Engineering Your Future

Engineering Your Future: The Professional Practice of Engineering, Third Edition

Stuart G. Walesh

John Wiley & Sons, Inc.

This book is printed on acid-free paper. ♾

Published by John Wiley & Sons, Inc., Hoboken, New Jersey.

Published simultaneously in Canada

Library of Congress Cataloging-in-Publication Data:

Walesh, S. G.
Engineering your future : the professional practice of engineering / Stuart G. Walesh. — 3rd ed.
p. cm.
Includes index.
ISBN 978-0-470-90044-4 (pbk.), ISBN 978-1-118-16043-5 (ebk.); ISBN 978-1-118-16044-2 (ebk.); ISBN 978-1-118-16045-9 (ebk.); ISBN 978-1-118-16300-9 (ebk.); ISBN 978-1-118-16301-6 (ebk.); ISBN 978-1-118-16302-3 (ebk.)
1. Project management. I. Title.
T56.8.W36 2012
658.4'04–dc23

2011028236

Printed in the United States of America

10 9 8 7 6 5 4 3 2 1

To that growing core of forward-looking engineers
committed to the reformation
of engineering education and early experience
in the U.S. and beyond.
You know who you are and your efforts are bearing fruit.

Contents

Preface to the Third Edition

Like the 1995 and 2000 editions, this third edition of *Engineering Your Future* offers students and recent graduates of engineering, or other technically-oriented academic programs, pragmatic management and leadership knowledge, skills, and attitudes (KSAs). This book is designed to be ***an insightful textbook for engineering and other technical program students* and a *practical reference book for young technical professionals***. Readers may be in the private, public, academic, or volunteer sectors. Emphasis is on professional, that is, non-technical topics that engineers and other technical professionals must master to fully realize their potential in the practice of their chosen profession. Presented in a results-oriented manner, the material in this book will be immediately useful to the student and the young practitioner. *Engineering Your Future* could also be helpful to experienced professionals who are in a review mode or moving from a primarily technical role to a mostly non-technical one.

TECHNICAL COMPETENCY: NECESSARY BUT NOT SUFFICIENT

Technical competency, although necessary, is not sufficient for those who wish to fully realize their potential in the consulting business, industry, government, academia, or volunteer organizations. They must supplement technical competency with basic management and leadership KSAs. Unfortunately, management concepts, knowledge, and skills are sometimes not introduced in undergraduate engineering and related curricula, and frequently nothing is taught about leading. As a result, aspiring professionals must learn managing and leading mostly by doing which is often inefficient and comes at high monetary cost to the employer, putting the young professional's career at risk.

This book does not describe how to transition from technical work early in your career to managing and leading later in your career. ***A premise of this book is that all engineers and other technical professionals can manage and lead from "day one,"*** that is, already as students. They should immediately manage their time, their assignments, their relationships with others, and their careers. And they should seek out leadership opportunities. Typically, the best managers and leaders in technical organizations are those who began to develop management and leadership KSAs as students. They learned, from the beginning, how to develop the "soft" as well as the "hard" side of their personal professional profiles.

Career management is increasingly important. The parents and grandparents of today's young professionals often entered into unwritten, but binding "contracts," with their employers. In that era, the young engineer or other technical professional would typically focus on technical matters and do them well. In turn, the employer would agree to provide long-term employment. Such "employment agreements" are vanishing, average periods of employment with a given employer are diminishing, and major organizational upheavals caused by financial difficulties, acquisitions, mergers, and globalization are increasing. Perceptive students and young professionals will recognize these changes, anticipate employment challenges, and prepare for employment opportunities. Perhaps a little further down the road, they will manage and lead their own businesses. This book will help you engineer your career.

AUDIENCES: STUDENTS AND PRACTITIONERS

I wrote this book with the assumption that readers are, or soon will be, graduates of an engineering or other technical program. The book assumes that readers want to take a pro-active approach and quickly build managing and leading capability on the foundations of their technically-oriented education. I hope that most readers of *Engineering Your Future* will see it as a means of earning career security, which is a much more viable life strategy than chasing job security.

Because of the two intended audiences, this book can be used as either a textbook for students or a reference book for young professionals. Instructors might use *Engineering Your Future* as the textbook for a capstone course. Another option is for students to use the book as a reference for many of their courses as they proceed through the curriculum from their freshman year to graduation with a baccalaureate or graduate degree.

Young practitioners can use *Engineering Your Future* in a just-in-time manner. For example, if you are offered a speaking opportunity or struggling with an ethical question, go immediately to Chapter 3, "Communicating to Make Things Happen" or Chapter 12, "Ethics: Dealing with Dilemmas" for guidance. Private, public, academic, and volunteer sector employers can use this book to support seminars, workshops, and webinars designed for young engineers and other technical personnel. Portions of the book can also be used as a textbook or reference book for young individuals who are not in technical fields.

Students will find much of the material in this book immediately useful. That is, while they are students, future technical professionals can draw on the presented information, tools, and techniques in areas such as goal setting, time management, communication, delegation, meeting facilitation, project management, business accounting, law, ethics, and marketing.

ORGANIZATION AND CONTENT

The book covers many aspects of managing and leading beginning with focusing on the individual and then moving into communication followed by developing productive relationships with others. The book's theme then shifts from personal development to professional practice topics including project management, achieving quality, design, construction and manufacturing, business accounting, law, ethics, the role and selection of consultants, and marketing professional services. *Engineering Your Future* concludes with a chapter that encourages the student and young practitioner to embrace and lead change.

Although the student or practitioner reader should follow some sequencing of chapters for effectiveness—e.g., Chapter 11, "Legal Framework" followed by Chapter 12, "Ethics: Dealing With Dilemmas"—reading or teaching all chapters in their order of presentation in the book is not necessary because each chapter is somewhat self-contained. Any given chapter begins with a very brief overview followed by the chapter's text which typically includes Personal, Historic Note, and/or Views of Others features. The first gives me an opportunity to reinforce a chapter's content with an anecdote and the other two use history and the thoughts of others to strengthen the chapter's message. The body of each chapter is immediately followed by a concluding statement, Cited Sources, and an Annotated Bibliography presenting a few carefully-selected resources which, although they were not specifically used in writing the chapter, are related to the chapter's topics and may be of interest to some readers.

Each chapter concludes with exercises which provide opportunities for readers to further explore or apply ideas, information, and techniques presented in the chapter. Some exercises are well-suited for modest to major team projects. Because teamwork is an integral part of managing and leading, those who use the book for teaching courses and for facilitating seminars and workshops are urged to assign some exercises as team projects. In that way, students and seminar or workshop participants will learn more about the subject matter while acquiring additional insight about being an effective team member and occasional team leader.

ADDITIONS AND IMPROVEMENTS

Since the 2000 publication of the second edition of *Engineering Your Future*, I have, as a result of study, teaching, and consulting, contemplated numerous refinements and additions to the book. Accordingly, the third edition provided the opportunity to make significant improvements. Examples of improvements are:

- Broadened the emphasis from primarily managing to both managing and leading as suggested, for example, by the Chapter 1 discussion of managing and leading and the last chapter's emphasis on effecting change.
- Made each chapter more self-standing, as explained above.
- Refined Chapters 2, 3 and 4 which focus on personal growth and developing relationships.
- Expanded the treatment of project management from one chapter to two (Chapters 5 and 6) with additions including a more detailed presentation of the project planning-execution-closing process and discussions of the Critical Path Method and scope creep prevention and resolution.
- Restructured the presentation of quality (Chapter 7) to include major sections describing how to develop a quality culture and ways to encourage creative and innovative thinking.
- Refined the design chapter (Chapter 8) including aligning it with the subsequent new constructing and manufacturing chapter.

- Added a constructing and manufacturing chapter (Chapter 9), to follow the design chapter and show how the results of design, the root of engineering, come to fruition in constructing and manufacturing, the fruit of engineering.
- Omitted the decision economics chapter because this increasingly complex topic is more effectively presented in a separate book intended for engineers and several are available (e.g., Blank and Tarquin 2005, Grigg 2010).
- Revised the business accounting chapter (Chapter 10) by adding a section that adds value to an understanding of accounting principles by showing how they can be used in career-long financial planning.
- Modified the legal chapter (Chapter 11) by focusing it even more on the role of the entry-level practitioner in helping to reduce liability exposure, both individual and organizational. The earlier organization of organizations chapter was removed and its essentials, from the perspective of students and young practitioners, were refined and moved into the legal chapter.
- Refined the treatment of ethics (Chapter 12) by broadening the discussion of ethics codes and including a case study.
- Refined the consulting chapter (Chapter 13) and expanded the treatment of qualifications-based and price-based selection of consulting firms.
- Expanded the breadth and depth of the marketing chapter (Chapter 14) which, while it continues to build on a simple, proven win-win marketing model, now includes an expanded discussion of marketing techniques and tools. The marketing model was retained, with added discussion, because of my even stronger belief in its value to those who offer and receive professional services.
- Finally, the last chapter has been almost completely rewritten. While retaining the previous paradigm discussion, the chapter now includes an in-depth treatment of change, more specifically, how the engineering student and young practitioner can and should anticipate, participate in, and ultimately effect change.

THIS BOOK AND ABET ENGINEERING ACCREDITATION CRITERIA

The first edition of *Engineering Your Future* (Walesh 1995) responded to the growing need for a systematic and comprehensive approach to providing engineers and other technical professionals with non-technical KSAs, although it did not use KSA terminology. After publication of the first edition, ABET, Inc. (formerly the Accreditation Board for Engineering and Technology), adopted *Engineering Criteria (EC) 2000*, a new statement of outcomes required of all engineering graduates beginning in 2001. *EC 2000* influenced the second edition (Walesh 2000) as well as this edition of *Engineering Your Future.*

ABET Basic Level Criterion 3, Program Outcomes and Assessment, which applies to all 27 engineering disciplines accredited by ABET (2011), states: "Engineering programs must demonstrate that their students attain the following outcomes:

a. an ability to apply knowledge of mathematics, science, and engineering

b. an ability to design and conduct experiments, as well as to analyze and interpret data

c. an ability to design a system, component, or process to meet desired needs within realistic constraints such as economic, environmental, social, political, ethical, health and safety, manufacturability, and sustainability

d. an ability to function on multidisciplinary teams

e. an ability to identify, formulate, and solve engineering problems

f. an understanding of professional and ethical responsibility

g. an ability to communicate effectively

h. the broad education necessary to understand the impact of engineering solutions in a global, economic, environmental, and societal context

i. a recognition of the need for, and an ability to engage in, life-long learning

j. a knowledge of contemporary issues

k. an ability to use the techniques, skills, and modern engineering tools necessary for engineering practice."

Clearly, engineering programs should be assessed on their demonstrated success in providing graduates with both technical and non-technical knowledge and skills. Six of the required 11 outcomes are non-technical (d, f, g, h, i, and j) and others have non-technical elements (c and e). When used as a textbook or reference book in an engineering program, *Engineering Your Future* provides content that supports these eight ABET's non-technical or partly non-technical outcomes as shown in Appendix A by a matrix of chapters versus outcomes.

ABET's basic-level accreditation criteria are supplemented, for most engineering programs, with program-specific criteria. These criteria include a variety of non-technical outcomes. For example, the Program Criteria for Civil and Similarly Named Engineering Programs state that "The program must demonstrate that graduates can... explain basic concepts in management, business, public policy, and leadership; and explain the importance of professional licensure." All of these non-technical topics are addressed in *Engineering Your Future* as shown in Appendix B with a matrix of chapters and the five topics. Other engineering programs that include non-technical topics in their program criteria are construction and engineering management and more programs will probably move in that direction.

In summary, this book provides some content that can be used by faculty members in helping their students attain the "a" through "k" outcomes and, as appropriate, satisfy program criteria. My hope is that this book's content will be useful to those who teach and advise students.

THIS BOOK AND THE BODY OF KNOWLEDGE MOVEMENT

Recognizing the need for major change, and under the leadership of the American Society of Civil Engineers (ASCE), U.S. civil engineering education and prelicensure experience are undergoing major reform. The effort is founded on the aspirational Civil Engineering Body of Knowledge (BOK) which is defined as "the necessary depth and breadth of knowledge, skills, and attitudes required of an individual entering the practice of civil engineering at the professional level [licensure] in the 21st century"

(ASCE 2008). Broadly speaking, the civil engineering BOK calls for individuals entering the profession to:

- Master more mathematics, science, and engineering science fundamentals
- Acquire appropriate technical breadth
- Attain broad exposure to humanities and social sciences
- Gain professional practice breadth
- Achieve greater technical depth in their chosen speciality area.

More specifically, the BOK includes 24 outcomes and desired levels of achievement for each outcome at up to three critical points: completion of the bachelor's degree, completion of the master's degree or equivalent, and completion of prelicensure experience. Fulfilling the civil engineering BOK requires a bachelor's degree plus a master's degree, or approximately 30 semester credits, and on-the-job experience.

Nine of the 24 outcomes are labelled Professional Outcomes, a term that is used in this book along with the term non-technical, which is similar. The names of the Professional Outcomes are: Communication, Public Policy, Business and Public Administration, Globalization, Leadership, Teamwork, Attitudes, Lifelong Learning, and Professional and Ethical Responsibility. The civil engineering BOK is aspirational whereas the previously-discussed ABET accreditation criteria are minimal. For civil engineering, the BOK and ABET criteria can be viewed as bracketing the range of possibilities.

Engineering Your Future, when used as a textbook or reference book in a civil engineering or similar program, provides content that supports achieving the portions of the Professional Outcomes to be fulfilled through the bachelor's degree. When used by employers of civil engineers, this book will help engineer interns achieve that portion of the Professional Outcomes to be fulfilled through prelicensure experience. The matrix in Appendix C relates the preceding nine non-technical BOK outcomes to the book's chapters.

Other U.S.-based engineering disciplines have mounted BOK efforts. For example, in 2009, the American Academy of Environmental Engineers (AAEE) published a BOK report (ASEE 2009). The chemical engineering profession conducted three workshops in 2003 that produced a vision and model for reform of undergraduate chemical engineering education (Armstrong 2006) and formed a BOK committee in 2009. The American Society of Mechanical Engineers (ASME) conducted a Global Summit on the Future of Mechanical Engineering in 2008 (ASME 2008) and ASME subsequently formed a vision/body of knowledge task force.

Partly as a result of the BOK movement, many members of the U.S. engineering academic community are considering changes to their programs. Some of these contemplated changes are meant to satisfy the minimum expectations of new or contemplated accreditation criteria. Other changes are more broadly based and are driven by the realization that major program changes—changes that go way beyond the minimum requirements—are desirable and possible. Many U.S. engineering educators, within the framework of each institution's traditions, mission, and aspirations, are open to and are searching for new approaches to the curricular, co-curricular, and extra-curricular elements of their programs.

The BOK movement extends beyond formal education. Engineer employers, in the business, government, academic, and volunteer sectors, who want to support licensure of their engineering personnel, are interested in helping those aspiring professionals fulfil that portion of the applicable BOK appropriate to the prelicensure experience.

Engineering Your Future is designed to assist academics and practitioners, in all engineering disciplines and beyond, as they work with students and employees to develop professional or non-technical KSAs. For civil and environmental engineers, this book will help students and young practitioners move toward licensure by fulfilling the aspirational BOKs. I trust this book's content will be useful to those who teach and advise students and to those who supervise, coach, and mentor young engineers and other technical professionals.

Acknowledgments

This book, in the form of its three editions, evolved over three decades as a result of my teaching a university management course; conducting seminars, workshops, and webinars; providing engineering and management consulting services; and researching for and writing all three editions. I acknowledge and sincerely appreciate what I have learned from and with former students; seminar, workshop, and webinar participants; clients; and colleagues.

I obtained and developed other materials and ideas, reflecting primarily management and leadership applications, over almost four decades while I was employed in the public and private sectors in engineering practice and in education and participated in volunteer efforts. During this period, I administered and was administered, managed and was managed, and lead and was lead and, in the process, witnessed some very enlightened and some very poor managing and leading. These excellent learning experiences constitute the personal experience base of this book.

I received a wealth of useful ideas and information from, and have interacted with and been positively influenced by, some exceptional individuals who I was privileged to work with in business, government, academic, and volunteer activities. More specifically, the following friends and colleagues kindly assisted in me in writing this book by suggesting and/or providing resources, outlining key ideas and information, clarifying and tightening text, and answering questions: Richard O. Andersen, PE; Wayne Bergstrom, PhD, PE; Brock E. Barry, PhD, PE; Vincent P. Drnevich, PhD, PE; John A. Hardwick, PE; William M. Hayden, Jr., PhD, PE; Douglas J. Hoover, CFP, ChFC; Jonathan E. Jones, PE; Merlin D. Kirschenman, PE; Chester F. Kochan, PE; David A. Lange, PhD, PE; Thomas A. Lenox, PhD; John W. MacDonald, Col USA Ret.; William D. Minor; Wayne P. Pferdehirt; Debra R. Reinhardt, PhD, PE; Jeffrey S. Russell, PhD, PE; William J. Schoech, PhD, PE; Clifford J. Schexnayder, PhD, PE; and Kimberly A. Walesh. While I am most appreciative of their assistance, I am totally responsible for the manner in which I have used their contributions.

My debt to other professionals is suggested, in part, by the extensive list of references cited in the book. I drew ideas, information, and reference materials from a wide range of sources. The resulting eclectic collection of cited and supplemental references suggests that engineers and other technical professionals can learn much about managing and leading by looking both within and outside of their particular disciplines.

I wrote much of the first edition of this book, which proved to be the most challenging edition and which laid the foundation for the subsequent two editions, in the early 1990's as my wife Jerrie and I traveled and worked for six months on our vessel Sabbatical while on sabbatical from Valparaiso University. This third edition of *Engineering Your Future* also reflects insight I gained as a result of authoring

Managing and Leading: 52 Lessons Learned for Engineers (Walesh 2004) and co-authoring with Paul W. Bush, *Managing and Leading: 44 Lessons Learned for Pharmacists* (Bush and Walesh 2008).

Vicki Farabaugh, owner of Creative Computing, helped draft the graphics. Her skills and responsiveness are most appreciated. The contributions of members of the Wiley-ASCE team are appreciated. Betsy Kulamer, ASCE Press Acquistions Editor, helped arrange for joint publication with Wiley and provided content resources. Daniel Magers, Senior Editorial Assistant at Wiley, provided guidance on using the publisher's standards and then Nancy Cintron, Senior Production Editor at Wiley, guided the book through production. Finally, Jerrie, my wife, did some of the word processing; meticulously proofed punctuation, spelling and grammar; critiqued content; and, as always, provided total support.

—Stuart G. Walesh
Cape Haze, FL
December 2011

CITED SOURCES

ABET, Inc. 2011. "Criteria for Accrediting Engineering Programs." (www.abet.org). October 7.

American Academy of Environmental Engineers. 2009. *Environmental Engineering Body of Knowledge*. AAEE: Annapolis, MD.

American Society of Mechanical Engineers. 2008. *2028 Vision for Mechanical Engineering*. ASME: New York, NY.

Armstrong, R. C. 2006. "A Vision of the Chemical Engineering Curriculum of the Future." *Chemical Engineering Education*, Vol. 40, No. 2, pp. 104–109.

American Society of Civil Engineers. 2008. *Civil Engineering Body of Knowledge for the 21st Century: Preparing the Civil Engineer for the Future-Second Edition*. ASCE Press: Reston, VA.

Blank, L. and A. Tarquin. 2005. *Engineering Economy-Sixth Edition*. McGraw Hill Higher Education, New York, NY.

Bush, P. W. and S. G. Walesh. 2008. *Managing and Leading: 44 Lessons Learned for Pharmacists*. American Society of Health-System Pharmacists: Bethesda, MD.

Grigg, N. S. 2010. *Economics and Finance for Engineers and Planners*. ASCE Press: Reston, VA.

Walesh, S. G. 1995. *Engineering Your Future: Launching a Successful Entry-Level Technical Career in Today's Business Environment*. Prentice Hall PTR: Englewood Cliffs, NJ.

Walesh, S. G. 2000. *Engineering Your Future: The Non-Technical Side of Professional Practice in Engineering and Other Technical Fields-Second Edition*. ASCE Press: Reston, VA.

Walesh, S. G. 2004. *Managing and Leading: 52 Lessons Learned for Engineers*. ASCE Press: Reston, VA.

List of Abbreviations

A	assets
AAEE	American Academy of Environmental Engineers
AAR	After-Action Review
AAWRE	American Academy of Water Resources Engineers
ABA	American Bar Association
ABET	ABET, Inc. (formerly Accreditation Board for Engineering and Technology)
ACEC	American Council of Engineering Companies (formerly American Consulting Engineers Council)
ACM	Association for Computing Machinery
ACOPNE	Academy of Coastal, Ocean, Port & Navigation Engineers
A/E	architect/engineer or architecture/engineering (see also E/A)
AGP	Academy of Geo-Professionals
AH HA	awareness-head-heart-action
AIA	American Institute of Architects
AIChE	American Institute of Chemical Engineers
AICP	American Institute of Certified Planners
AIPG	American Institute of Professional Geologists
APA	American Planning Association
APWA	American Public Works Association
ASCE	American Society of Civil Engineers
ASEE	American Society for Engineering Education
ASME	American Society of Mechanical Engineers
ASQ	American Society for Quality
AU$	Australian Dollars
BOK	body of knowledge
CAD	computer-aided drafting
CADD	computer-aided drafting and design
CASE	Council of American Structural Engineers
CE	civil engineering
CEO	Chief Executive Officer
CFP	Certified Financial Planner

ChFC	Chartered Financial Consultant
CII	Construction Industry Institute
CIP	capital improvement plan or capital improvement program
C of C	chamber of commerce
CP	critical path
CPM	Critical Path Method
CSC	compensated scope creep
CSI	Construction Industry Institute
CSO	combined sewer overflow
DAD	decide-announce-defend
DBE	Disadvantaged Business Enterprise
DFM	design for manufacturing
DIP	ductile iron pipe
DISC	Dominance, Influence, Steadiness, Conscientiousness
DVD	digital video disc
DWTSTWD	do what they say they will do
DWYSYWD	do what you said you would do
E	equity (same as NW) and expense
E/A	engineer/architect or engineering/architecture (see also A/E)
EC	Engineering Criteria, as in ABET's EC 2000
ECPD	Engineers Council for Professional Development (now ABET)
EFT	earliest finish time
EJCDC	Engineers Joint Contract Documents Committee
EST	earliest start time
EVM	Earned Value Method
FE	Fundamentals of Engineering (as in FE examination)
FEMA	Federal Emergency Management Agency
FPD	first professional degree
FTC	Federal Trade Commission
GNP	Gross National Product
HVAC	heating, ventilating, and air conditioning
I	income
IEEE	Institute of Electrical and Electronic Engineers
ISO	International Organization for Standardization
IT	Information Technology
KSA	knowledge, skills, and attitudes
L	liabilities
LCD	liquid crystal display
LFT	latest finish time

LL	lessons learned
LST	latest start time
M	multiplier
MBE	Minority Business Enterprise
mgd	million gallons per day
MOE	master's (degree) or equivalent
NCSEA	National Council of Structural Engineers Associations
NI	net income
NPDES	National Pollution Discharge Elimination System
NSPE	National Society of Professional Engineers
NW	net worth (same as E)
O	overhead ratio
O&M	operation and maintenance
P	total payroll cost
P′	non-billable payroll cost
P_w	total payroll cost per week
PBS	price-based selection (see QBS)
PE	professional engineer
PEPP	Professional Engineers in Professional Practice
PERT	Program Evaluation and Review Technique
PIN	personal identification number
PM	project management or project manager
PMI	Project Management Institute
POP	public owns project
POW	plan our work
POWWOP	plan our work - work our plan
PP	project plan
PVC	polyvinyl chloride, as in PVC pipe
Q&A	question and answer
QA	quality assurance
QC	quality control
QA/QC	quality assurance/quality control (same as QC/QA)
QBS	qualifications-based selection (see PBS)
QC/QA	quality control/quality assurance (same as QA/QC)
R	expense ratio (S/P)
R&R	rest and relaxation
RFI	request for information
RFP	request for proposal
RFQ	request for qualifications

ROI	return on investment
RPR	Resident Project Representative
S	non-salary costs that are non-billable
SECB	Structural Engineering Certification Board
SEI	Structural Engineering Institute
SMART	specific-measureable-achievable-relevant-time framed
SOP	standard operating procedure
SOQ	statement of qualifications
SPC	statistical process control
SWOT	strengths-weaknesses-opportunities-threats
U	utilization rate
URL	uniform resource locator
USA	U.S. Army
USACE	U.S. Army Corps of Engineers
USC	uncompensated scope creep
USMA	U.S. Military Academy
VCR	videocassette recorder
WBE	Women's Business Enterprise
WHO	World Health Organization
WOP	work our plan
WSDOT	Washington State Department of Transportation

Engineering Your Future

CHAPTER 1

INTRODUCTION: ENGINEERING AND THE ENGINEER

We recognize that we cannot survive on
meditation, poems, and sunsets.
We are restless.
We have an irresistible urge to dip our hands
into the stuff of the earth and do
something with it.

(Samuel C. Florman, engineer and author)

What do engineers and, by extension, many other technical professionals do? What roles and functions do they fulfill? This chapter uses several means to offer answers to those and similar questions and thus lay the foundation for the book's treatment of the professional or non-technical aspects of engineering. First, the roles of engineers are presented in the context of their frequent, dynamic interaction with clients, owners, customers, and constructors-manufacturers and other implementers. Then the chapter presents several definitions of engineering as another means of suggesting what engineers do within the framework of various constraints. These definitions also introduce the engineer's creative role. A discussion of leading, managing, and producing invites engineers to engage in all three roles early, beginning as college students. Included is a description of the seven qualities of effective leaders. Building, in the broadest sense, is discussed noting that this activity in its various forms is widely practiced across engineering. The wisdom of developing productive habits concludes the chapter.

THE PLAYING FIELD

Engineers and other technical professionals interact dynamically among themselves and with clients, owners, customers, constructors, manufacturers, and other implementers. The interaction process, as illustrated in Figure 1.1, typically begins with a client, owner,

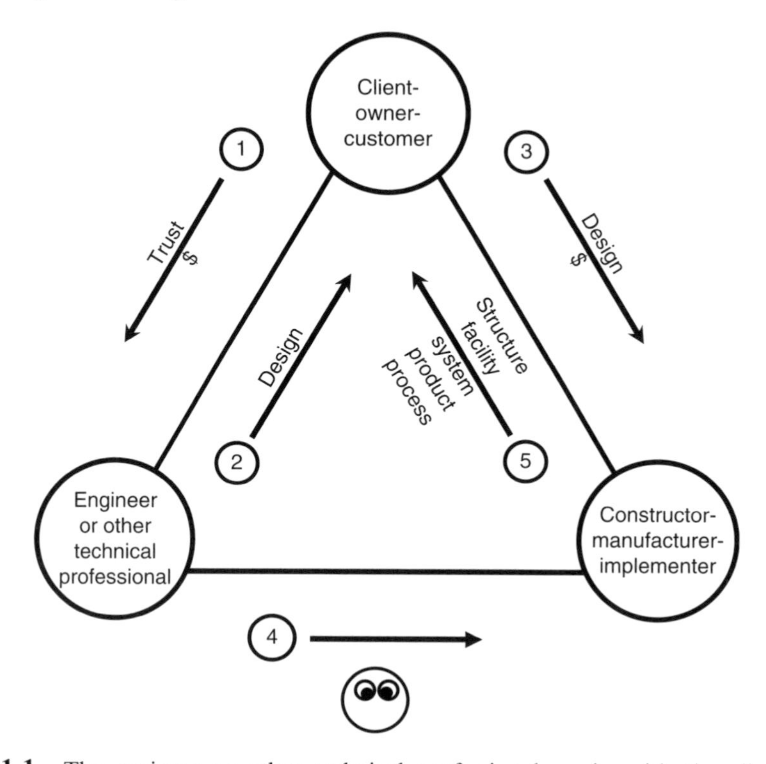

Figure 1.1 The engineer or other technical professional works with the client-owner-customer and the constructor-manufacturer-implementer to produce a useful structure, facility, system, product, or process.

or customer retaining a professional (e.g., engineer or architect) to conduct a study, perform preliminary designs, prepare a complete design, and deliver a contract package consisting of plans and specifications or other formal design. The client-owner-customer could be a private or public sector entity, such as a business or a municipality.

The client, owner, or customer then selects a constructor, manufacturer, or other implementer to produce the structure, facility, system, product, or process. The client, owner, or customer sometimes retains the professional to monitor the construction-manufacturing-implementation process so the final structure, facility, system, product, or process conforms to the original plans and specifications.

Design followed by manufacturing or construction can occur within a single organization. For a self-contained manufacturing organization, the bottom two vertices of the triangle shown in Figure 1.1 collapse into one point. Similarly the bottom two vertices become one in a design-build organization, that is, a single firm that both designs and builds structures, facilities, or systems.

Figure 1.1 is most likely to apply to engineers and other technical professionals who, at any time in their careers, might be at any one of the three vertices of the triangle. This book frequently refers to the triangular model and its variations. In one sense, managing and leading are the processes by which the various entities shown in Figure 1.1 interact with each other in the worlds of engineering and business.

DEFINITIONS OF ENGINEERING

Besides using the interactions shown in Figure 1.1 to show the role of engineers, we can also take a more fundamental approach to understanding "what engineers do," that is, examine some time-tested definitions of engineering. Aeronautical engineer Theodore von Karman said "Scientists explore what is, engineers create what never has been" (ECPD 1974). This succinct statement suggests how science and engineering differ; that is, creativity is essential in the latter.

ABET, Inc. (ABET 2002) offers this definition of engineering which focuses on a mathematics and science base, judgment, economic considerations, and the goal of benefiting society: "Engineering is the profession in which a knowledge of the mathematical and natural sciences gained by study, experience, and practice is applied with judgment to develop ways to utilize, economically, the materials and forces of nature for the benefit of mankind."

The creative and humanist dimensions of engineering were captured by Herbert Hoover, the 31st U.S. President, who had a long and distinguished engineering career. Stressing the thrill of creating and the satisfaction of enhancing the quality of life, he said (Fredrich 1989): "It is a great profession. There is the fascination of watching a figment of the imagination emerge, through the aid of science, to a plan on paper. Then it brings jobs and homes to men. Then it elevates the standards of living and adds to the comforts of life. That is the engineer's high privilege."

Finally, Professor Hardy Cross (1952), using direct, plain words, clearly captured the central, people-serving goal of engineering when he wrote: "It is not very important whether engineering is called a craft, a profession, or an art; under any name this study of man's needs and of God's gifts that they may be brought together is broad enough for a lifetime."

Based in part on the preceding definitions, the following six essential features of engineering appear:

- Science-based
- Systematic—However, except for trivial problems, judgment and other qualitative considerations always enter in
- Creative and innovative
- Goal-oriented—Satisfy the requirements and get the job done on time and within budget
- Dynamic—Technology, laws, public values, clients, owners, customers, stakeholders, and the physical environment continuously change
- People-oriented—Both in doing and in results in that engineering is essential to the survival of human communities and to the quality of life

LEADING, MANAGING, AND PRODUCING: DECIDING, DIRECTING, AND DOING

Leading, Managing, and Producing Defined

Another way of understanding what engineers and other technical personnel do, or could do, is examine their leading, managing, and producing roles. One paradigm for an organization, such as an engineering consulting firm, a manufacturing business, a government agency, an academic department, or a volunteer organization is that wholeness, vitality, and resiliency require attention to three different, but inextricably-related functions: leading, managing, and producing. The meaning of each of these terms is illustrated by the comparisons presented in Table 1.1. In a simplified sense, the leading, managing, and producing functions can also be represented by three Ds: deciding, directing, and doing.

Figure 1.2 uses the metaphor of a three-legged stool to suggest how attention to the leading, managing, and producing functions produces a stable organization—one that cannot easily be "knocked over." While an organization might temporarily survive balanced on two of the three legs, all three legs are needed for long-term survival. For example, an engineering consulting firm with only managing and producing legs may be precariously balanced. It believes that current success guarantees future success. It can be sailing along just fine. But, without leading, it fails to see the rocks versus the navigable waters ahead. Author Warren G. Bennis (1989) said, "Many an institution is very well-managed and very poorly-led. It may excel in the ability to handle each day all the routine inputs, yet may never ask whether the routine should be done at all."

With only leading and managing functions, exciting visions and careful operating plans are left lying on the table. Little or nothing happens—or what does happen fails to meet expectations. Finally, when only leading and producing are present, the organization lacks translation between the vision and the producing functions. This is sometimes the last stage of an organization started by a talented entrepreneur who cannot give up control. He or she hires managers who are expected to be clones—and, therefore, for all practical purposes, they add very little value. The organization produces, but very poorly. The vision, admirable and wise as it may be, does not get translated into services or products.

The Traditional Pyramidal, Segregated Organizational Model

Assuming you agree that each organization has leading (deciding), managing (directing), and producing (doing) responsibilities, consider the manner in which these corporate responsibilities might be met. More specifically, consider the matter of individual responsibility in achieving the three corporate responsibilities.

In what might be called the traditional pyramidal and segregated organizational model, as illustrated in Figure 1.3, the three functions reside in three separate groups of personnel. The vast majority of employees are the doers or producers, a distinctly different and much smaller group of managers are the directors, and one person, or perhaps a very small group, leads.

An aspiring and successful individual begins in a production mode and then passes serially or linearly maybe into managing and possibly into leading. Rather than being a

Table 1.1 Although leading, managing, and producing form a continuum, distinctions are drawn among them.

Leading	Managing	Producing	Source
Deciding where we want to go	Determining the best way to get there	Getting there	–
Deciding what ought to be done	Directing how things will be done, by whom, and when	Doing it	–
Moving forward to create something new	Taking care of what already exists	–	Kanter 1993
Selecting a jungle to conquer	Writing procedure manuals, setting up work schedules, sharpening machetes	Cutting through the jungle	Covey 1990
Deciding if the ladder is leaning against the right wall	Determining how to efficiently climb the ladder	Climbing the ladder and getting over the wall	Covey 1990
Dreaming during the day	Dreaming at night	–	Finzel 2000
Creating change	Reacting to change	–	Sanborn 2006
Using the right brain	Using the left brain	–	Covey 1990
Working through people and culture. Soft and messy	Working through hierarchy and systems. Hard and neat	–	Covey 1990
Influencing by permission	Influencing by position	–	Finzel 2000
Stressing people work over paper work	Stressing paperwork over people work	–	Finzel 2000
Causing people to want to do things	Causing people to have to do things	–	Badger 2007
Molding consensus	Searching for consensus	–	Martin Luther King in Goodale 2004
Creating teams	Directing groups	–	Sanborn 2006
Inspiring	Informing	–	Sanborn 2006
Having followers	Having employees	–	Sanborn 2006
Concentrating on what is right	Concentrating on who is right	–	Peyton 1991
Having very little formal education and training in this function	Having some formal education and training in this function	Having lots of formal education and training in this function	Finzel 2000
Very few personal models	Some personal models	Many personal models	Finzel 2000
Interruptions are the work	Work gets interrupted	–	Finzel 2000

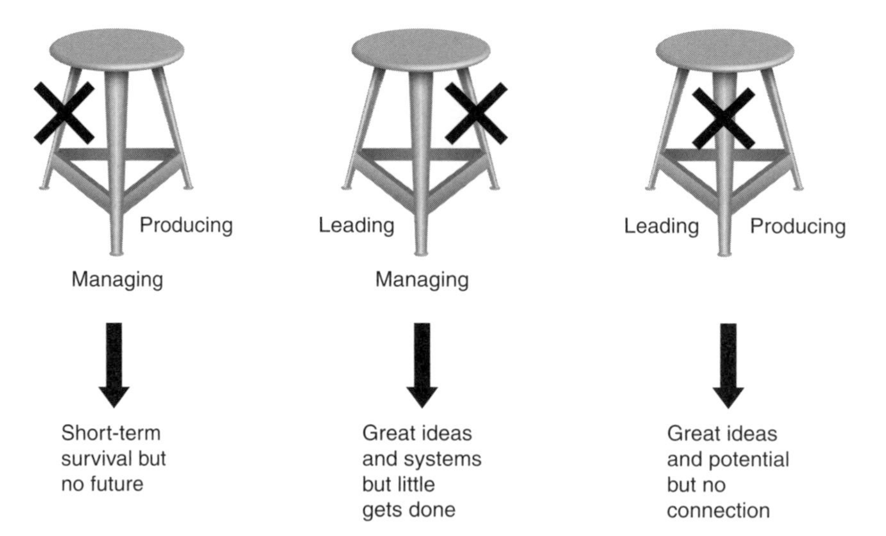

Figure 1.2 Organizational vitality requires a balance of leading, managing, and producing.

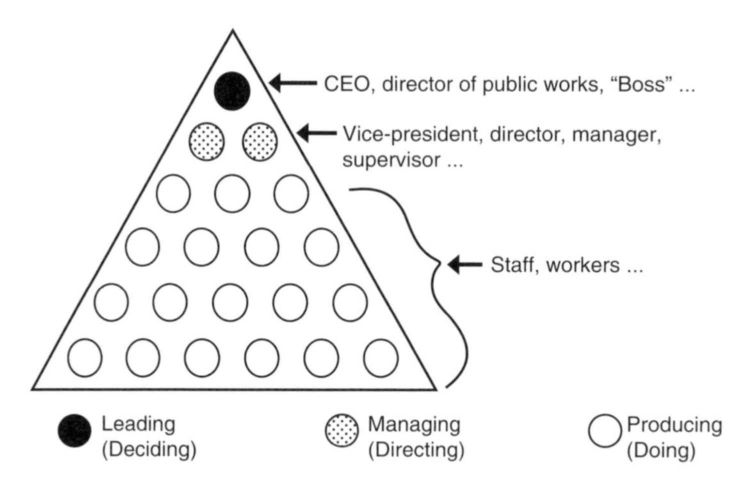

Figure 1.3 The traditional pyramidal and layered organizational model concentrates leading and managing at the top.

trait that many can possess, albeit to different degrees, leading is considered the end of the line or ultimate destination for a very few. But is this the optimum way for an organization to meet its leading, managing, and producing possibilities? I don't think so, although a command and control structure may be appropriate in certain situations.

The Shared Responsibility Organizational Model

An organization will be stronger if what I previously referred to as the three organizational responsibilities now also become individual responsibilities. The goal should

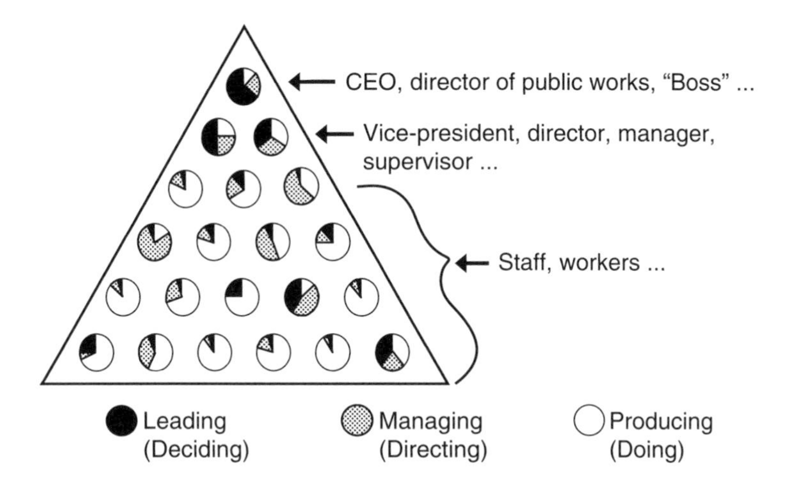

Figure 1.4 The shared organizational model enables dispersed leading, managing, and producing.

be to enable each member of the organization to be a decider, a director, and a doer as illustrated by Figure 1.4.

While the relative "amounts" of leading, managing, and producing will vary markedly among individuals in the organization, everyone should be enabled and expected to do all three in accordance with their individual characteristics. This shared responsibility organizational model, in contrast with the traditional segregated model, is much more likely to mine the organization's gold, that is, extract and benefit from the diverse aspirations, talents, and KSAs that are typically present within an organization.

Because essentially all members are fully involved, the shared responsibility organization is more likely to synergistically build on internal strengths, cooperatively diminish internal weaknesses, and learn about and be prepared to respond to external threats and opportunities. Striving to enable everyone creates confidence, results, and pride. As noted by Chinese philosopher Lao Tzu, "But of a good leader, who talks little, when his work is done, his aim fulfilled, they will say, we did this ourselves."

The Focus of This Book: Managing and Leading

The entry-level engineer or other technical person will, by definition, be well-prepared for and will spend a majority of his or her time producing, that is, carrying out the production function of the organization. An undergraduate education is typically a solid preparation for this function. The focus of this book is on managing and leading.

Your education may be preparing you, or may have prepared you, well for the managing and leading functions. For example, perhaps you are studying or have studied civil or environmental engineering in a department aligned with the Body of Knowledge (BOK) developed by the American Society of Civil Engineers (ASCE 2008), the BOK prepared by the American Academy of Environmental Engineers (AAEE 2009), or some similar BOK. As suggested by Appendix C in this book, current BOKs include managing and leading outcomes. If you are not in or did not study within that kind of environment, then you need to assume much more responsibility for developing your managing and leading KSAs.

Leading Misconceptions

Leading, which has been contrasted with managing and producing, warrants further discussion because aspects of it are often misunderstood. One misconception is that leading implies holding a high position within an organization. For example, in a city, state, or federal unit of government, do you view the chief engineer and department heads as leaders whereas individuals further down the "food chain" as not being leaders? Or in a manufacturing firm, do you think of vice-president and up as leaders and those with lesser ranks as not being leaders? Perhaps, as suggested by Sanborn (2006), leading is perceived as enjoying these four P's: position, power, prestige, and privilege. I reject this view, as clearly indicated by the previous discussion of the two organizational models. In my view and as reflected in this book, leading is a function that almost anyone can fulfill regardless of their position in the organizational hierarchy.

Another leading misconception is that one must be a charismatic extrovert in order to lead, that is, introverts need not apply. Not true. According to Jones (2006), Bill Gates, long-time Microsoft leader; Warren Buffet, Chairman and CEO of Berkshire Hathaway; and Steven Spielberg, film director and producer are all professed introverts and companies led by introverts do as well financially as those led by extroverts. Jim Collins, in his book *Good to Great* (Collins 2001), concludes this and dispels the notion that extroversion is needed to lead. Therefore, if you tend toward introversion—like me and the majority of engineers—that trait should not stop you from releasing the leader in you.

The third, and final misconception, is leaders are born, not made. You have probably heard expressions like "she is a born leader" and "his leading is natural." Is that true? I reject the idea that birth defines leading ability and dedicate some of this book to helping you learn how to lead, as well as manage and produce. If leading KSAs can be learned, then a discussion of the elements of leading, of what can be learned, is warranted and follows.

> **Views of Others**
>
> Famed football coach Vince Lombardi believed leaders are made. He said: "Leaders aren't born, they are made. And they are made just like anything else, through hard work." John C. Maxwell, a student of leading, says "Leadership is not an exclusive club for those who are born with it. The traits that are the raw materials of leadership can be acquired. Link them up with desire and nothing can keep you from becoming a leader."

THE SEVEN QUALITIES OF EFFECTIVE LEADERS

A review of books, papers, and articles on the topic and reflection on experience suggests seven important aspects of leading. These elements are discussed here for consideration by you, the student or entry-level practitioner. Think of these elements as an answer to the question: What do you look for in someone you might follow? For brevity, the word "leader" is used to mean the leader in each of us or the leadership part of each of us. Leader and leading do not refer to positions in an organization; anyone can be a leader anywhere in a public, private, academic, volunteer, or other organization.

Personal

After one of my webinars, a participant shared this thought with me: "I learned a long time ago that you can be perceived as a leader by others and not realize it yourself. Becoming a leader is something you can do without anyone's "permission." It is developed conscientiously through self-improvement. It's amazing when you realize others perceive you as a leader." If you possess, or work at possessing, the seven leadership qualities described here, others will perceive you as a leader. They will value you as an exceptional person. They will look to you when an unusual problem or opportunity arises.

Honesty and Integrity

The leader in each of us practices honesty and exhibits integrity, that is, tells the truth and keeps his/her word. On seeing this, some readers may be thinking, "oh sure, we've got to get this idealistic stuff" out of the way. Please understand, I am not offering this first leadership essential in some sort of pie-in-sky, ivory tower, got-to-say-it manner. On the contrary, honesty and integrity are, in my view, the first—and the most important—of the leading elements.

Regarding honesty, "a survey of several thousand people around the world and several hundred case studies—found that honesty was the most frequently cited trait of a good leader" (Woolfe 2002). Leaders are granted the privilege of leading by those who are prepared to be led. Honesty and integrity on your part are crucial to earning and retaining the privilege to lead. You must get people to buy into you before they buy into your vision (Maxwell 1993).

Although honesty and integrity are often used in a vague and even interchangeable manner, they warrant precise definitions. Covey (1990) says "Honesty is telling the truth—in other words, conforming our words to reality. Integrity is informing reality to our words—in other words, keeping promises and fulfilling expectations." Stated differently, honesty is retrospective and integrity is prospective; honesty is what you say about what you've done and integrity is what you do about what you've said. In high school, when you told your teacher that the family dog ate your homework, and that is exactly what happened, you exhibited honesty. Then, when you promised to redo the homework and hand it in at the next class, and did so, that was integrity

Vision: Reach and Teach

Leaders know where their organization is going; they believe in it and much of what they say and do reflects and supports the vision. Each organization should have a multi-year strategic plan and an annual business or operating plan. The strategic plan should contain and be based on brief, widely-understood mission and vision statements.

The mission statement explains the purpose of the organization—why it exists, what service or product it provides and to whom (Tompkins 1998). Tompkins offers this example of a mission statement for a consulting firm: "To provide our clients with total confidence in our solutions and to treat all employees with fairness and dignity."

In contrast with an organization's mission statement, its vision statement looks boldly into the future. "Vision . . . is mental, cognitive—not reality, or even close to reality, as we know it today. It is influenced, at least in part, by imagination, reflective of actual or desired values, and focused on "what," not "how." Finally, a vision is stimulating, energizing, engaging, and inclusive" (ASCE 2007). Consider the advice of city planner and architect Daniel Burnham: "Make no little plans, they have no magic to stir men's blood. Make big plans, aim high in hope and work and let your watchword be order and your beacon beauty."

The vision statement declares, as specifically as possible, what the organization intends to become. Like the mission statement, the vision statement must be clearly and widely understood throughout the organization. Tompkins (1998) provides these two examples of vision statements:

- For a consulting firm: "To be the best engineering-based consulting firm in the world while providing all employees a rewarding and satisfying work experience."
- For an appliance manufacturer: "To be the world's leading manufacturer of diversified laundry equipment by continuously improving customer satisfaction, employee motivation, and company profitability."

ASCE's Vision 2025 (ASCE 2007) provides another example of a vision. A companion document (ASCE 2009) illustrates a "roadmap" for implementing that vision. The ASCE – sponsored vision for civil engineering follows. "Entrusted by society to create a sustainable world and enhance the quality of life, civil engineers serve competently, collaboratively, and ethically as master:

- Planners, designers, constructors, and operators of society's economic and social engine – the built environment;
- Stewards of the natural environment and its resources;
- Innovators and integrators of ideas and technology across the public, private, and academic sectors;
- Managers of risk and uncertainty caused by natural events, accidents, and other threats; and
- Leaders in discussions and decisions shaping public environmental and infrastructure policy."

The Vision 2025 report explains the meaning of the word "master," which is a key adjective in the vision, as follows: "master means to possess widely recognized and valued knowledge and skills and other attributes acquired as a result of education, experience, and achievement. Individuals, within a profession, who have these characteristics, are often willing and able to serve society by orchestrating solutions to society's most pressing current needs while helping to create a more viable future."

To illustrate the influence of visions, the China Civil Engineering Society translated Vision 2025 and then endorsed it in May 2010. Spain's Association of Civil Engineers translated the two Vision 2025 reports into Spanish and, in June 2010, endorsed Vision 2025. Engineering societies in other countries have taken similar actions.

If the organization does not have a vision, a mission, and a plan, the leader takes action to get such statements articulated and a plan developed. But what if an individual's values and goals are in serious conflict with an organization's mission and vision? The answer is obvious—try to effect change and, if not successful, move on. Life is too short to dissipate yourself in a hostile environment or prostitute yourself by feigning allegiance to alien values.

The leader in you interprets opportunities and problems in the context of the vision and mission, always seeking ways to move the organization one more step in the direction of the vision. In the leader's mind, a constructive tension exists between many of the common, day-to-day occurrences and the organization's vision and mission. The tension pulls those occurrences in the direction of the vision and mission and is the force that enables the leader to take supportive steps. The leader in you seeks opportunities to communicate the vision recognizing that, as stated by Clark and Crossland (2002), "The difference between a vision and a hallucination is the number of people who see it."

Phillips (1992) describes President Abraham Lincoln as a visionary. Lincoln-based principles about vision include the following, quoted directly from the book:

- Provide a clear, concise statement of the direction of your organization and justify the actions you take
- Everywhere you go, at every conceivable opportunity, reaffirm, reassert, and remind everyone of the basic principles upon which your organization was founded
- Because effective visions can't be forced on the masses, set them in motion by means of persuasion
- When effecting renewal, call on the past, relate it to the present, and then use them both to provide a link to the future

Historic Note

During WWI, Herbert Hoover, engineer and later 31st U.S. President, was asked to lead an effort to save ten million people who were starving in Belgium. That country, which was heavily dependent on imports for food, had been overrun by the German army, which was intercepting imported food for its use. Hoover's response to the invitation: He would accept no salary but must be given "absolute authority" to do the job. He clearly thought he knew the magnitude of the task and had a vision of how it would be accomplished. Robertson Smith, who worked with Hoover on the Belgium effort, said this about Hoover's vision: "There is something almost terribly personal about it, in [his] desire that things shall change, that order shall be brought out of an existing chaos." Based on Hoover's heartfelt, personal vision and lots of work, the Belgium people did not starve (Ruth 2004).

The leader in us is like a TV or radio station that cannot turn off its transmitter. We are always "broadcasting" the vision. Receptive individuals hear what we say about the vision and watch what we do to move, with them, step-by-step, toward the vision. Consider some ways to broadcast your vision:

- When you read a relevant newspaper article or professional paper connected to your vision, share it with key individuals
- As opportunities arise during project and other meetings, mention your vision
- Write a "white paper" that describes your vision and the benefits it would provide
- Find others who share your vision and collaborate with them
- Strive to get your vision reflected in your organization's strategic plan
- Write an article about your vision for your organization's e-newsletter or for presentation at a conference

In summary, the leader in you enables others to understand, value, and commit to the vision and mission of the organization and, as a result, direct their aspirations, talents, and skills toward carrying out the mission and achieving the vision.

Strategies and Tactics to Achieve the Vision

Visions, as engaging as they may be, are just dreams until they are converted to strategies and tactics that will help to achieve them. Engineers, architects, planners, and other technical professionals know how to create plans that bring their projects to fruition. Plans, that is, strategies and tactics, are also needed to bring visions to fruition. The leader in you does not wait for someone to tell you what to do, when to do it, and how to do it. You act because you view the future as something you and others can make happen, not something that happens to you.

Drawing again on the example of Abraham Lincoln, Phillips (1992) says that Lincoln had "an almost uncontrollable obsession" to achieve. Phillips sets forth these action principles, all quoted, at the conclusion of a chapter titled, "Set Goals and Be Results-Oriented":

- Set specific short-term goals that can be focused on with intent and immediacy by subordinates
- Sometimes it is better to plow around obstacles rather than to waste time going through them
- Your war will not be won by strategy alone, but more by hard, desperate fighting

- Your task will neither be done nor attempted unless you watch it every day and hour, and force it.
- Remember that half-finished work generally proves to be labor lost

Consistent with the preceding, the leader in you sets goals and establishes strategies and tactics to achieve them.

Personal

I spoke with engineer Bill Ratliff who served as a Texas state legislator and Lt. Governor. He chaired committees and took on tasks in the legislature, such as education budgeting, clearly outside of his expertise. He gradually earned the reputation as "expert." As he explained, he just applied logic which is a characteristic of engineers and other technical personnel. Like Bill Ratliff, we engineers and other technical practitioners have the ability to deal with complex situations, like moving toward and achieving a goal, a series of goals, and, eventually, a vision.

Always a Student

Leaders develop and maintain, through formal study, self-study, and experience, their unique set of knowledge and skills. Maintaining one's expertise is a leadership element that is particularly important in a technical organization because rapid changes in science and technology drive the services and products the organization offers and produces.

Competency in knowledge and skills is crucial to an organization for three reasons. First, it contributes to what the organization can do or offer in serving its clients, owners, customers, and constituents. Second, maintaining currency at the individual level sets a positive example for others in the organization, earns respect from them, and encourages them to develop and maintain their expertise. Third, and finally, by being an expert, a person is much more likely to value expertise, people who possess it, and the wisdom of drawing on the proper mix of expertise in meeting needs of internal and external clients, owners, and customers.

Certainly the organization must support financially and in other ways the development and maintenance of individual skills and knowledge. However, the primary responsibility for maintaining expertise lies with the individual. The topic of managing personal professional assets is discussed in detail in Chapter 2.

The leader in you should also be a perpetual student of non-technical topics—of areas of concern and relevance outside of your area of technical expertise. Expertise implies depth of knowledge, in contrast with breadth of knowledge and understanding of context, which are also necessary. Harry S Truman, the 33rd U.S. President, advocated and exemplified gaining breadth of knowledge through reading. He said: "Readers of good books, particularly books of biography and history, are preparing themselves for leadership. Not all readers become leaders. But all leaders must become readers" (Poen 1982).

Personal

Near the end of my graduate studies, my grandmother, who never had an unpleasant word for me and had left school after the fifth grade, said "Stuart, what a shame! You are 27 years old and not working." I was so embarrassed! However, this and other events, led me to recognize that I was the studious type — not necessarily smart – and that I must seek employment and other activities that would enable me to contribute by continuing to be a student. And I have done that for decades and feel fortunate. The "always a student" advice works for me and may work for you. Consider one mechanism I've frequently used to "make the time" to study. Commit to something for which you will have to learn more. Examples: Sign-up to give a brown bag presentation in your office one month from now on a newly-developed technology, submit an abstract to present a paper at next year's geotechnical conference, or join a committee of professional or business society. When we do this, we are relying on a basic human characteristic: We do not want to embarrass ourselves. Works for me!

The notion of a leader who has extensive and largely sufficient knowledge based on education and experience and who uses that knowledge to direct the efforts of narrowly and sufficiently trained subordinates is dimming. The modern leader must continually seek and probe—and expect others to do the same. Rather than claiming to know in a static and superior sense, leaders increasingly focus on knowing how to learn and enabling others to learn as well. Continuous learning will increasingly characterize the world of work.

The perpetual student concept, so important to today's and tomorrow's leaders, is a common thread woven through history. Ancient works of fiction and nonfiction portray the heroes and the elite as relentless pursuers of knowledge and ideas—even when the resulting revelations threaten the seeker. For example, although Socrates probably protested too much, his claim that he did not know the truth but instead diligently searched for it can serve as a model for leaders. Directed discussion, used by Socrates and described by Plato, would seem to be the preferred modus operandi of viable organizations. Socrates convincingly expressed his faith in inquiry and discovery when he said (Plato 1981): "... but I would contend at all costs both in word and deed as far as I could, that we will be better men, braver and less idle, if we believe that one must search for the things that one does not know, rather than if we believe that it is not possible to find out what we do know and that we must not look for it."

Change will increasingly be the only certainty. Accordingly, leaders must create and support an intellectual environment in which discussion is directed toward identifying, interpreting, planning for, and in some cases, influencing the direction and shape of change. Given the increasing complexity of the world, no one person can possess sufficient knowledge to accommodate change. Interdisciplinary teams employing Socratic-style directed discussions are a promising alternative.

However, unlike Socrates' era when information was minimal, data now abound—usually to excess—and must enter into an organization's dialogues. Used in a modern organization, the Socratic method promises to elicit informed and relative contributions from many experts and other individuals, build the confidence of participants, sharpen critical thinking, encourage synthesis, and occasionally lead to serendipity. On the negative side, directed dialogue requires time and patience which are always precious resources in dynamic, action-oriented organizations. Dialogue may be difficult for the leader in some of us because the process may lead to challenges to our dearly-held notions and operating principles. Paradigm paralysis may afflict us. A leader's search for knowledge must be credible—it cannot be conditioned on the expected positive or negative impact of the findings. The truth must be determined and dealt with. The leader in you inquires continuously and is a perpetual student in keeping with the Arabian proverb: "Learning is a treasury whose keys are queries."

Personal

At one point in my career, I interviewed prospective engineering faculty members. I always asked this question: What are you studying? My reason: I did not want anyone in our classrooms who was not a perpetual student. During an even earlier period in my career, I interviewed prospective engineers for departments in a government agency and in an engineering-architectural firm. I did not routinely ask them that question. If I could do it over, I would ask each of them that question. My reason: I wouldn't want any professional serving our clients, owners, or customers who is not a perpetual student. By the way, if you are a perpetual student, that is, a nerd, geek, dweeb, or, as we used to say, an egghead or bookworm, you are in good company. Besides Harry Truman, history tells us that other studious types who led included George Washington, Thomas Jefferson, John Adams, Abraham Lincoln, Theodore Roosevelt, and Herbert Hoover (Phillips 1997; Ruth 2004).

Courageous

The fifth quality of the leader in us is courage. We see lots of physical courage: Mountain climbing, Olympics competition, and bungee jumping come to mind. These are commendable. However, the leader in us must pursue different kinds of courage. Writer and humorist Mark Twain wrote: "It is curious that physical courage should be so common in the world and moral courage should be so rare."

Leading requires courage to hold people accountable for carrying out their responsibilities and keeping their promises, to confront individuals exhibiting unacceptable behavior, to walk away from a project-client-owner-customer on ethical grounds, to speak up when someone is being treated unjustly, to aim high and risk apparent great failure, to apologize and ask for a second chance, and to persist when all others have given up. But, what constitutes courage and courageous people?

Aristotle (1987) offers a thoughtful and demanding perspective on courage. Aristotle defines courage as a precarious, difficult-to-prescribe balance between causes, motives, means, timing, and confidence. He says: "The man, then, who faces and who fears the right things and from the right motive, in the right way and at the right time, and who feels confidence under the corresponding conditions, is brave; for the brave man feels and acts according to the merits of the case and in whatever way the rule directs."

Aristotle goes on to say that courage is a mean between cowardice and rashness, confidence and fear. In summary, he defines courage as a fully informed, carefully-considered willingness to die for a noble cause. Aristotle refutes the notion that courage is reactive or instinctive. You might be tempted to say that Aristotle was not totally serious about his definitions of courage and courageous people—at least with respect to the "willingness to die" aspect. After all, he must have intended death as a metaphor for a willingness to incur great loss. Perhaps this interpretation is acceptable, at least for purposes of this book.

Aristotle outlines, in systematic and exhaustive fashion, five kinds of false courage. These might be referred to as lesser degrees of courage. They encompass much of what passes for courage in our society and help, by elimination, to define bonafide courage.

- The first type of courage is coercion courage or what Aristotle refers to as "the courage of the citizen-soldier." The possessor faces significant risks, but he or she has no choice. Leaders simply have to do many things—some of which are quite unpleasant and risky. Aristotle's coercion courage concept cautions the leader to maintain perspective and not to view these as courageous acts worthy of praise. These acts are part of the job—they come with the territory.
- What might be called high information or calculated courage is the second type. Aristotle uses the example of the professional soldier who seems brave in battle, but in fact entered the fray with far superior information and other resources that virtually guaranteed victory. The modern leader may be tempted to feign courage because he or she often has exclusive access to vital information or wields superior power by virtue of position.
- The third type of courage is passion courage. These reactionary acts conflict with the choice and motive elements clearly evident in Aristotle's model of courage. While the emotional outburst or sharp retort is often viewed as courage—as in "you sure told him/her"—these acts are often done without thought. Although passionate reactions may seem to immediately please onlookers, calm and reason in difficult circumstances may require more courage and lead to long-term benefits for all antagonists.
- Sanguine, to use Aristotle's word, or what might be called overly optimistic courage, is the fourth type of counterfeit courage. A string of business, government, academic, or other successes can lead to unrealistic optimism or even complacency, which may be viewed as courage. The U.S. global dominance in economic and military affairs during the four-decade post-World War II period is an illustration of Aristotle's sanguine courage. The modern leader must be alert and view expectations of continued success with suspicion. An earlier atmosphere of courage that enables an organization to achieve high levels of performance may gradually and unnoticeably give way to complacency.
- Aristotle's fifth and last type of false courage is the ignorance variety. As he bluntly says, "people who are ignorant of the danger also appear brave." As we

become an increasingly information-rich world, our private, public, academic, and volunteer organizations must devote appropriate resources to continuously sifting through new knowledge to identify and assess opportunities and threats.

Informed by Aristotle's ideas, the leader in you is more likely to recognize your and others' bravado. There will always be some pretense of bravery—particularly by people in high and prestigious positions. Recognizing this, the leader in you should place a premium on his or her acts and the acts of others that, in the face of risk and calamity, are carefully considered and indicate a willingness to sacrifice for the corporate or community cause. Courageous acts don't have to be extreme acts. When leaders take extreme positions, they may be less successful in defending a principle, advancing a cause, or achieving a worthy goal than when they assume courageous, but somewhat more moderate postures. The leader in you recognizes various types of false courage and seeks instead a courage that balances causes, motives, means, timing, and confidence.

Calm in a Crisis and Chaos

Whenever competent and committed people are involved in group efforts, often in competition with other organizations, difficult interpersonal and other serious conflicts and situations are inevitable. Confronted, usually unexpectedly, with such crises, the leader instills calm, seeks understanding, and does not make premature judgments. The leader should "Seek first to understand, then to be understood" (Covey 1990).

We've all heard the expression, "this is not a problem, this is an opportunity." The leader in us actually believes this—at least most of the time. When faced with a crisis or chaos, we look for that opportunity. Alexander Graham Bell, the inventor, offered this observation: "When one door closes another door opens; but we often look so long and regretfully upon the closed door that we do not see the ones which open for us."

Personal competency skills such as time management, listening, writing, speaking, delegating, and meeting facilitation are presented in Chapters 2, 3, and 4 of this book. These competencies enable you to effectively and efficiently carry out your day-to-day work. Of particular relevance here is that these skills are invaluable when faced with crises and chaos.

So much for getting your act together. The next way to prepare for handling crisis and chaos is to proactively seek a variety of personal and professional experiences. If you happen to be a civil engineer, please recognize that this profession is one of the few that provides many opportunities to have a career that includes employment in the government, business, and academic sectors. Travel is an excellent way to acquire personal and professional experiences.

Personal

During my first trip to Western Europe in the 1980s, I was astounded by the advanced level of recycling and the robust infrastructure. The fully integrated air-rail-bus-waterway transportation system amazed me. So did citizen concern with the environment. That and subsequent travel experiences influenced my thinking while working on various projects; I saw even more possibilities.

Actively participate in and sometimes lead ad hoc groups formed to carry out specific changes. These might be within your employer, in your professional society, or in your community. By working on these groups, you will rub shoulders with and learn much from individuals who know how to lead. Incidentally, this book's last chapter continues this change theme by offering advice on how you can effect change.

Varied experiences give you a wealth of knowledge, experiences, and contacts to draw on when you encounter crisis or chaos. They reflect the philosophy of regularly getting out of your comfort zone. As a result, when faced with crisis or chaos, you are more likely to say: "This situation reminds me of... "—you get the idea!

Having defused and perhaps even resolved the most recent crisis, the leader—regardless of his or her administrative level in an organization—is usually thrust back into the midst of a general chaotic situation typical of the dynamic organization striving to succeed in an ever-changing world. Engineering and other technical organizations must contend with rapid advances in science and technology; new environmental, personnel, and other laws and regulations; the globalization of business; client and customer demands or dissatisfaction; new competitors or old competitors offering new services in new ways; unexpected business or other opportunities; and turnover of professional and other personnel. Typically, the changes cannot be predicted and defy quantification, but they must be continuously confronted. The leader in you is calm in crisis and chaos.

Creative, Innovative, Collaborative, and Synergistic

Leaders seek ways to utilize the right hemisphere of the brain to complement the left hemisphere. Stephen Covey (1990) notes that decades of research have resulted in brain dominance theory. Covey explains: Essentially, the left hemisphere is the more logical, verbal one and the right hemisphere the more intuitive, creative one. The left deals with words, the right with pictures; the left with parts and specifics, the right with wholes and the relationship between the parts. The left deals with analysis, which means to break apart; the right with synthesis, which means to put together. The left deals with sequential thinking; the right with simultaneous and holistic thinking. The left is time bound; the right is time free.

Clearly, leadership qualities of creativity, innovation, collaboration, and synergism are products of the whole brain. Because of the nature of their college education, engineers and some other technical personnel are likely to be dominated by the left hemisphere of the brain. Cultivation and development of the right hemisphere is, therefore, especially important for engineers and other similar professionals who want to be creative, innovative, collaborative, and synergistic. The section of Chapter 7 titled "Tools and Techniques for Stimulating Creative and Innovative Thinking" can help you further develop these qualities of effective leaders.

According to Covey (1990), leaders create things twice—first in their minds and then in physical reality. Leaders have strong visualization capabilities. Their mental images are vivid and all-encompassing. What will the final structure, facility, system, product, process, organization, event, or thing look like? How will one feel to be in or around it? What will it smell like? How will it sound? The technical professional should resonate with the concept of creating things twice. This is the essence of first preparing plans and specifications—creating on paper—and then constructing, manufacturing, or otherwise implementing, that is, creating physically.

Covey argues that leaders apply the process of creating twice to all aspects of their lives. Somewhat ominously, Covey (1990) warns that "... there is a first creation to every part of our lives. We are either the second creation of our own proactive design, or we are the second creation of other people's agendas, of circumstances, or of past habits." Leaders view the future as something they make happen—not something that happens to them.

The leader in you leverages successes to produce even greater successes. The leader in you finds ways to invest money or other resources at the margin so as to yield larger returns at the margin (e.g., converting a successful project into a successful published professional paper). The leader in you is synergistic—seeking combinations such that the sum is greater than the parts. The leader in you searches for the silver lining in a black cloud. The leader in you believes—at least most of the time—that "this is not a problem, this is an opportunity."

Personal

I worked on an urban flood control project that started out as a technically feasible, but costly, single-purpose stormwater detention pond that would be funded by one source and would have been used only a few times each year. It ended up, as a result of creativity, innovation, collaboration, and synergism, being a cost-effective, multi-purpose facility funded by several sources and now used every day.

THE ENGINEER AS BUILDER

Engineering is an old profession; its roots can be traced back to the beginning of recorded history when nomads first came together and formed communities along what are now the Nile River in Egypt, the Tigris and Euphrates Rivers in Iraq, the Indus River in India, and the Yellow River in China. With the creation of communities came the need to provide basic infrastructure such as housing, transportation, defense, irrigation, water supply, and wastewater disposal. The work of the engineer had begun.

Besides being one of the oldest professions, engineering is one of the broadest. For example, ABET, Inc. the inter-engineering organization that accredits undergraduate engineering programs, recognizes 27 types of engineering programs, ranging from Aerospace Engineering to Systems Engineering, for purposes of undergraduate engineering accreditation (ABET 2011). Within any of the engineering types, engineers carry out a broad spectrum of functions such as research and development; planning; design; construction and manufacturing; operations; teaching; marketing; and management.

Throughout engineering's long history and within its great diversity, however, there is at least one widely-shared interest and function: building in the broadest sense. In the final analysis, whenever everything else is stripped away, the engineer is, at the core, a builder. Building is the glue that binds engineers together.

When civil engineers "build," they usually call the process construction. When mechanical engineers "build," they routinely refer to it as manufacturing. Whatever term you use, the ultimate end of the engineering process is to "build" something to meet human needs, usually something that never before existed. Examples include the water supply system "built" by the civil engineer, the energy-efficient and safe automobile "built" by the mechanical engineer, and the electrical power distribution system "built" by the electrical engineer. Some engineers "build" less concrete but nevertheless important things such as computer programs, better tools for performing engineering functions, and improved ways to organize engineering organizations.

"Builders" accrue great responsibility and liability as well as great satisfaction. U.S. President Herbert Hoover said (Fredrich 1989): "The great liability of the engineer compared to men of other professions is that his works are out in the open where all can see them. His acts, step-by-step, are in hard substance. He cannot bury his mistakes in the grave like the doctors. He cannot argue them into thin air and blame the judge like the lawyers. He cannot, like the architects, cover his failures with trees and vines. He cannot, like the politicians, screen his short-comings by blaming his opponents and hope the people will forget. The engineer simply cannot deny he did it. If his works do not work, he is damned... On the other hand, unlike the doctor, his is not a life among the weak. Unlike the soldier, destruction is not his purpose. Unlike the lawyer, quarrels are not his daily bread. To the engineer falls the job of clothing the bare bones of science with life, comfort, and hope."

As you begin your career, you will probably be increasingly cognizant of the diversity of engineers and the work they do. There is strength in diversity when that diversity is focused on a common and a meaningful interest. For engineering and other similar professions, that common bond is building for the benefit of society, that is "...clothing the bare bones of science with life, comfort, and hope." Building for that purpose is indeed a high calling.

CONCLUDING THOUGHTS: COMMON SENSE, COMMON PRACTICE, AND GOOD HABITS

Some of the content of this book might be correctly referred to as "common sense." An erroneous implication of this statement is the material is obvious and, therefore, does not warrant study or explicit disciplined application. However, experience teaches that which is common sense does not necessarily translate into common practice. Knowing something and using it are not the same. Knowledge is not power; knowledge applied is.

The student or entry-level engineer who is committed to high levels of achievement will take charge of his or her life. He or she will understand the need to translate common sense into common practice through study and self-discipline. When common sense ideas and approaches become common practice, that is, normal or ideally habitual, the young person will be well on the way to realizing his or her potential in the consulting business, industry, government, or academia. Or, as stated by author Og Mandino (1968): "In truth, the only difference between those who have failed and those who have succeeded lies in the difference of their habits. Good habits are the key to all success . . . I will form good habits and become their slave."

Studies by neurobiologists and cognitive psychologists conclude that the unconscious mind controls as much as 95 percent of human behavior (Martin 2008). We are on automatic pilot almost all the time. Have you ever driven your car a few blocks, or even a few miles, and suddenly realized you couldn't recall having done so? Your driving is largely habit and your unconscious mind was "at the wheel." We use our conscious mind for new situations while our unconscious mind—our habits, good or bad—take care of routine activities. My point: If so much of what you do is habitual, then improving any aspect of your life means you need to change your habits.

Covey (1990) describes what he considers to be the seven habits of highly effective people. These habits, which have a common sense tone, are: 1) Be proactive; 2) Begin with the end in mind; 3) Put first things first; 4) Think win/win; 5) Seek first to understand, then to be understood; 6) Synergize; and 7) Sharpen the saw (i.e., renewal of your physical, spiritual, mental, and social/emotional dimensions).

Playing off of Covey's ideas, Green (1995) presents the seven habits of highly ineffective people. They are: 1) Poor listening; 2) Negative thinking; 3) Disorganization; 4) Inappropriateness (i.e., there is a time and place for everything); 5) Decisions by default; 6) Randomization (i.e., performing tasks in random order rather than in logical sequence); and 7) Procrastination.

The opposing Covey-Green seven habits further emphasize the importance of good habits, as stressed by Mandino. You should decide what your most powerful habits are or will be. *Engineering Your Future* will help you learn about highly-effective managing and leading practices, some of which are common sense, so you can perhaps adopt many of them as habits. Let's conclude this introductory chapter with the following highly-relevant thought:

I am your constant companion.
I am your greatest helper or your heaviest burden.
I will push you onward or drag you down to failure.
I am completely at your command.
Half the things you do, you might just as well turn over to me,
and I will be able to do them quickly and correctly.
I am easily managed; you must merely be firm with me.
Show me exactly how you want something done,
and after a few lessons I will do it automatically.
I am the servant of all great men.
And, alas, of all failures as well.
Those who are great, I have made great.
Those who are failures, I have made failures.
I am not a machine,
though I work with all the precision of a machine.
Plus, the intelligence of a man.

You may run me for profit, or run me for ruin;
it makes no difference to me.
Take me, train me, be firm with me,
and I will put the world at your feet.
Be easy with me, and I will destroy you. Who am I?
I am habit!

(Ed Hirsch, baseball coach)

CITED SOURCES

ABET. 2002. *Accreditation Yearbook*. ABET, Inc.: Baltimore, MD.

ABET. 2011. "Criteria for Accrediting Engineering Programs." (www.abet.org/Linked%20Documents-UPDATE/Criteria%20and%20PP/E001%2010–11%20EAC%20Criteria%201-27–10.pdf). ABET, Inc: Baltimore, MD. April.

American Academy of Environmental Engineers. 2009. *Environmental Engineering Body of Knowledge*. AAEE: Annapolis, MD.

American Society of Civil Engineers. 2008. *Civil Engineering Body of Knowledge for the 21st Century: Preparing the Civil Engineer for the Future - Second Edition*. ASCE: Reston, VA.

American Society of Civil Engineers. 2007. *The Vision for Civil Engineering in 2025*. ASCE: Reston, VA.

American Society of Civil Engineers. 2009. *Achieving the Vision for Civil Engineering in 2025: A Roadmap for the Profession*. ASCE: Reston, VA.

Aristotle. 1987. *The Nicomachean Ethics*. Translated by D. Ross and revised by J. L. Ackrill and J. O. Urmson. Oxford University Press: Oxford, Great Britain.

Badger, W. W. 2007. Personal communication. Professor, Del E. Webb School of Construction, Arizona State University, August 25.

Bennis, W. G. 1989. *Why Leaders Can't Lead—The Unconscious Conspiracy Continues*. Jossey-Bass Publishers: San Francisco, CA.

Brown, T. L. 1990. "Leaders for the '90s." *Industry Week*, March 5, p. 34.

Clark, B. and R. Crossland. 2002. *The Leader's Voice*. SelectBooks: New York, NY.

Collins, J. 2001. *Good to Great: Why Some Companies Make the Leap and Others Don't*. HarperCollins: New York, NY.

Covey, S. R. 1990. *The 7 Habits of Highly Effective People*. Simon & Schuster: New York, NY.

Cross, H. 1952. *Engineers and Ivory Towers*. Edited by R. C. Goodpasture. McGraw-Hill: New York, NY.

Engineers Council for Professional Development. 1974. *Make Your Career Choice Engineering*.

Finzel, H. 2000. *The Top Ten Mistakes Leaders Make*. Cook Communications: Colorado Springs, CO.

Fredrich, A. J. (Ed.). 1989. *Sons of Martha: Civil Engineering Readings in Modern Literature*. American Society of Civil Engineers: New York, NY.

Goodale, M. J. 2004. "Leadership—Could You Be More Specific?" Focus on Strategic Planning, e-newsletter of Zweig White. September 5.

Green, L. 1995. "The 7 Habits of Highly Ineffective People." *American Way*. August 15: pp. 56–60.

Jones, D. 2006. "Not All Successful CEOs Are Extroverts." *USA Today*, June 7.

Kanter, R. M. 1993. Personal communication. February 22.

Mandino, O. 1968. *The Greatest Salesman in the World*. Bantam Books: New York, NY.

Martin, N. 2008. *Habit: The 95% of Behavior Marketers Ignore*. Pearson Education: Upper Saddle River, NJ.

Maxwell, J. C. 1993. *Developing the Leader Within You*. Nelson Business: Nashville, TN.

Peyton, J. D. 1991. *The Leadership Way—Management for the Nineties*. Davidson Manors, Inc.: Valparaiso, IN.

Phillips, D. T. 1992. *Lincoln on Leadership—Executive Strategies for Tough Times*. Warner Books: New York, NY.

Plato. 1981. *Five Dialogues: Euthyphro, Apology, Crito, Meno, Phaedo*. Translated by G. M. A. Grube. Hackett Publishing Company: Indianapolis, IN.

Poen, M. M. (Ed.). 1982. *Strictly Personal and Confidential—The Letters Harry Truman Never Mailed*. Little, Brown and Company: Boston, MA.

Ruth, A. 2004. *Herbert Hoover*. Lerner Publications: Minneapolis, MN.

Sanborn, M. 2006. *You Don't Need a Title to Be A Leader*. Currency Doubleday: New York, NY.

Tompkins, J. 1998. *Revolution: Take Charge Strategies for Business Success*. Tompkins Press: Raleigh, NC.

Woolfe, L. 2002. *Leadership Secrets from the Bible*. MJF Books: New York, NY.

ANNOTATED BIBLIOGRAPHY

Allen, J. 1983. *As A Man Thinketh*. DeVorss & Company: Marina Del Ray, CA. (Focusing on the mind, this short, uplifting book offers numerous thoughts such as "A man cannot directly choose his circumstances, but he can choose his thoughts, and so indirectly, yet surely, shape his circumstances" and "Thought allied fearlessly to purpose becomes creative force.")

Berson, B. R. and D. E. Benner. 2007. *Career Success in Engineering: A Guide for Students and New Professionals*. Kaplan Publishing: Chicago, IL. (Written by two engineer practitioners, this book offers advice starting with when the aspiring engineer makes the "transition from college to the real world." The focus is on nontechnical, that is, professional practice topics.)

Hatch, S. E. 2006. *Changing Our World: True Stories of Women Engineers*. ASCE Press: Reston, VA. (Using stories of 238 engineers, this book offers this woman's perspective: "engineering is an awesome way to change the world.")

Hill, N. 1960. *Think and Grow Rich*. Fawcett Crest: New York, NY. (Makes a case for having goals and a plan to achieve them and for the power of visualization and the subconscious mind.)

Leuba, C. J. 1971. *A Road to Creativity – Arthur Morgan – Engineer, Educator, Administrator*. Christopher Publishing House: North Quincy, MA. (Describes how Arthur Morgan, born of modest means, but with a stimulating home environment, creatively sought out and succeeded in a wide variety of professional roles. Describes his accomplishments as a water control engineer, founder of an engineering firm, creator of the Miami Conservancy District, rescuer of Antioch College, and organizer of the Tennessee Valley Authority.)

McCarthy, S. P. 2000. *Engineer Your Way to Success*. National Society of Professional Engineers: Alexandria, VA. ("Dozens of upper-level engineering executives, speaking to young engineers, explain what it takes to get ahead in the engineering profession today.")

McCullough, D. 1977. *The Path Between the Seas: The Creation of the Panama Canal 1870–1914*. Simon & Schuster Paperbacks: New York, NY. (A study of various approaches, successful and unsuccessful, to leading and managing.)

Walesh, S. G. 2004. *Managing and Leading: 52 Lessons Learned for Engineers*. ASCE Press: Reston, VA. ("The 52 lessons in this book present useful ideas for ways to more effectively approach the non-technical, soft-side aspects of working with colleagues, clients, customers, the public, and other stakeholders.")

Williams, F. M. and C. J. Emerson. 2008. *Becoming Leaders: A Practical Handbook for Women in Engineering, Science, and Technology*. ASCE Press: Reston, VA. (Structured in handbook style, this book offers practical actions that individuals and organizations can take to help women lead as students, career women, faculty members, and as managers in various sectors.)

EXERCISES

Note to Instructors: When this book is used as a course text, consider assigning Exercises 1.2, 1.3, 1.4, and/or 1.5 near the beginning of a semester or quarter so that students can work on them over a large portion of the semester or quarter. With this approach, students will be able to combine work on the exercises with material being presented in the course.

1.1 GOALS: The purpose of this exercise is to motivate you, the aspiring engineer or young practitioner, to think about your professional and other goals and how you plan to achieve them. The results of this exercise should be confidential, unless you want to share them with your instructor or other trusted persons. Suggested tasks are:

A. Write your goals, for 2 and 10 years from now, in each of the following areas: a) position (e.g., project engineer, project manager, instructor/professor, researcher, owner); b) annual salary and other income; c) function (e.g., design, marketing, construction, manufacturing, teaching, research, management); and d) other (e.g., international travel; present a paper; serve as officer in professional, community, or other organization; start your own business; hold elective office; earn a PhD; take a year off).

B. For each of the four goals, identify one specific thing you will do within the next year to move toward the goal. Note: You are in effect "planning a trip." How are you going to get to your destination? Do you have the necessary knowledge, skills, and attitudes or a way of obtaining them? Probably not, but you can obtain or develop them. Or are you going to let chance rule perhaps using the rationale that everything will come to you if you "just work hard?" If the chance route appeals to you, take a quick look at the contrary ideas presented at the beginning of Chapter 15.

1.2 ASSEMBLE PROJECT INFORMATION: This exercise's purpose is to provide a profile for engineering or engineering-related project, unique to you, for subsequent use in one or more assignments. Suggested tasks are:

A. Select a technical "project" you worked on or are working on. Examples are: a project you did during co-op, something you worked on during a summer job, a volunteer effort with some technical content, a design project in one of your courses, or your capstone project. The project must have at least 15 different tasks or steps.

B. Prepare a memorandum that includes: a) A description of your relationship to the project. That is, when, how, and why are or were you involved? b) A list of project tasks, in approximate chronological order, recognizing that some tasks may overlap. Assign a letter or number to each task.

1.3 BOOK REVIEW: The purpose of this exercise is to provide you with an opportunity to study, in depth, one professional/business author of your choice, subject to instructor approval, and to critique the thesis of the book. In so doing, you will be further introduced to the broad range of leadership and management literature with the hope that you will continue to read critically in this area. Suggested tasks are:

A. Select one business/professional book that addresses some aspect of leadership and/or management. The book could be recent or it might be old or even what you or others consider a classic. Some sources are books listed in the Cited Sources and Annotated Bibliography sections of this book's chapters, books reviewed in book sections of newspapers and magazines, books recommended by others, and those you find by searching under "leading," "managing," or similar words.

B. Request approval of the book from your instructor.

C. Read the book and prepare a review in which you do the following: a) cite your book (e.g., name, author, publisher, date), b) describe some of the key ideas and/or theses presented in the book, c) identify the evidence in support of the ideas/theses, and d) indicate whether or not you agree with the key ideas/theses. Refer to Chapter 3 of this book for writing guidance.

1.4 RESEARCH PAPER—INDIVIDUAL STUDENT VERSION: The intent of this exercise is to provide you with a means of studying, in depth, a leading/ managing topic of largely of your choice. This can broaden and deepen your knowledge, increase your awareness of the leadership and management literature, and strengthen your research and writing abilities. Suggested tasks are:

A. Select a leading/managing topic. To get you started, but at the risk of unintentionally confining your thinking, an array of varied topics follows to stimulate your thinking: cost control in engineering-construction-manufacturing, creativity and innovation, decision making, design-build, dual ladder organizational structure, engineering and/or other licensure laws, ethics, failures and learning from them, globalization, history of some aspect of engineering or technology, lean construction or manufacturing, liability, partnering, quality control and quality assurance, reengineering, risk, robotics, sustainability, and teamwork.

B. Request approval of the topic from the instructor.

C. Research your topic by drawing on a variety of sources. Consider using one or more personal sources that you contact in person, by telephone, e-mail, or letter. If you use a personal contact, cite them at the end of your paper using this format: Smith, J. A., personal communication, Director of Engineering, XYZ Company, Chicago, IL, January 26, 2012.

D. Write the paper. Assume your reader is an engineering or other technical profession major who knows little about your topic. Use the writing section of Chapter 3 for guidance.

1.5 RESEARCH PAPER—TEAM VERSION: Similar to the preceding exercise, except that the topic is to be selected and researched with the paper being written by a team. The purpose of this exercise is, in addition to the purpose stated for the preceding exercise, to simulate the team-oriented manner in which engineering work is done in practice. A possible approach for this exercise is to select a country and research the following aspects of engineering in the selected country: a) overview of country (e.g., location, size, population, topography, economy), b) engineering education, c) engineering licensing, d) engineering professional organizations, e) area(s) of science and technology for which the country is considered a leader, f) image and/or social status of engineers, and g) any other aspect of engineering that interests one or more members of the team. In addition to providing an instructive team experience, this approach could heighten student awareness of international similarities and differences in the approaches to engineering education and practice as well as variations in the status of engineers and the engineering profession recognizing that today's engineering students are very likely to work in a global engineering and business setting.

1.6 YOUR LEADER: The purpose of this exercise is to increase your sensitivity to the presence and influence of individuals who lead. As with Exercise 1.1, the results of this exercise should be confidential, unless you want to share them with your instructor or other trusted persons. Suggested tasks are:

A. Some of us are privileged to work for, or associate with, individuals who lead. Such situations are uplifting and instructional. Think of someone who led you or your group in a new direction, who enabled you or your group to see new possibilities and then achieve them.

B. Write about the experience noting, in particular, the qualities of the person who led. Compare them to the seven qualities of effective leaders presented in this chapter.

CHAPTER 2

LEADING AND MANAGING: GETTING YOUR PERSONAL HOUSE IN ORDER

The real contest is always between what you've done
and what you are capable of doing.
You measure yourself against yourself and nobody else.

(Geoffrey Gaberino, Olympic swimmer)

This chapter stresses the importance of getting your personal house in order as you move through your formal education and then into practice. It begins by discussing the roles-then goals idea which is followed by time management tips. The full-time employment versus full-time graduate school quandary is analyzed. The chapter then moves on to the culture shock that young technical professional may encounter during the transition from formal education to practice. Suggestions are offered on how to get off to a good start in your first professional position. Chapter 2 concludes with advice on three aspects of managing personal professional assets, namely continuing education, involvement in professional organizations, and professional licensure.

START WITH YOU

Before you, as a student of engineering or a similar profession or as an entry-level person, can be broadly involved in the managing and leading functions discussed in Chapter 1, you must lead and manage yourself. That is, whether on a day-by-day or year-by-year basis, what do you want to accomplish and how are you going to do it?

Consider a balance between success and significance as a way of thinking about your career. Define success as that which benefits you and perhaps your family or other dependents. Success indicators include income, net worth, and material possessions. In contrast, significance is your positive impact on others. Success is about you, or me, and our "stuff" while significance is mostly about others. British Prime Minister

Winston Churchill described the difference between success and significance when he said "We make a living by what we get and we make a life by what we give." Both success and significance are important. A comfortable, financially-secure lifestyle may be high on your agenda. But, you may also seek significance, may want your life to mean more than successfully accumulating "stuff." If you resonate with the success – significance career and beyond model, then begin pondering the success – significance balance that suits you.

Personal

I had a big corner office, a company car, and special perks. That was nice—that was success. At the same time I managed a project that resulted in the construction of a multi-purpose stormwater management-recreation facility. I see it often—it is heavily used—and hardly anyone remembers my role and that's not important. What is important is the positive impact that project has had, and will continue to have, on many people. That's my idea of significance.

Keeping your personal house in order requires self-discipline and most professional program students come a long way in that area during college. As illustrated in Figure 2.1, being able to manage yourself is the first step in a staircase of possibilities, an array of areas within which you can have a positive influence, be successful, and achieve significance within your profession and within society.

TIME MANAGEMENT: BUT FIRST ROLES AND GOALS

Time is a Resource

College students and young, harried practitioners often claim they "don't have the time" as though they have less time than other people. Each person has 24 hours a day and 365 days per year. Between college graduation and retirement at around age 65, you have about 400,000 hours at your disposal. Some engineering and other technical program graduates will use their time wisely to achieve much in their personal, family, financial, community, and professional lives. They will achieve their desired mix of the previously-mentioned success and significance. Others will fill many of those 400,000 hours with mediocrity. These individuals might live their last years regretting that they didn't do more with their gift of time. "We must all suffer one of two pains: the pain of discipline or the pain of regret," according to motivational speaker Jim Rohn, "discipline weighs ounces while regret weights tons." Without careful management, much of the time allotted to each professional is lost forever.

Roles, Goals, and Then, and Only Then, Time Management

The expression "time management" prompts images of a watch or clock. However, before thinking of a timepiece, consider a compass (Covey 1990). Before focusing on

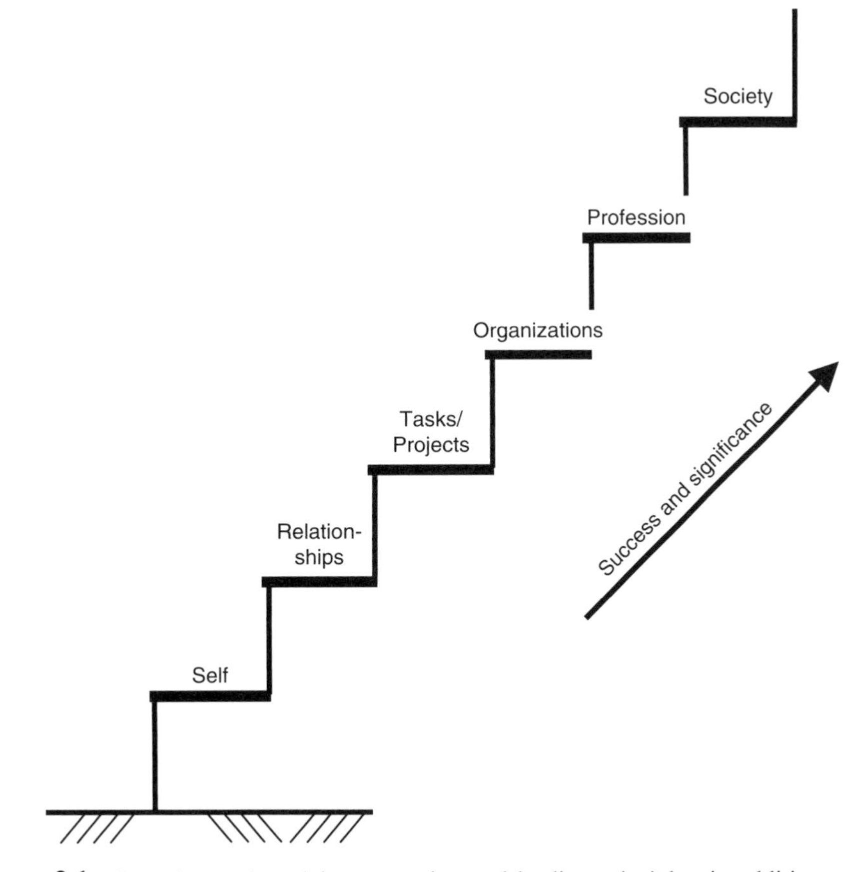

Figure 2.1 Learning and applying managing and leading principles, in addition to maintaining technical competence, will enable you to move up a staircase of career and life possibilities while enjoying continued success and significance.

the speed of your life, think about its direction. In other words, first choose your roles, then set goals for each, as illustrated in Figure 2.2. Then, and only then, focus on managing your time.

Adopt a holistic approach. Select your key roles in life, at least for the foreseeable future. Then establish goals and create action plans that will help you fulfill those valued roles. Mahatma Gandhi, Indian nationalistic leader and nonviolence advocate, stressed the need to identify key roles and act on them when he said: "One man cannot do right in one department of life whilst he is occupied doing wrong in other departments. Life is one indivisible whole."

A non-work role that many technical professionals are likely to share is parent. Other common non-work roles are daughter, son, wife, husband, neighbor, member of a religious group, athlete, and friend. Examples of profession-related roles are designer, administrator, marketer, project manager, partner, mentor, committee member, professor, and officer in a business, service, or academic society. Celebrate your goals and

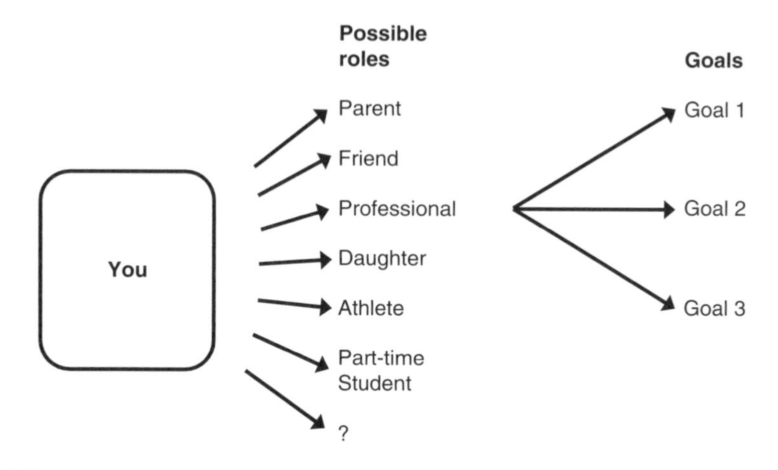

Figure 2.2 First choose your desired roles, set goals for each, and then focus on time management.

your efforts to achieve them. Your goals should be a source of pleasure in that they give added meaning to your life.

Clearly, you can establish goals without first defining roles. The danger is that you will inadvertently omit or diminish important segments of your being. You risk incurring deep regrets that cannot be remedied. In contrast, the suggested "roles—then goals" process leads to a balanced life (Walesh 1999).

Historic Note

As a teenager, Arthur Morgan wrote this: "I don't intend to be commonplace. I intend to make a great person of myself. . . great in having fulfilled my possibilities: great in having seen which of my possibilities are greatest." Are these the naïve thoughts of an arrogant young person or the serious reflection of a confident individual desiring to practice stewardship with his gifts? His life suggests the latter in that Morgan, an engineer who lived from 1878 to 1975, authored drainage laws for Minnesota, founded an engineering firm, created the Miami Conservancy District, rescued Antioch College, and organized the Tennessee Valley Authority (Leuba 1971).

Time Management: The Great Equalizer

While talent and intelligence will influence the student and the entry-level technical person's success or failure, how each uses his or her allotment of time will have a profound influence. As management expert Peter Drucker said, "time is our scarcest resource and unless it is managed, nothing else can be managed." Unlike talent, intelligence, wealth, and other personal attributes and assets, time is distributed equally. Accordingly, effective use of time can be the great equalizer.

Samuel Colt's pistol was the equalizer in the old west of the United States. Regardless of your size, you were equal to others if you knew how to use the Colt. Regardless of your talent, intelligence, and other personal attributes, you can be equal to others if you thoughtfully manage your use of time. You have the power, if not the responsibility, to carefully manage how you use your time in fulfilling your chosen roles and pursuing your chosen goals. Lives are built on the use of minutes, hours, and days. No one can get more time, but anyone can do more with what he or she gets.

In a narrow sense, time management means getting more productivity out of a person's allotment of time by efficiently handling communication, telephone calls, correspondence, meetings, and various tasks and responsibilities. In a broad sense, time management means going well beyond these practical tools and techniques and building a meaningful life in accordance with your desired roles and supporting goals. How the young professional decides to use his or her minutes, hours, and days determines how that young professional will spend his or her life. If you don't decide how to use your time, be assured that others will decide for you.

Time Management Tips: The ABCs

Table 2.1 lists 26 time management tools and techniques. Because there are 26 tips, they can be viewed as the management ABCs. Although some are self-explanatory, all are discussed because of the crucial importance of time management for the student and young professional.

Table 2.1 Choose from among these 26 time management tips—ABCs—to help you fulfill your roles and achieve your goals.

A. Define roles, set goals, and determine action items	N. Use discretionary time wisely
B. Plan each day	O. Avoid telephone and email tag
C. Act immediately and constructively	P. Delegate
D. Bring solutions—Accompany each problem with at least one solution	Q. Keep door closed but access open
E. Identify and use your best time	R. Write it down—Document-document-document
F. Organize your space—Create an efficient work environment	S. Network
G. "Chain" yourself to the task until it is done	T. Minimize "toxic" situations
H. Distinguish between efficiency and effectiveness	U. Use travel and waiting time productively
I. Create professional files	V. Meet with yourself
J. Keep related materials together	W. Log your time
K. Meet only when necessary	X. Adopt a holistic philosophy
L. Apply the 20/80 rule	Y. Guarantee small successes
M. Break projects into parts	Z. "Goof off"

Personal

A participant in one of my time management webinars said that if anyone did all the things suggested in the webinar—all of these ABCs—there would be insufficient time to get any work done! I don't expect you to do everything. Select those tips that seem new, or remind you of something valuable you once did, and apply them. If useful, they will become habits—habitual actions or procedures take little or no thinking or time. I habitually practice most of the tips.

Tip A–Define Roles, Set Goals, and Determine Action Items: Tip A repeats the earlier admonition to select roles and set goals. It is the foundation of your time management and warrants repetition. Conceptualize, refine, and write-out monthly, annual, and multi-year goals for personal, family, financial, community, and professional areas and affairs. Clear goals, consistent with your selected roles and quantified to the extent feasible, are crucial to charting and navigating the seas of a business and professional career.

Think of your daily activities as falling into one of the four quadrants shown in Table 2.2. As suggested by the four quadrants, you can easily get diverted, goals keep you on track. So does consciously thinking about what you are doing, how productively you are using your time.

Consider these examples of one possible activity for each of the four quadrants:

- Example of I-U: "Boss" says do this now.
- Example of NI-U: The office "whiner" drops in to complain, to play "ain't it terrible" or "did you hear what they are doing now?"
- Example of NI-NU: Reading junk snail mail or junk email.
- Example of I-NU: Completing an action item related to one of your goals.

Clearly, you should try to spend most of your time professionally, and perhaps beyond, in the first, that is, important row.

Perhaps the acronym SMART will help you formulate your goals. Each goal should be:

- *S*pecific
- *M*easurable—cast in quantitative terms if at all possible
- *A*chievable—while the goal will stretch you, you must be able to accomplish it

Table 2.2 Time can be categorized into four quadrants on an urgency and importance basis.

	Urgent (U)	Not Urgent (NU)
Important (I)		
Not Important (NI)		

(Source: Adapted from Covey 1990)

- ***R*elevant**—the goal is appropriate for your chosen roles and, as appropriate, your organization's current circumstances
- ***T*ime-framed**—you establish a schedule for achieving the goal or its components.

"Manage a big project" might be a worthy goal, but it lacks specificity. Use the SMART approach to articulate a more specific goal such as: Profitably manage a transportation project with a contract amount of at least $100,000 by the end of next year.

Practice "proper selfishness" (Handy 1998), that is, regularly get into the Important—Not Urgent quadrant shown in Table 2.2. Listen to Scottish novelist and poet Robert Stevenson: "Don't judge each day by the harvest you reap, but by the seeds you plant." When you are in the upper right quadrant, you are practicing proper selfishness. You are planting seeds—for yourself, your family, your organization, those you serve, and your community.

Consider selectively sharing your goals with family, friends, selected colleagues, the "boss," and some subordinates as a means of enlisting their cooperation and support. Most people won't ask about an individual's goals because they think the topic is too personal or because they don't think in terms of goals. But they are likely to be interested if the information is volunteered and they are often, with that knowledge in hand, in a position to help you achieve your goals.

Assume, for example, that one of an aspiring professional's goals is to present a technical paper at a regional or national conference. Obviously, the individual would review the calls for papers published by professional organizations to find an appropriate opportunity. However, by also sharing this goal with colleagues, particularly those who are involved in professional organization activities, the young engineer or other technical person, in effect, enlists the potential help of many individuals and is more likely to achieve the goal.

Personal

At one point in my career, I set a goal of being the manager of one of our firm's offices. I had worked hard and had good ideas that I was eager to implement. However, when the new manager was named, I was not selected. To make matters worse, I learned later that I wasn't even considered? One reason: I had foolishly failed to share my goal.

Don't assume that friends, colleagues, professors, supervisors, supervisees, and others know your goals by virtue of working with or near you. Similarly, thinking that all good things will come your way if you simply "put your nose to the grindstone" is naive. Rather than being timid or overly modest about professional and other goals, adopt a proactive communicative approach and then be prepared to follow through as opportunities arise.

Tip B–Plan Each Day: Plan each study or work day, prioritizing tasks, in writing, according to importance. Such planning, which might be done the evening before or first thing in the morning, will require less than 15 minutes of time but be highly productive. You will realize a great return on the investment of your planning time. Victor Hugo, the French writer and romantic who turned to politics in later life, said:

He who every morning plans the transactions of the day
and follows that plan carries a thread that will guide him
through the labyrinth of the most busy life.
The orderly arrangement of his time is like a ray of light
which darts itself through all his occupations.
But where no plan is laid, where the disposal of time is
surrendered merely to the chance of incidence,
chaos will soon reign.

In other words, if you do not plan the use of your time, you will probably fail to use it effectively.

Others suggest (e.g., Covey 1990) that, in addition to planning each day, active professionals should also plan each week. Lee Iacocca earned an industrial engineering degree and rose to be the chief executive officer of the Chrysler Corporation. His autobiography (Iacocca 1984) indicates that he devoted Sunday evenings to planning his week's activities. He did this partly as a matter of organization and partly as a way of focusing his energies. He said "every Sunday night I get the adrenaline going again by making an outline of what I want to accomplish during the upcoming week."

Tip C–Act Immediately and Constructively: After interacting with someone, try to immediately do something constructive. For example, after reading an e-mail, hanging up after a telephone conversation, or saying good-bye to someone who visited you, take a specific action such as scheduling a meeting, drafting an email or memorandum, obtaining a file, or asking a colleague for information. Rather than only making a note to do some tasks and adding that note to your "to do" pile, try to take one meaningful step.

With specific reference to materials and requests that come to you, try to respond immediately as shown in Figure 2.3. Act, that is, do something useful, file the input electronically or in hard copy, or discard or delete. To reiterate, setting things aside somewhere in your office or work area until you "have the time" will simply result in a net accumulation of materials.

You may have trouble acting immediately and constructively because you are a procrastinator. "Procrastination is opportunity's natural assassin" according to Victor Kiam, Remington Shaver Company owner. Have you ever hesitated too long—procrastinated—and lost an opportunity? You may be controlled by fear of imperfection, the unknown, success, change, responsibility, finishing, rejection, and mistakes (Emmett 2000). Consider imperfection. Maybe you are reluctant to finish a task, or maybe even start it, because the project or result will not be perfect. Some ways to tackle procrastination are: give yourself deadlines, don't duck difficult problems, and avoid perfectionism paralysis (Alessandra 2004).

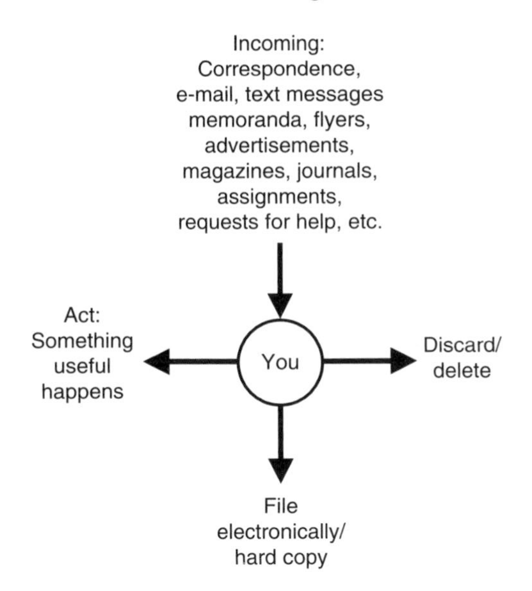

Figure 2.3 Act immediately and constructively.

Tip D–Bring Solutions—Accompany Each Problem with at Least One Solution: Make sure that you provide solutions when you present a problem to your supervisor. Insist that subordinates bring solutions when they bring problems. Anyone who is perceptive enough to identify a problem is capable of suggesting at least one solution to the problem.

Writing a description of a problem is a disciplining activity that focuses thinking and enhances understanding of a problem's facets. Writing encourages you and others to think deeper and wider. This, in turn, often leads to solutions in that a problem well defined is partially solved. Whatever your personal style, don't rush to your supervisor immediately upon discovering a problem. Think about it, perhaps write about the problem, and identify at least one feasible solution. Then talk to your supervisor. This will enable both you and your supervisor to make the best use of your collective time. Expect the same of people who report to you.

Tip E–Identify and Use Your Best Time: Identify your best time of day in terms of energy level and intellectual and creative ability, and try to schedule your most challenging study or work tasks into that time period. Some individuals are morning people or larks, some are night people or owls (Zaslove 2004). Others are most productive and creative at other times of the day. Recognize this in yourself and others and use it to encourage effective time utilization.

Tip F–Organize Your Space—Create an Efficient Work Environment: Examine the arrangement of your office, cubicle, or other study or work space. Small changes can result, over time, in time saved. For example, consider moving the printer closer to you, putting frequently used reference books and materials within arm's reach, and/or using your desk drawers as "in and out baskets." The elements of an effective and efficient office are highly subjective. Contemplate what enhances your outlook such as lighting, the view, photos, and other memorabilia. What increases the likelihood that you will do the right things and do them well?

Try keeping your desk clean of all but your current project. This suggestion is offered because most people are easily distracted by materials in their area of peripheral vision, particularly if those materials remind them of other tasks or projects that need attention.

Personal

I've had several opportunities to design a "perfect" office. In one case, my office overlooked a park with ponds. A bright yellow, functional desktop ran down the center. Seventeen two-drawer file cabinets supported the desk and provided counter space for office equipment, books, and other materials. Worked for me! Try to tailor your study or work space to work for you.

Tip G–"Chain" Yourself to the Task Until it is Done: Commit to studying, analyzing, or writing for an hour; calculating for half an hour, or emailing for 30 minutes. You may resist this chaining tip (Wetmore 2003) arguing that it won't work because you are frequently interrupted. And who is interrupting who? Pattison (2008) claims that half of the interruptions experienced by U.S. office workers in high-tech companies are self-interruptions, that is, "jumping from task to task." We frequently interrupt ourselves! The author also notes that a significant time period is needed to get back "on task." See the book by Jackson (2008) for a similar message.

Or you may object to this "chaining" suggestion because you are a multi-tasker. Like a grasshopper, you jump from task to task. You google, you blog, you email, you text—and you did all of that in just the last five minutes! You are certainly busy, but are you effective and efficient? Sources, such as those cited above, remind us that activity is not necessarily progress. The bottom line: Prioritize your tasks and then focus on one task for a significant period of time.

Multitasking is inefficient. In an intriguing and informative article titled "E-mail Is Making You Stupid," Robinson (2010) says "People may be able to chew gum and walk at the same time, but they can't do two or more thinking tasks simultaneously." About 2000 years earlier, Publilius Syrus, the Latin writer of maxims, said "To do two things at once is to do neither." Multitasking, which is really jumping from thinking task to thinking task to thinking task, etc. is very inefficient because of the time, perhaps unnoticed, but nevertheless needed, to resume a task.

When the task or a series of tasks is finished, reward yourself! Kick back, grab a cup of coffee, enjoy one of those candy bars hidden in your desk, or take a walk around the office or the block. "Stressed" spelled backwards is desserts! Offset some of that intense work with one or more pleasurable "desserts."

Speaking of "chaining" yourself to task and the clock, a related tactic is to group similar activities such as analyzing, designing, writing, telephoning, and emailing. Grouping increases efficiency. What I'm calling grouping is also referred to as blocking, as in dedicating a block of time to one type of activity. The idea is for you to "get in a groove" on one activity during one sitting and be more efficient (Zaslove, 2004).

Tip H–Distinguish Between Efficiency and Effectiveness: Distinguish between doing things right and doing the right things, giving preference to the latter. Avoid doing useless tasks efficiently. A situation like that often arises because the original need for a report or other action has vanished but the process, which has been well-established, is blindly adhered to.

Consider the following story. An ambitious and inquisitive worker starts a new production line job at a tire manufacturing plant. While working diligently during her first week on the job, she asks many questions. As the work week winds down, she asks her supervisor a final question. "Why do we wrap each finished tire in brown paper?" His answer: "To protect the whitewalls." Her response: "But I learned that this plant stopped making whitewalls ten years ago." The factory workers were simply doing the wrong thing. They may have been efficient, but they were not effective.

Personal

Consider a personal experience that illustrates the need to first be effective. For many consecutive Fridays, John, the Assistant Office Manager, and I, a department head, wrote a brief status report about all of the wastewater projects being conducted in our office of an engineering firm. We immediately sent the report to another of the firm's offices. This task was performed quickly every Friday. We were efficient! One Friday I asked John why we prepared and sent the status report. I thought he knew. He didn't either. In a somewhat flippant manner, I suggested we not prepare the report and see if anyone noticed—somewhat reluctantly, John agreed. You guessed it; no one noticed. We had been doing efficiently something that did not need doing! We were efficient but not effective.

Remain vigilant. Ask this question of yourself, other students, and your colleagues: are we doing the right things or only doing things right? The ideal is to do both. "Doing the right things well" embodies effectiveness and efficiency. Avoid the folly of doing well that which should not be done.

Tip I–Create Professional Files: Recall the three possible actions with "incoming," that is, act-file-discard/delete. Some of that "incoming" is potentially valuable data and information. Therefore, you should have a storage and retrieval system. Please respect the expectations and rules regarding confidential information associated with your employment.

Therefore, beginning in college and continuing into professional practice, develop personal professional files. These files, which will start modestly and are likely to include digital and hard copy items, might be set up based on an initial set of technical and non-technical categories. A systematic file system that can be expanded, as needed, is a partial answer to the problem of trying to read all the material that you receive. Be selective with what you file. Don't use that file cabinet or electronic file as a trash compactor (Zaslove 2004).

Tip J–Keep Related Materials Together: Work out a system to keep all the physical parts of a task or project together as it moves from person to person within an office or between offices. For example, when you finish with your responsibility for a small project such as writing a memorandum, producing a spreadsheet, or preparing a table for a report, and it is time to pass the project on to someone else, ensure all project materials remain together.

Personal

I once worked with an administrative assistant who recognized that, at any time, we were working on many, mostly little tasks or projects. She had an idea for keeping related, in-process materials together: Use a zip lock freezer bag for each project. My initial reaction was one of skepticism. However, it worked very well and I have used this time management method every since. The bags work well as temporary containers because they accommodate 8½″ × 11″ materials and because the contents of the bags are readily visible. Here's one way it worked back then: A letter request would come into our office. The administrative assistant would open it, make an initial assessment, write a note on the letter and/or put something with it, put everything in a bag, and give it all to me. I would review the contents, add some content and/or request some action, put it all in the bag and give it all back to the administrative assistant. The project would go back and forth, using the bag, until finished.

Tip K–Meet Only When Necessary: To the extent feasible, ask or insist that all meetings, begging with your college days, are carefully planned and conducted and that follow-up responsibilities are clearly assigned. This guideline applies to all meetings—including conference calls. Meeting management is so important to personal time management and overall organizational effectiveness and efficiency that a major section of Chapter 4, "Developing Relationships," is devoted to the topic.

Tip L–Apply the 20/80 Rule: Most activities involve input and output, work and results, effort and accomplishments, or other cause and effect relationships. But each unit of input, work, effort, and cause does not necessarily lead to the same relative result. Some kinds of input, work, effort, and cause are more productive than others. This is the concept portrayed by the 20/80 rule shown in Figure 2.4. Simply stated, 20 percent of the total input produces 80 percent of the total results. While there is no strong substantiation for the rule in a quantitative sense, the 20/80 rule—also called the Pareto Principle (Covey 1990)—makes intuitive sense.

Effective time management suggests that input, study, work, effort, or cause be partitioned and that each piece be examined for its individual contribution to the output, results, accomplishments, and effects. Input, which requires time and energy, should not just be considered as a homogenous whole, but rather as a heterogeneous mix of components, some of which are more effective than others. Less effective input should be replaced with more effective input.

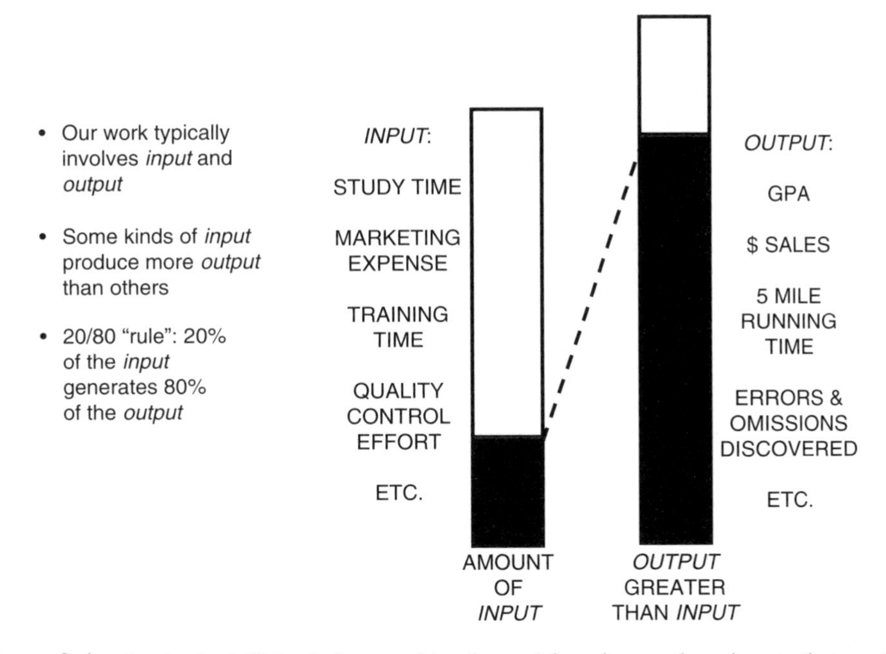

Figure 2.4 Apply the 20/80 rule by searching for and focusing on those inputs that produce the greatest outputs.

College students or graduates should have an intuitive sense of the 20/80, rule based on what they learned about their own personal learning style, that certain types of studying are more effective for any given student than are others. Some students prepare for an examination by reviewing and perhaps even rewriting class notes, going over completed homework, studying with friends, and meeting with their professor. While all means of examination preparation can be used, that means which seems to have the most impact on test performance should be emphasized.

Use an organization's marketing effort as another example of the 20/80 Rule. Marketing inputs or expenses include, but are not limited to, relationship building, advertising, open houses, and cold calls. The 20/80 rule leads you and your organization to identify those marketing inputs that produce a disproportionately larger output and do more of them and less of other inputs. Incidentally, Chapter 14 provides an introduction to marketing of professional services. More broadly, if you accept the 20/80 idea, then continuously search for small shifts in emphasis in essentially everything you do because those little adjustments can lead to large shifts in results (Connellan 2011).

Tip M–Break Projects into Parts: As a result of an engineering or other technical education, most entry-level professionals have developed an ability to examine complex mechanical, electrical, and other physical systems; identify their components; and understand the interrelationships between them. This analytic skill is transferable and applicable to any complex system regardless of whether it is primarily technical or non-technical in nature.

Do not be overwhelmed, as a student or practitioner, by the next big task or project that comes your way. Instead, first break it into parts so that the time invested in the task or project is managed effectively and efficiently. Writer and humorist Mark Twain advocated the "break it into parts" approach by saying "The secret of getting ahead is getting started. The secret of getting started is breaking your complex overwhelming tasks into small manageable tasks, and then starting the first one."

By visualizing the components of a task or project and understanding the relationships between them, you will be prepared to apply the Critical Path Method, which is discussed in Chapter 6, or other network methods. In addition, partitioning a task or project into its component parts provides the basis for effective delegation, which is discussed in Chapter 4 and is another way to make effective use of your time.

Tip N–Use Discretionary Time Wisely: Let's take another look at the time-use quadrants (Table 2.2) introduced in the Tip A discussion. One way of looking at the ways you use time in your work environment is to view it as:

- "Boss"-imposed or professor-imposed, such as doing what he or she asks you to do
- System-imposed, such as completing time sheets and other forms
- Self-imposed, such as getting into the upper right-hand quadrant while carrying out your various student and employee responsibilities

To reiterate the advice offered earlier, invest some time, if not daily at least weekly, in that third bullet which is another way of describing that upper right important, but not urgent quadrant. Ralph Waldo Emerson, schoolmaster, minister, lecturer, and writer offered this advice: "Guard your spare moments. They are like uncut diamonds. Discard them and their value will never be known. Improve them and they will become the brightest gems in a useful life."

Understandably, your "boss" needs to carry out his/her responsibilities. And those needs tend to put you in the upper left, important-urgent quadrant. Understandably, and legitimately, you have your goals. You want to invest some time in the upper right, important—not urgent quadrant.

Inevitably, you may experience conflict between what your "boss" wants you to do and what you want to do. Given this tension, how can you regularly get into the upper right quadrant? Consider these ideas:

- Carefully select your "boss," that is the person you will report to. This advice is especially important early in your career when your "boss" is most likely to influence your attitude and your acquisition of knowledge and skills.

Personal

When I was a professor, I recall senior engineering students coming back from job interviews. They described employment opportunities in terms of geographic locations, types of projects, technology they would use, and salary and benefits. Rarely did they say anything about their potential or likely supervisor. I suggested to them that the characteristics of the individual they would report to was of prime importance.

- Make sure you understand your responsibilities. Maybe your "boss" is overbearing and highly-intrusive because you are not voluntarily doing what he or she expects you to do, that is, you do not fully understand your responsibilities.
- Try to understand your "boss's" situation, that is, his or her responsibility, style, and goals.

Personal

I worked for a government agency and learned that the "boss" wanted national recognition for the organization. My upper right-hand quadrant goals included preparing, presenting, and publishing professional papers and expanding my network of professionals. Because my goals were aligned with the goals of my "boss," I received support and we enjoyed a "win-win" situation.

- Share your goals. Your "boss" and others can help you achieve your goals only when they know them. Recall the earlier personal story of not even being considered for the office manager's position because of failing to selectively share that goal.

Tip O–Avoid Telephone and Email Tag: Avoid "telephone tag," the wasteful practice by which two individuals repeatedly place calls to each other but do not connect or in any way exchange useful information. Try to make an appointment for a telephone meeting that you will initiate by calling the other party at the agreed-upon time. Leave a specific voice mail message, ask to speak to someone else who might be able to assist, or leave an intriguing voicemail message that might induce the elusive person at the other end of the line to call you. Email or texting tag can also occur resulting in similar frustration.

Given the difficulty of contacting some people by telephone, ensure the call will be productive by preparing a list of topics you would like to discuss, questions you need answered, and information you want to share. This written agenda could also serve as the basis for your written documentation of the telephone conversation.

Leave a specific voice mail message or send a specific email or text request. Be more specific than "Mary: Need to talk to you." "Mary: Need to talk to you about reinforced concrete" is better. An even more effective voice mail, email message, or text message would be "Mary: Need to talk to you about using reinforced concrete for pilings and decks in a saltwater marina"—not long, but specific and even more likely to raise Mary's interest.

Tip P–Delegate: Delegate appropriate tasks, along with the necessary authority, to other capable individuals. Effective delegation, while challenging, can greatly increase your time utilization effectiveness and efficiency and offer organization-strengthening benefits. Many professionals do not delegate either because they have personal "reasons" for not delegating and/or they are in a culture that does not encourage and support delegation. Reluctance or inability to delegate results in wasteful use of your time and talents and the time and talents of others. Delegation is discussed in detail in Chapter 4.

Personal

I led a department in an engineering firm. One of our departmental policies was that engineers, including entry level engineers, should be responsible for at least half the time of one paraprofessional, such as a technician or technologist. Benefits of such a delegation-oriented policy include: 1) Reduced task cost because work is "pushed" down to the lowest cost-per-hour person capable of competently performing the work, 2) Increased morale because most personnel are routinely working at their higher levels of competence and/or are being challenged to learn more, and 3) Development of planning and management knowledge and skills among personnel because of the need to be an effective and efficient delegator.

Tip Q–Keep Door Closed but Access Open: Consider adopting, where feasible based on the physical situation and the corporate culture, a closed-door procedure with respect to your office, cubicle, or workspace while implementing an open-door policy with respect to access to you. The closed door procedure doesn't mean that, if you have a door, you literally should keep it closed. It means you should discourage unnecessary interruptions. If you work in a cubicle, discouraging unnecessary interruptions may be as simple as how you orient your desk and chair relative to the cubicle's entrance.

Consider keeping an "Interruptions Log" for a week (Wetmore, 1999). Look for patterns such as "Bill came to me six times this week looking for documents he could have found just as easy." Explain to Bill how electronic and/or hard copy documents are filed.

Tip R–Write it Down—Document-Document-Document: Be sure that you or others document activities such as telephone or face-to-face conversations, meetings, and field reconnaissance. The time you or others invest in documentation, may seem, when you do it, like a poor or marginal use of valuable time. However, going back later and trying to remember and reconstruct what happened days, weeks, and months earlier is a greater time sink. For many people, the process of describing, in one's own words the essence of a just completed telephone conversation, meeting, or presentation adds immeasurably to understanding and retention.

Personal

My practice is to create skeleton notes during a discussion with a client, potential client, or other person. I usually ask their permission to do so. Immediately after, and before getting back to the office, I detail the documentation while the conversation is fresh in my mind. This practice has proven to be an effective and efficient use of my time. Another thought about the importance of documentation: One of the largest expert fees I ever earned resulted from failure of a project manager to document meetings at which critical decisions were made. This and similar experiences convince me that documentation reduces the likelihood of litigation and other interpersonal or inter-organizational conflicts.

Tip S–Network: Recall Tip A, which suggests setting goals consistent with your desired roles. That tip includes a recommendation to selectively share goals by reaching out to your network. The same idea applies when you have a need for data, information, and ideas. Use your network to make the best use of your time. Reach out for help to various subsets of your global (figuratively and literally) network. Sometimes responses will help meet your immediate need. At other times, someone in your network will connect you with someone in their network thus possibly helping you and expanding your network.

Using a network assumes you have a network. Having a network—and the relationships it implies—is essential, not just for time management purposes, but for your overall well-being. Members of your network are likely to:

- Share material that they think resonates with your interests
- Encourage you when they think you are "down"
- Ask about you when you've been out of touch
- Open doors for you—sometimes even when you do not ask
- Share lessons learned with you

Wetmore (2006) states that "probably 50 percent of your personal productivity has to do with effective relationships with other people." Build, cultivate, and cherish your network for time management and other benefits. Chapter 4 offers relationship-building advice.

Tip T–Minimize "Toxic" Situations: While networks, or more specifically, the people within them, can be uplifting, other individuals and groups with whom we interact may sap our enthusiasm and drain our energy. We should try to resolve "toxic" situations and/or avoid the people who cause them.

Regardless of how much we may try, we inevitably encounter "bad chemistry" situations where we are in frequent or continuous conflict with certain individuals. "Toxic" situations are relevant to time management because they diminish individual and group effectiveness and efficiency. "Toxic" situations also require that we invest time and energy with people who are not likely to help us fulfill our roles and move toward our goals.

Assuming you are not the cause of the toxic situation and that you cannot "fix" it, extricate yourself or at least reduce your involvement. Take action such as resign from the committee, ask for a transfer to another department or office, request a new academic advisor, find a new position, or "fire" the client, owner, or customer.

Tip U–Use Travel and Waiting Time Productively: Many professionals, even entry-level personnel and some students, spend considerable time traveling. They also often find themselves waiting for a meeting or a travel connection or having spare time between appointments. Travel and waiting time offers an excellent opportunity to be productive. Work on whatever project prompted the travel and also move forward on other projects. Besides taking materials to read or projects to work on, the traveling professional should also take appropriate tools and supplies.

The ready availability of email, texting, cell phones, websites, facsimile machines, and overnight mail, have enhanced the ability to effectively use travel and waiting time. For many of the functions they carry out, the location of the professional person is becoming irrelevant because of these communication devices and links.

Tip V–Meet With Yourself: Making appointments with yourself" is a legitimate and effective means of managing your time. Consider meeting with yourself in two modes. One is the prospective mode. Plan your day, week, or next few months. Update roles, goals, and action items. The other mode is retrospective. Reflect on recent events and experiences—good and bad. Possibly write about them. Learn what you can from events, activities, and experiences such as attending a meeting, giving a presentation in a college class, having a proposal rejected, or interviewing for employment.

Reflective thinking or, what the Japanese call Hansei (Liker 2004, Maxwell 2003) is necessary for Kaizen, that is, the Japanese word for continuous improvement. Improvement results from reflection followed by intentional, small improvement steps. For example, reflect on a presentation. Did you get irritated, and reveal it, when someone in the audience reacted in knee-jerk fashion to your suggested change? Don't do that next time. How did the audience respond to your use of props? If positively, try it again.

Writing, while reflecting, offers two benefits. First, you are more likely to think deeply and broadly about the event, activity, or experience as you try to describe it and what you learned. Second, you have documentation to review later.

Views of Others

Perhaps you feel that reflecting isn't productive. Consider this thought by consultant Alan Weiss: "Many of us have the... belief that we must be doing something or we're somehow nonproductive. Thinking, reflecting, meditation, observing, remembering, visualizing, absorbing, resting, recharging, and enjoying are all doing something." Or how about philosopher Raoul Vaneigem's thought?: "There are more truths in 24 hours of a [person's] life than in all the philosophies." Self improvement speaker Dennis Waitley advocates the "meet with ourselves" advice this way: "The most important meetings you'll ever attend are the meetings you have with yourself."

Tip W–Log Your Time: Keep a time log, at about 15 to 30-minute intervals, for several days to a week as a way to identify undesirable patterns or trends in time utilization. Deduce ways to be more effective in meeting your personal goals and the goals of your organization. For example, in keeping with Tip A, a time log may reveal that you are frustrating achievement of your goals by not devoting enough time to important—not urgent tasks.

Author and poet Carl Sandburg said: "Time is the coin of your life. It is the only coin you have, and only you can determine how it will be spent. Be careful lest you let other people spend it for you." How are you spending your coin? Every now and then, and especially if you do not formally log your time, stop and reflect.

Tip X–Adopt a Holistic Philosophy: Strive to maintain an intellectual, physical, emotional, and spiritual balance. Creative and innovative ideas tend to occur during transitions from one focus to another, such as from work to play. Adopt a holistic philosophy and resist the sometimes strong tendency, in response to the pressures of the work place, to focus excessively on intellectual activities. Maintain the intellectual, physical, emotional, and spiritual dimensions of your being and the balance between them.

Personal

When I get bogged down at my desk, I take a walk, ride my bike, or work out at the gym. Often, while doing one of these, the answer to my dilemma appears. Henry David Thoreau, essayist and poet, said, "Me thinks that the moment my legs begin to move, my thoughts begin to flow."

Tip Y–Guarantee Small Successes: "Set yourself up" for at least one significant success each day. Plan each day to include one or more work-related or other activity that is both enjoyable and likely to be accomplished. This may seem trivial, but failures and other disappointments are inevitable. You can anticipate their negative impact and, to some extent ameliorate it, by giving yourself the opportunity to enjoy at least one study, work, or other success each day.

Tip Z–Goof Off!: Life goes fast—be sure to take some time "off the top" for yourself, the things you like to do, and the people you love to be with. For example, when you attend that next conference, participate in the sessions and meetings. And also, depending on the geographic location, visit the Taj Mahal, walk the Shanghai Bund, or walk into the Lincoln Memorial. Simultaneously recharge your batteries and learn about people and places!

A Time Management System

Assuming that you, as a student or practitioner, intend to practice some or most of the preceding time management tips, how will you keep track of those things you plan to do and have done? Adopt one system, that is, a single tool for your study, work, family, community, and other activities. Your time management system could be the traditional hard copy type or an electronic device. Its form, which depends on your preference, is secondary. The primary factor is your self-discipline in using the system. Commit to proactively and repeatedly using effective time management tools and techniques so that they become habitual.

Key Ideas about Time Management

You have all the time there is. While it is certainly influenced by intelligence, talent, other personal attributes, and luck, the difference between the accomplishments of people is heavily influenced by how those people use their time. Time management techniques and tools are intended to help you get more things done and, even more important, to get the right things done. Your approach to time management should be guided by a compatible set of chosen roles and supporting goals.

EMPLOYMENT OR GRADUATE SCHOOL?

As you near completion of your undergraduate studies, consider your options which include, but are not limited to, full-time graduate study and full-time employment. If you have a solid academic record (B or better), consider attending graduate school in engineering or a related field on a full-time basis immediately after earning your first professional degree. Colleges and universities operate placement offices that help senior students secure full-time employment. Most also offer helpful resources to engineering and other students who wish to explore the graduate school option. Engineering students interested in graduate study should read Hardy Cross's (1952) timeless book *Engineers and Ivory Towers*.

Table 2.3 summarizes positive and negative aspects of full-time graduate school and full-time employment immediately after earning a bachelor's degree in engineering or similar technical field. Another option is full-time employment supplemented with part-time graduate study leading to a graduate degree in engineering, another technical discipline, business, law, or some other profession. Often the young person's employer reimburses the cost of this form of graduate study in whole or in part. However, the demands of professional practice will sometimes conflict with classes and study. Increased availability of quality web-based education and training will offset some of the disadvantages of part-time graduate study. For purpose of this discussion, full-time graduate study and full-time employment are compared.

Table 2.3 Consider the pros and cons of full-time graduate study and full-time employment options.

	Full-Time Graduate Study	Full-Time Employment
PRO	• In-depth knowledge • More career choices and enhanced autonomy • Financial support • "Buy" time	• Real-world perspective • Income
CON	• Study "burnout" • Uncertainty of area of specialization • Short-term cost • Reduced number of employment positions for those who earn PhDs	• Technical obsolescence • Low probability of returning to school for full-time graduate study

Full-Time Graduate Study

The best reason to enter full-time graduate study immediately upon completion of the first professional degree in engineering or other technical field is the opportunity to study a field in depth by pursuing a master's degree and possibly a doctorate. Because they are stimulated by their college education, many young people learn to love their new-found discipline and want to immediately continue to study the depth and breadth of that discipline. Full-time graduate study can help satisfy that need. Furthermore, the resulting in-depth knowledge will be valued by some employers who require graduate degrees for certain positions. As explained in the Preface section titled "This Book and the Body of Knowledge Movement," engineering education is in the early stages of reform. The intent of the reform is to improve the education and pre-licensure experience of engineers which includes earning a technical specialization master's degree or courses equivalent to that degree.

More career choices and enhanced career autonomy, including access to higher-level positions, are other reasons to consider acquiring a graduate degree including a PhD. Graduate schools often offer promising potential graduate students attractive financial packages. While financial assistance is not likely to be the primary reason for graduate study, it can help offset the financial burden. Finally, some imminent college graduates are simply uncertain as to what the future holds for them in terms of type of employment (e.g., private practice vs. government) or function (e.g., planning, design, construction, or manufacturing.) A year or so in graduate school, during which they acquire a graduate degree, might be a productive way to "buy time" as they work through the decision-making process.

On the negative side, full-time graduate study immediately after the first professional degree might lead to, or raise the fear of leading to, "study burnout." By its very nature, graduate study in engineering or other professions tends to be specialized and, therefore, is not advised for the recent graduate who is uncertain about his or her area of preferred specialization. Another negative aspect of full-time graduate study is that, even with excellent financial support, there will be a short-term net cost. Finally, graduate study leading to a terminal degree, such as a PhD, may reduce the number of positions available to you. However, the opportunities offered by the smaller pool of positions may be superior to opportunities available to holders of bachelor's or master's degrees and the number of professionals seeking the positions will be much smaller.

Personal

I was advised by a well-meaning professor that once I earned my PhD, I should select one employment sector, such as academia or private practice, and devote my career to it. I decided otherwise and found that my advanced degree enabled me to work in academia, private practice, and government and, later in my career, to function as an independent consultant and author. Graduate education opened many doors for me.

Full-Time Employment

Just as the opportunity to continue in-depth study is probably the primary attraction for full-time graduate study, the opportunity to participate in real-world engineering and other technical applications is the primary driving force for entering full-time employment immediately after earning an undergraduate degree. As an imminent new baccalaureate graduate, you may have a strong desire to prove yourself in the technical practice world and see a product, structure, facility, or system come to fruition as a result of applying your studies. This desire may be very strong if you have not been involved in cooperative education or summer employment related to your chosen technical discipline.

The probability of a high and secure salary is also a major attraction to the soon-to-graduate engineering or other technical profession student. Many students acquire significant debt as a part of their college education and, understandably, want to begin repaying that obligation. The young professional should not be too enamored by what appear to be high starting salaries. Your income over a several decade career is not likely to be strongly correlated with a starting salary immediately after college. Other factors, including higher compensation typically associated with graduate degrees, may offset the apparent advantage of a high starting salary immediately after earning a bachelor's degree in engineering or other technical area.

Perhaps the principal negative aspect of full-time employment immediately after earning an engineering or similar undergraduate degree is the possibility of technical obsolescence. Unless you are able to continue a carefully-crafted program of personal professional development, your value to the employer may gradually diminish and you may increasingly be assigned routine work, while more sophisticated assignments are given to recent graduates who bring newer and higher technologies they learned in their bachelor's or graduate programs. Furthermore, compensation may flatten over time while new hires receive salaries that approach yours. Stated differently, your engineering or similar education, in both technical and non-technical areas begins with—does not end with—your bachelor's degree. The crucial importance of managing your personal professional assets, of continuing your education, and of being involved in professional organizations is discussed later in this chapter.

Consider another potential negative aspect of full-time employment. The possibility of, at sometime, engaging in full-time graduate study intrigues you, even though you are now inclined to enter full-time employment. Given the realities of financial and other responsibilities and becoming used to a comfortable life style, returning to full-time graduate study after several or more years of full-time employment will be very difficult.

Learn From Potential Employers

The employment search process, which most engineering and other technical program students undertake near the end of their formal education, provides an excellent opportunity to learn about the value of graduate study. In their zeal to fill positions, recruitment representatives of private and public engineering and similar organizations may disparage full-time graduate study, arguing that advanced degrees are not really needed and that you can always earn a graduate degree on a part-time basis. So that you may gain a comprehensive and balanced view of each recruiter's organization, request a list of top people or a chart showing the organization's personnel structure. Then find out how many of the organization's upper-level personnel hold advanced degrees. Draw your own conclusions regarding the importance of graduate education.

Much more insight into a particular organization's real position on post-graduate study can be gained during cooperative education assignments, internships, and summer jobs. Note who holds graduate degrees and the disciplines represented by those degrees. As you get to know the graduate degree holders, ask them to share their experiences and views about formal education beyond the baccalaureate degree.

THE NEW WORK ENVIRONMENT: CULTURE SHOCK?

Demands and expectations change as you move from the world of study to the world of practice. Completing tasks and projects correctly and on time is further complicated by the expectations that the tasks and projects be done within a budget. You may encounter not only new demands and expectations, but find that they are presented with greater intensity. Finally, the rate of change will probably increase.

No Partial Credit

Although partial credit is routinely granted on examinations and assignments at the university to encourage the learning process, the partial credit paradigm is much less applicable in professional practice. To be successful, each of the structural, mechanical, electrical, aesthetic, operation and maintenance, and cost features of an engineered product, structure, facility, or system must meet the owner's, client's, or customer's requirements. Neither the design professional nor his or her organization are likely to receive partial credit if the product lack lacks any of the required attributes. The college tactic of banking quality points in easy courses to offset poor performance in difficult courses won't work on projects. Meeting all requirements is the foundation of Chapter 7, " Quality: What is It and How Do We Achieve It?"

Doing especially well on one feature of a project, as admirable as that may be, is not likely to offset the failure to meet requirements on another part of the project. In fact, striving for perfection—going well beyond some of a project's requirements—may result in excessive labor and expense charges against the project budget and cast you in a negative light.

Little Tolerance for Tardiness

Although a proposal for an engineering or other professional services contract may be creative in its approach and handsome in its presentation, the proposal is likely to be rejected by the potential client if it is delivered after the stated deadline. The excellent job you do on an assignment from your supervisor is likely to be negated if you complete the assignment late. Young professionals who frequently arrive late at meetings, even if they are well-prepared, risk antagonizing their colleagues, many of whom place a high value on their time.

Personal

As a young engineer, I was part of a team that prepared a proposal to do a preliminary design for a marina on one of the U.S. Great Lakes. Unfortunately, our team leader delivered the proposal late—less than one hour late—and the potential client correctly declined to accept it. My disappointment lasted for years because I frequently traveled past the project site and saw the preliminary design move through final design and then into construction of an impressive marina. Perhaps, if we had respected the client's proposal schedule, our team could have had the satisfaction of doing the project.

Assignments are Not Graded

Students grow accustomed to professors reading, critiquing, and grading their work. In the professional world, careful reading of all submitted materials is uncommon and constructive written critiques are rare. Teaching, coaching, and mentoring are simply not the principal business of engineering and similar organizations.

A word of caution is in order. You may have derived satisfaction from grades and other academic recognition you received in college rather than from the intrinsic value of the work. Therefore, you may feel unappreciated as you begin working in the profession because the absence of continuous, positive feedback. Worse yet, your self-confidence may waver because you are not receiving frequent direction and affirmation. Recognize that many supervisors and managers follow a "no news is good news" approach. That is, your good work may be appreciated but you are not hearing about it.

Schedules are More Complicated

The daily and weekly schedule of a typical university student is relatively simple in that so much time is blocked out by predetermined activities and events such as classes and meetings or activities of professional, student government, and other campus organizations. Daily, weekly, and monthly schedules in the engineering and business world are much less repetitive than those of the academic world and are more likely to change quickly and dramatically in response to owner, client, customer, and stakeholder needs. If you haven't already done so, develop a systematic time management philosophy and system, as suggested earlier in this chapter.

Higher Grooming and Dress Expectations

Engineering and other technical program faculty members should serve as role models and mentors for their students in all aspects of engineering and business including grooming and dress. Unfortunately, the appearance of some professors and instructors falls below the standard of the world of professional practice. This exacerbates the grooming and dress problem for some professional program students. The professional and business community has little tolerance for inappropriate grooming and dress, although its expectations are rarely explicitly communicated. Accordingly, the young professional may lose opportunities and be relegated to secondary tasks because he or she does not understand the importance of personal appearance. Additional ideas about appropriate dress are presented later in this chapter in the section titled "Dress Appropriately."

Some engineers complain, Rodney Dangerfield style, that "we get no respect"—have less status than and earn less than attorneys, bankers, stockbrokers, medical doctors, and other professionals. Then they show up for meetings with these individuals and/or for presentations to them dressed and behaving as though they worked for them rather than with them. Their PowerPoint presentation might be first rate while their personal presentation is second rate. The professional work they describe or provide when meeting with other professionals may be conscientiously and competently prepared, but is judged as being less than that—it is devalued—because of the perception created by their dress, grooming, and demeanor. The good news is that if anyone sees a need to improve how they present themselves, the solution lies largely within them. Start by searching "business dress" on the internet and for additional thoughts, see Walesh (2008).

Teamwork is Standard Operating Procedure

Teamwork is becoming standard operating procedure (SOP) in professional practice. Except for the most trivial planning and design projects, interdisciplinary work and often formal interdisciplinary teams are required. Unfortunately, academia sometimes gives little attention to developing team skills. Students are often pitted against each other in competition for high grade point averages. The resulting individualistic paradigm is not likely to be well-received in the professional practice community. Chapter 4 includes a discussion of teamwork.

Expect and Embrace Change

Change is inevitable throughout one's career. Failure to frequently welcome and sometimes lead change will frustrate desires to realize one's potential in the consulting business, industry, government, or academic sectors. Chapter 15, "The Future and You," provides an in-depth treatment of the kind of changes likely to occur in the 21st Century and suggests ways in which you can embrace change and, in some situations, help to lead it.

THE FIRST FEW MONTHS OF PRACTICE: MAKE OR BREAK TIME

As noted earlier, demands and expectations will change abruptly as you move from the world of study to the world of practice. Success during the first few months of practice will not depend only on your technical knowledge. As part of the process of recruiting new graduates to fill a particular position, employers explicitly or implicitly define the minimum range and level of technical competence. A candidate's ability to satisfy the technical requirements of an available position can be determined largely by examining the candidate's resume and academic transcript. Employers will also assess professional or non-technical attributes of candidates. Success in your first position will depend on a blend of technical and non-technical attributes; you will need to draw on a broad range of KSAs.

Recognize and Draw on Generic Qualities

Fortunately, as part of their engineering or other technical education, young professionals are usually given ample opportunity to learn and develop widely-applicable qualities. Assuming you saw and seized those opportunities, now you can draw on those generic resources as you navigate the first few months of full-time employment. Consider adopting an attitude of quiet confidence based in part on what you have already accomplished by successfully completing a demanding education program. Appreciate and draw on the valuable generic qualities you should have developed as part of the process of earning an engineering or other technical degree. These qualities are:

- The ability to work hard and exert intense effort. For example, the typical engineering student is required to successfully complete significantly more—and more demanding—courses or credits per year than the majority of the students at a university. Completion of an engineering or similar rigorous degree suggests the student's strong work ethic.
- Persistence—continuing in spite of difficulty, being resourceful and ingenious, and having the ability to see opportunities where others see problems.
- A high degree of analytic ability including skills such as understanding complex processes and systems, identifying components of those systems, understanding the relationships between components, determining the cause of problems or failures, conceptualizing and developing alternative solutions, comparing options, and selecting the best course of action, and implementing the solution.
- Broad and effective communication skills including listening, speaking, writing, and using mathematics and graphics.

Florman (1987) discusses qualities that engineers, by virtue of education and experience, bring to their work and to society. He cites belief in scientific truth, ability to work hard, risk-taking, dependability, belief in order, pragmatic orientation, a democratic tendency, creativity, and openness to change.

Incidentally, most recent engineering and other technical program graduates will have opportunities to be involved in neighborhood, community, and religious organization activities such as service projects, recruitment efforts, political processes, and fund-raising campaigns. Generic qualities developed from the technical education experience will be valuable in these non-technical endeavors.

Suggestions for enhancing your effectiveness during the first few months of full-time professional employment are presented here. This discussion assumes that you are adopting the roles-goals approach and many of the time management tools and techniques presented in the previous section of the chapter. Time management skills, especially those that become habitual, provide one way to become productive during the first few months of employment as you move to achieve success and significance.

Guard Your Reputation

Unlike craftsmen, who are typically judged on material products such as paintings and pottery, engineers and other technical professionals are judged by the credibility of their advice, which is closely tied to their reputation. A client of a technical professional is often not qualified to judge the advice or recommendations of the professional; however, he or she is capable of judging the quality of the professional. Minister and writer Ralph Waldo Emerson said: "What you are . . . thunders so that, I cannot hear what you say . . . ." Clients, owners, customers, and the public often cannot "hear" or fully understand what engineers and other technical professionals say, but they can and do judge character and use it to value and trust—or devalue and mistrust—the professional's advice and recommendations.

Personal reputation, like a hand-crafted crystal vase, takes a long time to create and once damaged, might never be repaired. Tell the truth. Keep your word. Give credit for ideas and information. Do your share. Don't blame others. Accept responsibility for your errors and, to the extent feasible, correct them. Chapter 12 discusses the ethical aspects of studying and practicing engineering and other professions.

As bluntly stated by the Roman statesman Cicero, "A liar is not believed even when he tells the truth." One seemingly harmless way in which a young person's personal reputation may be tarnished is failing to keep what appear to be small promises. For example, you meet someone at a local meeting of your professional society, you exchange business cards, and you promise to send him or her an article about a common interest. But you forget. Or you run into an acquaintance whom you have not seen for some time, talk briefly, and agree you should get together for lunch. You offer to make arrangements. But you forget. Although individual instances like the two cited here might be considered harmless oversights, a series of them will damage your credibility. A time management system is an effective way to help you keep your promises. By keeping small promises, you will build big relationships.

Learn and Respect Administrative Procedures and Structure

In the first few days of employment, you are likely to be deluged with forms, written procedures, policy statements, and information about how the organization is structured and how it functions. There isn't a form that can't be refined or a procedure that can't be improved. Perhaps, after you are well-established in the organization, you will want to make constructive comments about administrative policies, structure, and processes. At the outset, however, you should focus on learning and respecting the established policies, structure, and processes. Focus on doing your assignments well.

Complete Assignments in Accordance with Expectations

Regardless of how unimportant or trivial initial assignments may seem, assume that your supervisor knows what he or she is doing. Some of the simple mechanical tasks you are asked to do may, in fact, be tasks typically assigned to technologists, technicians, or other paraprofessional or support staff. Your supervisor may give you these assignments simply because the work has to be done and you are readily available. Or, your supervisor may be assigning routine, simple tasks to help you develop a comprehensive understanding of the variety of work done in the organization. He or she may be grooming you to assume responsibility for managing that work.

During the first few months of employment, give your supervisor the benefit of the doubt regarding the appropriateness of your assignments. Focus your energies on understanding the context of and expected results for each assignment, learning how to do the work, and appreciating the constraints under which you are to carry out each assignment. Do your assignments in accordance with those expectations and constraints and, in the process, learn all you can.

Get Things Done

What and whom you know are secondary to how you utilize what and whom you know to make good things happen. Take the initiative to start an assignment and keep it going. Don't wait for someone to tell you what to do next—decide for yourself or ask. Be resourceful by seeing opportunities in problems and, at minimum, learning from them. Be persistent; don't let setbacks become roadblocks.

As you gradually earn a reputation for meeting expectations and getting things done, reflect on the KSAs that have enabled you to be successful. Also analyze what KSAs need work. Then, in the spirit of continuous personal growth and ongoing contribution to your organization, ask for even more varied and/or challenging tasks.

Trim Your Hedges

Develop the habit of answering questions in a positive manner and stating your findings without excessive qualifications. What you write and say should be in the context of the expected or actual audience. For example, do not begin the answer a question about the required size of an electric motor with, "If I did the calculations correctly . . . ." You are responsible for doing the calculations correctly. A qualified answer such as "Based on the limited field data, I believe that there will be no foundation problems," might be an acceptable qualification in a conversation with professional peers who understand the complexity of your work, but is not likely to be appropriate in a presentation to a non-technical client. Overly qualifying statements and responses on technical matters beyond your audience's area of expertise is nonproductive. Listeners may perceive your hedges as a lack of competence, confidence, or commitment.

The tendency to overly-qualify statements and responses suggests inadequate preparation, lack of ability, low self-confidence, or insensitivity to colleagues, clients, owners, customers, and stakeholders and detracts from the performance of the young technical professional. If not rectified, this tendency will interfere with professional advancement within the organization. The fact that you are well-prepared, have ability, and are confident is irrelevant if you are perceived to be otherwise. Perception is fact. When explaining or reporting the results to others, be very sensitive to the nature and interest of the audience. Speak in a simple, declarative, and brief fashion unencumbered with inappropriate caveats.

The preceding focuses on how what you say influences others. What you say and how you say it also influences you. Consider the more positive effect on your subsequent performance when you say, "I will get the draft report to you by Friday noon," rather than "I will try to get the report to you by Friday noon." Less hedging leads to more commitment.

Keep Your Supervisor Informed

Given the pressure of their responsibilities, many supervisors manage by exception. That is, you are unlikely to hear from them unless your performance is unsatisfactory, is exemplary, or if they have a new assignment for you. These types of management-by-exception individuals will probably expect you to function in a similar fashion, especially in your reporting to them. Determine their preferred mode of operation and function accordingly. If you are working with a management-by-exception individual, keep that person informed of the status of your assignments, particularly if the task or project is encountering problems that may have consequences for your supervisor and others.

Speak Up and Speak Positively

In addition to asking questions to help you quickly become a productive member of the organization, also gradually begin to offer suggestions when you see what appears to be a better way of doing things. Being new to an organization may be considered a disadvantage, as you have much to learn about the organization and the tasks you are given. But, in a sense, as a newcomer especially from another employer, you have an advantage because you are able to take a fresh, relatively unencumbered look at the organization and your assignments. This phenomenon, which is also discussed in Chapter 7, is referred to as the novice effect (Gross 1991). Accordingly, you may have valuable insights that could be shared with your colleagues and supervisor.

Regardless of the situation, speak thoughtfully. Think before you speak and, if needed, do some research. Consider this observation by my high school drafting teacher Walter A. Johnson: "He who thinketh by the inch and talketh by the yard should be kicketh by the foot."

When you speak up, whether it is in a conversation in the office or in a more formal setting such as a meeting, consider some of Benton's (1992) suggestions. Avoid a "bored room" voice, that is, a monotone. Do not speak too softly or raise your voice at the end of a declarative sentence, in effect changing it to a question. Women are more prone to doing these things. Talk at a moderate speed—people tend to listen more carefully if they think you are thinking while you speak. Talk a little but say a lot. Finally, speak honestly and positively. Be a "glass is half full" not a "glass is half empty" person.

Personal

While I served as an engineering dean, prospective students and their parents would frequently ask: "What engineering programs do you have?" One department head almost always responded negatively. He would say "We only have civil, computer, electrical, and mechanical engineering" or "We don't have . . . " My typical answer was "We have four strong, accredited programs—civil, computer, electrical, and mechanical engineering." Put yourself "in the shoes" of the inquirer. Do you see the sharp differences among these answers on the negative-positive scale?

Some final advice on the topic of speaking up: talk to strangers. RoAne (1988) challenges us to "work the world." Adopt the philosophy that you are surrounded by opportunities to learn and make contacts. But, you often have to take the initiative, whether you are at your workplace, doing personal errands in your community, or sitting in an airport between flights. Will talking to strangers always yield useful information or a new contact? Certainly not. However, Roane says: "That's not the point. The point is to extend yourself to people, be open to whatever comes your way, and have a good time in the process. You never know . . . the rewards go to the risk-takers, those who are willing to put their egos on line and reach out—to other people and to a richer and fuller life for themselves." As hockey great Wayne Gretzky said, "You miss 100 percent of the shots you never take."

Dress Appropriately

Dress and grooming significantly affect the professional success of any young professional. Strive to be well groomed and attractive at all times from the top of your head (e.g., clean, trimmed hair) to the bottom of your feet (e.g., clean, polished shoes).

Appropriate dress and immaculate grooming are, of course, not sufficient, but they are absolutely necessary. While it is true that a person has a right to dress as he or she pleases, it is also true that others have a right to react as they wish to that person's dress and grooming. Scottish poet Robert Burns, in his poem *To a Louse*, wrote "O wad some power the giftie gie us to see oursels as ithers see us!"

The usually unwritten definition of what constitutes acceptable dress for advancing young professionals varies from organization to organization. The extremes range from dark-colored, traditional style suits for men and women to jeans and sport shirts. To determine the appropriate dress for your organization, observe individuals one or two levels above you in the organizational structure.

If you work directly with people outside your organization, you must also be sensitive to their perceptions of you and the organization you represent, especially early in a new relationship. As noted by humorist and social commentator Will Rogers, "You never get a second chance to make a good first impression." Clothing that might be appropriate within your office may not be suitable within your client's or customer's workplace.

Personal

I recall an engineering firm executive telling me he was reluctant to have one of his project managers participate in certain outside meetings because the project manager did not dress appropriately. And, to my knowledge, no one told the project manager about that liability.

While dress styles vary from organization to organization, the range of acceptable grooming is likely to be much narrower. Incidentally, good posture is essential to achieving the full benefit of the professional clothing you wear. As noted by Benton (1992), "It's not what you wear but how you wear it that gets you to the top and keeps you there." Benton goes on to note that good posture also helps you appear energetic, improves health by reducing undue pressure on internal organs, and enhances voice quality. Benton also argues that all professionals should stand up for a variety of reasons, including showing respect and signaling the end of a meeting.

Hone Communication Ability

Look for opportunities to develop your communication knowledge and skills. For example, offer to write the minutes for a meeting. As a result, you will probably find that, in addition to honing your writing skills, you obtained the best understanding of the ideas and information exchanged at the meeting. For similar reasons, offer to draft a letter, e-mail, or report that will be eventually sent to one of your organization's clients, customers, or stakeholders. Pursue opportunities to write papers about your work and then present them at meetings and conferences.

Volunteer to make oral presentations to colleagues; clients; customers; and student, community, and professional groups. Every time you prepare for and deliver an oral presentation, you have the opportunity to improve on a skill that is highly valued within the engineering profession and society. Most of Chapter 3 is devoted to helping you write effective documents and make convincing oral presentations.

Seize Opportunities for You and Your Organization

Luck results when opportunity meets preparation. You certainly will not be successful solely on the basis of luck. On the other hand, luck opens windows of opportunities, often for only a fleeting moment. As written by Shakespeare in *Julius Caesar*, "There is a tide in the affairs of men, which, taken at the flood, leads onto fortune; omitted, all the voyage of their life is bound in the shallows and miseries. On such a full sea are we now afloat; we must take the current when it serves or lose our ventures."

If you see those opportunities and have the courage to seize them, you and your organization may benefit. For example, as a result of your initiative at a local meeting of an engineering society, you meet a representative of an organization and learn that the organization needs specialized services your firm provides. You immediately follow-up by informing your company's marketing group and your firm obtains a very attractive contract. As someone said, "Don't be afraid to go out on a limb. That's where the fruit is."

Choose To Be a Winner

Most externally-imposed situations are neutral, that is, they are neither inherently "good" nor inherently "bad." They are what we make of them. While we cannot control much of what happens to us, we can choose the attitude of our response.

Many apparent problems are actually opportunities in disguise. For example, your failure to be selected to work on an exciting new project might prompt you to suggest that a similar project be undertaken with a different client. Or, assume you are asked to work long hours for several days to correct calculation errors made by a recently-released employee. You could choose to view this as an unfair imposition, or you may choose to develop a spreadsheet or computer program that could be used not only for this assignment, but also in the future to minimize errors and reduce the amount of time necessary to do calculations. Consider the powerful attitude advice of pastor and author Charles R. Swindoll:

The longer I live, the more I realize the import of attitude on life.
Attitude, to me, is more important than facts.
It is more important than the past, than education,
than money, than circumstances,
than failure, than successes,
than what other people think or say or do.
It is more important than appearance, giftedness, or skill.
It will make or break a company . . . a church . . . a home.
The remarkable thing is we have a choice every day
regarding the attitude we will embrace for that day.
We cannot change our past,
we cannot change the fact that people will act in a certain way.
We cannot change the inevitable.
The only thing we can is play on the one string we have,
and that is our attitude.
I am convinced that life is 10 percent what happens to me and
90 percent how I react to it.
And so it is with you . . . we are in charge of our attitudes.

Recognize that individual and group attitudes, whether they are predominately positive or negative, are contagious in organizations. Unfortunately, negativism appears to move through an organization with greater ease and speed than positivism. However, positivism can permeate an organization if a few people at all levels choose to take a winning, rather than losing, perspective and course of action.

Personal

I couldn't help but notice the sole flight attendant on the evening flight from Indianapolis to DC. She was attentive, articulate, energetic, positive, and helpful. Before leaving the plane, I quietly said to her something like "you really enjoy your job, don't you?" The essence of her reply: "I love my job! If I didn't, I'd work somewhere else; there are so many great opportunities." What a great attitude! Music industry executive Bruce Flohr stressed the importance of attitude this way: "Hire attitude and train for skill. The most talented person will destroy a business is there is an attitude problem, and fellow employees are the first to know it."

Be known as an action-prone person rather than just a talk-prone person. Experience suggests that for every 10 to 100 people who say something like "you know what we ought to do?" no more than one will actually do it. Be that one.

Summing it Up

The world of professional practice is different than the world of engineering and technical education. In the former, technical knowledge is often assumed and success always depends on performance in professional, that is, non-technical areas. Forewarned is forearmed. Appreciate and use your generic qualities and characteristics. During the first few months of your first full-time professional employment, guard your reputation, learn and respect administrative procedure and structure, do all assignments well, get things done, trim your hedges, keep your supervisor informed, speak up and speak positively, dress appropriately, hone your communication skills, seize opportunities, and be a winner.

MANAGING PERSONAL PROFESSIONAL ASSETS: BUILDING INDIVIDUAL EQUITY

A downturn in the stock market is disconcerting for many individuals, even the young professional who is just beginning to build his or her balance sheet. Suddenly, the value of retirement accounts, mutual funds, and other investments plummets. Individual net worth may drop sharply. There is a gnawing fear of the negative long-term effect on the material well-being of a spouse and dependents. The young person vows to be more careful about investments. After all, prudence requires careful management of personal financial assets.

Personal Professional Assets

What about the status of and attention to personal professional assets? By virtue of individual talents, education, and experience, each young engineer or other technical professional has significant value to society. In contrast with your financial assets, many of which can be measured to the penny, the value of your professional assets defies quantification. In a narrow sense, individual professional assets or personal professional equity might be valued as the present worth of the projected stream of future earnings. In a broader and more accurate sense, the young technical professional's assets are measured by the actual good accomplished and by all the good a person has the potential to accomplish through conscientious use of his or her talents, education, and experience.

Although the true value of personal professional assets defies quantification, it nevertheless is very real. Furthermore, the value of professional assets, even if narrowly defined, is likely to exceed the value of the entry-level person's financial assets, except perhaps for those few fortunate people who are independently wealthy. Like personal financial assets, personal professional assets can appreciate, remain level, or decline.

Annual Accounting

Beginning as a student, appraise your professional assets at least once a year, perhaps as part of a resume update exercise. What new areas of technology have been mastered? What new managing and leading techniques were used? What new concepts, ideas, or principles were studied? What new skills were acquired? What new challenges and responsibilities were accepted? What new opportunities were seized? What new risks were taken? What knowledge was shared with professional colleagues? What new contributions were made? In what ways have you been a "good and faithful servant" with talents?

While experience is valuable, too much of one kind of experience can hamper your growth. As you review annual accountings of your professional assets, will you find several years filled with new experiences or one year of experience repeated several times? Resist the temptation to settle into the comfort of routine. Listen to Mandino's (1968) warning about excessive experience:

I will commence my journey unencumbered
with either the weight of unnecessary knowledge
or the handicap of meaningless experience...
In truth, experience teaches thoroughly
yet her course of instruction devours men's years
so the value of her lessons diminishes
with the time necessary
to acquire her special wisdom.

If an annual accounting of your personal professional assets reveals a loss or no increase in value, you have experienced a personally devastating "stock market crash." You have lost a year of growth and increased contribution, neither of which can ever be redone or perhaps regained. As a result of mismanagement of your professional personal equity, you have failed yourself, dependents, your employer, your clients, your profession, and society.

Careful Management of Personal Professional Equity

Each young professional is gifted with a unique combination of talents. A challenge in the early years of professional life is to discover and develop through reflection, education, and experience, a special set of KSAs and then to dedicate and direct them to meaningful professional work and service. You should commit to managing personal professional assets at least as well as you manage your financial assets. Some ways of managing personal assets were presented earlier in this chapter in the section titled "The First Few Months of Practice: Make or Break Time." The following sections discuss three major additional ways to manage your personal professional assets: continuing education, involvement in professional organizations, and licensing.

Continuing Education

Continuing education is an important mechanism for maintaining your personal professional equity. It is also a requirement for continued licensure as a professional engineer in over three-fourths of the U.S. licensing jurisdictions (Casey 2010). Entry-level engineers or other technical professionals have many means available for immediately beginning their individualized continuing education and professional development programs. Examples are internal and external workshops, seminars, and webinars; university classes, offered in the traditional classroom manner or in a distance education mode and possibly leading to one or more graduate degrees; and, for the very disciplined person, self-study. "Read widely and eclectically, including articles, books, newspapers, and other publications that address a range of topics—technical, historical, economic, social, and contemporary. Consider the goal of reading a book a month" (ASCE 2008). A series of varied and challenging work assignments is another effective mechanism for enhancing your personal professional equity. While many means are available, you have the primary responsibility for your continuing education and professional development.

Most employers will support your continuing education by means such as offering financial assistance and providing released time during normal office hours. However, organizational education and training is sometimes conducted on a hit-or-miss basis. Some engineering and related organizations maintain highly-focused internal education and training units referred to as corporate universities (e.g., Meister 1998, Roesner and Walesh 1998). Other organizations have created formal mentoring programs (e.g., Bonar and Walesh 1998, Finchum 2003, Jackson 1999). Nevertheless, if you are not in charge of your continuing education, chances are no one is.

Continuing education is vital to you for one or two reasons. First, rapid changes in technology require constant learning to be current. You probably noticed changes in technology during the short time you were in college. If entry-level technical professionals seek success, they must remain current. A second reason to continue your education is to prepare yourself to function in areas other than the technical ones—areas such as research, marketing, administration, finance, and teaching. Continuing education is essential if you want to make functional changes within your current employment situation or with a new employer or want to start your own business (Walesh 2000).

Think about your commencement ceremonies. On the surface, commencements celebrate the beginning of a career. Experience indicates, however, that such occasions also mark the commencement of the rest of your education. Although it may not be as formal as the first part, it will be just as important and it should last much longer—for the rest of your life.

Involvement in Professional Organizations: Taking and Giving

In addition to varied and challenging work assignments and continuing education, active involvement in professional organizations is the third way to increase the value of your personal professional equity. Entry-level professionals should realize that they will derive a satisfying and prosperous living from their profession and, accordingly, ought to give something back to their profession. English philosopher and statesman, Francis Bacon said: "I hold every man a debtor to this profession; from that which man has a course to seek countenance and profit, so ought they of duty to endeavor themselves, by way of amends, to be a help and ornament there unto."

Note the large number of materials that have been produced for your use, usually in the context of professional organizations, such as books, papers, conference proceedings, and manuals of practice. Most of these are created by volunteers who gave something back. How could you not do the same?

The call to be actively involved in professional organizations goes beyond maintaining one's currency and meeting an obligation. Such participation provides an opportunity for you to enjoy and benefit from the company of leaders. Engineering and other technical professions and their various subdivisions typically have many members but very few doers. The doers are usually committed, creative, ambitious, and accomplished people. The young professional can learn much from associating with them and the "ticket" is a commitment to being actively involved in the work of professional organizations.

Upon joining such an organization, or moving from the student membership status you had in college to the practicing professional status, select one or more types of activities for your involvement and contribution. Besides attending meetings, consider presenting and publishing papers, serving on and chairing technical and non-technical committees, helping arrange and run meetings and conferences, and serving as an officer. By investing your time and talent in one or more professional organizations, you will realize the significant return on your investment in terms of knowledge gained, satisfaction of contribution, and the association with the leaders of your profession. As you contemplate involvement in professional organizations, consider these thoughts from British Prime Minister Winston Churchill:

There is a place for everyone, man and woman,
old and young, hale and halt;
service in a thousand forms is open.
There is no room for the dilettante, the weakling,
for the shirker, or the sluggard.
From the highest to the humblest tasks,
all are of equal honor; all have their part to play.

As noted, presenting and publishing papers is one way to be actively involved in a professional organization. Personal and organizational benefits of individual or co-authored papers include:

- Improved writing and speaking ability, which is directly and immediately transferable to many aspects of your professional and personal life

- Increased confidence as a result of interacting with peers
- Expanded visibility for your organization with emphasis on its accomplishments and abilities
- Earned membership in networks of leaders, which provides quick access to assistance when needed
- Returning something to the profession

Having considered some of the benefits of presenting and publishing papers, you may argue that you simply don't have the time. This may be a valid argument if you have to first complete special work that will serve as the subject for the paper. Instead, you and co-authors should write about the good work that you have already done. The extra investment in time and effort, on top of what you have already done, will typically be small compared to the total effort expended. Equally important, the extra investment in time and effort can be small compared to the extra benefit that will accrue to you, co-authors, and your organization.

Figure 2.5 illustrates this concept of great return on a marginal investment. Cost on the horizontal axis might include labor, expenses, research, creativity, and/or energy required to complete a project for a client/owner/stakeholder. Benefit on the vertical axis could include fee, satisfaction, learning, and/or recognition resulting from the completed project. The cost increment on the horizontal axis shows the additional effort required to write and speak about the project. The somewhat longer benefit increment on the vertical axis presents the additional benefit realized as a result of the paper such as: increased confidence, network membership, exposure for organization, and contribution to the profession. To reiterate, you do not have to create work to find something to write and speak about. Instead, write and speak about the work you and colleagues have done or are doing.

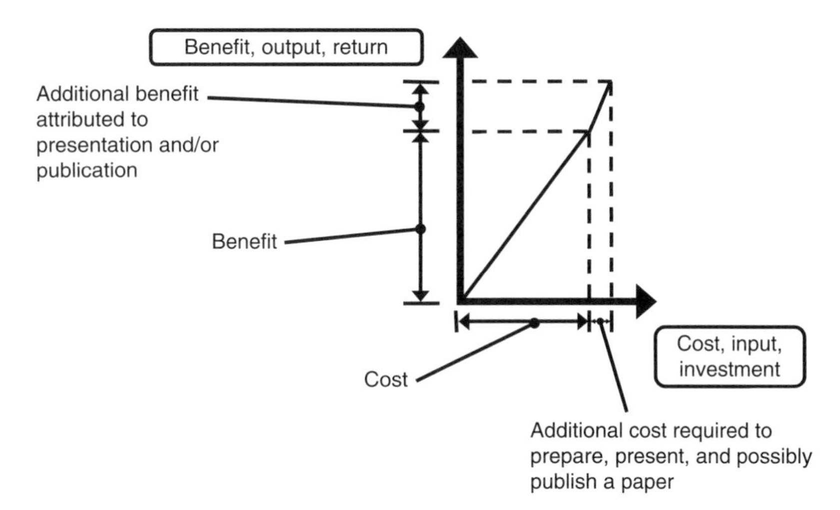

Figure 2.5 The small incremental cost required to prepare a presentation can result in a large incremental benefit.

Licensing

A system of licensing engineers and certain other technical professionals, such as architects, has been established in the United States and elsewhere primarily to protect the public by establishing minimum requirements for individuals who plan and design public facilities. In the U.S., for example,"...engineers are licensed in 50 states plus the District of Columbia and four U.S. territories (Guam, Puerto Rico, Northern Mariana Islands, and the Virgin Islands) for a total of 55 licensing jurisdictions. Illinois has a separate board for structural engineering. Therefore, there are 56 boards that license engineers" (ASCE 2008). McGuirt (2007) provides an historical account of U.S. engineering licensure while Nelson and Price (2007) address licensure's future.

Various federal, state, local, and other laws and regulations specify when engineering work must be done under the direction of a licensed engineer. Licensing laws focus on protecting public health, safety, and welfare. Because civil engineers, in comparison to most other engineers, work most closely with the public, licensing is a virtual necessity for them. However, as stressed later in this licensing discussion, engineers in all disciplines should strive to become licensed.

One benefit of engineers obtaining one or more Professional Engineering (PE) licenses is the availability of more engineering opportunities. Without the license, the engineer will most likely always do engineering work for or under the direction of someone else. With the license, the engineer will be able to do higher level work, be responsible for more engineering projects, have access to more favorable employment opportunities, and be in a position to someday own and operate his or her consulting engineering or other engineering-based business. Holding of one or more engineering licenses is also a mark of achievement.

The Evolving Licensing Process

As illustrated by the upper portion of Figure 2.6, today's engineering professional track in the U.S. typically consists of this sequence:

- Earning a four-year baccalaureate degree in engineering from a program accredited by ABET, Inc. and successfully completing the Fundamentals of Engineering (FE) examination.
- Obtaining four years of progressively responsible experience as an engineer intern. Some U.S. licensing jurisdictions allow some graduate study and/or cooperative education to count toward the four years.
- Passing the Principles and Practices of Engineering Examination and becoming licensed
- Practicing engineering while continuing life-long learning

The engineering professional track is changing in the U.S. As outlined in this book's Preface, some engineering disciplines are reforming education and prelicensure experience. Specialty certification, which is recognition for post-licensure professional achievement, is gradually expanding to more engineering specialties.

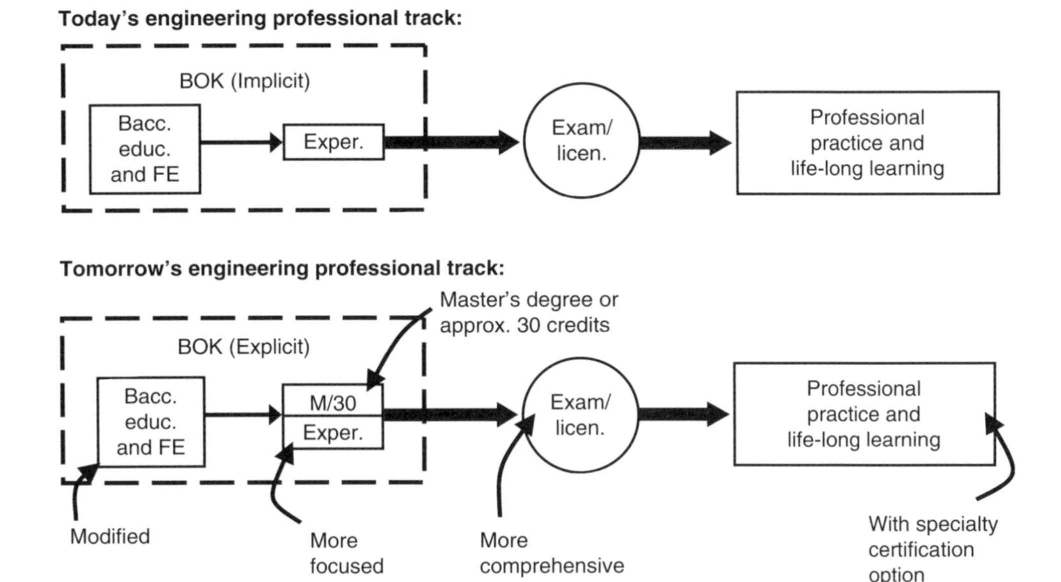

Figure 2.6 Today's engineering students and young practitioners should consider preparing for tomorrow's engineering professional track. (Source: Adapted with permission of ASCE from ASCE 2008)

Accordingly, the U.S. engineering professional track is gradually shifting to that shown in the lower part of Figure 2.7. Major changes from "today's" track to "tomorrow's" track include providing an explicit definition of an engineering discipline's BOK, modifying the baccalaureate program, requiring a master's degree or its equivalent (denoted by "30" for 30 credits), a more comprehensive licensure examination, and the availability of specialty certification.

Taking the Fundamentals Examination While in School

A few additional words are in order regarding the wisdom of taking the FE Examination while in engineering school. Most civil engineering majors take the examination at that time because they learn through faculty and others that holding an engineering license, or at least being in the licensing process, is expected by most civil engineering employers.

Engineering students in other disciplines, however, are likely to hear about what is called the "industrial exemption." Under the industrial exemption, which is in effect in most U.S. jurisdictions, state registration laws indicate that an engineer does not necessarily need to have an engineering license to practice engineering.

"Some words of caution: Be wary of arguments—sometimes very self-serving—against licensure. Someone may say that you are working in an employment sector that is under the industrial exemption and, therefore, that you do not need a license. Will you always want to work in that sector? Others will oppose licensure because it

Figure 2.7 Earning an engineering degree and then not taking the Fundamentals of Engineering Examination is like buying a classic sports car and leaving it in the garage. (Source: Author, his car)

results in having to pay higher compensation to licensed engineers. While they will not make that argument directly to you, if you are employed in their organization and are not licensed, you are likely to incur a penalty in compensation and opportunities" (ASCE 2008).

"Others will say that licensure is merely a shallow "prestige" credential and that your employment with them—and perhaps even others—is secure as long as you maintain your technical competence. After all, that's what really counts. But what if, someday, you want to start your own business—perhaps first as an individual proprietor and then later as the leader of a small and growing engineering firm? Can you exercise that option without a PE? Even if you never start your own firm, but choose instead to spend your professional career as an employee of an engineering organization, state laws require that the engineer in responsible charge of engineering work be licensed. Are you willing to relinquish this opportunity? That is very unlikely, so keep your options open by proactively seeking licensure" (ASCE 2008).

Except for extreme circumstances, earning an engineering degree and then not taking the FE Examination in college is like taking a driver's education course in high school but not getting a driver's license, running a marathon and quitting 10 yards short of the finish line, or buying a sports car and leaving it in the garage, as illustrated in Figure 2.7. You should think long and hard before you "buy" the industrial exemption argument. Wise people keep options open.

While many engineering faculty members will support the engineering licensing process, especially for the engineers who intend to practice engineering, some will be neutral or negative toward licensing. They may base this partly on the industrial exemption provision, as already discussed. They may not value an engineering license because, in most situations, engineering faculty members do not need licenses to practice their profession, that is, teaching and research. While a PE license may not be appropriate for them, it may be appropriate, if not essential, for you.

Comity

Comity is the process through which the board of registration for engineers of one state or other jurisdiction may license a person to practice engineering in that state or jurisdiction based on a license issued by another one. Under comity, an engineer can usually have the FE Examination results transferred from one state to another or obtain an engineering license in one state as a result of holding a valid license in another state. While there are some exceptions (such as special structural engineering licensing procedures in some U.S. states), comity is common throughout the United States.

License Renewal

While renewal of engineering licenses is required, retesting is not required as a condition of a license renewal. However, continuing education is a requirement of license renewal in a majority of U.S. jurisdictions and there is a continuing movement in that direction. In the U.S., boards administer provisions of the state and territorial licensing laws for PEs. Sanctions for violations range from letters of reprimand to permanent loss of a license to practice engineering in a given jurisdiction.

CONCLUDING THOUGHTS: GETTING YOUR PERSONAL HOUSE IN ORDER

"People don't want to be managed. They want to be lead," according to John C. Maxwell, a student of managing and leading. If you want to lead, you must first manage yourself. This includes selecting roles, setting goals, managing your use of time, deciding if graduate study is for you, proactively approaching and performing in your first employment, and managing your professional assets. Just as the condition of a house is determined primarily by the actions, or inactions, of the owner so your condition – your readiness to achieve success and significance and to lead – will be determined primarily by you.

> Up to a point a man's life is shaped by environment, heredity, and movements and changes in the world about him. Then there comes a time when it lies within his grasp to shape the clay of his life into the sort of thing he wishes to be.
> Only the weak blame their parents, their race, their times, lack of good fortune, or the quirks of fate. Everyone has it within his power to say this I am today; that I will be tomorrow.
>
> *(Louis L'Amour, author)*

CITED SOURCES

Alessandra, T. 2004. "Procrastination," *Dr. T's Timely Tips*. e-newsletter, August 18.

American Academy of Environmental Engineers. 2009. *Environmental Engineering Body of Knowledge.* AAEE: Annapolis, MD.

American Society of Civil Engineers. 2008. *Civil Engineering Body of Knowledge for the 21st Century: Preparing the Civil Engineer for the Future-Second Edition.* ASCE: Reston, VA.

Benton, D. A. 1992. *Lions Don't Need to Roar: Using the Leadership Power of Professional Presence to Stand Out, Fit In, and Move Ahead.* Warner Books: New York, NY.

Bonar, R. and S. G. Walesh. 1998. "Mentoring—Investing in People: A Case Study." *Compendium of Educational Material—Fall Conference*. American Consulting Engineers Council. November.

Casey, T. M. 2010. "Lifelong Learning to Meet Engineering Challenges." *CE News*. August, pp. 20–24.

Connellan, T. 2011. *The 1% Solution for Work and Life.* Peak Performance Press: Chelsea, MI.

Covey, S. R. 1990. *The 7 Habits of Highly Effective People*. Simon & Schuster: New York, NY.

Cross, H. 1952. *Engineers and Ivory Towers.* Edited by R. C. Goodpasture. McGraw Hill: New York, NY.

Emmett, R. 2000. *The Procrastinator's Handbook: Mastering the Art of Doing It Now.* Walker & Company: New York, NY.

Finchum, M. J. 2003. "Cultivating the Next Crop of Leaders." *Leadership and Management in Engineering – ASCE*, July, pp. 150–152.

Fredrich, A. J. (Editor). 1989. *Sons of Martha: Civil Engineering Readings in Modern Literature*. American Society of Civil Engineers: New York, NY.

Florman, S. C. 1987. *The Civilized Engineer.* St. Martin's Press: New York, NY.

Gross, F. 1991. *Peak Learning*. Jeremy P. Tarcher, Inc.: Los Angeles, CA.

Handy, C. 1998. *The Hungry Spirit: Beyond Capitalism: A Question for Purpose in the Modern World.* Broadway Books: New York, NY.

Hatch, S. E. 2006. *Changing Our World: True Stories of Women Engineers.* ASCE Press: Reston, VA.

Iacocca, L. with William Novak. 1984. *Iacocca—An Autobiography*. Bantam Books: Toronto, Canada.

Jackson, L. 1999. "Putting People First." *Civil Engineering - ASCE,* July, p.124.

Jackson, M. 2008. *Distracted: The Erosion of Attention and the Coming Dark Age.* Promethus: Amherst, NY.

Layne, M. E. 2009. *Women in Engineering: Pioneers and Trailblazers.* ASCE Press: Reston, VA.

Liker, J. K. 2004. *The Toyota Way: 14 Management Principles from the World's Greatest Manufacturer.* McGraw-Hill: New York, NY.

Leuba, C. J. 1971. *A Road to Creativity – Arthur Morgan – Engineer, Educator, Administrator.* Christopher Publishing House: North Quincy, MA.

Mandino, O. 1968. *The Greatest Salesman in the World.* Bantam Books: New York, NY.

Maxwell, J. C. 2003. *Thinking for a Change.* Warner Books: New York, NY.

McGuirt, D. 2007. "The Professional Engineering Century." *PE.* NSPE: Alexandria, V A. June, pp. 24–29.

Meister, J. C. 1998. *Corporate Universities: Lessons in Building a World Class Work Force.* McGraw-Hill: New York, NY.

Nelson, J. D., and B. E. Price. 2007. "The Future of Professional Engineering Licensure." *PE.* NSPE: Alexandria, VA. June, pp. 30–34.

Pattison, J. 2008. "Worker Interrupted: The Cost of Task Switching." *Fast Company.com*, July 28.

RoAne, S. 1988. *How to Work a Room: a Guide to Successfully Managing the Mingling*. Shapolsky Publishers: New York, NY.

Robinson, J. 2010. "E-mail Is Making You Stupid." *Entrepreneur,* March, pp. 60–63.

Roesner, L. A. and S. G. Walesh. 1998. "Corporate University: A Consulting Firm Case Study." *Journal of Management in Engineering - ASCE.* March/April. pp. 56–63.

Walesh, S. G. 2008. "First Impressions." *Indiana Professional Engineer.* January/February, p.3.

Walesh, S. G. 2000. *Flying Solo: How to Start an Individual Practitioner Consulting Business.* Hannah Publishing: Valparaiso, IN.

Walesh, S. G. 1999. "Roles—Then Goals." *Journal of Management in Engineering - ASCE.* March/April. p. 3.

Weingardt, R. G. 2005. *Engineering Legends: Great American Civil Engineers.* ASCE Press: Reston, VA.

Wetmore, D. 1999. "How to Plug the Big Hole in your Day." Productivity Institute: Stratford, CT.

Wetmore, D. 1999b. "The Paper Blizzard." Productivity Institute: Stratford, CT.

Wetmore, D. 2003. "Time Tip – Chain Yourself." *Timely Time Management Tips.* e-newsletter. July 15.

Wetmore, D. 2006. "Top Four Time Management Issues." *Time Tip,* e-newsletter. January 17.

Zaslove, M. O. 2004. *The Successful Physician: A Productivity Handbook for Practitioners.* Jones and Bartlett Publishers: Sudany, MA.

ANNOTATED BIBLIOGRAPHY

Allen, J. 1983. *As A Man Thinketh.* DeVorss & Company: Marina Del Ray, CA. (Focusing on the mind, this short, uplifting book offers numerous thoughts such as "A man cannot directly choose his circumstances, but he can choose his thoughts, and so indirectly, yet surely, shape his circumstances" and "Thought allied fearlessly to purpose becomes creative force.")

Hill, N. 1960. *Think and Grow Rich.* Fawcett Crest: New York, NY. (Makes a case for having goals and a plan to achieve them and for the power of visualization and the subconscious mind.)

Leuba, C. J. 1971. *A Road to Creativity – Arthur Morgan – Engineer, Educator, Administrator.* Christopher Publishing House: North Quincy, MA. (Describes how engineer Arthur Morgan, born of modest means but with a stimulating home environment, creatively sought out and succeeded in a wide variety of, professional roles. Describes his accomplishments as a water control engineer, founder of an engineering firm, creator of the Miami Conservancy District, rescuer of Antioch College, and organizer of the Tennessee Valley Authority.)

Phillips-Jones, L. 2003. *Strategies for Getting the Mentoring You Need: A Look at Best Practices of Successful Mentees.* The Mentoring Group: Grass Valley, CA. (Stressing that "you own your development," this booklet offers advice for individuals who seek mentoring but are not involved in a formal mentoring program.)

Urban, H. 2003. *Life's Greatest Lessons: 20 Things That Matter.* Simon & Schuster: New York, N.Y. (Chapter 11, "Real Motivation Comes From Within," urges us to create mental pictures of the success [and significance?] we desire, noting that we don't think in words, we think in pictures.)

Walesh, S. G. 2010. "Remember When You Almost..." *Leadership and Management in Engineering - ASCE,* October, p. 197. (Cautions against "rationalizing away one inspirational possibility after another" and, as a result, experiencing late-life regret.)

EXERCISES

2.1 TIME MANAGEMENT TIPS: The purpose of this exercise is to share personal time management tips among members of the class and with the instructor. While 26 time management ideas are presented in the chapter, they reflect my ideas. You and others have also found ways to make more effective and efficient use of your time and this exercise, because of the sharing aspect, could benefit many. Suggested tasks are:

- **A.** Decide if this exercise is to be done as a group effort or individually. If the former, meet and share ways group members make the best use of their time. Ask one person to draft a summary of the results, in the form of a memorandum to your instructor, for review by the others.
- **B.** If this exercise is done individually, then briefly interview a cross-section of individuals such as students, faculty members, and administrators to learn how they optimize their use of time. Present what you learn in a memorandum to your instructor.

2.2 SELF-DISCIPLINE: This exercise focuses on the importance of self-discipline, or what is referred to in this chapter as "getting your personal house in order." This objective is accomplished by gaining in-depth knowledge and understanding of one accomplished "engineer." Suggested tasks are:

- **A.** Select an accomplished engineer or other technical professional, or someone who made significant accomplishments in what is now considered a technical profession. Examples are Leonardo Da Vinci, A. G. Eiffel, Catherine Anselm Gleason, Sextus Julius Frontinus, Herbert Hoover, Margaret S. Peterson, Ellen Swallow Richards, Emily Warren Roebling, Joseph Strauss, and Orville and Wilbur Wright. Books by Fredrich (1989), Hatch (2006), Layne (2009), and Weingardt (2005) are excellent sources of biographies.
- **B.** Request approval of your instructor.
- **C.** Read the book and prepare a report for your instructor in which you a) cite your book (e.g., name, author, publisher, date) and b) describe some of the ways in which the individual exercised self-discipline and the results of that effort. Refer to Chapter 3 of this book for writing guidance.

2.3 ENGINEERING BODY OF KNOWLEDGE: This exercise's purpose is to increase your familiarity with a relevant engineering body of knowledge (BOK) to help you guide your continued professional development. The exercise assumes that you can find a BOK relevant to your discipline. For example, see ASCE (2008) for the civil engineering BOK and ASEE (2009) for the environmental BOK. As noted in this book's Preface, an engineering BOK movement is underway in the U.S. and, therefore, other engineering disciplines may be developing or have developed a BOK. Suggested tasks are:

A. Study the BOK for your engineering or related discipline.

B. Analyze the BOK and, more specifically, a) compare it to your current or recently-completed engineering program, b) note any differences and their relevance to your plans and aspirations, and c) indicate in what way, if any, the BOK will influence your continued professional development. Prepare a report for your instructor in which you discuss the preceding three points. Refer to Chapter 3 for writing assistance.

CHAPTER 3

COMMUNICATING TO MAKE THINGS HAPPEN

With public sentiment, nothing can fail;
without it, nothing can succeed.
Consequently, he who molds sentiment
goes deeper than he who
enacts statutes or pronounces decisions.

(*Abraham Lincoln, 16th U.S. President*)

Starting with the premise that we should be proficient with five forms of communication—listening, writing, speaking, visuals, and mathematics, this chapter draws sharp distinctions between writing and speaking when considered in the context of the interaction between the writer and the reader and the speaker and his or her audience. The chapter then provides in-depth discussions of listening, writing, and speaking. Visuals and mathematics, the other two communication modes, are woven into the listening, writing, and speaking sections. Ideas and information presented here will allow you to build on the communication foundation provided, or being provided, as part of your formal education. If you have graduated from college and your formal education did not stress the importance of communication and require you to develop basic communication skills, you probably have a serious liability. Ideas and information set forth in this chapter, coupled with your desire and self-discipline, will help you overcome that deficit.

FIVE FORMS OF COMMUNICATION

Communication forms, as defined here, consist of listening, speaking, writing, and using graphics and mathematics. The unity and completeness suggested by the circle showing in Figure 3.1 indicates that a technical professional needs knowledge and skills in all five areas to be a complete communicator.

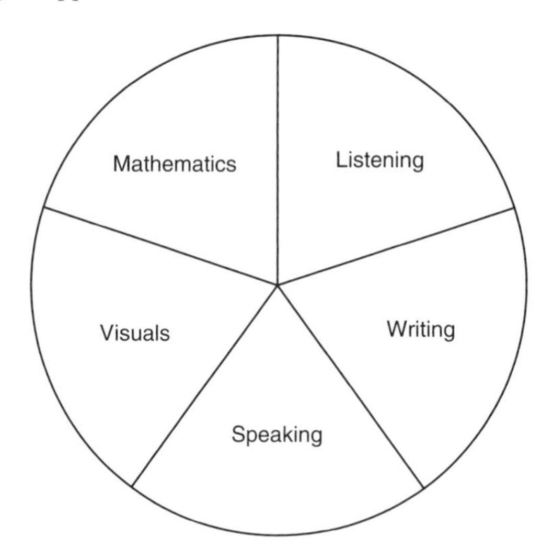

Figure 3.1 The effective communicator is able to draw on combinations of up to five forms of communication.

If you agree that all five forms of communication are important, consider the instruction you probably received, or are receiving, in each of them. Begin with mathematics in Figure 3.1 and proceed counterclockwise. Engineering or other technical education typically includes a strong emphasis on the use of mathematics and perhaps visuals or graphics, although neither may have been oriented toward communicating with largely non-technical audiences. Writing instruction and critiquing typically receive moderate attention in technical education programs. Even less emphasis is usually placed on developing speaking skills and explicit instruction on building listening skills is rare. Accordingly, you may have some communication deficiencies and, listed in order of decreasing priority, they are likely to be in the areas of listening, speaking, and writing.

Personal

My communication defining moment, or turning point, occurred during the fall semester of my freshmen year. The engineering dean abruptly entered our graphics class, apologized, and said something like this: "While you are here at the university, develop your communication abilities. Learn how to write and speak and how to use mathematics and graphics." I do not remember his exact words. However, like it was yesterday, I vividly remember the message. I began work on my communication knowledge and skills in the dean's four areas plus one I added, that being listening. That commitment has taken me to many places, that is, assignments within organizations and countries around the globe. Whatever success and significance I've achieved, it is based in part on continuously improving my communication knowledge and skills.

Your concepts, ideas, discoveries, creations, and opinions will contribute to making things happen only if you effectively communicate them to others. Effective communication, in the context of a particular situation, means using one or more of the five communication forms to understand colleagues and others and to accurately and convincingly convey your thoughts to them. The most exciting vision, the most thoughtful insight, the most elegant solution, or the most creative design are all for naught unless they are effectively communicated to others. Lacking such communication, the intellectual and other seeds that you plant within your organization and within your professional circles are not likely to sprout and bear fruit. You and your colleagues will be denied the bounty of your labors.

Stated differently, effective communication is necessary, but not sufficient to achieving success and significance in engineering and other technical professions. There may be the rare exceptions, such as the non-communicative geniuses tolerated by others because of their extraordinary intellectual or creative gifts. Then there is the occasional recent technical program graduate who also happens to be the owner's daughter or son and who lacks communication and other interpersonal skills, but is foisted on the other members of the organization. Unless you are a genius, are inextricably linked to the organization's owner, or enjoy some other rare privilege, you need effective communication skills to realize your potential.

You might be tempted to agree with the importance of communication, but argue that as a college student or an entry-level professional in a new environment, your "plate is full" and, therefore, you will defer developing effective communication skills for a few years. The fallacy of your position, if it persists, will gradually become evident as you find yourself thinking thoughts such as:

- I know I had the best solution to the design problem but we decided on an inferior course of action
- Mary and I started at about the same time, but she is increasingly having out-of-office contact with clients while "they" never let me out of this place
- Juan was selected to attend a valuable two-day seminar based on the "brown bag" summary he gave the other day, but my request was turned down
- I'm just not appreciated around here. I think I'll start looking around for other opportunities

THREE DISTINCTIONS BETWEEN WRITING AND SPEAKING

Three distinctions can be drawn between writing and speaking, two of the five available communication modes. Effective communicators understand and capitalize on these differences.

Single-Channel versus Multi-Channel

First, the reader receives written communication as "linear, single-channel input" (Decker 1992). The writer sends and the reader receives written words, one after another. Words only, with the possible exception of supporting tables or figures, must carry the written message. Therefore, the words must be carefully chosen and arranged.

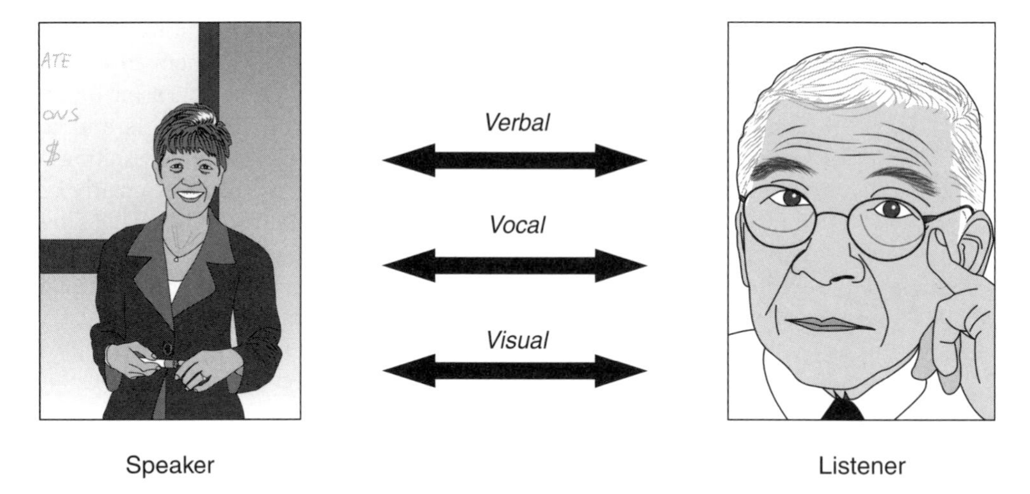

Figure 3.2 Speaking, contrasted with writing is multi-channel and two-way.

In contrast, the listener receives spoken communication as "multi-channel input," as Figure 3.2 illustrates. The speaker sends, intentionally or unintentionally, and the listener receives a three-component message (Decker 1992). The components of your spoken message are: 1) verbal, the words you use; 2) vocal, how you use your voice to say the words; and 3) visual, your facial and body expressions and motion and the visual aides you use as you speak. What is the relative input of the three "Vs" of spoken communication? Research suggests (e.g., Atkinson 2007, Restak and Kim 2010) that the visual is at least as important as the verbal. Plan your spoken presentations accordingly. Explicitly plan for and use the verbal, vocal, and visual components of spoken communication, and place appropriate emphasis on each.

One-Directional versus Two-Directional

A second distinction between writing and speaking is that writing is essentially one-directional and speaking is two-directional. As the two-directional arrows in Figure 3.2 suggest, as soon as speakers send a verbal-vocal-visual message, they may receive verbal-vocal-visual feedback. For example, individuals receiving the message respond verbally in the form of comments or questions. The tone of their voices is the vocal feedback. Finally, they respond visually by the facial expressions, posture, and other aspects of body language. Verbal, vocal, and visual feedback facilitates mid-course corrections. In contrast, writers don't know how their message is being received.

Conveying versus Convincing

The third and last writing-vs.-speaking distinction is that writing is much more effective in communicating "facts, data, and details" than is speaking (Decker 1992). In contrast, speaking clearly holds the power of persuasion. Depending on your primary objective—information transfer or persuasion—choose wisely and carefully execute your medium.

> **Personal**
>
> Consider an example when I unfortunately reversed the modes of communication while working as a middle manager in an engineering and architectural firm. I prepared a detailed written proposal advocating one-month sabbaticals for senior professionals under certain conditions and sent it "cold" to the firm's president. Response? None! Maybe I would have been able to make sabbaticals happen if I had first spoken with the president, assessed the level of his receptivity, and then followed up with a written proposal that, among other things, addressed his concerns.

LISTENING: USING EARS AND EYES

Of the five forms of communication, listening might appear to be the easiest. On the contrary, listening effectively, that is, listening to understand what others mean and how they feel, is difficult. Hearing, one of the body's five senses, is necessary to listen, but it is not listening in and of itself. Linver (1978), said that hearing "... is a natural, passive, involuntary activity. Anyone with a normally functioning ear and brain will involuntarily hear sounds of certain intensity." Effective listening goes well beyond hearing because it requires you to be attentive, verify understanding, and use what is heard. Good things come to those who know how to listen. Epictetus, the Greek philosopher, suggested that we are designed to listen effectively when he said "Nature has given to men one tongue, but two ears, that we may hear from others twice as much as we speak."

Be Attentive

Covey (1990) identifies five levels or degrees of attentiveness in how a person uses his or her hearing ability. As Figure 3.3 shows, they are: ignoring, pretending to listen, using selective listening, using attentive listening, and using empathetic listening. Ignoring

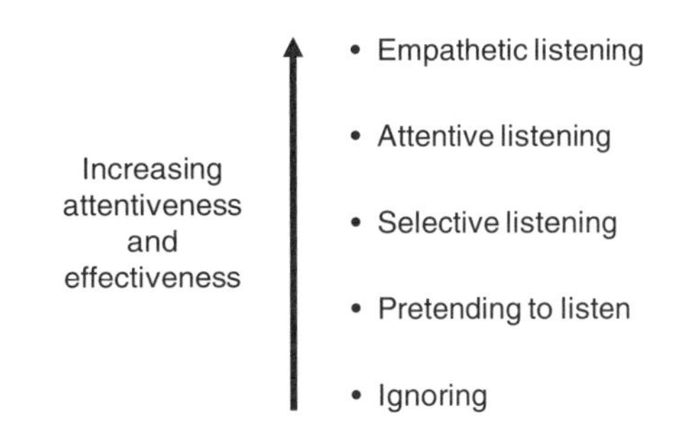

Figure 3.3 Listening can be categorized into five levels, the most effective of which are attentive and empathic listening.

the speaker, the lowest level of listening, may seem unusually rude, especially in a small group setting; however, most people have had the humbling experience of realizing that no one was listening to them.

While appearing to be attentive—pretending to listen—may help, actually being attentive though selective listening or attentive listening will accomplish much more. Incidentally, Covey distinguishes between selective and attentive listening, levels three and four, by explaining that the former means hearing only parts of a person's message, perhaps by design, and the latter means hearing all of a person's words.

Empathetic listening, the highest level according to Covey, means "listening with the intent to understand" or trying to see the situation as the other person sees it. Decker (1992) calls this "feeling listening" and distinguishes it from "fact listening." Empathetic listening is not necessarily sympathetic listening. "The essence of empathetic listening is not that you agree with someone; it's that you fully, deeply understand that person, emotionally as well as intellectually" (Covey 1990). Empathetic listening is "risky" for the listener. If you truly achieve empathetic listening, then you have discovered how another person thinks and feels about something. As a result, and as noted by Covey, you may be profoundly influenced and your thinking and feeling may change accordingly.

The Value of Facts and Feelings

Attentive and empathetic listening, especially in response to your questions, is a vital part of the process used to understand client, owner, and customer wants and needs as described in Chapters 7 and 14. You may wonder why empathetic listening, in addition to attentive listening, is important in various aspects of professional practice. The decisions we and others make and the resulting actions are typically driven by a combination of facts and feelings. Therefore, our success in understanding and influencing colleagues, clients, owners, customers, stakeholders, and others will be enhanced if we know the facts and understand the feelings. Engineer and author Samuel Florman offers this ominous observation: "One of the failings of engineers is they overestimate the power of logic and underestimate the power of emotion."

What techniques can you use to achieve the highest two levels of listening? Benton (1992) suggests that you "Silence all internal dialogue. Let extraneous thoughts pass through your mind without dwelling on them or allowing them to take your attention away from the speaker." Comfortable eye contact is important, according to Roane (1988), who suggests avoiding the extremes of glaring continuously at the speaker and looking everywhere but at the speaker. Other body language messages are also important, according to Benton who cautions against looking bored, intimidated, or intimidating. She suggests encouraging the speaker by signaling, "tell me more" in one way or another.

Body Language: The Silent Messenger

Having mentioned body language, it warrants additional attention. View seeing and interpreting body language as part of listening as in "listening" with your eyes. Listening is markedly enhanced or diminished by body language—yours and his, hers, or theirs. Body language is non-verbal communication such as posture, facial expression, arm positions, handshake, eye contact, and dress. In 1959, anthropologist Edward T. Hall used the expression "the silent language" for what we now typically call body language. According to him, silent language functions "in juxtaposition to words" by conveying feelings, attitudes, reactions, and judgments (Bauerlein 2008). "When the eyes say one thing and the tongue another," according to schoolmaster and minister, Ralph Waldo Emerson, "a practiced [person] relies on the language of the first."

Riggenbach (1986) discussed the role of body language in various types of negotiations and claimed that 95 percent of the non-verbal gestures being received during negotiations are not used by the receiver. According to the author, the nature and interpretations of body language varies among countries and cultures. Finally, Riggenbach says, "Body language shows the inner feelings and attitudes of a person—actions do speak louder than words!"

As extreme and irrational as these results and claims may seem, review some of your recent positive and negative transactions with one or more people. To what extent was your speaking effectiveness influenced by your body language? How much were you affected, as you spoke and as you listened, by theirs? Some examples of body language as interpreted in some western cultures (Quilliam 2009, Wang 2009) are:

- Arms crossed on chest: Resists your message, closed mind
- Touching nose: Thinks your words are deceptive
- Hand on back of neck: Has questions/concerns
- Raised eyebrow(s): Does not believe you
- Looking at ceiling: Deciding
- Relaxed and smiling expression, eye contact: Good decision

Look for repetition and consistency. A single type of body language exhibited for an instant must not be interpreted out of context with other types of body language. Someone might touch their nose because it itches or look at the ceiling because they admire the light fixture. You get the idea!

By the way, when you read body language, you are using primarily the right side of your brain. Your brain's right side operates or thinks in a "visual, perceptual, and simultaneous" mode contrasted with your brain's left side which thinks in a "verbal, analytic, and sequential" mode (Edwards 1999). Refer to the Chapter 7 section titled "Mind Mapping" for additional discussion of how the brain's hemispheres function.

As a college student or young practitioner, you are in a group that tends to be very adept at faceless communication such as using cell phones, texting, email, blogging, and tweeting. That's fine—these electronic communication tools can be both effective and efficient. On the other hand, taken to extreme, you may be part of what Mark Bauerlein (2009) calls "The avalanche of all-verbal communication" and, therefore, you may be diminishing your communication ability. You may be less likely to understand the importance and meaning of body language, that silent language that is part of total interpersonal communication. You increasingly risk missing non-verbal clues, which could harm you and your organization.

Verify Understanding

Recall that the goal is attentive and empathetic listening—understanding the other person intellectually and emotionally. You may need to verify your understanding of speakers and their messages. Obvious techniques include asking questions and paraphrasing or summarizing what you think the speakers are saying and seeking their confirmation. You may have to courteously interrupt the speaker to verify your understanding, but this will probably be viewed positively because it indicates you are listening attentively and perhaps even with empathy. Engineers, and other individuals with technical and scientific abilities, sometimes use impromptu sketches as a way of verifying understanding.

Use What Is Learned

The most obvious way to demonstrate that you correctly and fully—intellectually and emotionally—heard what the speaker said is to reflect your new understanding of facts and feelings in subsequent written and oral statements and in your actions. Your statements and actions won't necessarily reflect the speaker's wishes, but they will be informed by the speaker's perspective. No one can ask for more of you.

WRITING TIPS: HOW TO WRITE TO MAKE THINGS HAPPEN

A document is sometimes the most used and influential product of a technical investigation. Depending on the situation, that document might be more specifically called a formal report, manual, memorandum, article, letter, or email. Assuming that the data and information are available, the smorgasbord of over 20 ideas or tips presented here will assist engineers or other technical professionals, beginning as students, in producing all or portions of a document with minimal effort and maximum communication potential. The intent is also to help you efficiently create documents that will move people to action–that will make things happen. As publisher Malcolm Forbes noted, "Putting pen to paper lights more fire than matches ever will."

We spend relatively little time writing documents. Why? Because we spend most of our time doing the work! However, our work will very often be judged by our documents, that is, how they are viewed—not by the work itself. As suggested by Figure 3.4, the document that explains the work, not the work itself, will, at least initially, impact the client, owner, customer, or stakeholder. Therefore, if you want to make things happen, your documents, especially those sent to those you serve, need to be well done—they need to look and read well. Don't produce a mediocre document about your and your team's excellent work.

Define the Purpose

The first thing to do, if you want to write to make things happen, is to articulate your purpose. What do you, as the writer or lead writer of anything from an email to a formal report, want someone or others to do? Before going any further, draft one or more statements describing what you want your readers to do and begin each statement with an active verb such as understand, support, commit to, or oppose.

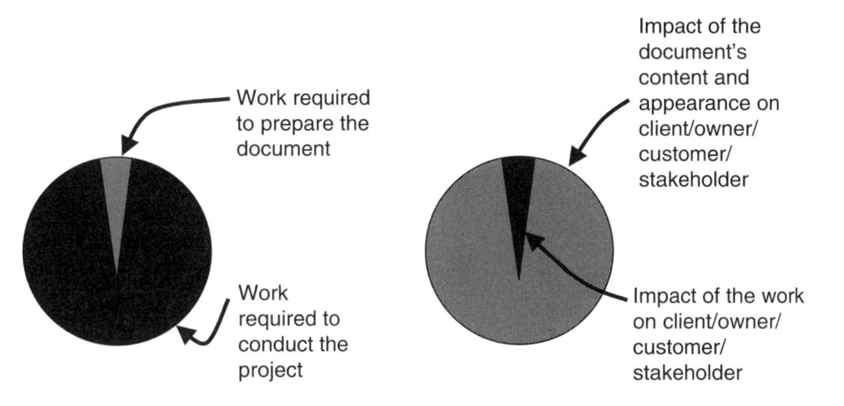

Figure 3.4 A document's content and appearance may have more impact on those being served than the work itself.

Consider, as an example, a project I worked on. The principal message for this project's audience was: Understand that the watershed can be developed without adverse downstream water quantity – quality effects. The project progressed through two documents and then design and construction—and the purpose remained constant and visible.

Profile the Audience

Technical professionals often write documents intended for technical and non-technical individuals and organizations. The same item must be intelligible and useful to multiple audiences. For example, a document produced by design engineers in a manufacturing organization may be directed to technical personnel such as other engineers guiding the production process and to non-technical personnel such as staff in marketing and finance departments. Similarly, a document written by a civil engineering consulting firm may be directed to technical personnel such as the city engineer or director of public works and non-technical individuals such as the mayor, city council members, business leaders, and citizens.

Accordingly, the first step in writing to make things happen is to profile or define the audience. Think about what you know about the individual, or individuals, likely to read your document. For example, what do you know or can you find out about their functions, responsibilities, goals, disciplines, specialties, education, experience, age, gender, ethnicity, and religion. One thing you can be sure of is that your audience is not exactly like you. Write for them, not you.

This profiling task is simple if you are composing an email to a person you know well. However, we often write documents to people we don't know well. Some disciplines, such as civil engineering, tend to have large, diverse audiences as illustrated in Table 3.1. Such audiences are not necessarily big audiences in terms of the number of likely readers, but they are diverse audiences. You should recognize that your documents will not be best sellers, but they can be effective sellers. They can help to make things happen if you define the breadth and diversity of your likely audience and try to write to all of them.

Table 3.1 The audience for some engineering documents, while it may be small, can be very diverse.

Chamber of Commerce member
Chief Engineer
Chief Executive Officer
City Council member
City Engineer
City Planner
College faculty and students
County Board member
Department of Natural Resources representative
Director of Public Works
Land developer
Mayor
Newspaper reporter
Plant engineer
Sierra Club member
United States Environmental Protection Agency representative
Village Board member

Structure the Document to Reflect the Audience Profile

The pyramid structure shown in Figure 3.5 is an effective way to write a major report to an audience with highly-varied members, including technical and non-technical individuals. The areas associated with Executive Summary, Body, and Appendices very roughly represent the relative number of pages to be devoted to each of those three sections of a major document. More specifically:

- The Executive Summary is designed for individuals such as the mayor, chief executive officer, and council and board members. Write it last, focus on forward-looking essentials, and keep it short.
- The Body of the document serves those who want a comprehensive understanding such as the city engineer, the chief engineer in an industry, or other technical professionals.
- Appendices are designed for scientific/technical experts, special interest groups, and others with interests in detailed data and information.

The preceding structure suggestions refer primarily to full-fledged, formal reports. However, the fundamental ideas apply, in mini-form, to basic written communications such as memoranda, letters, and emails. For example, if you are writing a memorandum, make your point and attach something that provides additional documentation for those who may want to go into further depth. This way not all readers will have to wade through everything to get to the "bottom line."

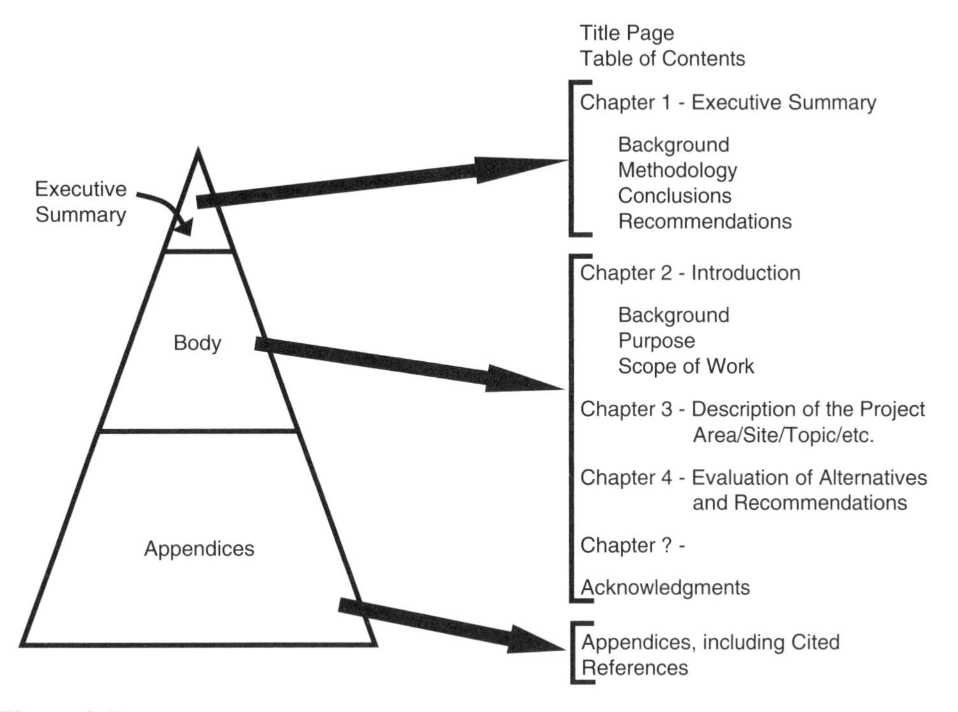

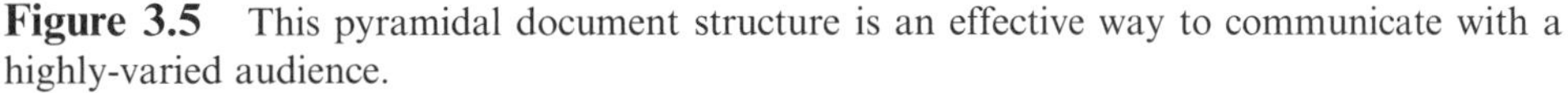

Figure 3.5 This pyramidal document structure is an effective way to communicate with a highly-varied audience.

Personal

While working as a project manager and a lead document writer for an engineering firm, I learned that the company's leadership was not pleased with the effectiveness of writing across the organization. Accordingly, they hired a "burned-out" high school English teacher to help us. He reviewed documents I had written and one of his comments hit home. He said I was writing for me, not readers. My documents were chronological accounts of all that we did on a project. Readers had to wade through the entire document, even if they just wanted the essentials. Clearly the pyramid structure illustrated above was the answer. I enthusiastically embraced it then and have used it ever since. The English teacher also helped our firm develop a very useful style guide. More about this tool later.

Students and young professionals may be troubled by the inverted model. Presenting the conclusions and recommendations before describing the data collection and analysis procedures is certainly contrary to the way the work was actually done. But, most readers will not care what was done, other than to have the confidence that it was done correctly. They will focus on their areas of interest and the recommendations and implications for them or their constituencies.

As careful as you may be, you never know for sure where your document will go, therefore, put "wheels" on it, make it travel well. Depending on its size and complexity, consider providing user-friendly features such as an executive summary, a glossary, a list of abbreviations, a table of contents, informative headings and subheadings, tabs, graphics, generous white space, wide margins, and an index. The Spanish writer Enrigue Jardiel Poncela said this about carefully-designed written products "When something can be read without effort, great effort has gone into its writing."

Ask About Document-Writing Guidelines

Some public and private organizations have style-guides and procedural manuals for writing documents. These helpful guidelines address topics such as abbreviation, capitalization, citation of references, gender-neutral writing, punctuation, typestyles, graphics, common mistakes, and overall format. Inasmuch as engineering documents are often written as a team effort, style-guides and procedure manuals help achieve internal consistency within any given document. They also contribute to inter-document consistency within an organization. Finally, writing guides and manuals save time and minimize frustration by reducing the need to make numerous decisions about the mechanics of a document. Determine if your organization has guides or manuals. If not, perhaps you will be in a position, after having written various documents, to suggest that a style-guide be developed and to contribute to its preparation.

At minimum, refer to widely available guides such as the Chicago Manual of Style (University of Chicago 2010) and Strunk and White (2000). The preparation of a style-guide or manual to meet your organization's needs can be simplified by making reference to specific provisions in readily available manuals and guides such as these.

Personal

I once served as a consultant to a team working on state department of natural resources project. One of my responsibilities was to facilitate the preparation of a final comprehensive document that summarized the group's work and recommendations. From the outset I saw inconsistencies as various writers weighed in. For example, the sponsoring organization was referred to in a half dozen ways. Therefore, I led the preparation of a project-specific style guide for this project. It was effective and serves as an example of how a project-specific style guide, which is easy to do upfront, saves much time and frustration later. Incidentally, I also prepared and used a style guide when writing this book.

Start Writing on "Day 1"

If a document is one of the project deliverables, start writing it when the project starts (Walesh 2004c). Do I really mean writing on the first day of the project? No, not exactly. But I do mean during the first week of a project having duration of months.

The document should be written while work is in progress and, as draft portions of it are produced, they should be provided to members of the project team; the client, owner, customer; and possibly other interested parties. Experience suggests three benefits of writing parallel to working:

- Improved communication within the project team. Writing and reading draft descriptions of project features such as background, purpose, scope, and approach is likely to raise intra-team concerns, questions, and suggestions. Those who write drafts will clarify previously vague aspects of the project in keeping with playwright Edward Albee's statement, "I write to find out what I'm thinking."
- Improved communication with clients, owners, customers, and stakeholders in areas such as overall document format and content, level of document detail, data and information sources, assumptions made, approaches taken, alternatives developed, and recommendations considered.
- Avoidance of "last minute" rushing and resulting errors and omissions which diminish the perceived quality of the project. Sadly, excellent work is sometimes delivered in a mediocre package and, as noted earlier in this chapter, the quality of the effort is judged by the appearance of the package.

A common objection to starting the writing on the first day is that there is nothing to write about because no work has been done. The partial answer to this is that a draft document outline can be prepared immediately. Furthermore, background, purpose, and scope subsections can be drafted. A glossary, a list of abbreviations, and a list of cited references can be started. If, after the project has passed the ten percent mark, there is nothing to write about, then maybe nothing useful is being accomplished. Nagle (1998) offers similar thoughts on paralleling working on and writing about a project.

Consider a scenario for starting the writing on "Day 1." During the first week of a multi-month project, the project manager drafts an outline or table of contents of the document and shares it with the project team. This provides a means to further communicate his or her ideas of where the project is going and to solicit input.

During the second week of the project, the project manager sends the revised outline to the client, owner, customer, and possibly some stakeholders asking them to review the outline or table of contents in light of their understanding of the project and its deliverables.

As the project progresses, drafts of various document sections or chapters are shared first with the project team and then, as refined versions, with the client, owner, customer, and stakeholders. The process improves communication between the project manager and the project team, within the project team, and between the team and others.

Get Started: Overcome Writer's Block

Assume you are at the very beginning of writing a document. The blank page—or the blank screen—can be intimidating. You may suffer from "writer's block." Author Peter De Vries said "I love being a writer. What I can't stand is the paperwork." Consider a process for getting started—it works for me and might work for you.

Brainstorm

Place a blank sheet of paper on your desk or bring up the word processor on your computer. Quickly write single or a few words placed anywhere on the paper/monitor. Avoid writing sentences. Fill the medium with ideas and information. This is your initial "brain dump!"

If you still can't get started, consider taking a walk—around the office or around the block. The German novelist Thomas Mann said "Thoughts come clearly while one walks." William Wordsworth, the English poet, composed most of his poetry while walking. He memorized it and put it in writing once he returned home (Johnson 2009).

Engage Your Subconscious Mind

Set aside the results—do something else for a few hours or even a day. Consultant Debra Benton suggests "Plant subjects you are considering in the garden of your subconscious" (Benton 1992). Then return to the brainstorming and discover that additional ideas will appear. A participant in one of my webinars shared this thought about use of his subconscious: "I long ago developed the habit of rough drafting particularly sensitive letters or memoranda in the late evening; then when I get back to them first thing the next morning, my subconscious often will have done a pretty good job of overnight editing."

As an alternative to the preceding two steps, consider using mind mapping as a "get started" tool. Chapter 7 includes a detailed description of mind mapping within the major section of the chapter titled "Tools and Techniques for Stimulating Creative and Innovative Thinking."

Group, Outline, Share, and Refine

Brainstorming is typically productive, especially if done in two or more sessions. Now begin to bring some order. Provide some structure by arranging ideas and information into clusters or groups of similar ideas, topics, or information. William Zinsser, writer, editor, and teacher observes that "Writing is thinking on paper." Now your efforts begin to suggest some resemblance of order. Use the groups to create an outline. You might do this in two or more settings in order to give your subconscious mind additional incubation time.

Share your outline with friends, team members, or clients/owners/customers/stakeholders. Give them and their subconscious minds time to work on the outline. Use their input to refine the outline. Resist being prematurely bound by the outline. It's just words on paper, that is, freely delete, insert, and rearrange.

Draft the Document

Draft the document using sentences and paragraphs. Notice, we invested significant effort to get here, that is, to begin writing sentences. The point? Don't start here. Don't start by writing full sentences and paragraphs. Instead, start with the "brain dump" and then engage your subconscious mind, group and outline, and share and refine the outline. Reason: Optimize use of your subconscious mind and those of your colleagues.

"Listen" to your writing by reading your draft out loud. Poet Robert Frost said: "the ear is the only true writer and the only true reader." This suggestion will be revisited in the next subsection of chapter as part of the tin ear discussion. Look for ample use of metaphors and similes, that is, images to help your reader. Gabriele Rico (2000) says that a metaphor "consists of images connected to something they literally cannot be." For example, "he ate a mountain of ice cream for dessert" or "the proposal she wrote nailed the project for our firm." He didn't eat a mountain and she didn't use a hammer. However, as illustrated by these two examples, metaphors bring additional dimensions to our writing and, as a result, make more connections with readers. In other words, when we write, we do not have to always be purely literal.

"Both metaphor and similes originate in the ability to perceive similarity in dissimilarity," according to Rico, "the difference is that simile has "pointers"—like or as—to explicitly signal that we are joining logically unjoinable entities. Metaphor dispenses with these pointers, simply asserting a likeness between two unlike things, thus making a metaphor richer, more open-ended, and more resonant." If we converted the mountain metaphor to a simile it might read like this: "The amount of ice cream he ate last night for dessert was like a mountain." Not quite as descriptive as the metaphor mode.

You may, as the designated writer of all or portions of a document, be most familiar with or enthusiastic about certain aspects of the project. For example, perhaps you developed a computer program and used it to design a structure. Maybe you supervised data collection in a laboratory or carried out field studies. Go after the low-hanging fruit first so that you get some words on paper and continue to beat down writer's block. Documents do not have to be written in chronological fashion! A glossary might be an early writing product when a team is working in a new area because drafting and refining the glossary will align team members on their understanding of terminology.

Avoid Tin Ear

Do you "hear" what you write? If not, you may be suffering from "tin ear" which means failing to see or, more precisely, hear excessively repeated words. At first read, the following text may appear acceptable, that is, it looks "OK" (Alexander and Rivett 1998). But how does it sound? Read the text out loud.

The objective of the study is to identify all trends
in the markets of the company.
The study will identify products which offer
added opportunities to the company.
In addition, the study will identify the requirements
which must be met by the company.

If you had not already noticed, you have now heard "study" three times, "identify" three times, and "company" three times. This repetition of those words within three sentences is aesthetically undesirable. A more pleasing version follows (Alexander and Rivett 1998):

The objective of the study is to identify trends
in the company's market,
products which it might add to its line,
and
the requirements imposed on the company
by these products.

In this more pleasing version, the frequency of "study" goes from three to one, "identify" from three to one, and "company" from three to two with "company" appearing once as a possessive and once as a noun. And, as with many of these writing tips, the frosting on the cake is that the revised version is shorter—43 words to 31 words.

Reading text out loud will help you find other ways to improve your writing. Examples of deficiencies that can be discovered are confusing syntax, excessively long sentences or too many sentences of the same length, passive structure, using technical terms that have not been defined, and the need of additional punctuation.

Retain Some of the Outline in the Document

Carefully selected, placed, and spaced first, second, and perhaps third order headings are useful to the reader who skims a document for topics or areas of particular interest. The writer can also use those headings, perhaps down to the second order level, to construct a table of contents for a document that further enables the selective reader to focus on the critical parts. A study of engineer's reading preferences concluded that they preferred text structured in subtitled sections as opposed to large blocks of text (Silyn-Roberts 2002). Subheadings, and the related white space, make the text more inviting. This book makes ample use of white space and first, second, and third order headings to assist you, the reader.

A final thought about getting started. Recognize that you, the author of the document, may be asked, often on short notice or in an impromptu situation, to summarize the report. By "turning the pages" of a document that has carefully-designed headings and sub-headings, you can make an effective presentation.

Write Major Documents in Third Person: Mostly

Most major engineering and other technical documents are written in the third person. Let's review first, second, and third person writing. Consider this example of a sentence written in third person: "A geotechnical survey of the site revealed bedrock 10 to 20 feet below the ground surface." A second person version could be: "You conducted a geotechnical survey of the site and found bedrock 10 to 20 feet below the surface." In first person, this might be: "We conducted a geotechnical survey of the site and found bedrock 10 to 20 feet below the surface."

Exceptions to the dominance of third person writing are appropriate where the overall effectiveness of the document will be improved. For example, you might decide to use a first person structure for that part of the document that presents recommendations. Each recommendation might be preceded with the first-person statement, "We recommend. . . ." By switching abruptly from third person to first person, the report structure draws the reader's attention to the importance of the recommendations. The style change also reminds the reader that recommendations are based on the data and analysis presented, combined with the experience and judgment of members of the group or organization that prepared the document. Irish and Weiss (2009) reinforce this suggestion.

Personal

As part of a book proposal, I sent the draft first chapter of the book to the publisher. It was written in third person. The publisher's representative, noting that the book would be written to offer advice to a young audience, suggested that I mix second and third person. He argued that this use of two persons would speak more directly and informally to the target audience. I followed his advice, wrote the book (Walesh 2000) in second and third person, and believe that this use of second and third person was successful. You are reading the third edition of this book which uses first, second, and third persons to further personalize it.

Employ a Gender-Neutral Style

Unless the expected audience is composed of one gender, draft the documents to treat genders equally. If the basic style of your document, or the accompanying oral presentation, is offensive to part of the audience, your message may not be "heard."

Besides using the now common "he or she" pronouns, Walters and Kern (1991) offer two useful suggestions. First, avoid or reduce gender-specific pronouns. For example, instead of writing "When a person needs assistance with software, he or she should contact the Office of Electronic Information Services," use the gender-neutral pronoun "anyone." Rewrite the sentence to read, "Anyone needing assistance with software should contact the Office of Electronic Information Services." The second approach Walters and Kern suggest is to write in a plural format rather than a singular format. For example, instead of writing, "If a member of the staff needs assistance in making travel arrangements, he or she should call the Human Resources Office," write "Staff members needing assistance with travel arrangements should call the Human Resources Office."

Write in an Active, Direct Manner Rather Than a Passive, Indirect Manner

Active writing, or writing in the active voice, is direct, rigorous, definite, and shorter. Active voice typically uses a simple subject-verb-object structure. In contrast, passive writing, or writing in the passive voice, is indirect, tame, and sometimes indefinite.

Let's begin with this passive voice example: "It was determined from application of computer modeling that the existing freeway system could not accommodate rush-hour traffic." Note that the example begins with "it," a common feature of passive writing. Therefore, be wary of sentences that begin with "it"—for two reasons. First, the "it" is likely to indicate a passive structure. Second, casual use of "it" may lead to what Vesilind (2007) refers to as "hidden antecedents"—not knowing what preceding noun is referenced by "it." Expressions like "it was" and "it is" are known technically as the "it-cleft" which is defined as "a construction that has no intrinsic meaning" (Irish and Weiss 2009). Go through some of your writing, if you find sentences beginning with "it," where "it" does not refer to something definite, you are probably writing in the passive voice.

Let's rewrite the passive sentence to convert it to the active voice: "Computer modeling indicated that the freeway system will not carry rush-hour traffic." And here is a "bonus"—fewer words. The passive voice version has 19 words while the active voice requires only 13 words.

"An active verb makes somebody or something responsible for an action," according to writer and editor Patricia T. O'Conner. Instead of the passive "It was suggested that . . . " write "Frank suggested that . . . " In place of "The Plans have been called deficient," write "Betty called the plans deficient." In other words, identify the subject, the one doing the suggesting, calling, etc. (O'Conner 1999).

Here is another example of converting from passive to active writing. Passive version: "It should be noted that Hydrologic Type C soils are most common in the watershed." Convert to an active style: "Hydrologic Type C soils are most common in the watershed." In this case, the text drops from 15 words to 10 words.

Having stressed the desirability of the active voice, situations arise in which the passive voice is more desirable. Consider this example (Vesilind 2007):

- Active: "The falling rock hit the geologist."
- Passive: "The geologist was hit by the falling rock."

The latter is preferred because it emphasizes the unfortunate geologist, not the rock. Another reason to occasionally use the passive voice: Introduce variety into your text.

You may still have "hang-ups" on length of your written products. You think length is good. Perhaps you are still being influenced by that pressure you felt in high school or college when the teacher or professor said something like "write a 25 page paper on . . ." How would you ever find 25 pages of material? Most of your clients, owners, customers, and stakeholders value brevity. As someone anonymously said: "I didn't have time to write a short document, that's why this one is so long." Extra effort is required to produce writing that is brief and to the point. However, brevity, with clarity, is appreciated—and more active voice will automatically do this!

Recognize that Less Is More

As noted, strive for brevity, that is, documents that are no longer than they have to be. Engineers and other technical professionals value efficient facilities and equipment. Let's also strive for efficient writing. While active voice writing reduces document length, other means are available to tighten up your writing.

Table 3.2 Replace two or more words with one word or eliminate redundant words to reduce the length and improve the clarity of a document.

More	Less
During the course of the study, we learned...	During the study, we learned that...
The treatment process exhibits the ability to...	This treatment process can...
The team is currently defining a vision	The team is defining a vision
We recommend that diesel be used for fuel purposes	We recommend that diesel be used for fuel
The question as to whether we choose Option A or Option B depends on...	Whether we choose Option A or Option B depends on...
Owing to the fact that pier scour occurred, we...	Because pier scour occurred, we...
in favor of	for
due to the fact that	because
in order to	to
in the near future	soon
in close proximity to	near
lacked the ability to	couldn't
take into consideration	consider
in addition to the above	also
has the capability	can
perform an analysis of	analyze
pointed to the fact	noted
held a meeting	met
proved of benefit to	benefited
reached an agreement	agreed
present a conclusion	conclude

One approach, as illustrated in Table 3.2 which draws on ideas offered by Berthouex (1996), Iacone (2004), Marketing Sherpa (2003), and Strunk and White (2000), is to replace two or more words with one word or to eliminate words. The third example involves the word "currently," which usually adds nothing, and is widely used by engineers. Perhaps it is in their genes. For example, what are you doing now? You might answer "I am currently reading this book." The word "currently" adds nothing.

Another way to prune our writing is to drop commonly used redundant words as illustrated in Table 3.3. Consider the first example "PIN number." Would you say "Please give me your personal identification number number?" The others in the list, while not repeating a word, all contain unnecessary or redundant words—words that add no new information.

Consider one more example of pruning, in this case a sentence (Walesh 2007a and 2007b). The first 38 word version is: "During the course of the study, we were able to determine that the wastewater treatment plant exhibits the ability to accommodate only up to 2.5 mgd of flow owing to the fact that the pumps have that capacity." Although acceptable, the sentence can be tightened up for the reader's benefit to this 23 word version: "During the study, we determined that the wastewater treatment plant

Table 3.3 Omit redundant words to reduce the length and enhance the clarity of documents.

More	Less
PIN Number	PIN
advanced planning	planning
blue colored	blue
totally destroyed	destroyed
work tasks	tasks
revert back	revert
continue on	continue
temporarily suspended	suspended
basic fundamentals	fundamentals
illegal crime	crime
most unique	unique
past history	history
true facts	facts
equal halves	halves
adequate enough	enough
completely full	full

(Source: Adapted from Vesilind 2007.)

can accommodate only up to 2.5 mgd because the pumps have that capacity." Note the changes in going from the first to the second version:

- "The course of" is simply dropped
- "Were able to determine" is replaced with "determined"
- "Exhibits the ability to" is replaced with "can"
- "Of flow owing to the fact that" is replaced with "because"

Yes, you want your writing to be complete and correct. Strive to do that while using only the necessary words. Prune your writing—it will yield more fruit. Thomas Jefferson, the 3rd U.S. President, said "The most valuable of all talents is that of never using two words when one will do." And, as the principal author of the U.S. Declaration of Independence, he was an accomplished writer!

Apply Rhetorical Techniques

Rhetoric is the art of using words effectively. Conger (1991) argues that the most effective communicators use carefully selected rhetorical techniques. He identifies and gives examples such as repetition, rhythm, metaphors, analogies, brief stories, and alliteration. Martin Luther King's "I Have A Dream" speech, in which he says "I have a dream" eight times, illustrates the power of repetition and rhythm. Incorporate carefully selected rhetorical techniques in your writing.

Alliteration, that is, repetition of two or more words, as in the Martin Luther King speech, or repetition of a sound, usually a consonant, is a rhetorical technique that can be used in the titles of documents to attract readers. Consider these examples:

- Report: Engineering the Future of Civil Engineering (ASCE 2001)
- Professional Paper: "Leading Lessons Learned" (Walesh 2004b)

Alliteration can also be used within the text of a document. For example, alliteration in the form of three pairs of words, each pair beginning with the same letter, is used in the following sentence which appears later in this chapter: "Perhaps your document needs a sentence shortened, a table tightened, or a figure fixed."

Format writing is another form of rhetoric. Consider, for example, a report that will repeatedly describe similar alternatives or computer simulations. Develop a generic format for alternatives or simulations and use it repeatedly, with minor variations for interest, to describe all alternatives or simulations. The generic format for describing each of a series of alternatives might include the following sub-topics in the indicated order: location, function, cost, positive attributes, negative characteristics, and implementation sequence. The format writing helps the writer write and the reader read.

Finally, metaphors and similes, which are discussed earlier in this chapter, are examples of rhetorical techniques. Irish and Weiss (2009) provide a detailed discussion of using rhetorical techniques to enhance document effectiveness.

Adopt a Flexible Format for Identifying Tables, Figures, and Sources

As a document evolves, particularly in an interdisciplinary environment, writers and readers will suggest numerous changes and many of them will be implemented. The writers will expand and delete text; modify, expand, or delete tables and figures; introduce new chapters; and cite additional reports, papers, books, and other sources.

Recognizing the dynamic, evolving nature of the document, consider adopting techniques to minimize the effort required to incorporate the inevitable changes. For example, in the text, cite sources such as reports, papers, books and other sources of information using a format like "Smith (1984)" rather than "Reference 3." As references are added, the latter format usually requires frequent renumbering of the reference list and of referenced citations in the text. In contrast, the former format, which is used in this book, simply refers to an alphabetized list of references at the end of a chapter, or perhaps as an appendix to the document. The list can be gradually expanded, without requiring numbering changes in the text.

If figures and tables will be included, consider numbering them within each chapter (e.g., as in this chapter, Figure 3.1, Figure 3.2, etc., and Table 3.1, 3.2 etc.) rather than consecutively from the beginning of the document. With this approach, the addition of a figure or table in the chapter will limit the re-numbering process only to that chapter. Pages can be numbered in a similar fashion. Again, the extent of renumbering will be less if this procedure is used.

Explicitly mention all tables, figures, appendices, and attachments at least once in the text with some explanation of their significance. These supporting items should appear immediately after or close to where they are first mentioned. Exceptions to this placement advice are appendices and other special supporting materials, which appear at the end of the document, or perhaps in a separate document available on special request.

Use Lists

Add to the variety, clarity, and attractiveness of your document by occasionally using lists as an alternative to narrative in paragraph form. An example is the list used in the earlier "Structure the Document to Reflect the Audience Profile" section of this chapter. Chan and Lutovich (1994) use a list to offer the following suggestions about using lists:

- Explain the list by introducing it to the reader with an appropriate statement.
- Provide ample white space by widening the margins and/or placing extra space between lines. (A wider left margin is used in this list.)
- Mark items with neutral symbols, that is, use numbers and letters only if necessary, such as to indicate priority or to provide for ease of subsequent reference to selected items in the list. (This list provides an example of the use of symbols.)
- Utilize end punctuation for each item only when one of the listed items contains more than one complete sentence. If punctuation is used at the end of any item, it should be used at the end of all items. (This suggestion is followed in this list.)
- Use a similar construction for all items in the list. (For example, all items in this list begin with an active verb.)

Design a Standard Base Map or Diagram

Your effort in preparing graphics will be minimized if you carefully design base maps, plan sheets, diagrams, and other graphics at the beginning of the document preparation process. Computer-based maps, plan sheets, diagrams, and other graphics offer more flexibility and are more easily modified than manually-based materials. However, the anticipated use of computers does not diminish the need to carefully design graphics.

Assume that a consultant's document proposes a rapid transit plan for a growing metropolitan area. The base map on which the existing system, various alternatives, and the recommended system will be shown should be carefully designed at the beginning of the report preparation process. In fact, the anticipated document aside, base map decisions should be made early in project work. Similarly, assume that an engineering project involves retrofitting a heating, ventilating, and air conditioning system into a large, older building. Develop a set of plans and elevations of the building early in the project so you can repeatedly use the plans and elevations when displaying the existing facilities, the recommended system, and alternatives. This approach avoids major re-drafting of maps, plans, elevations, and other visuals.

Compose Informative Titles

Assume that you prepared Table 3.4 for inclusion in a document. You originally titled it "Gravity sewer construction costs." Because you wanted the table to be even more effective for readers, you subsequently selected the indicated longer title. The longer

Table 3.4 Gravity sewer construction costs are relatively insensitive to conveyance capacity.

Discharge ratio	Construction cost ratio
1.5	1.16
2.0[a]	1.30[a]
4.0	1.68
5.0	1.83
6.0	1.96

[a]For example, doubling the flow capacity of the sewer would increase the cost by only 30 percent.

title is, in effect, a short declarative sentence. It is more reader-friendly than the original title because it explains the table's message. Informative titles, like that used in Table 3.4, are used throughout this book. An option, if you prefer short titles, is to use captions or notes within a table, figure, or other visual to explain its principal message.

Establish Milestones

Recall the earlier suggestion to write a major document parallel with the project work. Milestones are dates by which key parts of the writing effort will be completed. For example, milestones might be established for completing the outline of the document, each of the chapters, tables and graphics, and appendices. Such milestones are important because there is no limit to the improvements and refinements that can be made to any document. There are always better ways to present data, describe analysis, and present recommendations. Establishing milestones will offset the open-endedness of writing projects and focus the project team on producing results.

Produce an Attractive and Appealing Document

Word processing and other software capabilities such as underlining, right justification, bold-faced type, special fonts, color, icons, photographs, and other digitized materials provide an opportunity to produce an attractive and appealing document. In fact, failure to do so suggests that the authoring organization is not well-equipped in its document production capabilities and perhaps its other capabilities. Give special consideration to the cover of a document because it is all that some potential readers may ever choose to look at. Rarely does anyone have to read your document—make them want to read it by virtue of its appearance and structure. The attractiveness idea also applies to much more modest documents such as a one-page memorandum. View it objectively—how does it "look" as determined by font, headings and subheadings, white space, and margins.

Cite All Sources

Give credit for all ideas, information, and text created by others and used in preparing your document. Respect the intellectual and creative works of others. Cite the references in the text and provide a list of the cited references, as illustrated in this book by the Cited Sources at the end of each chapter. Some writers prefer to use footnotes, that is, show cited sources at the bottom of each page. This might be even more reader friendly—the reader does not have go to the end of a chapter or the back of a document.

Personal

I was reading an article and came to a particular paragraph. It looked familiar—it was familiar—I had written and published it earlier. The author had used it without attribution. I contacted him and he blamed his assistant. My point: I felt violated because I write for a living. Using my text without attribution or my permission is the same as breaking into my home and stealing personal property.

An acknowledgments section provides an opportunity to recognize the assistance and support of private and public sector individuals and organizations. Be gracious—err on the side of acknowledging the assistance of individuals and organizations even though some of them may have only been marginally involved in the work culminating in the written document.

An acknowledgments section, typically placed near the end of a document, also provides a means by which the author or authors indicate that, although they appreciate assistance provided by others, they are responsible for the essence of the final product. The acknowledgments section might read, in part, as follows: "The authors gratefully acknowledge the assistance received from the indicated individuals and organizations. However, the authors are solely responsible for the opinions expressed, conclusions drawn, and recommendations made in this document."

Read One More Time

As noted at the beginning of this document-writing section, the focus is on production—generating a document with minimal effort and maximum communication potential. A small diversion is in order at this point in the process. When your document is finished, and after you have used the spell checker on your word processor, take one last critical look at it. Table 3.5 presents suggested proofing tips based, in part, on Alexander (2001) and Walesh (2000).

Search for errors in syntax, grammar, punctuation, spelling, and facts and find omissions. At a higher level, consider refinements you could use to enhance communication with your readers. Perhaps your document needs a sentence shortened, a table tightened, or a figure fixed. Then perhaps ask your spouse, significant other, a friend, or colleague to critically read the document. They offer a fresh perspective. Make whatever refinements are needed to yield a final polished product.

Humorist and fiction writer Mark Twain said this: "The difference between the right word and the almost right word is the difference between lightning and the lightning bug." Charles Dickens exemplified the editing and "read one more time" advice. In early versions of *A Christmas Carol*, Tiny Tim was Small Sam, Little Larry, and Puny Pete. And the now widely-recognized "Bah, Humbug!" was originally just "Humbug!" (O'Conner 1999).

Table 3.5 Use some of these tips when proofreading your writing.

Print in landscape mode, different font, or double spaced to increase your awareness
Check overall look—hold at arm's length
Check consistency of format, headings, and subheadings
Cross-check Table of Contents headings, subheadings, and page numbers with text
Read backwards
Try a change of scenery
Read out loud
Let it sit so your subconscious can work
Do not rely solely on word processor's spell check (e.g., did you mean "chose" or "choose," "heard" or "herd"?)

SPEAKING TIPS: HOW TO SPEAK TO MAKE THINGS HAPPEN

Engineering students and entry-level professionals have many opportunities to make, or help to make, formal and semi-formal presentations. Typical audiences are fellow students, colleagues within their organization, clients and potential clients, professional associations, service clubs, and K-12 students.

Much may be at stake in some of these oral presentations. For example, a technical professional may be trying to convince colleagues of the validity of his or her ideas in order to see them implemented. Or, a young engineer may be urging representatives of a client, owner, or customer to accept a recommendation based on an engineering investigation. A potential client's, owner's, or customer's decision to retain your firm may hinge on the effectiveness of the presentation a representative of your firm makes. The stakes are moderate for other presentations. Nevertheless, much good could be accomplished and much satisfaction achieved if the presentation is excellent. More specifically, effective speaking leads to a win-win-win-win result because:

- You "win" in that you learn as a result of preparing, include researching and thinking. English novelist E.M. Foster said: "How can I tell what I think till I see what I say?" You also "win" by gaining more confidence each time you speak.
- Your audience "wins" in that they learn based on your study and preparation. In any given week you could, in effect, be teaching and your audience learning at a professional society meeting, in a fifth grade class room, or in a "brown bag" session in your office.
- Your profession "wins" in that you give back to your profession by sharing your knowledge and experience. This is especially true if your speaking is eventually captured in conference proceedings, published articles and papers, and books.
- Your employer "wins" because when you stand in front of that audience, you are, at least temporarily, your organization. Your presentation enhances your organization's reputation.

As an aside, look for opportunities to test the hypothesis that people's perceptions of all types of organizations are often based on only one contact with a representative of that organization. Engage people in conversation as opportunities arise. Ask them what they think about a particular organization, such as a consulting firm, a manufacturing company, a government agency, or an educational institution, and then ask them how they know. You might be surprised to find the basis for their perspectives. Often the impression, whether it is positive or negative, will be based on a presentation by one representative of the organization.

Therefore, the community of technical professionals places a high premium on the ability to make effective presentations. Individuals who develop good to excellent speaking skills are well-rewarded in terms of span of influence, promotion, compensation, perquisites, added opportunities, and perhaps most important of all, personal satisfaction for a difficult task well done.

Conquer Reluctance to Speak: Commit to Competence

Although numerous opportunities arise within the engineering and business world to make effective presentations, and although effective presentations are needed and valued, relatively few professionals, young or otherwise, voluntarily seek out such opportunities. Furthermore, many of the presentations that do occur are mediocre to poor—even some of those made at regional and national conferences.

Anyone can learn effective speaking, like any other skill, through study and practice. However, because they must practice this particular skill in front of other human beings, and because they may fear failure, relatively few students and practitioners make the commitment to become effective speakers. Their speaking is usually infrequent, by request, and under duress. They react, out of fear, to the need to speak rather than seek the opportunity to do so. Perhaps baseball great Yogi Berra said it best: "Public speaking is one of the best things I hate."

As a student or entry-level professional, commit to continuously improving your presentation ability. The specific and pragmatic ideas presented here will help you prepare for, present, and follow up on presentations within and outside of your professional activities.

No matter where you, as a student or young practitioner, are today as a speaker, you can get much better if you combine knowledge of speaking fundamentals with speaking practice. Even if you are fearful of speaking and/or are a poor speaker—I was both—you can become a good to excellent speaker. As schoolmaster, minister, lecturer, and writer, Ralph Waldo Emerson said, "All great speakers were bad speakers at first."

Personal

I heard an effective sermon. After, during a luncheon, I happened to sit next to the pastor. I asked him if he would explain his sermon preparation process. He described how he worked on each sermon over an entire week. Total preparation time: 20 hours. That's 20 hours of preparation for 20 minutes of delivery for a preparation to delivery ratio of 60. On the heavy side, but not by much. My rule of thumb for a new presentation: 20/1 or higher.

Prepare the Presentation

Break the speaking project into three parts: Prepare, present, and follow-up. The first part requires, by far, the most effort. However, all three parts require careful attention if you want to use speaking to make things happen. The following discussion addresses each of the three parts.

Define the Purpose and Profile the Audience

Review the earlier writing sections of this chapter titled "Define the Purpose" and "Profile the Audience." The same principles apply, supplemented with some additional thoughts about the audience. You might ask for names of likely participants. You may know some of them and that can help you shape your message. For example, you are a consultant and will speak to the local Kiwanis club. You learn that Mayor Susan Jones will be in the audience. As a result, you include in your presentation a comment about your firm's award-winning municipal projects.

As with a written document, especially a major document, your audience may consist of many audiences. However, unlike a document that can be written at various levels and divided into different sections so the readers can choose which portions they want to read, you must make the entire oral presentation to the entire audience. This is a major challenge. Try to have something to say to each audience segment.

If you are one of a group of speakers, find out as much as you can about the other speakers, such as their topics and the chronology of the presentations. Also, determine the total time allotted for your presentation and post-presentation questions and discussion and respect those limits. Determine if you will be permitted to entertain the questions immediately after your presentation or if all questions will be deferred until all speakers have completed their formal remarks.

Personal

Weeks before a presentation to structural engineers, I asked the session organizer to help me profile the audience. He provided this description "The audience will be about five percent students, 40 percent owners, 50 percent designers and five percent other, including architects. We are guessing that the audience will be about 50/50 for public and private work." That was very helpful! A very diverse audience. I needed to speak to all audience segments.

Select the Topic and Title

What are you going to talk about? Sometimes the topic is selected for you or, as when you respond to an invitation to speak about a specific topic or when you respond to a call for papers that specifies a topic or a theme. Other times, you select the topic. A natural, initial thought is to select a topic about which you are an expert. However, another basis for topic selection is something you'd like to learn more about. This is one of the reasons I've enjoyed full-time and part-time teaching throughout my career—helps me remain a student, a learner, and study topics that interest me.

Look objectively at the work you and your organization are doing. Might others be interested? Sometimes we are too close to what we do to see the value others may see. As an example, a client and I gave two well-received presentations on the role of mentoring in ownership transition. All we did was describe the process we used in his firm. What's going on in your "shop" that might interest others?

Having decided on the topic, create a "catchy" title. This may increase the probability of co-workers coming to your brown bag presentation in the office or participants coming to your session at a conference. Which presentation would you like to hear, "Converting Disasters into Opportunities" or "Government Assistance Program"? As with writing, employ rhetorical techniques, such as alliteration in "Leading: Lessons Learned" (Walesh 2004b). Another approach is to suggest benefits, as in the preceding example, or promise to show "how" as in "Flying Solo: Some Thoughts on Starting an Individual Practitioner Consulting Business" (Walesh 2004a). Maybe use an intriguing title such as "DAD is Out, POP is In" (Walesh 1999). You may dismiss these little ideas as gimmicks or not worth the effort. Think again—the total is the sum of the parts. Therefore, attend to the parts, the details.

With a topic and title in mind, begin to think about the overall structure of your presentation. Consider the T^3 approach:

- Tell them what you are going to tell them
- Tell them
- Tell them what you told them

This structure uses repetition because repetition helps the audience follow your presentation and remember the principal points. Individuals who hear your presentation, unlike individuals who read your document, cannot easily refer back to what you "said" earlier. Unless they are conscientious listeners and note takers, you will need to reiterate your principal points. The "tell-tell-tell" approach is an effective model for meeting this need. In the introductory part of your presentation you might briefly tell the audience what you are going to tell them. Then, in the major part of your presentation, you tell them. At the end, you wrap up by briefly telling them what you told them. Now for the substance. What will you say?

Outline the Presentation

The process of preparing the material from which you will speak is similar to the previously presented process of writing a document. Refer again to the following topics within this chapter in the major portion titled "Report Writing Tips: How to Write to Make Things Happen":

- "Structure the Document to Reflect the Audience Profile"
- "Get Started: Overcome Writer's Block"
- "Avoid Tin Ear"
- "Employ a Gender-Neutral Style"
- "Recognize that Less is More"
- "Include Rhetorical Techniques"

- "Use Lists"
- "Compose Informative Titles"
- "Cite All Sources"

You can readily use advice offered in these sections to help you prepare the outline for your presentation. Assemble and organize, in your script, more ideas and information than you will need during your presentation. Adopt the approach that you know or will learn much more about your topic than you will actually share during your formal remarks. Plan to ask the audience for something and to ask it early. Doing so reduces audience anxiety and gets you and they aligned.

Personal

Your request to the audience does not have to be major. For example, I've spoken many times about a then controversial policy that called for increasing the formal education of civil engineers, who want to become licensed, to a master's degree or 30 credits. My request was just this—I asked audience members to try to understand why this policy was adopted. I did not ask for agreement or support. That request reduced audience anxiety and enabled them to focus on my message.

Use discretion as you select material, particularly data, to share with your audience. Don't, in your zeal to support your principal points, overwhelm your listeners with quantitative material so that they miss your message. Don't allow them to "drown in data while thirsting for knowledge" (Decker 1992).

As you begin to flesh out the content of your presentation, you are very likely to go to the internet. How can you evaluate the credibility of various internet sources? Author Cella Jaffe (2001) offers this advice: "Even before you click onto a document, begin evaluating its source by looking at its URL. A document with a .gov or .edu sponsor is sponsored and maintained by an institution with a reputation to uphold, .com sites are commercial, and .org web pages are sponsored by organizations with varying reputations." She goes on to suggest that you may want to determine the author of a web page and learn about and possibly contact him or her.

If your presentation advocates major change, consider using the past. Look for historic changes that are similar to what you are advocating for the future. For example, as explained in this book's preface, some U.S. engineering societies are leading reform of the formal education and pre-licensure experience of tomorrow's engineers. Achieving this will require many changes, one of which is revisions to the licensure laws/regulations/procedures in 55 U.S. states and territories. Sound impossible? No, look at the historic explosion of continuing education requirements for licensed engineers that started with the Iowa licensing board in 1979. As of 2011, continuing education was a condition of continued licensure in over three-fourths of U.S. licensing jurisdictions (Casey 2010).

As already noted, if you are one of a group of speakers, you should become familiar with the general themes or content of the other presentations and the order in which the presentations will be made. Consider contacting one or more of the other speakers to learn more about their presentations and how you can coordinate your efforts. Craft your presentation so it relates to, supports, or contrasts with the other presentations. Whenever you are part of a group of speakers, you are not speaking in a vacuum; therefore, you are obliged to try to coordinate your efforts with theirs.

You have an outline—a thoughtful and perhaps collaboratively-prepared outline. Now you are ready to finish the structure or to "put some meat on the bones."

Recognize Preferred Ways of Understanding

As you think about proceeding, recognize varied preferred ways of learning. The idea is to design sensitivity to the different preferred ways of learning into your presentation. The three preferred ways of learning are:

- Auditory—Understand mainly by hearing. For auditory learners, words are critical—slides and other visuals supplement their understanding of your message.
- Visual—Understand principally by seeing. Visual learners rely on slides and other visuals. Albert Einstein was a visual learn. He said: "If I cannot picture it, I cannot understand it."
- Kinesthetic—Understand primarily by touching and doing. If you are introducing software to an audience, the kinesthetic learners can't wait to get their hands—or precisely—fingers on it.

So how do you accommodate the varied learning styles of your audience? You usually won't know and/or can't control which types of learners will be in your audience. Therefore, expect all three types to be present. Think of the room you will speak in to be full of ears, eyes, and fingers. Then consider:

- Using careful word selection and definition for the auditory learners. They will be "all ears" and "hanging on your every word." Later, we consider the importance of practice partly for the benefit of the auditory learners.
- Creating visuals such as photographs, drawings, graphs, cartoons, icons, and videos for the visual learners.
- Preparing handouts for the kinesthetic and visual learners. Make sure that they are stand-along documents, that is, they include the title of your presentation, your name, affiliation, contact information, date, and the event.
- Arranging for props for the benefit of the visual and kinesthetic learners. The word "prop" comes from the theatrical world and is a shortened form of property, which means any object handled by an actor during a performance. When you or I give a presentation we are like actors giving a performance during which we strive to communicate with our audiences. Let's use whatever works, including props. Professionals, such as engineers and architects analyze, plan, design, construct, manufacture, fabricate, and operate material things

for the benefit of society. Accordingly, they have many opportunities to use props. For example, assume you are an architect and will be describing the proposed City Hall addition to the City Council. Bring samples of bricks that could be used on the addition's exterior. Given the choice between talking about "it" and bringing "it," if feasible bring "it," whatever "it" may be (Walesh 2009).

Historic Note

After decades of using cast iron and then ductile iron pipe (DIP), a water utility was asked by a member of the local developer community to allow use of polyvinyl chloride (PVC) pipe. The principal argument for the PVC pipe was less initial material cost. In response, the utility's board created an advisory committee to investigate and report back with recommendations.

The utility's construction manager opposed allowing PVC pipe. Accordingly, he made a presentation to the advisory committee during which he stressed what he viewed as negative aspects of PVC pipe and positive features of DIP. The construction manager referred to a 20-foot long, eight-inch-diameter PVC pipe that he had placed in the meeting room. He noted that when PVC pipe failed, the failure tended to affect the entire length of pipe and require a 20 foot long excavation. In contrast, DIP failures tended to be localized. They were repaired by excavating a small hole down to and around the pipe and clamping a saddle over the failure. He showed the audience a saddle and observers could not help but note the small size of the saddle compared to the 20-foot length of PVC pipe. The large size differential between the two objects also reinforced the relative size of the repair excavation and, by extension, the relative repair costs.

Clearly, the speaker could have simply described the high cost of repairing PVC failures relative to the cost of repairing DIP. However, the two props greatly enhanced his argument. Note: The request to allow use of PVC pipe was ultimately denied as a result of many factors, only one of which was differential repair costs (Pipe Panel 2009).

Prepare Visual Aids: Nine Tips

Let's take a "time out" before offering visual aid tips. Note that while we are well into preparing the presentation, we have not yet discussed use of PowerPoint, Keynote, or other presentation slideware. That is intentional. I urge you not to be concerned with the presentation medium you will use until completing preparation basics such as defining purpose, profiling the audience, selecting the topic and title, outlining, and recognizing preferred ways of understanding.

Before automatically starting to use slideware, the most common visual aid, recognize the alternatives. Examples include the already-mentioned props plus writing on newsprint pads, relying solely on handouts, and using prepared display boards.

Personal

I once participated, as part of a team, in the official release of a report at the National Academy of Engineering in Washington, D.C. All we used were five boards. We displayed them as needed and left all of them up once they were used to facilitate the question and answer period. As another example, I've spoken on the topic "10 Tips for Success and Significance." I use ten props that I bring out, one at a time, from a box. Each one represents one of the 10 tips.

Nine tips for preparing and using visual aids follow. Most apply to slideware because that medium is the most common, although not always the most effective.

Tip 1–Revisit the Value of Visuals: Recall the earlier reference to visual learners. For them, as the Chinese say, "One seeing is worth a thousand tellings." Visuals may becoming even more important, perhaps at the expense of verbal expression or text. Futurist John Naisbitt (2006) says: "A visual culture is taking over the world... Our literacy, and with it our verbal communication skills, are in decline."

Tip 2–Keep Each Slide Simple: If you use slides, each projected image should clearly portray one or maybe two ideas. Remember that your audience is simultaneously listening to you, looking at your visual aid, and presumably trying to understand and assimilate new concepts, ideas, or information. Material that is familiar to you is new to them. Use carefully-designed graphics to help your audience understand what you are trying to say. Simple, familiar visual analogies are helpful. Assume you are trying to convey an acre of area or an acre-foot of volume. Show an aerial photograph of a football stadium and explain that one acre is approximately the size of the playing field.

Historic Note

The U.S. city of Milwaukee, Wisconsin was considering use of a deep tunnel bored in bedrock to temporarily store and convey combined sewage prior to treatment. Citizens were concerned with leakage of combined sewage from the tunnel into the surrounding aquifer. The tunnel was to be located about 200 feet below the water table. Someone on the project team suggested using a submarine to convey the fact that leakage, if it occurred, would be inward—not outward. This analogy proved to be an effective means of communication.

Tip 3–Use the 1/30 Rule: Size alphanumeric characters on slides to be at least 1/30 the height of the slide image, otherwise some audience members will have difficulty reading the slide. A 28 pt. font achieves this. Sometimes the audience has the advantage of a handout which they can look at if the projected image is too small. However, this is not always the case and moving from the screen to a handout is distracting. Because of this size limitation, do not indiscriminately use a graphic, containing alphanumeric characters, taken from a document, book, or published paper. This will typically violate the 1/30 rule approach, and while it may ease your preparation effort, much or all of the visual will be useless to your audience. As an alternative to the 1/30 rule, print your draft slides, place them on the floor, stand up and try to read them. If you can't read them your audience won't be able to read them (Guetig 2011).

Tip 4–Use Sans Serif Fonts: A sans serif font, such as Arial, is easy to read on a slide (Dyrud 1995, Irish and Weiss 2009), while serif fonts tend to clutter the image. Not everyone agrees with this advice. If you are one of them, then consider not mixing fonts, especially on a given slide.

Tip 5–Design Slide Content in Accordance with Research Findings: Many presentation slides use the format illustrated in Figure 3.6, that is, a statement supported by bulleted points. The speaker thinks this is helpful, at least for him or her, because it provides notes to speak from.

However, research indicates that this common word only or word-intensive format is not an effective way to communicate with an audience during a presentation, although it could be effective as part of a handout. Reason: Audience members are confused by word inconsistency as they see one set of words on the screen and hear another set as the speaker paraphrases the slide. Or, worse yet, the speaker reads the slide to the audience which raises this question: Why is the speaker needed (Atkinson 2007)?

Research further reveals that the most effective slide contains a declarative statement with a supporting image as shown in Figure 3.7. Audience members tend to learn better from words and pictures rather than from words alone. This approach virtually eliminates the previously mentioned confusion between what the slide says and what the speaker says.

Taking this slide design tip a step further, consider the custom of English being read from top to bottom and left to right. That which appears first on the slide, when being viewed by English speakers, will be given highest priority. Therefore, if you are using

Figure 3.6 Research reveals that this frequently-used slide format is not an effective way to communicate with an audience during a presentation.

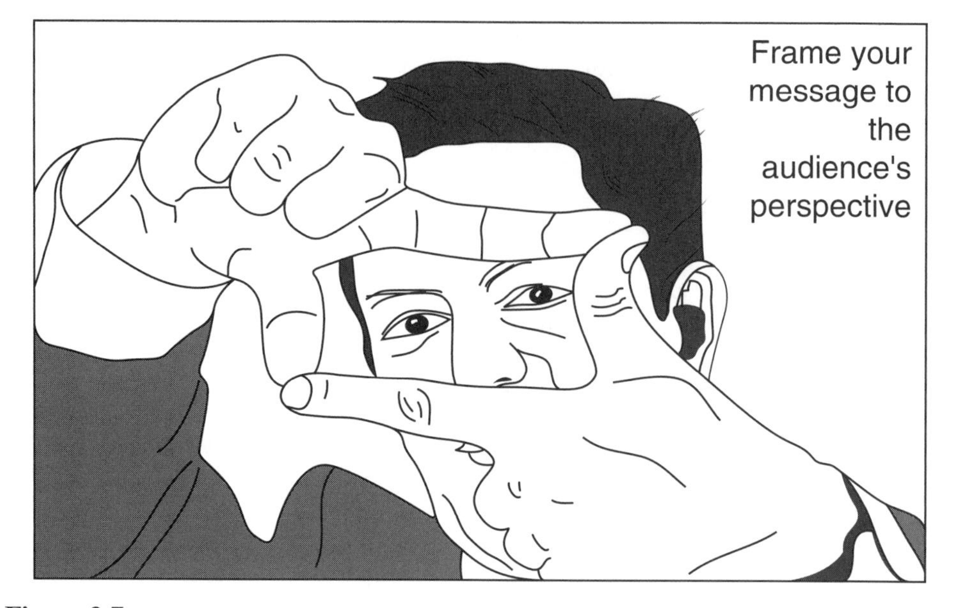

Figure 3.7 Studies reveal that the most effective presentation slide is a short declarative sentence coupled with a supporting image.

slides in the declarative statement-image format and if the declarative statement is more important than the image, the statement should appear at the top of or on the left side of the slide. Do the opposite if the image is more important (Irish and Weiss 2009).

"Sometimes, the most effective way to convey a message is through multiple images. This is a form of collage," according to Irish and Weiss (2009). "Often these multiple images will get sequenced by a speaker so that the audience receives one image at a time but still gets the overall effect of juxtaposed information. In such instances, the set of images, working together, makes a point far stronger than words alone are capable of."

Tip 6–Title for Maximum Impact: Select a title for each slide, board, or other visual that states the principal idea, like suggested in the writing section of this chapter. Informative titles will tend to be longer, but much more informative. For example, instead of the title "Dam Location" use "The Dam Will Be Located at the Narrowest Part of the Valley." Or, rather than saying "Annual Production of Widgets," use the title, "Widget Production Increased During Each of the Last Five Years." The subject, verb, and object approach is used in this book for tables and figures. Another similar titling approach is the feature, verb, and benefits sequence. An example is "Computer Modeling Leads to a Less Costly Solution."

Tip 7–Select Colors Carefully: Some color combinations do not work well for most viewers. An example is red on dark blue. Test your slides—preferably by projecting them on a screen. Color selection is important in another way, an admittedly subjective way, as illustrated in Table 3.6 which is based on Raskin (2002) and Mittelsteadt (2011).

A related thought: Use one or more of your sponsoring organization's or your audience's organizations colors. For example, when I speak on behalf of or at a

Table 3.6 One person's view of how colors communicate.

Colors	Communicate
Black	Elegance, sophistication, intimidation
Blue	Trust, authority, security, serenity
Brown	Security, stability
Green	Money, growth, environment
Orange	Movement, construction, energy
Pink	Femininity, calm
Purple	Royalty, spirituality
Red	Power, aggression, activity, rescue
White	Cleanliness, class
Yellow	Light, nostalgia, future, philosophy

meeting of the American Society of Civil Engineers, I emphasize blue because blue is the dominant color in ASCE's logo and publication. And, as noted in the list, blue also suggests trust, authority, and security.

Also consider using the logo of the sponsoring organization, as well as the logo of your organization, on some of your slides or other visuals. Alignment with audience and/or the sponsoring organization helps you communicate. Using their logo suggests that you were thinking of the audience and their concerns as you prepared the presentation materials. Perhaps the sponsoring organization's logo and your organization's logo could appear on a title slide and a summary slide.

Tip 8–Prepare a Handout: A carefully-crafted handout enhances your presentation in these four ways:

- As you prepare the handout, you will think even broader and deeper about your topic. As a result, you put more and better thoughts into words and visuals.
- A handout provides a back-up in the event of an equipment failure during your presentation. Number the pages of your handout so that you can refer to specific pages if the equipment fails.
- Handouts are likely to be appreciated by audience members because they reduce the need to take notes and provide a "take away" for later referral.
- You can leverage your presentation effort by having something to send to interested individuals who could not attend. The handout connects you and your presentation to anyone who receives it. Accordingly, and as noted earlier, make your handout a "stand alone" document. That is, include on the title page, your name, affiliation, and contact information plus the name, location, sponsor, and date of the event at which you spoke. In other words, enable anyone who receives the handout, whether or not they were at the presentation, to easily contact you.

Don't necessarily use your complete slide presentation as your handout. The reason: Some slides in a presentation are meant to maintain attention, serve as transitions, or make a temporary point. Such slides do not necessarily convey useful information when viewed without the benefit of the oral presentation. Furthermore, a copy of the slide presentation may be unnecessarily voluminous. A short, but dense and detailed handout, may be a much more useful "take away" for participants. It could be developed, at least in part, from your slide presentation.

"When to hand out the handout?" is a question that often arises. I usually distribute it immediately after my presentation, otherwise many audience members will be distracted by the handout during the presentation. There are exceptions. For example, you may want the audience to refer to a detailed table that cannot be conveniently projected on the screen. Accordingly, you may provide the handout at the beginning of your presentation or during your presentation. Or you might provide just the table immediately before your presentation and the complete handout after.

Recall the cite sources admonition in the writing section of this chapter. The same advice applies to a slide presentation and to handouts. Give credit where credit is due.

Tip 9–You Are the Principal Means of Communication: Visual aids such as slides, boards, props, and handouts, are important, but never more important that you. Avoid the trap of thinking that visual aids, even sophisticated, high-technology graphics, can lead or carry your presentation. You cannot "PowerPoint" an audience into: Being aware of . . . , understanding . . . , committing to . . . , buying . . . , and/or doing/not doing something. You must convey the message:

- Verbally—the words you use
- Vocally—how you use your voice
- Visually—your appearance, supplemented by the visuals aids you use

You better be more excited about the topic than about the technology you use to present it. As someone said, if you can be replaced by slides, you should be!

Personal

Prior to presenting a speaking webinar, I and my then 11-year old grandson had this email exchange:

Me: "Do you make Power Point slides at school? I think you told me that you did. If you do, do you use the standard backgrounds? They are also called "designs." I am trying to learn more about how elementary students use Power Point. Thank you, Grandpa."

Grandson: "Hey Grandpa—we use the standard background for social studies and math. We used it a lot in the fourth grade, but none at all yet in the fifth grade."

Do you, as a professional, want to rely heavily on visual features being used by fourth graders? When preparing slides, are you smarter than a fifth grader?

Arrange On-Site Logistics

The success of your presentation will depend partly on the physical arrangements. Examples are: projector, remote control, pointer, screen, podium and/or table, flip chart, white board, sound system with fixed or lavaliere microphone, and special seating arrangements such as classroom style, semi-circular, or tables.

Practice Out Loud

You are almost ready to present—however, you have one last preparation task: Practicing out loud. By practice "out loud" I mean stand up and talk and gesture through your entire presentation at least twice, sometimes in front of a mirror. If you have never done this, try it. Ask someone to listen or, if that is not feasible or makes you uncomfortable, record your voice and, if feasible, your image. However, do not memorize or read your presentation. Include use or manipulation of your visuals in the rehearsals, whether they are slides, props, or boards. As they say, "timing is everything." You want to effectively transition from one visual to another (Jaffe 2001). "Out loud" practice offers you three major benefits:

- Timing is the first reason to practice "out loud." It is the only way to know how long your presentation will last. Reading or speaking the presentation "to yourself" provides a poor measure of the actual presentation time, and as a result, you are very likely to underestimate the delivery time. Most presenters are given a specific time allotment. If you underestimate the time required to deliver your presentation, you risk agitating the audience, offending other speakers, and embarrassing yourself and your organization. Consider this rule of thumb if you have text and plan to speak from it: The average person speaks 120–150 words per minute (Detz 2000).
- The second reason to practice "out loud" is to discover and rectify distractions such as "ah," "you know," fiddling with your eye glasses or keys, rocking back and forth, and focusing on just a portion of the imagined audience.
- Achieving effective vocabulary, pronunciation, cadence, and emphasis are the third reason to practice "out loud." You are less likely to be searching for words or stumble over them because when you actually make the presentation to your audience you will already have given a similar speech two or more times. Many presentations that have been well-prepared in terms of content and organization are less than they could be because the speaker seems to be searching for the right words. This probably reflects the lack of "out loud" practice, not a lack of substantive content or familiarity with material. "Out loud" practice also helps you develop an appropriate cadence (rise, fall, rhythm), and emphasis. Musician, Arthur Schnable, said: "The notes I handle no better than many pianists. But the pauses between the notes—ah, that's where the art resides." If you do not practice out loud, you will not have "heard" how you will use cadence, volume, and flow (Cooper 2004).

Segment your presentation and practice each segment and, as you do, note changes you want to make. For example, use different terminology, change the order, or revise a visual. Make the changes to the segment and then practice it again. Maybe do this one more time on that segment if you are not satisfied. Then move to the next segment and repeat the process. I suggest this segment-by-segment approach because, by using it, you are much more likely to remember and act on needed improvements, and to focus on those parts of your presentation that need the most improvement. I realize this process of practicing by segments may seem cumbersome. However, it works for me. At least give it a try. By the way, Geoff Colvin in his book *Talent Is Overrated* (Colvin 2008) refers to this as "deliberate practice." For even more thoughts about the importance of "out loud" practice and tips on how to do it, refer to Irish and Weiss (2009).

Views of Others

Practice might be a key to your speaking success, as it has for those who speak or do other things well. Vince Lombardi, professional football coach, said: "Practice does not make perfect. Only perfect practice makes perfect." Martina Navratilova, nine-time Wimbledon singles winner, said this about practice: "Every great shot you hit, you've already hit a bunch of times in practice." Winston Churchill prepared meticulously for his speeches—he practiced one hour for every minute of the speech (Selbert 2006). And then we have the late Steve Jobs, the former Apple CEO, who gave those casual, conversational, friendly, and passionate presentations when Apple releases new products. Now he was a natural speaker! No he wasn't. As noted by Garr Reynolds (2008), Steve Jobs "and his team [prepared] and [practiced] like mad to make sure it looks easy." Perhaps you, too, could find the time to practice your presentation "out loud" a few times.

A final thought about preparing to speak. Do you know how your voice sounds to others? Unless you've listened to and studied recordings of your voice, you don't know. Decker (1992) explains that, when you talk, the voice you hear is conducted largely through the bones in your head while the voice others hear is transmitted through the air. He says that the recorded voice is the "real" voice, at least as far as your audiences are concerned.

Unless you plan to talk only to yourself, you ought to know how you sound to others and, if you don't like it, change it. This is another reason to consider recording all or part of your presentation as part of your preparation. A small recording device that you turn on and place on the lectern or near you at the beginning of your practice session and/or your actual talk will capture the verbal and vocal components of your presentation. Study the recording and be prepared to be shocked by the strengths and weaknesses of your spoken communication. Although more difficult to arrange, an audio-video recording is even more valuable because it will capture all three components of spoken communication: verbal, vocal, and visual.

Use your organization's voice mail system to hear how you sound to others in the normal course of the business day (Decker 1992). For example, send a voice mail message to your voice mail box. Critically listen to the verbal and vocal components of your spoken messages. Identify strengths and weaknesses and explicitly build on the former while you diminish the latter.

Deliver the Presentation

So much for ideas and tips on preparing. Imagine you are one hour from the presentation. What might you do? Recall the previously-discussed three channel-two way characteristic of speaking. You have many things going for you because you planned your presentation with this phenomenon in mind. Many presentation ideas follow.

Check Out Equipment and Facilities

Arrive early. Sit in all corners of the room to see what audience members will be able to see, or not see. Find and test light switches. Get comfortable with the physical set up and test the audio-visual equipment, or have someone test it. Don't assume that something is functioning simply because it is there. Audio-visual personnel who sometimes set up presentation rooms typically do not give presentations. Therefore, the equipment may work but it may not be positioned correctly. I always move or adjust something. Locate the nearest restrooms—someone may ask you.

In many cultures, people read from left to right. Therefore, you should try to speak from left of the screen as viewed by the audience (Cooper 2004). Audience members will look at you as you say something and then they will look slightly right at the screen which illustrates what you just said. You are aligned with the audience.

Confront Fear

If you applied most of the ideas and tips offered in the preceding speech preparation discussion, any presentation fear you may have had should be greatly diminished. Why? Because you are well-prepared. Accomplished speaker Jack Valenti, President of the Motion Picture Association of America, said: "The most effective antidote to stage fright and other calamities of speech making is total, slavish, monkish preparation."

Consider visualization as another fear confrontation technique. First, assume you have already given a successful presentation. Use your imagination to visualize various aspects of the speech and how it was received. Picture how confidently and knowledgeably you looked and sounded during the speech, how positively the audience responded, and how satisfied you felt afterward. Athletes use visualization. We hear about this practice at the Olympics and in college and professional sports. Sports psychologist Bob Rotella says "The secret to great performances—in golf or the boardroom—is in the mind." He goes on to urge aspiring performers to focus their minds on what they want to have happen and sublimate what they want to avoid (Burke 2007). Psychologist William James was the first to describe the operable "as if" principle. Think or act "as if" you are confident and you will be confident. Your body tends to do what your mind expects (Detz 2000).

Sometimes our presentation is in a less formal setting—maybe at a meeting with participants sitting around a large table. In this situation, even if and especially if you are nervous, stand up—maybe simply stand behind your chair. This enables others to see and hear you. As noted by Timothy Koegel, communications consultant: "We're more persuasive on our feet than in our seat."

On the way to becoming a better speaker, you will occasionally fail, in some small or major way. I once fell off the side of the speaker's platform prior to a presentation to hundreds of people. When something like that happens, remember that to have failed does not make you a failure. Learn from it. And note that we can learn as much, if not more, from failure than from success, as nicely noted by Henry Ford who said "failure is only the opportunity begin again more intelligently."

Stage fright or nervousness is natural. Comedian Johnny Carson said he was nervous prior to every monologue. Johnny was nervous over 4000 times! (Koegel 2002). Maybe instead of calling it "stage fright," we should refer to it, in a more positive way, as "speech excitement" (Laskowski 2001).

Suggest an Introduction

To the extent feasible, ask the host or session chair to give you a brief but proper professional introduction. Audiences appreciate a synopsis of the speaker's education and experience. Even a young professional has a "background" that is of interest to the audience. In anticipation of being properly introduced, prepare a brief (one or two short paragraphs) biographical sketch that hits the high points and offer this to the person who will introduce you. The introduction might include a brief summary of post-high school education, your employer, and perhaps reference to some project experience.

Connect With the Audience

Once you are satisfied with the logistics, engage the audience, even before you are introduced. "Work" the room, that is, greet people one-on-one as they arrive—you may know some of them—and introduce yourself to others.

When you start to speak, connect to the audience. For example, if you haven't been introduced, briefly introduce yourself. Tell a humorous personal story that illustrates your humanity. Say hello to someone you know in the audience. Share a relevant personal anecdote. If I was talking to elementary teachers, I would mention Miss Blaha, the elementary school teacher that introduced me to geography and history. You might share a favorite quote, refer to a relevant current event, tell a joke, show a short video, play a brief audio, or share a startling number (Jaffe 2001). Distribute your handout, or part of it, or indicate that it will be available immediately after the presentation and perhaps before the question and answer (Q&A) session.

Conduct a survey. For example, how many audience members have visited a construction site? What's one benefit individuals hope to receive from your presentation? Invite participation, if that is your intent, by saying something like "please feel free to ask questions at any time."

> **Personal**
>
> As an engineering dean, I and some of our faculty members often hosted groups of prospective students and their parents. We always invited some of our current students to join us—and referred as many questions as possible to them. The students were experts on many topics of interest to our visitors and student presence helped engage the audience.

Do Not Apologize

Do not apologize for any aspect of your presentation including, but not limited to, the amount of time you spent preparing for it, the content, the organization, the visibility of visual aids, your experience, or lack of it. All of these things are within your realm of control and, out of respect for your audience, you should have taken care of all of them before you were introduced. A possible exception is something that happens at the last minute which, in spite of all of your and the organizers' planning, you cannot control. A simple apology might be appropriate. Then, take feasible corrective or offsetting action and get on with your presentation.

Speak to the Audience

As you speak, remember where your audience is located. Your audience is not the wall to your left or to the right, they are not on the ceiling, they are not in your notes, and they are certainly not in the images projected on the screen behind you. Your audience consists of those people seated directly in front of you. Try to speak to them—each and every one of them. Do this explicitly by gradually moving your gaze around the room. Make eye contact of up to about five seconds with specific individuals. These "eye-fives" will make your message more personal and help you avoid speaking to just one portion of the audience.

If you are using slideware, place the computer between you and the audience. Look at the computer monitor, not the screen. The advice may seem obvious. However, too many speakers spend too much time looking at the screen. If you speak to the screen, rather than the audience, they may have difficulty hearing you, they will not see your expressions, and the audience may be insulted by being forced to look at your back.

Notes are certainly acceptable. However, stay connected with the audience. Talk to them, not just in front of them (Anholt 2006). Use your notes unobtrusively (Jaffe 2001). My point: The screen, your notes, and the ceiling never made a decision, never retained a consultant, and never approved a budget (Koegel 2002). People in your audience do those things. Talk to them!

Maintain Audience Attention with Enthusiasm and Variety

Better to be very enthusiastic, than to look as though you do not care. Repeat words or phrases for effect. For example, say "The cost of Alternative B is only 70 percent—70 percent of the cost of Alternative A!" Consistent with your "out loud" practice, vary volume and pause to provide variety. Avoid filler words such as "Ah," "you know," "like," and "currently," as in, we are currently considering the second alternative. Use a remote control so you are not "glued" to the computer, and, instead, can move around in front of and/or into the audience.

To sum up the preceding provide-variety suggestions, consider Aristotle's advice: "It is not enough to know what we ought to say; we must also say it as we ought... There are three things—volume of sound, modulation of pitch, and rhythm—that a speaker bears in mind." Remember, you are the principal visual. While body language is subjective, as discusssed earlier in this chapter, some aspects of body language messages are widely understood.

Conclude Definitively

Consider briefly summarizing your principal points and/or state the take-home message. Clearly indicate that you have finished, you are done. And hopefully, in some way, you and your audience have won. People tend to remember what they hear last. Anholt (2006) calls this the "stress position" in a sentence—and your last formal words are the "stress position" in your presentation. Irish and Weiss (2009) expand this to the primacy or recency effect saying that "people remember the beginnings and the ends of things better than the middle." Avoid diluting your principal message. Dilution examples are ending with credits, ideas for future studies, and secondary issues. Any of these may be appropriate, but if they are secondary topics, don't present them at the very end of your presentation. These items can be addressed earlier or in a different manner.

Possibly signal or accentuate the imminent end of your formal comments with some physical action. For example, step away from the podium, lower or raise your voice, walk toward the audience, use a prop, or project a dramatic slide that summarizes key points and stimulates questions or other post-presentation activity (Jaffe 2001).

Or perhaps echo, that is, refer to or repeat something you said at the outset of your presentation (Jaffe 2001). For example, you are speaking about the need to control erosion within the watershed of the U.S. Great Lakes. At the end of your presentation you might echo your initial concern with erosion by saying "Assuming the average rate of sediment flow into the Great Lakes, about X tons of sediment has entered the lakes during my presentation."

Prompt Post-Speaking Discussion

The only thing more satisfying than delivering an effective speech is having it followed by a Q&A session characterized by a stimulating exchange of ideas and information and active participation by many audience members. Such an exchange is one indication that you have succeeded in making something happen. However, starting that exchange is often difficult.

If there is a lull after the conclusion of your presentation, consider asking a general question of the audience such as, "Would anyone care to share a similar or different experience?" Share a previously-asked question as in "The last time I spoke about this topic, some asked . . . ? My response was . . . " Or, try to find someone in the audience who has an inquisitive or doubtful expression on his or her face and diplomatically "challenge" the person on the chance that he or she will be compelled to speak. For example, look at the person and say, "You look skeptical about the usefulness of what I just described. How do you feel about it?" Then remain silent. Most people will feel compelled to fill the silence with a comment. The comment is likely to "break the ice" for other members of the audience, and in domino fashion, cause others to offer a comment or ask a question.

You might even consider "planting" a person in the audience who agrees to ask a question, of his or her preference, if no one else does. This may be the catalyst needed to stimulate participation by many members of the audience. When listening to and responding to questions, stay put or move toward the asker. Do not back up—it is likely to be interpreted as a lack of confidence—as weakness. Instead, focus on and engage the questioner.

During the question and answer session, consider repeating the question or perhaps rephrase it for clarity. Although you may be able to hear the question clearly because the questioner is facing you, members of the audience, especially those behind the questioner, may not hear the question. Therefore, they may not be able to benefit from your answer. Restate a negative question positively. For example, say that the negatively-stated question is "Why are your service's priced so high?" Restate it more positively as "You want to know how we arrived at the cost of our services?" and then answer that question.

Speaking of questions, you might want to be prepared for these kinds of questions (Jaffe 2001):

- Loaded questions—Questions intended to put you on the defensive, as illustrated by the preceding negative question.
- Closed questions—Requests for brief answers, that is, questions that can be answered with yes, no, or a fact.
- Open questions—Questions that invite you to explain or elaborate on some aspect of your topic.

To conclude this thought, you might ask yourself prior to your presentation what loaded, closed, and open questions might be asked.

A participant in one of my seminars noted that his big fear is that someone in the audience will know more than he does about some aspect of the topic. Very likely—one of the purposes of speaking, and the subsequent discussion, is to learn—and that includes you! Perhaps one of your fears is not "knowing all the answers." Who does? Your responsibility as the speaker, during the post-presentation discussion is to help find the answers. If you do not know the answer, ask if someone in the audience does. Your audience may include relevant experts. Or promise to find the answer and get back to the questioner. Ask for the asker's business card. The subsequent exchange could lead to some mutually-beneficial results.

"Hang Around" After the Event

Individuals reluctant to comment in a group setting or those with what they consider sensitive issues are likely to approach you after your presentation and the Q&A. These post-session discussions can lead to productive exchanges and possibly new professional relationships. I secured a multi-year consulting assignment as the result of such a post-session discussion. I complimented the speaker and shared some common interests. One thing led to another.

Obtain Input So You Can Improve

View each presentation as a learning experience. Unobtrusively place a small recorder on the podium and record all or a portion of your presentation for later review. Organizers sometimes provide an evaluation form or you can bring your own. Upon receiving the evaluation, search for ways to improve while expecting contradictions. Your spouse, significant other, a family member, or a trusted friend can share their views and report on what they heard audience members say or do during and after your presentation.

Personal

Speaking of contradictions, I presented a webinar about speaking. One anonymous reviewer wrote that what he/she disliked were "all of the quotes by famous people, there were too many and quite unnecessary." Another reviewer said that one of the features he/she liked was "the quotes that were included throughout the talks." I often see such wide-ranging reviews. They illustrate the previously-discussed different learning preferences of members of an audience. Accordingly, you need to throw out many different "hooks" during your presentation so as to eventually "catch" each member of the audience. You also need thick skin. While you cannot connect with everyone in the audience for your entire presentation, strive to connect with each person for a part of your presentation.

Follow-Up the Presentation

Your presentation is completed. Glad that's over with! You did well! Aren't you finished? No! Consider some important "loose ends" to take care of within a few days after the presentation, especially if, in keeping with the theme of this speaking section of this chapter, you want "to make things happen." Recall that our discussion of speaking has three parts. We've worked through preparing and delivering the presentation and now conclude with following up on the presentation.

Say Thank You

In preparing for and delivering your presentation, you probably didn't do it all by yourself. Most good-to-excellent presentations are team efforts. Members of your team might include students or colleagues that constructively critiqued your draft materials, graphics experts in your organization, on-site audio-video personnel, your host or moderator, and your "boss"—who supported your participation in the first place. Therefore, as may be appropriate, thank them and perhaps share some of what you learned with them or others.

Commit to Improvements

Based on evaluations and other input, you have information on how you might improve your presentation knowledge and skills. Commit to doing so. Assume for example, that you let too much verbal graffiti (Koegel 2002), such as "ah," "you know," and "um," creep into your presentation. Then practice reducing verbal graffiti from your conversations and telephone calls. You don't have to wait until you are preparing for a presentation.

As you strive to become a better speaker, you are unlikely to experience dramatic breakthroughs. Continuous improvement is the more realistic road to success. And this requires the self-discipline to act on information you receive. Getting better is not "rocket science" in that most of us can all do simple things to greatly improve our presentation effectiveness. But, and this is a big but, we need to know what needs improvement. Maybe your simple thing is less verbal graffiti, looking at the audience more, improved visuals, or using props. Find it and fix it.

Act on Promised Follow-Ups

Recall the first "Define the Purpose and Profile the Audience" section in this discussion of preparing the presentation. You may have connected with one or a few audience members who can advance that purpose. Pursue that opportunity by following up. What you do or don't do, now that the presentation is over, may determine the success or failure of your speaking effort.

For example, if you promised to send a copy of your handout or some other document, do so shortly after the meeting transmitting it with a letter or email that references the presentation, reminds the reader of the request, and asks the person to share the handout with others and to report any critiques back to you. If you were referred to a paper, some other publication, or another person, obtain the paper or publication or contact the person. Consider writing to the referrer and thanking him or her for the referrals. By properly following up on post-presentation conversations, you advance your stated purpose by learning more about your subject and widening your network of professional contacts. Learning more about your subject and getting to know more people are two significant personal and organizational benefits of making effective presentations. Make sure you get this return on your investment.

When people ask you to follow up after your presentation, ask for their business cards so you have the necessary information. Afterwards, examine the business card to determine the person's organization and to deduce their responsibility in that organization. This may prompt you to send them individual items of interest, such as your organization's website link, copies of materials that may be in your personal professional files, and names of individuals whom you think the person might want to know. Consider adding the individual to your organization's newsletter or other mailing or emailing list.

Leverage the Presentation

You have invested considerable time and energy in preparing and delivering your successful presentation. How can you earn an even bigger return on that investment?

- Post the handout on your organization's website, use it to conduct an in-house brown bag presentation, or send the handout to potentially-interested individuals.
- Present the same basic materials to other audiences, tailored, of course, to their profiles.
- Speak again, about a different topic, to the group you just spoke to if you resonate with them.
- Convert your handout into a draft paper or article. This can be a very effective way to gain much larger exposure for you and your organization. Perhaps you gave your presentation to a live audience of 50 people. Now, as a publication, your audience could be tens of thousands of readers.

CONCLUDING THOUGHTS ABOUT WRITING AND SPEAKING

Frankly, excellent documents and presentations are few and far between. They are often "thrown together" when time is tight and/or the budget is blown. An excellent document is attractive, logically structured, well-written, informative, useable by the targeted audience or audiences, and convincing. An effective presentation—a presentation likely to achieve its purpose and make things happen—has these three carefully orchestrated parts: preparation, delivery, and follow-up. While the first is by far the most time consuming, all are essential.

Producing an excellent document or presentation about a project is a project in itself. Sadly, clients, owners, customers, and stakeholders are sometimes denied the value of excellent technical work and the participants in that project are denied the satisfaction of seeing their efforts come to fruition because the document and/or the presentation, the crucial link between the professionals who performed the work and those who could implement it, is weak. The quality of the document or presentation should be at least as high as the quality of the technical work described.

How much time—absolute or elapsed—have you invested in acquiring your technical expertise? Whatever your answer, the relative time required to learn writing and speaking fundamentals is miniscule. The writing tips and speaking tips presented in this chapter provide the fundamentals. Leverage those fundamentals by proactively seeking opportunities to practice-practice-practice them. The ball is in your court!

Some of the most accomplished engineers of all time
have paid as much attention
to their words as their numbers,
to their sentences as to their equations,
and to their reports as to their designs.

(Henry Petroski, engineering professor and writer)

CITED SOURCES

Alexander, A. 2001. "Keeping Errors at Bay." *IBD's* 10 Secrets of Success column, *Investor's Business Daily*, May 15.

Alexander, D. and A. Rivett. 1998. "The Tin Ear." *Journal of Management in Consulting*. November, p. 55.

Anholt, R. R. H. 2006. *Dazzle 'em With Style: The Art of Oral Scientific Presentation Second Edition*. Elsevier: Amsterdam.

ASCE Task Committee on the First Professional Degree. 2001. *Engineering the Future of Civil Engineering*. October. ASCE: Reston, VA.

Atkinson, C. 2007. *Beyond Bullet Points: Using Microsoft PowerPoint 2007 to Create Presentations That Inform, Motivate, and Inspire*. Microsoft Press: Redmond, WA.

Bauerlein, M. 2008. *The Dumbest Generation: How the Digital Age Stupefies Young Americans and Jeopardizes Our Future*. Jeremy P. Tarcher/Penguin: New York, NY.

Benton, D. A. 1992. *Lions Don't Need to Roar: Using the Leadership Power of Professional Presence to Stand Out, Fit In, and Move Ahead.* New York: Warner Books.

Berthouex, P. M. 1996. "Honing the Writing Skills of Engineers." *Journal of Professional Issues in Engineering Education and Practice—ASCE*, July, pp. 107–110.

Burke, M. 2007. "How Not to Choke." *Fortune*.

Casey, T. M. 2010. "Lifelong Learning to Meet Engineering Challenges." *CE News*. August, pp. 20–24.

Chan, J. F. and D. Lutovich. 1994. "Using Lists . . . a Simple Solution to a Complex Problem." *HYDATA*, March, pp. 9–10.

Colvin, Geoff. 2008. *Talent Is Overrated: What Really Separates World-Class Performers from Everybody Else*. Penguin Group: New York, NY.

Conger, J. A. 1991. "Communicating a Vision That Inspires." *Engineering Management Review*, Winter, pp. 69–77.

Cooper, C. 2004. "A Winning Presentation." *Investors Business Daily*, July 1.

Covey, S. R. 1990. *The 7 Habits of Highly Effective People*. Simon & Schuster: New York, NY.

Decker, B. 1992. *You've Got to be Believed to be Heard*. St. Martins Press: New York, NY.

Detz, J. 2000. *It's Not What You Say, It's How You Say It*. Bristol Park Books: New York, NY.

Dyrud, M. A. 1995. "The Visual Aspect." *Annual Conference Proceedings*. American Society for Engineering Education.

Edwards, B. 1999. *Drawing on the Right Side of the Brain*. Jeremy P. Tarcher/Putnam: New York, NY.

Guetig, M. 2011. "Harness the Power of PowerPoint." *PM NETWORK*, September, pp. 50–54.

Iacone, S. J. 2004. *Write to the Point: How to Communicate in Business With Style and Purpose*. Barnes & Noble: New York, NY.

Irish, R. and P. E. Weiss. 2009. *Engineering Communication: From Principles to Practice*. Oxford University Press: Don Mills: Ontario, CA.

Jaffe, C. 2001. *Public Speaking: Concepts and Skills for a Diverse Society - Third Edition*. Wadsworth/Thomson Learning: Belmont, CA.

Johnson, P. 2009. "Walking Our Way Out of Recession." *FORBES*, September 21, p. 17.

Koegel, T. J. 2002. *The Exceptional Presenter*. Greenleaf Book Group Press: Austin, TX.

Laskowski, L. 2001. *10 Days to More Confident Public Speaking*. Princeton Language Institute, Warner Books: New York, NY.

Linver, S. 1978. *Speak Easy*. Summit Books: New York, NY.

MarketingSherpa, Inc. 2003. "High-Impact Email Writing Part I: Useful Lists of Short Words, Strong Verbs, and Blah Words." e-newsletter, August 13.

Mittelsteadt, S. 2011. "Healthy Hues." *Vim & Vigor*, Fall, pp. 10–13.

Nagle, J. G. 1998. "Seven Habits of Effective Communications." *Today's Engineer*, Summer, pp. 22–25.

Naisbitt, J. 2006. *Mind Set! Reset Your Thinking and See the Future*. Collins: New York, NY.

O'Conner, P. T. 1999. *Words Fail Me: What Everyone Who Writes Should Know About Writing*. Harcourt: San Diego, CA.

Pipe Panel. 2009. "Report of the Advisory Pipe Panel to the Valparaiso City Utilities Board." Valparaiso, IN, January 8.

Quilliam, S. 2009. *Body Language: Actions Speak Louder Than Words*. Fall River Press: New York, NY.

Raskin, A. 2002. "The Color of Cool." *Business 2.0*, November, pp. 49–52.

Restak, R. and S. Kim. 2010. *The Playful Brain: The Surprising Science of How Puzzles Improve Your Mind*. Riverhead Books: New York, NY.

Reynolds, G. 2008. *Presentations: Simple Ideas on Presentation Design and Delivery*. New Riders: Berkeley, CA.

Rico, G. 2000. *Writing the Natural Way: Using Right-Brain Techniques to Release Your Expressive Powers*. Jeremy P. Tarcher/Putnam: New York, NY.

Riggenbach, J. A. 1986. "Silent Negotiations: Listen With Your Eyes." *Journal of Management in Engineering - ASCE*. Reston, VA.

Roane, S. 1988. *How to Work a Room: A Guide to Successfully Managing the Mingling*. Shapolsky Publishers: New York, NY.

Selbert, P. 2006. "Missouri Museum Recalls Historic Churchill Address." *Chicago Tribune*, May 21.

Silyn-Roberts, H. 2002. "Document Structure and Its Effect on Engineers' Reading Strategies." *Journal of Professional Issues in Engineering Education and Practice - ASCE*, July, pp. 115–119.

Strunk, Jr., W. and E. B. White. 2000. *The Elements of Style*. Fourth Edition. Longman: New York, NY.

University of Chicago. 2010. *The Chicago Manual of Style-16th Edition*. University of Chicago Press: Chicago, IL.

Vesilind, P. A. 2007. *Public Speaking and Technical Writing Skills for Engineering Students*. Lakeshore Press: Woodsville, NH.

Walesh, S. G. 1999. "DAD is Out, POP is In." *Journal of the American Water Resources Association*, June, pp. 535–544.

Walesh, S. G. 2000. *Engineering Your Future: The Non-Technical Side of Professional Practice in Engineering and Other Technical Fields*. ASCE Press: Reston, VA.

Walesh, S. G. 2004a. "Flying Solo: Some Thoughts on Starting an Individual Practitioner Consulting Business." presented at the University of Wisconsin Reunion Conference on International Issues in Environmental Engineering and Education, Madison, WI, August 10–12.

Walesh, S. G. 2004b. "Leading: Lessons Learned." Structural Engineering Conference. ASCE-Iowa Section. Iowa State University, Ames, IA. November 8.

Walesh, S. G. 2004c. *Managing and Leading: 52 Lessons Learned for Engineers*. "Lesson 21: Start Writing on Day 1," ASCE Press: Reston, VA.

Walesh, S. G. 2007a. "Writing – Less Is More – Part 1." *Indiana Professional Engineer*, March/April, p. 3.

Walesh, S. G. 2007b. "Writing – Less Is More – Part II." *Indiana Professional Engineer*, May/June, p. 8.

Walesh, S. G. 2009. "Prop Up Your Presentation." *Leadership and Management in Engineering - ASCE*. October, p. 215.

Walters, R. and T. H. Kern. 1991. "How to Eschew Weasel Words." *Johns Hopkins Magazine*, December, pp. 25–32.

Wang, J. 2009. "7 Non-verbal Cues and What They (Probably) Mean." *Entrepreneur*, May, p. 15.

ANNOTATED BIBLIOGRAPHY

Advanced Public Speaking Institute. 2003. "Public Speaking: Stage Fright." *Leadership and Management in Engineering – ASCE*, January, pp. 4–5. (Recognizes stage fright, noting that some fear is helpful. Offers 16 visualization tactics that can be used anytime and 36 tactics for possible use just before you speak.)

Clarke, B. and R. Crossland. 2002. *The Leader's Voice*, Select Books: New York, NY. (Argues that "the most effective communicators use three essential channels to convey important leadership messages... These channels are factual, emotional, symbolic.")

Karp, J. 2006. "Raytheon Penalizes CEO Financially After Plagiarism." *Wall Street Journal*, May 4, p. B7. (Reports that the Raytheon Company's board of directors decided not to raise

Chairman and CEO William Swanson's salary and decided to reduce his incentive-stock compensation "because he plagiarized parts of a management booklet." He used, without indicating the sources, almost verbatim passages from W. J. King's 1944 book *The Unwritten Laws of Engineering*. Approximate cost to Swanson: $1 million.)

Lublin, J. S. 2004. "To Win Advancement You Need to Clean Up Any Bad Speech Habits," *Wall Street Journal*, October 5. (Cautions young people to avoid "teen speak," such as "like." Suggests that women avoid "uptalk" which includes ending declarative sentences with a rising inflection thus making the speaker seem tentative or uncertain.)

Pagan, A. R. 2006. "An Ethical Dilemma Regarding Report Writing." *CE News*, February, p. 14. (Raises the issue of when the consultant should and should not make changes to a draft document as requested by a client or other stakeholder.)

Pagan, A. R. 2006. "Report Writing – Part 2." *CE News*, April, p. 18. (Readers share their thoughts on when a consultant should and should not make changes to a draft document when requested to by a client or other stakeholder.)

Roam, D. 2009. *The Back of the Napkin: Solving Problems and Selling Ideas with Pictures*. Portfolio-Penguin Group: New York, NY. (This book's freshness and its possibilities for you are suggested by this quote: "Visually representing someone or something – regardless of actual likeness or detail – always triggers insights that writing a list cannot achieve.")

Truss, L. 2003. *Eats, Shoots, and Leaves*. Profile Books: London, U.K. (British author stresses the importance of commas, colons, dashes, and other punctuation.)

Tufte, E. R. 1997. *Visual Explanations: Images and Quantities, Evidence and Narrative*. Graphic Press: Chelshire, CT. (Includes highly-varied, extraordinary visuals many of which, as suggested by the title, communicate quantitative information.)

Witt, C. with D. Fetherling. 2000. *Real Leaders Don't Use PowerPoint*. Crown Business: New York, NY. (The book's theme, in the author's words, is "Just because everyone else shies away from giving speeches or relies too much on PowerPoint is no reason for you to.")

Zinsser, W. 1988. *Writing to Learn*. Harper Resource: New York, NY. ("Writing enables us to find out what we know—and don't know—about whatever we're trying to learn." The author's view suggests that that if we select a topic and start writing about it, in a disciplined manner, we will become an expert on that topic.)

EXERCISES

3.1 APPLICATION OF WRITING TIPS: The purpose of this exercise is to suggest the value of writing tips presented in this chapter. Select something you have recently written, which is either in draft form or was completed and submitted or sent, such as a research paper for a class, a laboratory report, a capstone course report, a book review, a design memorandum, or a letter and resume sent to a potential employer. Save a digital copy of it for later reference. Suggested tasks are:

A. How thoroughly and explicitly did you define the document's purpose, that is, how carefully did you define, before you started writing, what you wanted the reader(s) to do or not do?

B. How carefully did you define the audience(s)?

C. What aspects of the document are user-friendly and which are user-unfriendly?

D. Read the document out loud. Do you hear excessive repetition of some words?

E. Scan the document's headings and subheadings. From the reader's perspective, are they sufficient and helpful?

F. What person(s) did you use? On reflection, was that appropriate?

G. Is the document gender neutral and, if not, is that justified?

H. Is the writing style mostly active and direct or passive and indirect?

I. What words and/or content could you eliminate?

J. To what extent did you employ rhetorical techniques?

K. How effectively did you use tables, figures, other visuals, and sources to support the text and how efficiently did you reference them from within the text?

L. Do the titles of tables, figures, and other visuals describe the principal message of each?

M. Did you thoroughly cite all sources?

N. Are you pleased with the appearance of the document?

O. After answering each of the preceding questions and, in effect, thoroughly critiquing your document, revise all or some of it.

P. Compare the original document to the revised version or revised portion of it. In terms of improving your writing ability as you move forward, comment on the value, if any, of this exercise.

3.2 APPLICATION OF SPEAKING TIPS: The purpose of this exercise is to suggest the value of speaking tips presented in this chapter. Select something you have recently prepared, which is either in draft form or was completed and already presented. Save a digital copy of it for later reference. Suggested tasks are:

A. Conduct a point-by-point analysis of the presentation material, and the actual presentation and follow-up, as appropriate, similar to that called for in Exercise 3.1. However, the points of reference are the applicable speaking tips presented in this chapter.

B. After thoroughly critiquing your materials and, as appropriate, the presentation and follow up, revise the presentation materials or a portion of them. If you actually made the presentation, think how you would change your actions during the presentation and the follow-up.

C. Compare the original "package" to the revised version. In terms of improving your future speaking ability, comment on the value, if any, of this exercise.

CHAPTER 4

DEVELOPING RELATIONSHIPS

> No matter how ambitious, capable, clear-thinking, competent,
> decisive, dependable, educated, energetic, responsible, serious,
> shrewd, sophisticated, wise, and witty you are,
> if you don't relate well to other people, you won't make it.
> No matter how professionally competent, financially adept,
> and physically solid you are, without an understanding
> of human nature, a genuine interest in the people around you,
> and the ability to establish personal bonds with them,
> you are severely limited in what you can achieve.
>
> (*Debra A. Benton, consultant and author*)

This chapter offers guidance on how to develop relationships with the wide variety of people you will typically interact with within and outside of your organization. It begins by describing the range of attitudes and perspectives that you are likely to encounter. Major sections of the chapter are devoted to the art and science of delegation, orchestrating meetings, interacting with paraprofessionals and support personnel, selecting co-workers, and working with your boss. The chapter concludes with discussions of caring for colleagues, coaching, teamwork, and making the most of attending conferences.

TAKING THE NEXT CAREER STEP

Going back to the Chapter 2 theme, you are getting your personal house in order. You've chosen roles, set goals, and implemented an effective time-management system; you are conscious of the importance of personal integrity; you function well within the administrative structure of your organization; and assignments are being completed well and on time. Although you watch what you say and how you say it, you are not reluctant to speak up when you have a question or suggestion. You dress appropriately, are developing communication skills, and have a positive outlook. In addition, you passed the Fundamentals of Engineering or similar examination and are beginning to gain valuable experience toward your professional license. Finally, you are involved in continuing education and are active in at least one professional organization.

You decided on full-time professional employment rather than full-time graduate school, made the initial transition from the world of study to the world of practice, and are in your first full-time employment situation. Or maybe you are in graduate school.

Regardless of your situation, including still being a university student, you are now ready to focus more of your time and energy on developing relationships with others. Early success in education or practice, as satisfying as it can be, may result in young technical person's failure to grasp the importance of the next step in their professional development. That step is continued growth of interpersonal knowledge and skills and development of relationships with others.

If employed, you are increasingly interacting professionally with colleagues, research personnel, clerical staff, technicians, computer-aided drafting and design technicians, IT personnel, sub-contractors, data processing staff, surveyors, vendors, clients, owners, customers, and stakeholders. Perhaps you are also being asked to assist with the orientation of the new entry-level professional staff and to direct some of the efforts of those staff on technical projects.

Technically and scientifically-oriented people, especially young people, are probably most prone to thinking that "it's what you know that counts." While "what you know" is certainly necessary, it is not sufficient in the world of engineering and other technical practices. Patterson (1991) advises young professionals to "take time to smell the social roses." He goes on to observe, "It's just that the real shots are called by those who find a way to fit within the social scheme" or, as stated anonymously, "people will forget what you said, people will forget what you did, but people will never forget how you made them feel."

PERSONALITY PROFILES

At the risk of over-simplifying people's personalities, a brief review of a variety of simple models is useful to stimulate your thinking about you and others. Within your organization, among people you serve as a result of your professional work, and within your community, you are likely to encounter widely-varying personality profiles such as:

- People who make things happen, watch things happen, or ask what happened?
- Risk takers, caretakers, and undertakers.
- Leaders, laborers, and losers.
- Acceptors, undecided, and rejectors (Brenner 2009).
- Fighters, followers, and "fleers."
- Individuals who are on top or on tap.
- Winners and losers (Harris 1968).

- Those who see "it" as an opportunity or see "it" as a problem.
- Movers, movables, and immovables (Annunzio and Liesse 2001).
- Proactive and reactive.

The last item is exemplified by typical expressions used by proactive and reactive individuals. For example, a reactive person is likely to say, "There's nothing we can do," while a proactive individual would probably say, "Let's look at our alternatives." Other reactive-proactive contrasts are "I must" versus "I prefer" and "They make me angry" versus "I control my feelings" (Covey 1990).

These simple personality profile models suggest that individuals vary markedly in their outlook and perspective. Wide variations often occur even in presumably homogeneous groups such as civil engineers employed in the small consulting firm or mechanical engineers involved in design and manufacturing in an automobile plant. Of course, individuals can change. Although each person has natural tendencies, any of us can recognize the available range of outlooks and perspectives and work toward those that are more likely to be productive. An end-of-chapter exercise provides you with an opportunity to further explore your personality profile.

As a student or entry-level professional, try to understand the needs and motivations of the people around you, particularly those with whom you must work to carry out your responsibilities. Although you may not share much of what you hear and see in terms of personal philosophy, respect and try to understand the individuals who hold those philosophies. Recall the powerful idea of empathetic listening discussed in Chapter 3. One of Covey's (1990) seven habits of highly effective people is appropriate: "seek first to understand, then to be understood."

MASLOW'S HIERARCHY OF NEEDS

One way to better understand why the people with whom we interact exhibit such a variety of outlooks and perspectives is to consider the work of psychologist Abraham Maslow. He developed one of the first models of people's needs (Dychtwald and Kadlec 2009, Martin 1986, and McQuillen 1986.) Maslow's work, which was done in the 1940s, followed Frederick W. Taylor's scientific management, which emphasized training workers for efficiency and essentially ignored human factors (Hopp and Spearman (2001). Maslow's efforts also followed Mayo's human relations approach, which was published in 1945 and pointed to the uniqueness of individuals and suggested that workers should have some degree of group control over their work (Martin 1986).

The Hierarchy

Maslow's hierarchical model helps us understand the basic drive or motivation of people around us, including students, colleagues, clients, owners, customers, those we report to, and those who report to us. As Figure 4.1 shows, Maslow envisioned five levels of needs, any one of which may apply to any individual at any time. The most basic needs are physiological and include basics such as food and shelter. The second level addresses safety and security and includes being physically, psychologically, and economically safe and secure. Assuming the first two levels are satisfied, Maslow

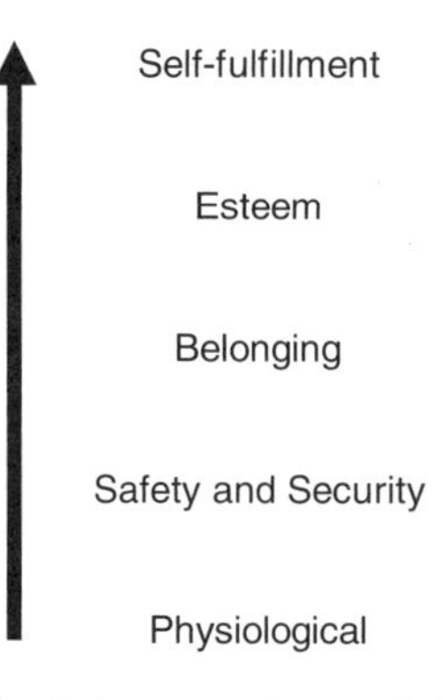

Figure 4.1 Maslow's hierarchy helps us understand the basic motivations of others.

identified belonging as the third level. Individuals want to be accepted as part of a group, feel wanted and appreciated, and give and receive affection.

Maslow referred to the first three levels as lower order needs. The upper two levels he called the ego needs. The first of the upper needs is esteem—having self-respect and the respect and recognition from others. The highest level need is self-fulfillment or self-actualization—thinking and feeling that one has fully realized one's potential. This is the upper end of the success and significance staircase discussed in Chapter 2.

Application

Awareness of needs and the satisfaction of those needs do not necessarily occur in sequence, and rapid, temporary changes can occur. An example is an unexpected life-threatening situation that could take someone you normally study or work with from a level four dominance down to a level one dominance. Consider another situation. After several months of employment, your physiological needs are met and you feel safe and secure, certainly in an economic sense. But, you have a strong desire for more indications that you belong, that you are accepted as part of a group and that you are wanted and appreciated.

While on the belonging theme, look for appropriate opportunities to participate in social and other activities that, while they are not part of your formal job responsibilities, are connected to your place of employment. Examples include attending holiday parties, assisting with fund raising to meet a community need, helping to host open houses, and occasional after work socializing. By engaging in these kinds of "extracurricular" activities, you enable your employer to carry out its mission while you benefit by connecting with co-workers, learning more about your employer and the immediate community, and enhancing your communication and other knowledge and skills.

However, a word of caution is warranted. Be wary of regular, intense socializing with peers, supervisees, and supervisors. Information exchanged and views expressed, by you and others, in those informal social settings can complicate carrying out your and their work-related responsibilities.

Refer again to Figure 4.1. Just as you may be at one point in Maslow's hierarchy, imagine that any one of your colleagues could be at any one of the other levels at any given time. Maslow's needs model helps you understand the actions and reactions of others by providing insight into what motivates them. Consider, for example, the technician who does what he or she is told and appears to have no other aspirations. Perhaps this person is under extreme financial pressure and is focusing primarily on economic security. Being employed is more important than advancing up the organizational ladder.

You may know an upper-level manager or executive who works long hours, appears "driven," and is involved in many extra-curricular projects. This individual may be focusing on level four with emphasis on seeking the respect and recognition from others. Consider also other young professionals in the office who are seeking other employment opportunities. While they feel safe and secure in their present positions and find the work challenging, they are at level three and are not finding the feelings of acceptance and appreciation that they need within the present employment situation.

THEORIES X AND Y

In understanding and honing their attitudes toward study and work and in interacting with others and their attitudes, students and young practitioners should be aware of two fundamentally different perspectives—Theory X and Theory Y. These two perspectives, which were developed by Douglas McGregor (1960), bracket the range of attitudes toward work. Most college students or recent college graduates have enough part-time or full-time work experience to recognize these two very different work styles, although they may not know them by the names Theory X and Theory Y.

Definitions

Theory X people dislike and avoid physical and mental work. They hate work and want security—the kind of security represented by Maslow's level one and level two needs. Theory X people work in response to threats, coercion, control, and close direction, and must be reminded constantly that their security is at risk. Theory X people must be watched and controlled.

In stark contrast, Theory Y persons assume that physical and mental work are natural, as natural as play and leisure. They want to contribute by assuming responsibility, exercising self-discipline, working hard and smart, and putting something of themselves into the product or service. Theory Y individuals want to understand, support, and become active participants in the mission, objectives, and efforts of the group. Accordingly, Theory Y people respond positively to rewards for and recognition of achievement. Theory Y assumes that most people possess a wealth of intelligence, creativity, imagination, and energy. When released in the work environment, these resources bring individual satisfaction and organizational success.

Applications of Theory X and Theory Y Knowledge

Assuming you see some validity in the Theory X-Theory Y model, consider some hypothetical examples of its usefulness as applied to individuals and even organizations. Theories X and Y can help you understand the responsiveness or non-responsiveness of others to typical interpersonal situations. For example, a Theory X manager, because he or she is not satisfied with the progress being made by a supervisee who happens to be a Theory Y adherent, threatens to discharge the supervisee. The Theory Y individual does not respond with the expected fear, but instead is insulted and disappointed.

Or a Theory Y-oriented young engineer, seeking a first permanent employment opportunity, interviews with what he or she immediately determines to be a Theory X-run organization. The young person is perceptive enough to not pursue the position further, even though an attractive salary and benefit package is offered. Finally, consider this situation. A new Theory Y manager is meeting individually with all of his or her employees. While meeting with one of the supervisees, a Theory X adherent, the manager asks for ideas and inquires about the supervisee's aspirations. Later, after the employee leaves the manager's office, the Theory X employee tells a co-worker to "watch out for the new manager," noting that the manager talks strange and must "have something up his or her sleeve."

Dominance of Theory Ys

John Mole, an educated and experienced manager, became a successful writer, quit his management position, and went underground for two years as a "management mole." During that time, Mole had temporary jobs in 11 organizations—jobs he obtained without revealing his educational and business background. He wrote about his experience in the book *Management Mole: Lessons from Office Life* (Bredin 1988; Mole 1988).

Mole reported that the majority of junior-level staff he encountered wanted to learn, work, contribute, and succeed, that is, they were primarily adherents of Theory Y. Unfortunately, much of what junior-level staff offered and aspired to was wasted because of poor organizational managing and leading. On-the-job training and orientation were virtually non-existent in the organizations on which the book was based. As a result, the "blind led the blind," and young employees who had little or no training, education, or orientation informally oriented new personnel. Mole's book suggests that, even today, many Theory Y people are an unused resource. This resource should be tapped and enabled to the benefit of the individuals and their organization.

Views of Others

Expect to encounter an amazing, sometimes uplifting and sometimes discouraging, variety of people as nicely explained by this anonymous quote: "We could learn a lot from crayons. Some are sharp, some are pretty, and some are dull. Some have weird names and all are different colors, but they all have to live in the same box." In the interest of keeping our relations with others in perspective, columnist Earl Wilson says "One way to get high blood pressure is to go mountain climbing over molehills." Finally, when relationship problems arise, cleric Charles L. Bromley says "Why not make the best of things? Any fool can make the worst of them."

DELEGATION: WHY PUT OFF UNTIL TOMORROW WHAT SOMEONE ELSE CAN DO TODAY?

You may be experiencing an ever-growing work load and looking for some relief, a "life preserver." One approach is to simply stick with the status quo and hope the situation improves. That approach rarely works. As someone said "If you do what you did, you will get what you got." Then there's the work harder option. You are probably willing to do this, to make the extra effort. However, there's a limit to how much harder you can work. Delegation, or more effective delegation, may be the answer to your "drowning in work" dilemma.

Delegating means legitimately assigning part of your assigned task to someone else. Properly done, the delegator retains responsibility for the entire task because the original assignment, part of which has now been delegated, was given to him or her. From the perspective of the delegator's "boss," the delegator still has the responsibility. The delegator should never blame the delegatee for a deficiency in the overall assignment.

Delegation is not giving orders, that is, holding back authority and regarding the delegate as an object or robot with no initiative. Nor is delegation dumping, that is getting rid of responsibility, especially when the going gets tough. Another form of "dirty delegation" (Finzel 2000) is you delegate much of your task to others, they contribute, and you take all the credit. The discussion of delegation here is based, in part, on ideas derived from Covey (1990) and Culp and Smith (1997).

Reasons to Delegate

In spite of compelling arguments in support of delegation, engineers and other technical professionals are often reluctant to practice delegation. Before exploring resistance to delegation, however, consider the following reasons to practice effective delegation:

- ***Leveraging***: As illustrated in Figure 4.2, delegation leverages the delegator's effort, that is, his or her knowledge and skills. When you delegate, you move the fulcrum to the right and now one unit of your input yields more, often much more, than one unit of output.
- ***Building the bench***: Delegating gives other members of the organization an opportunity to learn, grow, and contribute in new ways to the organization's work. Within a delegation environment, everyone, regardless of position, rank, or salary, is challenged, stretched, and pushed to learn and contribute more. The organization becomes more resilient because more people know how to do more things. This is analogous to the concept of "strengthening the bench" on an athletic team.
- ***Reducing task costs***: Delegation reduces tasks costs, as illustrated in Figure 4.3. Task costs are reduced when they are delegated to less costly (per hour) personnel. This assumes that the less costly personnel can do the tasks as well as the more costly personnel and not require significantly more time. Whether or not that happens depends, in part, on your effectiveness as a delegator. Through effective delegation, typical organizational tasks such as drafting a memorandum, calling on a potential client, making a presentation to a potential client or

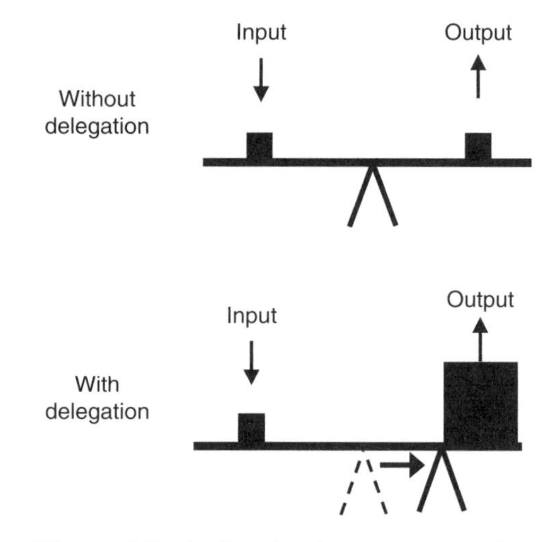

Figure 4.2 Delegation leverages your effort.

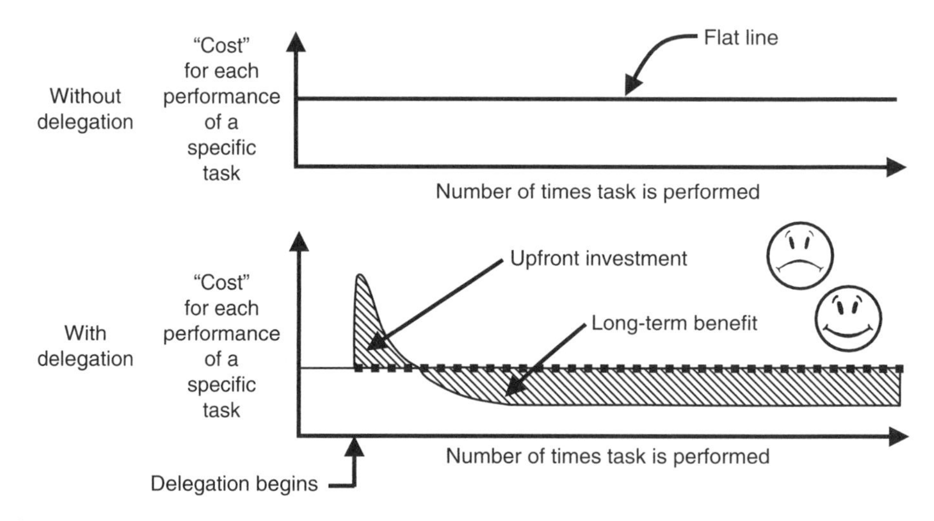

Figure 4.3 The up-front investment in delegation typically yields larger long-term benefits.

customer, managing a project, preparing a poster, performing calculations, and preparing a proposal are all done in whole or part by competent individuals who have lower labor rates.

Views of Others

A participant in one of my webinars shared this observation: "A percentage of division managers are not bottom-line aware of tasks. They have $45/hr. people doing $22/hr. work even when the task is done more efficiently [and] accurately by the $22/hr. workers." "The best executive is the one who has the sense to pick good [people] to do what [needs to be] done," according to Theodore Roosevelt, 26th U.S. President, " and self-restraint to keep from meddling with them while they do it."

- ***Invoking the novice effect***: The novice effect is the sometimes surprising and powerful result of giving" a task normally done by an "expert" to an "amateur" (Gross 1991). The "amateur" may have an idea for a better way to do the task. This is one result of giving up some authority, that is, you may "turn on" the novice effect. State differently, delegation helps individuals learn from others and sometimes the learning flows from the delegatees to the delegators. Because of the novice effect, whenever an "old" or routine task is given to a new person, the result may be fresh and improved approaches.

Reluctance to Delegate

In spite of all the good reasons to delegate, some technical professionals never learn how to do it effectively. This failure hampers the success of the organization and hinders the individual's advancement. Experience reveals the following seven reasons, or more precisely, rationales for this reluctance.

- ***No time***: Delegation, as already noted, usually requires an up-front investment of time and energy and some individuals believe they simply don't have the time to provide basic information required to delegate tasks to others. This is contrary to the previously-discussed return on investment associated with delegation.
- ***Excessive pride of ownership***: Young professionals sometimes oppose delegation because of excessive pride of authorship or ownership or perhaps even arrogance. That is, they believe they are the only ones who know how to do the work correctly. However, anyone who has moved from one position to another, whether in the same organization or another, realizes that somehow, once they are gone, others are quite capable of doing what they did.
- ***No one to delegate to***: Delegation may seem attractive to you, but you may feel that you have no one to delegate to. As a young professional, you probably do not have supervisory responsibility for full-time personnel in your organization. However, you do have access to and can seek assistance from your organization's paraprofessionals and other support personnel (see "Working With Technologists, Technicians, and Other Team Members" later in this chapter), consultants, subcontractors, your supervisor, and other individuals above you in the organization's structure. Of course, the preceding assumes that your organization's culture promotes collaboration and teamwork. My definition of organizational culture is "how things really work around here," not the values talked about, but the values practiced. We are assuming that the organization has a culture in which delegation can occur outside of the supervisor-supervisee structure.
- ***Poor interpersonal abilities***: The potential delegator's poor organizational and communication knowledge and skills can frustrate delegation. Although you

thoroughly understand the tasks that need to be done because you have done them many times, and although you realize the value of delegating some of those tasks to others, you are not able to effectively explain what needs to be done and to suggest ways to do it. If this situation applies to you, immediately commit to improving your organizational and communication knowledge and skills so that you can delegate and, in many other ways, contribute to your organization and advance your career. Poor communication ability will not only frustrate your attempts to delegate; it will also impede your advancement.

- ***Loss of job security***: Fear of losing knowledge-based job security may frustrate delegation. The young professional may be hovering around Maslow's second level and believe that his or her job security is linked in part to the exclusive ability to carry out certain tasks. High pressure to maintain billable time within professional service firms raises the specter of job insecurity and frustrates delegation. Accordingly, some personnel "hang onto" tasks that could and should be delegated "down" to lower cost per hour personnel. They maintain their expected utilization while driving down project profitability and work satisfaction and denying others personal development opportunities. The answer to the job security issue lies largely in organizational culture. That culture determines the extent to which fear of losing knowledge-based job security will frustrate delegation and thus deny an organization the many benefits of delegation. Progressive organizations value sharing and communication and, therefore, are wary of individuals who keep things "close to the chest."
- ***Appearing lazy***: Fear of appearing lazy or incompetent may inhibit delegation (Raudsepp 1978). For example, you just received an assignment and may think "If I give part of the assignment to someone else, my "boss" may think I'm lazy or don't know how to do the work." Again, the answer to this dilemma lies in the organization's culture. Is delegation valued, in fact, expected because it tends to benefit everyone? If delegation is valued, then senior personnel should exemplify it, not just talk about it, and when delegating a task to someone, encourage them to in turn delegate portions of the task to others and provide an appropriate array of professional, paraprofessional, and other support personnel to facilitate delegation.
- ***Fear of advancement***: Finally, you may be hesitant to delegate because you fear advancement or you are concerned that advancement may occur too rapidly as you delegate tasks to others and free up your time to take on new responsibilities. The fear could be offset by reviewing the desirability of your roles and goals and the extent to which delegation will facilitate achieving those goals.

Personal

A personal experience illustrates the effectiveness of creating a delegation culture. Coming from outside of an organization, I was hired as its executive. Shortly after arriving, I saw a de facto, two-class organizational structure. The "professionals" were "highly educated." The "support group" did what they were told to do, no more, no less. The "professionals" tended to hold onto what they did, even though some were "overworked," and the "support group" was underutilized and somewhat demoralized. Some of the "professionals" were rude when interacting with the "support group." After first asking "support group" personnel not to tolerate rude behavior by anyone I then encouraged everyone to delegate more work and I set an example. Some did delegate, the "support group" ended up working harder, doing more, and they and some "professionals" were happier. Most of us, regardless of the level in an organization, want to learn, grow, and contribute. A delegation culture produces a positive result.

Delegation Isn't Always Down

Delegation is usually downward in the organizational structure—that is, from a supervisor to a supervisee or from someone "high" in the organization to someone "low" in the administrative structure. However, as suggested by the preceding discussion of no one to delegate to, lateral and upward delegation can be an effective way to effectively use an organization's human resources.

Assume, for example, you are responsible for drafting a report on a subject that is new to you. To start with an overall report structure that incorporates experience already within the organization, you might ask a senior person to prepare an outline based on that senior person's experience. Similarly, you might delegate the task of preparing minutes of a meeting to someone at your level who has talent for preparing effective minutes. All members of the organization should be receptive to being delegatees, regardless of position and rank. By doing so, they demonstrate their ability to carry out the important "doing" or "producing" function which is discussed in Chapter 1 in the context of the leading-managing-producing continuum. A word of caution: In an environment with great pressure on maintaining a high percent of billable time, delegating upward may open the door for more than desired involvement by a "high-priced" expert.

Delegation Tips

Assume that the delegator is a Theory Y person and is about to delegate a task to another person in the organization—ideally a Theory Y person. Provide the delegatee with the context for the task. Explain how the task is one small, but important part of a major effort. For example, you may be delegating water sampling work to one of the organization's paraprofessionals. You should provide an overview of the project, indicate the expected outcome of the project, and clearly explain how the water quality data will be used in project tasks that will culminate in the desired outcome. Explain the desired results of the delegated task in appropriate terms, including when it is needed and the quality, size, format, location, and documentation. Identify the resources that are available to carry out the task such as budget, personnel, information, and data.

Personal

Consider a delegation situation in which I failed to provide context. As project manager in an engineering firm, I was delegating a surveying task, namely field surveying of channel-floodplain cross-sections to be used in a water resources engineering project. During a break, I overheard two surveyors indicate that they would read and record elevations in the cross-sections to the nearest 0.01 feet. That accuracy greatly exceeded what was needed and would have resulted in unnecessary labor expense. I had failed to provide the full context, that is, the manner in which the surveyed cross-sections would be used. Fortunately, having heard the surveyor's comments, I was able to rectify the situation.

Resist the temptation to over-prescribe "how" the delegated task should be carried out. That is, focus on the desired outcome and the resources available to achieve the outcome. Depending on what you know about the delegatee, delegate the task in such a way that you, the delegatee, the project, and the organization benefit from the potential novice effect.

Consider providing the delegatee with milestones. For example, you might suggest that the delegatee prepare an outline of how he or she will go about doing the delegated task and that the two of you will discuss the outline within three days. Clearly give authority commensurate with what you are asking the delegatee to do.

Three Possible Outcomes

When you assign a task, recognize that there are only three possible outcomes. Expect the first and most desirable outcome to occur most often, but recognize that the second and third outcomes will occasionally occur. The three possible outcomes are:

- Results are delivered as needed
- Results will not be provided as needed but the delegator is so advised by the delegatee well before the deadline
- Results are not going to be completed as expected and the delegator learns about the deficiency at or after the time the work was to be completed

The first result, as already noted, is by far the most common. While the second outcome is unfortunate, it is much better than the third outcome because the second provides the delegator with an opportunity to seek an alternative course of action. If a delegated task is not going to be successful, the delegator should be informed immediately.

In reviewing the results of delegated assignments with a delegatee, remember to critique the work, not the person. Avoid "you" messages, especially when the work is deficient. Instead, note which aspects of the work satisfy the requirements and which parts are deficient. Identify results that are acceptable and those that are not. Say "thank you," in person or in writing, especially when the work has been carried out in accordance with requirements

Recognize that not adopting a delegation mode of operation within an organization will likely relegate you to the slow track. Through delegation, you demonstrate important interpersonal and management knowledge and skills such as communicating, planning, coaching, and teamwork. Failure to delegate is likely to label the entry-level engineer or other technical person as not being a "people person" and stymie both technical and managerial advancement.

ORCHESTRATING MEETINGS

Comedian Milton Berle said "A committee is a group that keeps minutes and loses hours." Although he may have found humor in committees and their meetings, many personnel at all levels in organizations find little to smile about as they sit through long, non-productive, and sometimes unruly and highly stressful sessions that are often unduly dignified by being referred to as "meetings." A meeting is defined here as three or more people sitting face-to-face or connected electronically, discussing business or professional work. Meetings are often a waste of time and excessively frustrating because of poor planning, execution, and follow-up that sometimes indicates lack of respect for the time, talent, and feelings of others. The negative aspects of meetings become more frustrating as people progress in their careers because the percentage of time devoted to meetings tends to increase with increased levels of responsibility.

Some meetings are absolutely necessary because they are the best way to enable groups of people to accomplish certain things. Because so much time is devoted to meetings and because so much important work can be accomplished at meetings, careful meeting management is mandatory. This section offers tips for successfully planning and executing meetings and dealing with difficult people and situations at the meetings. These suggestions should be useful to you already as student when you participate in or lead meetings of student groups. Incidentally, engineers and other technical professionals occasionally participate in public meetings or hearings. Some of the material presented in this section is applicable to managing public meetings.

Reasons to Meet

Only two legitimate types of meetings are consistent with the preceding definition of meetings. The first is a working or highly-participative session devoted to defining problems, hearing status reports, brainstorming, conceptualizing courses of action, creating alternatives, and selecting and beginning to implement solutions. The principal feature of the working meeting is informed, positive participation by all attendees.

The second legitimate reason to meet is the critical information meeting which focuses on explaining non-routine, crucial topics such as personnel issues, acquisition of an organization, and serious financial matters. Although little discussion is expected, in contrast with the working meeting, the meeting is justified because of the serious implications of the information shared and the need to fully inform all invited participants. A memorandum or other form of written communication is not appropriate, although it these documents may follow an information meeting.

Personal

I was a manager within an engineering firm that was experiencing a continuing decline in backlog, that is, contracted work. We managers had received a series of messages from the company president indicating that we needed to "fix" the situation. He was not satisfied with our response so he convened an emergency evening meeting of all managers at the corporate office. He stressed the seriousness of our situation at the beginning of the meeting by indicating that our salaries would immediately be reduced by five percent and there would be no end-of-year bonuses until the declining situation was reversed and resolved. He got our attention and we responded accordingly.

When Not To Call a Meeting

Interestingly, and perhaps this is the cause of meeting problems, there seem to be more reasons not to call a meeting than to call a meeting. Consider the following three reasons not to call a meeting (Walesh 2000b):

1. Your mind is made up but you want others to think you are seeking their input; you are patronizing them.
2. While you know what needs to be done, you are reluctant to take full responsibility. Therefore, you call a meeting, indicate what should be done, obtain formal approval or informal acquiescence, and then you can "spread the blame" if needed while taking all the credit if the decision is correct.
3. You don't know what needs to be done but you have the necessary responsibility and authority. However, you want somebody else to make the decision.

Tips for Successful Meetings

To the extent possible, try to give attention to the tips suggested here. They are generally applicable to all types of meetings, that is internal and external meetings in the business, government, academic, and volunteer sectors.

Planning the Meeting

1. ***Confirm the Need to Meet:*** This is one advantage of ad hoc committees and similar groups in that there is less danger of falling into a pattern of periodic, but unnecessary meetings. Permanent groups that meet as a matter of practice tend to encourage the creation of agenda items and marginal discussion to fill the time available.
2. ***Invite Only the Necessary People:*** Avoid the "it would be nice to have them there" thinking. Those who need to know the results of the meeting, but do not need to participate in the meeting, can receive copies of the minutes.

3. ***Bring Geographically-Dispersed Individuals Together As Soon As Possible:*** Physical separation, in spite of electronic communication, can lead to too little communication or miscommunication because it is not informed by the richness of face-to-face encounters. Accordingly, consider doing the following very early in the life of a new group or team composed of geographically-dispersed individuals who have not worked together: "Bite the bullet" on cost and bring the new team together for a face-to-face social and work gathering. Then rely primarily, if not exclusively, on electronic meetings that are carefully planned, executed, and followed up. The project team productivity you desire ultimately comes from connectivity, not proximity. The upfront, face-to-face meeting is the foundation of that connectivity.
4. ***Provide an Agenda Prior to the Meeting:*** The following memorandum is an example of an agenda that would be distributed prior to a meeting. Note the following features:

   - All invitees are identified in the "To" portion of the memorandum. In the example, names are listed in alphabetical order. In a more formal environment, the custom might be to list each person by name, followed by position titles. By listing all the attendees, everyone knows who else will attend the meeting. This information provides an opportunity to informally discuss other matters with one or more individuals before or after the meeting.
   - Both the date and the day of the meeting are indicated. This way, invitees are much less likely to miss the meeting by arriving on either the wrong day or the wrong date.

MEMORANDUM

Date: April 17, 2012

To: Members of the Design Team – H.O.T. Air, B. Careful, O.U. Kidd, B. Level, and U.R. Liable

From: I.M. Boss, Project Manager

Re: Agenda for Meeting 9, Friday, April 20, 2012
8:00 – 9:00 A.M., Conference Room 2

1. Welcome to new members
2. Additional agenda items?
3. Surveying (see memorandum, included as Attachment A, for alternatives and some pros and cons) (B. Level).

a. Discuss alternatives.
b. Select/follow-up.

4. Design criteria (see pages from State code included as Attachment B) (U.R. Liable).

a. Recommended course of action.
b. Decision/follow-up.

5. Alleged design error

a. Summary of 4/6/12 meeting (see memorandum included as Attachment C) (B. Careful).
b. Discussion.
c. Follow-up.

6. Next meeting

Enclosure: Attachments A, B, and C
c: Vice President I.M. Bizee (without enclosures)

- The meeting starting time and ending time are shown on the agenda, enabling attendees to plan their day.
- The heading "Additional agenda items?" appears near the beginning of the agenda. This provides the chairperson with the opportunity to add topics for discussion and facilitates input from attendees, some of whom may not have been involved in preparing the original agenda.
- Individuals who have reporting or other responsibilities are identified by name. This helps ensure they will be prepared to report on their efforts or lead a discussion on the indicated topic.
- Background, support, and other materials are provided as attachments. This extra effort by the person managing the meeting arrangements provides participants with the opportunity to be prepared so the meeting time can be used effectively.
- An action-oriented theme is established by using words and expressions such as "select," "follow-up," and "recommended course of action." The agenda is structured to encourage and expect action.
- For each agenda item, the invitees are informed, via the agenda and the enclosures, what the committee is expected or encouraged to do. For example, the person arranging the meeting clearly indicates if the committee is to develop, discuss, and select alternative solutions to a problem (e.g., agenda item 3) or respond to a recommended course of action developed prior to the meeting by a member of the committee or by others (e.g., agenda item 4). That is, the person managing the arrangements for the meeting focuses the committee members' energy. Recognize that the committee may elect to broaden its response to the agenda, but is unlikely to do this if trust is established between the chairperson and the committee members.

- Depending on the nature of the group and/or the meeting, an agenda item titled "Good news" may be added under which positive happenings are briefly shared (Walesh 2000a). "Bad news" need not be shown on the agenda—it will take care of itself.

Time spent by the person responsible for meeting arrangements, possibly assisted by a few group members, prior to the meeting can save considerable time for the entire group. In contrast, a brief agenda—or none at all—and little time spent preparing for a meeting may result in a long and unproductive session. If you are invited to a meeting and a written agenda is not provided, ask if an agenda will follow. If there is not to be an agenda, insist on knowing what will be discussed so you can prepare or arrange to have some other, more appropriate person attend in your place.

Assume there is an important point you, as an invited participant, want to make at a meeting, but there is no directly related agenda item, because either you did not have an opportunity to get your concern on the agenda prior to the meeting or the chair was not receptive to adding items at the meeting. Carefully prepare your idea or information. Look for opportunities at the meeting to make your point, perhaps in answer to a question posed to you or to someone else.

5. ***Give Every Invited Participant a "Job":*** I offer this suggestion for two reasons. First, having a "job" connected with upcoming meetings gets a person thinking about the meeting. Each person is more likely to be engaged. Second, having "jobs" spreads and accelerates the work. Everyone is more likely to carry their share if the group culture expects it from the "get go." Furthermore, the group will have some work done before they meet. Examples of small to large tasks that could be assigned to meeting participants are bring refreshments, arrange for equipment, host a new participant, lead discussion of an agenda item, and prepare a white paper about a key topic.

Personal

For the first meeting of an improve profitability group that I led, I asked a member to speak about off-shoring and its impact on profitability. He did a great job researching and speaking about this topic. Twelve people attended that first meeting with 11, including me, having pre-meeting assignments that required presentations at the meeting. The 12th person agreed to record action and parking lot items as the meeting progressed. We accomplished a lot—we hit the ground running.

6. ***Arrange Logistics:*** The devil is in the details. If you are the meeting facilitator, attend to the details. Go beyond handouts and extra agendas. Address the physical facilities which may include a whiteboard, markers, flip chart, LCD projector, overhead projector, refreshments, and access to copy machines and restrooms. Sometimes logistics might include a special meeting site. Perhaps a project site or a

neutral place for discussion of a contentious issue. Avoid conducting a meeting on a critical topic within a setting where participants can be easily interrupted or are uncomfortable.

Recognize that your client/owner/customer/stakeholder may want to meet in your offices. Don't jump to the conclusion that they always want to meet in their offices—because that is more convenient for them. They may welcome the opportunity to "get out," to see where you work. Bottom line: Make participants feel special when they walk into the meeting room or place which will, in turn, encourage a productive session.

Conducting the Meeting

1. ***Establish Minutes Responsibility:*** Meeting minutes or a meeting summary are essential. One person could volunteer or be asked to provide this service, or possibly all could "take turns" over a series of meetings. When encouraging individuals to prepare the meeting minutes or summaries, stress, besides their service to the group, the benefits to them. Personal benefits include enhanced listening and writing skills and improved organizational knowledge. We learn via the mental-physical process of writing.

   Some individuals are reluctant to volunteer to draft meeting summaries because of writer's block and/or concern with appearance or elegance. Offset this by being flexible on the format (e.g., "bullets are OK") or by providing an example. Stress the need to record decisions, action items, and individuals responsible for action items.

   Interim minutes are real time minutes prepared as the meeting progresses and displayed for all to see, such as on whiteboards or on newsprint sheets taped to the walls of the meeting room. This approach is useful for complex discussions where participants would like to have everything that has transpired so far during the meeting in front of them. After the meeting, these interim minutes can serve as the basis for the actual minutes or they could be photographed, distributed via email, and become the minutes.

2. ***Start and Stop on Time:*** To show respect for the people who have been invited to the meeting, do everything you can to start and stop on time. If a meeting is to be the first of a series of meetings of a group, the pattern established at the first or early meetings will tend to prevail. Consider scheduling meetings late in the morning (e.g., 11:00 AM) or late in the afternoon (e.g., 4:00 PM) so impending lunch or dinner encourages focus and brevity. Occasionally schedule a meeting very early in the morning prior to the start of normal office hours, during lunch, during dinner, or in the evening to emphasize urgency or provide variety. Consider setting time limits for each agenda item as a further means of encouraging focus and brevity. Some meeting participants may only need to attend a portion of the meeting. Out of consideration to them, the chairperson could schedule their appearance at a specific time and excuse them when they are finished.

A group that meets frequently should establish a fixed meeting time and then cancel a meeting if it is not needed. This practice enables participants to plan their schedules. A group that meets occasionally should schedule its next meeting at the current meeting. For this, and other reasons that may arise during the meeting, all attendees should bring their time management systems to each meeting. One of the most wasteful practices is to expect a committee member or someone on the support staff to try to schedule a meeting among people with varying responsibilities and already complicated schedules.

3. ***Adopt Meeting Protocol:*** Early in the "life" of a new group that expects to meet frequently, the group should adopt meeting protocol or "rules of the road." Examples are:
   - Prepare agendas and minutes
   - Turn off cell phones
   - Expect everyone to participate
   - Never exceed one hour
   - Bring time management systems
   - Honor action items
   - Use "parking lot," that is, log topics that arise during deliberations but are not an agenda for possible consideration at the end of the meeting or the next meeting

   I emphasize that these are just examples because each group should create meeting protocol consistent with its characteristics and needs. One of my clients adopted "rules of the road" for their management group, framed them, brought them to meetings, and set them in the middle of the conference table. You could also consider somewhat higher level protocols than those listed above (Tompkins 1998). One example is transparency—issues get discussed here, involving the entire committee, team, or group, and decisions get made here. Another example is no repercussions—honest views are expected and those who express them will not be penalized.

4. ***Seek the Minority View:*** You and others should want to hear the minority position on any topic or issue, and the reasoning behind it. Initially, in addressing an issue or solving a problem, the majority view is offered and it isn't necessarily the best. It often reflects "how we've always done things around here." Therefore, draw out all views—especially minority views.

   To reinforce encouraging everyone to contribute, consider the results of a study of approximately 10,000 projects at 35 *Fortune 500* firms (*PM Network* 2006). Seventy percent of the projects failed because project participants simply did not speak up. Stated differently, while most of the group members and their leaders may be surprised by a failed project or other effort, one or more persons within the group "saw it coming" but, for whatever reason, did not speak up! Create a meeting culture in which individuals feel comfortable in expressing their views—especially contrary views.

Personal

I recall serving on a committee and, from the "get go," one person pressed for an approach to solving a problem that was contrary to everyone else's preference. She presented her view in a rational manner. Committee members heard her out—she persisted, she prevailed, she was correct.

What if you as project manager, department head, office manager, or other group leader always get only "I agree" at your meetings? View that as a danger signal. For example, you may have successfully cloned yourself; you may have successfully sent the message that you are the "boss," you know what's "best," and maybe you and your group are not doing anything exciting or forward-looking.

5. ***Deal with Difficult Behavior During Meetings:*** Some individuals may occasionally exhibit difficult behavior at meetings you lead. Give those individuals the benefit of doubt. Most of us exhibit difficult behavior at times—and it is not intentional. Furthermore, most people who occasionally exhibit such behavior are not, at the core, difficult people. However, as chair of a committee or manager of a project, you will have to deal with difficult behavior when it surfaces at meetings.

   If the difficult behavior is growing or frequent, take time to first look inward. Might you be the principal cause of the "difficult behavior" of others. Perhaps you are inadvertently offending or intimidating others by actions such as snickering, rolling your eyes, acting bored, shrugging, talking too much, interrupting, and checking email. Maybe your meetings are poorly planned and/or conducted, including exceeding the designated ending time. After introspection, consider the suggestions presented in Table 4.1 for addressing the difficult behavior of others.

6. ***Maintain Perspective:*** C. Northcote Parkinson, the British author and professor, said: "Time spent on any item of the agenda will be in inverse proportion to the sum involved." Some of us tend to focus and dwell on the relatively unimportant agenda items because they are easy to deal with. Incidentally, this is the Parkinson credited with Parkinson's Law which says: "Work expands to fill the time available for its completion" (Parkinson 1957).

   Steer your group to the most important items. For example: "I know refreshments for our next meeting are important, but let's move to discussing our department's marketing goals for next year." One way to maintain perspective is to assign time periods to agenda items. For example, refreshments for next week's meeting (5 minutes) and marketing goals for next year (45 minutes).

7. ***Seek Consensus:*** I suggest trying to seek consensus is in contrast with relying on a formal decision system such as Robert's Rules of Order. View seeking consensus, after a deep and broad discussion, as a worthy goal. You are more likely

Table 4.1 Many means are available for dealing with difficult behavior during meetings.

Person's Behavior	Possible Solution
Makes clearly erroneous statement	• Offer a correction. • Ask others to comment.
Talks too much	• Interrupt, thank, and then engage others.
Speaks, in a side conversation, in low, possibly negative tone	• Pause to enable others to listen. • Ask the person to repeat the comment for the benefit of the group.
Obstinate, presses for a course of action opposed by the group	• Ask to "put it in writing" for the record and/or future consideration
Uses poor choice of words or pronounces words incorrectly	• Help them by restating
Habitually arrives late	• Start on time. • Brief late comers, if appropriate. • Discuss privately with the offender.
Does not contribute	• Appeal to experience. • Appeal to position. • Stop inviting.

to reach consensus on any issue if you focus on the vision - mission - values of your organization and the purpose of your meeting. Ask questions such as what are we trying to do at this meeting? How does that relate to our core values and our mission and vision? Individuals should detach themselves from their ideas and permit them to be modified as a condition of having their ideas accepted by the group. Individual ideas that the group adopts and commits to implementing should become group ideas.

Personal

While working at a university, I led meetings of a group of department heads for eight years. We never voted on anything. We made decisions by consensus and accomplished much.

8. ***"Recap" Action Items:*** One way to "recap" action items is to go around the room and say: "Most of you committed to action items—please review them for the group before we adjourn." Another approach is to develop, as the meeting progresses, a visible, running list of action items and who will do them.

   Visit the "parking lot," that is, the list of topics that arose during the meeting and did not fit the agenda. Perhaps they were written on newsprint or on a whiteboard as the meeting progressed. If time permits near the end of the meeting, discuss the topics. If not, put them on the agenda for the next meeting. Other U.S. terms for "parking lot" are "issues bin" and "issues list." The British are comfortable with "parking lot." Other Europeans, in their languages, say

"Marmalade jar" (Danish), "Parking" (French), "Aside for now" (Dutch), and "Can we talk about this later?" (Italians) (Brenner 2007).

Following-Up After the Meeting

1. ***Practice Confidentiality:*** Practicing confidentiality means that debate and discussion should end at the meeting, especially for sensitive issues. Respect the group's decision, even if your view was different, and move on.

Personal

I recall an experience I had as a consultant to an engineering - architectural (E/A) firm. The president told me that he observed unrest within their executive group and morale problems beyond. He asked me to attend a meeting of the executive group. I did and immediately felt tension, cynicism, and reluctance to speak their minds. I asked permission to anonymously interview a cross-section of personnel within and outside of the executive group. Some personnel outside of the executive group said that members of the executive group were the laughing stock of the company. They would oppose an action by the executive group and then reveal, in a biased manner, confidential information to personnel outside of the executive group. My simple recommended solution to the executive group: Keep discussion of sensitive issues within the group. They did this and took other actions to rectify the situation.

2. ***Promptly Distribute Meeting Minutes:*** The meeting isn't over until the paperwork is done. Therefore, always quickly create and distribute meeting minutes or summaries. The incremental cost of preparing minutes is typically miniscule relative to the time already invested in planning and conducting the meeting. Two kinds of problems arise as a result of undocumented meetings:
   - Participants forget, more specifically, they forget what they agreed to do, their action items. English author and critic Virginia Woolf said this which is very relevant to meeting minutes: "Nothing has really happened until it has been recorded."
   - You and others will have to deal with as many versions of the meeting as there were participants. Given the opportunity, we tend to "rewrite history" in our favor and to remember things as we wanted them to be, rather than as they were. And this leads to later conflict. Englishmen Jonathan Lynn and Anthony Jay observed the following: "It is characteristic of committee discussions and decisions that every member has a vivid recollection of them and that every member's recollection differs violently from every other member's recollection."

   When you or others distribute meeting minutes you may, for the sake of accuracy and completeness, want to use a statement like the following to provide an opportunity for meeting participants to comment: "If you have questions about or corrections or additions to these minutes, please contact_______________. We will consider the minutes to be accurate unless comments are received within two working days of the date of the minutes."

Distribute meeting summaries quickly, to impart a sense of urgency and to build on the meeting's momentum. Don't wait to prepare and distribute meeting minutes immediately before or at the next meeting. This causes people to spend time making excuses regarding what they were supposed to do rather than spend time doing and reporting on what they were supposed to do.

Personal

When I am responsible for minutes, my goal is to distribute a draft within three business days. Some individuals do much better. I've participated in many conference calls after which the summary is distributed via email within one hour.

3. ***Set an Example by Doing Your Action Items:*** Back to you if you led the meeting —be the example for meeting follow-up. Reinforce and maintain your credibility. Others will be watching you! Author William Dean Howells put it this way: "An acre of performance is worth a whole world of promises." Be what you want others to be.

Additional Meeting Thoughts

If a manager calls a meeting to "discuss" a topic, the manager gives up the option to act unilaterally on that topic or area. To do otherwise is to risk losing credibility. Meet standing up—I once had had a "boss" who did this.

Planned and unplanned supervisor-supervisee interactions and discussions with colleagues constitute special kinds of meetings—certainly informal ones. While such "meetings" usually do not follow a formal agenda that is provided ahead of time, try to bring a list of discussion items and indicate to the other person that there are specific topics you would like to discuss. These informal meetings often lead to various kinds of follow-ups. Rather than a single document, such as minutes, the follow-up might be in the form of documents that are related to topics discussed and decisions made at the meeting with your supervisor or supervisee. An example is an email to a third party with a copy to the person with whom you met.

WORKING WITH TECHNOLOGISTS, TECHNICIANS, AND OTHER TEAM MEMBERS

Consider some definitions to set the scene for this aspect of developing relationships. An ASCE (2010) task committee developed definitions of professional, technologist, and technician and they are quoted here:

- ***Civil Engineering Professional (CE Professional):*** A person who holds a professional engineering license. A person initially obtains status as a CE Professional by professional engineering (PE) licensure obtained through the completion of requisite formal education, engineering experiences, examinations, and other requirements as specified by an appropriate board of licensure. A person working as a CE Professional is qualified to be professionally responsible for engineering work through the exercise of direct control and personal supervision of engineering activities and can comprehend and apply an advanced knowledge of widely-applied engineering principles in the solution of complex problems.
- ***Civil Engineering Technologist (CE Technologist):*** A person who exerts a high level of judgment in the performance of engineering work, while working under the direct control and personal supervision of a CE Professional. A person initially obtains status as a CE Technologist through the completion of requisite formal education and engineering experiences and may include examination and other requirements as specified by a credentialing body. A person working as a CE Technologist can comprehend and apply knowledge of engineering principles in the solution of broadly-defined problems.
- ***Civil Engineering Technician (CE Technician):*** A person typically performing task-oriented scientific or engineering related activities and exercising technical judgments commensurate with those specific tasks. A person working as a CE Technician works under the direct control and personal supervision of a CE Professional or direction of a CE Technologist. A person initially obtains status as a CE Technician through the completion of requisite formal education, technical experiences, examination(s), and/or other requirements as specified by an appropriate credentialing body. A person working as a CE Technician is expected to comprehend and apply knowledge of engineering principles toward the solution of well-defined problems.

While the preceding definitions of some likely team members are specifically for the civil engineering discipline they could, in principle, apply to other engineering disciplines as well as other technical areas. Presumably you as an engineering student or entry-level practitioner aspire to be a professional, as generally defined above. Prior to and after reaching that goal, you will frequently work, within and outside of your organization, with technologists and technicians, as generally defined above, as well as team members such as administrative assistants, secretaries, clerical workers, and data-entry personnel. You and they will constitute vital teams.

Essential Members of the Organization

The entry-level engineer or other technical person should view the other team members as essential because of their special knowledge and skills and their lower labor cost. These personnel should be brought into a project or task whenever and wherever their combination of expertise and cost per unit of production is more favorable than having the tasks performed by professionals. Even if the professional has the necessary knowledge and skills, the cost per unit of production should govern involvement assuming the other personnel have the necessary knowledge and skills. Just because a professional could perform a task does not mean a professional should do the task, as noted in the earlier delegation section of this chapter.

Entry-level technical persons should recognize that they need the other team members as much as the other team members need them. You cannot be productive without significant assistance in various specialized areas. While entry-level technical persons might be able to do some functions, they will not be able to do them as efficiently, as measured by productivity (labor and other costs per unit of work completed) and as well as some other team members.

Challenges Unique to Working with Varied Team Members

Working with the many and varied individuals who populate the engineering and business work place is a challenge. Some of those challenges are discussed here.

Communication

One challenge is communication—mainly writing, speaking, and listening—between aspiring professionals and other team members. The former routinely communicate with each other using terms, concepts, acronyms, abbreviations, and other expressions known only to them. Other team members, because of their narrower and more pragmatic or applied training and interests, may not be as widely conversant. Therefore, the aspiring professional should focus on communication if he or she is to be productive—to effectively work with the people and use the resources available in the organization.

Age Differential

Most team members will be older than the aspiring professional. The 22-year-old, entry-level engineer may rely on support from the 45-year-old chief CADD operator and the 32-year-old surveying party chief. Inherent in the age differentials is a wealth of knowledge and skills in areas such as "how this place really runs" and where you can find something. Aspiring professionals should conduct themselves in a manner that enables them to benefit from that knowledge and those skills. One indication that the entry-level person has earned the confidence and respect of other team members is when the latter make comments to the young person such as, "I'm sure you know how you want this done, but if I were you I would consider doing it this way. . . ."

Devaluation

Unfortunately, a few aspiring professionals "look down on" some other team members. As these engineers survey an organization, particularly the larger, diverse public or private entity, they are likely to see people in a variety of positions, some of which appear undesirable. For example, from the perspective of the recent college graduate, some jobs may be characterized as simple, boring, dirty, noisy, and hot. Because the entry-level professional would not want to do those jobs, he or she might devalue those who do.

This devaluing may be indicated by the use of the terms "boy" or "girl" in referring to men and women, as in "I'll have one of the boys enter the data" or "let's have one of the girls in the field check this out." Another example is the uninvited use of first names. A final, and particularly disturbing example, is being impolite to and surly with some team members but polite to and friendly with professionals. The overall impact is negative.

People's inherent value is not determined by the positions they hold on an organization chart. While the economic value of people's work varies widely, reflecting their education, experience, and responsibility, their inherent value as individuals is an independent matter. Aspiring professionals are advised to take the position, unless they know otherwise, that everyone in the organization is doing basically what they want to do. Assume that each person has consciously made a series of education and work decisions that lead to their current position. No one should be embarrassed by or have to apologize for the position they hold or the work they do unless the work does not meet the expected quality and quantity for the position.

A Dozen Tips for the Entry-Level Technical Person

Working effectively with the very diverse personnel that typically compose teams within business, government, academic, volunteer, and other organizations presents challenges. The following suggestions should help the aspiring professional develop effective working relationships.

1. Identify the organization's technologists, technicians, and other team members learn how they are organized. Who are they, where are they, what services do they perform, and to whom do they report? Sources of information include administrative charts, the directory, personnel handbooks, maps or drawings of the physical facilities, and colleagues.
2. Introduce yourself to other team members and the individuals who direct their work. If you are not introduced to these personnel during your initial orientation to the organization, take the initiative to do so on your own. Try to complete this process before you begin to request or offer assistance.
3. Unless they volunteer the information, ask people how they want to be addressed or observe how they are addressed by others. This is particularly important when dealing with senior personnel who may be offended by excessive familiarity on your part.
4. Always be polite. Words and expressions like "please," "excuse me," and "thank you" show respect for individuals and maintain civility in the civilized work place and encourage it in the uncivilized one.
5. Find out how to request support. Expect procedures to vary widely from organization to organization and even within an organization. The standard operating procedure may be very informal or there may be a formal, written work request form. Respect the established work request procedure. However, if the requesting procedure is informal, such as a verbal request, accompany or follow the verbal request with written instructions. There are three reasons for

this. First, you are more likely to think through your needs if you write them down; that is, your request will be clearer and more complete. Second, there is a much lower probability that other team members will misunderstand your needs if your request is written rather than just spoken. Finally, when the requested work is completed, your written instructions will remind you how the product is to be used or what is to be done next. Ask that your written instructions be returned to you preferably with a sign-off by the team members who completed the task or tasks.

6. Always indicate what you need—that is, the product you want—and when you need it. With respect to the desired deliverable, try to provide an example such as a copy of a previous letter, memorandum, drawing, survey, or data sheet. With respect to the desired completion time, avoid saying things like "as soon as possible," "no hurry," or "when you get around to it" unless there is a clear understanding of what these terms mean.
7. Explain, as appropriate, the context or purpose of the work or product you are requesting, as discussed earlier in the delegation section of this chapter. Team members are much more likely to produce work or products that meet your needs if they understand the context. They may even suggest an alternative approach.
8. Minimize interruptions. Others cannot make progress on your work or the work of others if you are frequently talking to them or interrupting them. The previously-mentioned written work request procedures minimize interruptions. Consider grouping verbal requests to minimize the number of interruptions and scheduling a brief working meeting at a mutually-agreeable time.
9. Do not, through your poor planning or as a means of getting attention, allow too many requests to be in a crisis or panic mode. If you "cry wolf" too many times, you will lose credibility within the team. Try to develop a sense of the absolute and elapsed time required to complete certain tasks. You will occasionally encounter bona fide emergencies and, if you have credibility with team members, they will respond favorably. Remember, in keeping with Theory Y, most people want to be helpful.
10. Be prepared to prioritize. You may not be able to get everything you want when you want it. If necessary, establish priorities and look for alternative ways to accomplish certain tasks.
11. Insist that you are informed as soon as possible if the product will not be done in the manner requested and/or completed on time. Recall the earlier "three outcomes" discussion under the topic of delegation.
12. Provided you get the necessary product on time, be slow to criticize how team members do their work. As you develop an effective working relationship, you might ask questions about why they do things the way they do them and perhaps eventually make some suggestions. You will learn much in the processes. Ask how you could be more effective in your joint efforts. And, as noted elsewhere in this chapter, say thank you for the service you receive and for what you may learn in the process.

SELECTING CO-WORKERS AND "MANAGING YOUR BOSS"

The title of this section may initially sound manipulative or presumptuous. However, it emphasizes the need to actively approach relationships with personnel all around you in the organizational hierarchy—up, down, and laterally.

Carefully Select Your "Boss" and Co-workers

Assume you have high career aspirations. As a soon-to-graduate student, you just learned about an employment opportunity. Or, as a practicing engineer, you just learned about an open position that could be within or outside your current employer. Which of the following factors—and in what order of importance—would you consider in making your decision to seize the opportunity or stay where you are: job location, salary, computer and other equipment you would use, benefits, consistency with goals, style/condition of office building and/or your work area, and the people you would work for and with. In my view, at or close to the top of your list should be the last item: People you will work for/with!

Carefully choose your co-workers and your "boss," especially in the early, more formative part of your career. These frequent associations in a variety of settings are crucial because they will influence your attitude toward your profession, will affect the knowledge and skills you acquire, and will determine the expansion of your network. Unless you are an unusually independent, highly self-disciplined person, you will be shaped by the people environment within which you work. Consider this advice of humorist and writer, Mark Twain: "Keep away from people who try to belittle your ambitions. Small people always do that, but the really great make you feel that you, too, can become great." To reiterate, if you have high aspirations, carefully choose your co-workers and your "boss."

Seek a Mutually-Beneficial Relationship

Gabarro and Kotter (1993) define managing your "boss" as "... the process of consciously working with your superior to obtain the best possible results for you, your boss, and the company." Using the word "company" in the definition certainly does not limit the importance of the "boss" management process to business. Clearly, "managing" your supervisor is applicable in all employment sectors, including government, academia, and volunteer organizations. You are encouraged to have high expectations for both you and your immediate supervisor while remembering that you and your supervisor are, as noted by Gabarro and Kotter, in a situation of "mutual dependence between two fallible human beings." In other words, you need each other and neither of you is perfect.

Gabarro and Kotter offer some specifics. Try to understand your supervisor's goals and objectives, pressures, and preferred work style. Look within yourself and assess your strengths, weaknesses, preferred work style, and predisposition toward authority. Use the preceding to "manage your boss" or, more specifically, your part of what ideally will be a mutually-beneficial relationship.

Avoid Being a "Yes" Man/Woman

Consider this conversation in Shakespeare's Hamlet, Act III, Scene 2 as quoted by Norton (1994) who notes that "Polonius is the one who gets stabbed to death behind the curtain:"

Hamlet: Do you see yonder cloud that's
almost in the shape of a camel?
Polonius: By the mass, and 'tis like a camel,
indeed.
Hamlet: Methinks it is like a weasel.
Polonius: It is backed like a weasel.
Hamlet: Or like a whale?
Polonius: Very like a whale.

Sometimes it's easier to just go along. Say and do what your "boss" wants, or you think he or she wants. Keep your supervisor happy rather than developing an informed, mutually-dependent, beneficial relationship in which you give the best that you have to offer.

The "yes" man or woman approach is easiest in the short run but harmful in the long run for you, your supervisor, your organization, and those your organization serves. As Norton (1994) noted, "... the culture of yes men hurts the bottom line by corrupting the information flow and depriving managers of the best options." He goes on to say that in "yes" man and woman cultures, "... fewer ideas are generated by groups than by individuals working alone..." and thus the organization is denied synergism.

As you move into supervisory positions, note how much your supervisees disagree with you. If not at all or rarely, you, your supervisees, and the organization are all short-changed. Your actions are creating a serious problem and only your actions will correct it. "Where all have the right to speak, some foolish speaking is done," according to Frank L. Weil, lawyer and social worker, "but where, as in a dictatorship, all speak alike, little thinking is done."

CARING ISN'T CODDLING

As you develop relationships with your supervisor, your supervisees, and others, an element of caring should be evident in your actions toward others and in their actions toward you. Caring, as used here, does not mean coddling. If caring isn't coddling, what is it?

Think about those former teachers or professors whom you believe really cared about you. They probably demonstrated their concern for you narrowly as a student and broadly as a person through a variety of meaningful interactions, such as: delivering well-prepared lectures, making regular and demanding assignments intended to deepen and broaden understanding of the course material, providing opportunities for independent study such as a research paper or laboratory project, encouraging you to participate in co-curricular and extra-curricular leadership and service activities, offering an encouraging word at a discouraging time, and praising when nobody else seemed to notice what you had accomplished. The preceding were not offered in a paternalistic, condescending, ostentatious manner. Instead, these actions were part of a high-expectations and high-support environment intended to stretch without snapping, provide example without expecting cloning, and build confidence without imparting arrogance.

Caring is also exemplified by the parent who said, "If it's worth doing, it's worth doing well" and the colleague who, at a meeting in your firm, has the courage to ask the awkward question or raise the sensitive issue that almost everyone knows must be addressed. Caring is also shown by the manager who says no to the pleading employee who did not strive to meet the established requirements and now wants to avoid the adverse consequences.

You may also recall, with disdain, those teachers, supervisors, colleagues, and others who were generally "nice," but didn't expect all that much of you. Often, you delivered in accordance with their expectations. You and they could have done so much more. Perhaps they didn't really care about you, or even themselves.

Caring isn't coddling. Caring is pushing, pulling, admonishing, stretching, demanding, encouraging, urging, challenging, cajoling. Caring is high expectations coupled with high support. Caring helps individuals and organizations meet their goals and realize their full potentials (Walesh 2004).

COACHING

As you gain experience, others can benefit from your experience. Consider coaching as one way of leveraging your experience for the benefit of others (Walesh 2004). Coaching means occasional, ad hoc, one-on-one focused interactions between a caring senior employee and a receptive junior employee. Coaching is distinguished from mentoring, which typically requires a major ongoing effort over an extended period of a year or more. Coaching is accomplished intermittently and opportunistically during the course of ongoing projects and activities. It is not a systematic education and separate training process.

Coaching Tips

How can you, as an increasingly-experienced professional, coach less experienced engineers and other personnel? How will you have time to fit this additional activity into your already busy schedule? The answer is to look, in the normal course of your work, for specific coaching opportunities that will take the receptive young technical professional up the knowledge, skills, and attitudes learning curve. Tailor the coaching situations to the level, needs, and receptivity of the young person. For example:

- **In your role as project manager:** Invite a young engineer to accompany you to a public meeting and observe the proceedings. On the way to the meeting, describe what appear to be the key issues, how they were determined, and how they have been or will be discussed with stakeholders. Outline the anticipated outcomes of the meeting. After the meeting, on the way back to the office, constructively discuss the meeting and critique the tactics and effectiveness of the various participants.

- **In your role as manager of all or a portion of a project:** Invest some quality time with a young technical person by walking him or her through an in-process report on a project. Discuss and illustrate the critical, client-owner-customer friendly features of the report, such as an attractive cover, documentation of all alternatives and their pros and cons, ample white space, effective graphics, proper citation of sources, a list of abbreviations, a glossary, and a bottom line-oriented executive summary with a brief description of issues and recommendations. Explain how the implementation of the project is likely to depend on a carefully-crafted combination of quality technical work and the polished manner in which that work is presented in the report.
- **In your role as proposal and interview manager for a potential project:** Ask an entry-level person to sit in on an initial strategy session, some subsequent conversations with graphics and other personnel, a practice session prior to the actual interview, and the actual interview. Use these interactions as opportunities to share aspects of the organizational philosophy, such as focusing on developing relationships with clients-owners-customers vis-à-vis pursuing projects. Emphasize the importance of client-owner-customer and project research and illustrate the advantages of teamwork as exemplified by clerical, engineering, graphic, and marketing personnel working together to identify and meet client and project requirements.

Concluding Thought

Aspiring young engineers and other technical personnel are usually bright, but they don't always know what they don't know, especially if it is outside the technical or hard-science arena. They need to be told, by word and by example, that a judicious blend of technical and non-technical knowledge and skills is essential to securing and completing engineering assignments. Coaching is a way to share what you know about soft skills. You may have been fortunate to have been coached. If so, seize opportunities to pass on the experience. Consider this advice of minister and writer, Ralph Waldo Emerson: "It is one of the most beautiful compensations of this life that no man can seriously try to help another without helping himself."

TEAMWORK

As noted in Chapter 2, teamwork is becoming standard operating procedure (SOP) in successful business, government, academic, and volunteer organizations. Examples of teams are departments in various types of organizations, a student group working on a project, a team of government personnel preparing a grant application, and an ad hoc professional society committee. You will have many opportunities to serve on teams, initially as a team member and later as a team leader. As used here, teams include groups that meet and work mostly face-to-face as well as virtual teams, that is, groups that are always or usually physically separated.

Much of what has already been presented in Chapters 2 and 3 and in this chapter should help you be an effective team member. Examples are managing personal time, developing communication skills, understanding body language, recognizing types of people and their needs, delegating, orchestrating meetings, appreciating and working with all team members, and coaching.

Personal

As a college freshman, my new-found, naïve friends and I joined the same fraternity. Our choice, as we soon discovered, was not a good one—at least in the short run. The fraternity had not distinguished itself. Within the first year of our joining, and largely by coincidence, a musically-inclined, two-year member of the fraternity presented a vision. He argued that we could begin to rebuild our group's role and reputation on campus by winning the fraternity portion of the next annual Greek Songfest. (In retrospect, this singing competition seems trivial. But, then and there, it was the big event and winning it was the big prize. Please indulge me.)

The vision—although it initially seemed ludicrous—became contagious; it spread throughout our fraternity and with more and more of us beginning to think that we might be able to achieve it. Commitment to the vision led to individual singing auditions. Strength in diversity emerged. We had a good mix of basses, baritones, tenors, altos, and sopranos. A trusting organizational and operational structure soon emerged. Its foundation was a rigid schedule of frequent practices; everyone was needed and held accountable. We won! We also won the next two years. By the time my class graduated, and as a result of the catalytic effect of the first Songfest win, our fraternity was one of the most highly-regarded on campus excelling in academics and leadership, as well as singing. The "secret" of the team's success?—vision, diversity, and trusting structure.

Three Teamwork Essentials

Based on participating in various successful team efforts in the business, government, academic, and volunteer sectors, I am convinced that the three key elements of successful teamwork are:

- **Vision:** A strong and shared commitment to an ambitious vision; the bolder the better. "Far better it is to dare mighty things, to win glorious triumphs, even though checkered by failure," according to Theodore Roosevelt, 26th U.S. President, "than to rank with those poor spirits who neither enjoy much nor suffer much, because they live in the gray twilight that knows neither victory nor defeat." The litmus test: The vision should initially appear highly desirable but unachievable. Sometimes teams are carrying out more routine tasks.

In these situations, the word vision may be too lofty, may not be appropriate. Instead use an end point term like objective or goal and, as with vision, seek a strong and shared commitment to it.

- **Diversity:** Assemble an optimum mix of players. All the necessary bases must be covered. Factors to consider in forming a team are to seek individuals who, besides sharing the vision, collectively bring the necessary knowledge, skills, attitudes, connections, and time availability. Consider this thought, offered years ago by John W. Gardner, former U.S. Secretary of Health, Education, and Welfare, about honoring and valuing diversity: "The society which scorns excellence in plumbing as a humble activity, and tolerates shoddiness in philosophy because it is an exalted activity will have neither good plumbing nor good philosophy: neither its pipes nor its theories will hold water."
- **Trusting, communicative structure:** Trust and open, on-going, intra-team communication are essential. Now the emphasis moves from the team members to the relationships among team members. Those relationships must be respectful and trusting. "The way a team plays as a whole determines its success," according to Hall of Fame Baseball Player Babe Ruth, "you may have the greatest bunch of individual stars in the world, but if they don't play together, the club won't be worth a dime." Football coach Vince Lombardi stressed commitment when he said: "Individual commitment to a group effort—that is what makes a team work, a company work, a society work, a civilization work." Commitment is the "glue" that holds a team together.

All three elements are needed, none is sufficient. An exciting vision without diverse players is a dream. A talented team toiling in a vision void is poor stewardship. Engineer and educator Arthur E. Morgan said "Lack of something to feel important about is almost the greatest tragedy a man may have." A superb organizational structure without talented players degenerates into bureaucracy.

Creating a Team

You, as a student or as a young practitioner, will eventually have an opportunity to create an ad hoc or relatively permanent team to accomplish a purpose or drive toward a vision. Or you will be asked to be member of a team created by someone else. Consider some ideas, as illustrated in Figure 4.4, that might be used by an individual, or small group, to enable them to create and launch an effective team.

Begin by stating, in writing, the team's purpose or vision. We are tempted to say "Oh, you know what I mean." How will they know if you don't know? Most of us don't really know what we mean until we express it in writing. Then draft a charge for the proposed team. Besides the team's purpose or vision, typical elements in a charge are background and context, deliverables, available monetary and other resources, schedule, and stakeholders that should be involved. Think of the charge as the definition of the issue to be addressed, the problem to be solved, or the opportunity to be pursued and the overall framework for carrying out the stated challenge. The written charge reflects the idea that an issue, problem, or opportunity well defined is half resolved, solved, or achieved.

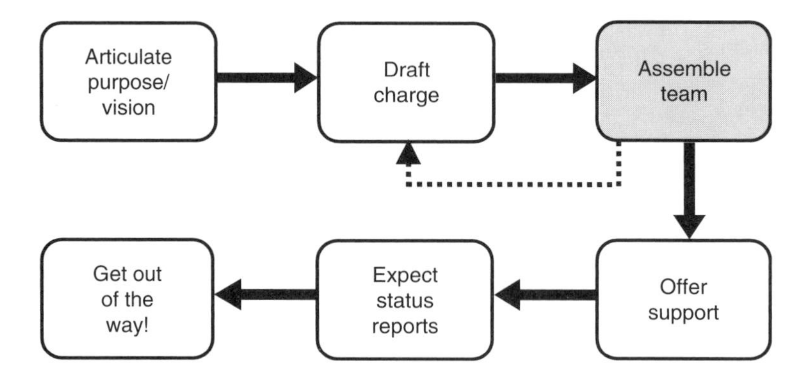

Figure 4.4 This process will help you create and launch an effective team.

Identify potential team members with diversity of knowledge, skills, and attitudes being the key consideration. Team creators should resist the temptation to clone themselves. Sure, that would be more comfortable, but the lack of diversity would also greatly limit the team's potential. Secondarily, find individuals with basic personal and interpersonal skills like those discussed in Chapters 2, 3, and this chapter. That is, to use the Chapter 2 metaphor, seek team members who have their "personal house in order" in that they know how to manage their time, set and achieve goals, delegate, listen, write, speak, and do what they say they will do (DWTSTWD).

Stated differently, avoid dysfunctional individuals. Behavioral "red flags" include cynicism, aloofness, an I win-you lose mentality, discomfort with ambiguity, impatience with indecisiveness, defensiveness in response to criticism, and intolerance of other people's mistakes. By the way, this list is taken from an article by consultant Thompson (1996) titled "Engineers Don't Always Make the Best Team Players." Perhaps we could also substitute "geologist," "architect," "planner," "accountant" and so on for "engineer?" Time for a little introspection?

Sometimes teams, and their members, are collectively so dysfunctional that they do not recognize that they are dysfunctional. They think that the following are "normal":

- Combative, win-lose environment at team meetings.
- "Commitments" made at meetings routinely ignored.
- Superficial discussions during team meetings with real decisions occurring off-line, behind the scenes.

The advice about the appropriateness of potential team members is directed in two directions the first being to those who are forming, or helping to form, a team and want the team to be successful. Ideally, they seek a diverse group with each person possessing basic personal and interpersonal attributes. Second, if you are invited to join a team, and have the option of declining, find out who will be on the team and who will lead it. This advice may be hard to follow, especially in employment situations. That is, you may be assigned to a team, have nothing to say about its formation, and not have the option of declining.

Assume the team has been assembled and begins to meet. Encourage them to study the charge and possibly ask that it be modified to expand or reduce the scope, change the schedule, or make other change. The goal is to enable the team to own the charge. If the individual or small group that formed the team is not to be part of the team, he, she, or they should offer support, indicate that status reports are expected, and then get out of the way. If you or others have assembled a trusting, communicative group of appropriately diverse individuals who share a purpose or vision, they can accomplish almost anything. I've experienced this—I commend the above process to you.

The Forming-Storming-Norming-Performing Process

The start-up of a new permanent or ad hoc team typically presents challenges. One model for understanding the evolution of a newly created team is the sequential forming-storming-norming-performing process. According to Brown (1992) and Martin and Tate (1999), these four stages of team development are characterized as follows:

- **Forming:** Politeness, inquiry, waiting to see what will happen, no or very little productivity
- **Storming:** Disagreement, confusion, conflict, factions, some productivity
- **Norming:** Conflict resolution, goal setting, decision making, establishment of protocols, ownership, accountability, moderate productivity
- **Performing:** Teamwork, adjustments, deliverables, full ownership and accountability, can and will attitude, high productivity, satisfaction, celebration

A newly-formed team, especially a diverse one, is likely to exhibit some "storming." That step in the team's evolution is not necessarily negative and team leaders and sponsors can take some actions to minimize and work through this stage. First, follow the advice offered earlier in this chapter to bring geographically-dispersed individuals together as soon as possible recognizing that while the team may be virtual, its internal relationships must be real (King 2007). Second, provide context by discussing and answering questions such as how did we get here? and, looking forward, who cares about the results of our work and why?

Third, agree on terminology recognizing that misunderstanding of key words is often a major stumbling block. Agree on terminology and put it in writing. Maybe start a glossary especially if one team deliverable is a report. Fourth, encourage everyone to speak up; leverage the diversity that influenced the selection of team members. Fifth, define some roles that logically follow from diversity. For example, assume someone volunteers to edit the team's report, possibly because of their writing knowledge and skill. Then generally defer to that person on routine editing decisions. Sixth, and last, refer again to the tips for successful meetings presented earlier in this chapter.

Closing Thoughts about Teams

My hope is that the preceding discussion of the three key teamwork essentials, the suggestions for creating a team, and the forming-storming-norming-performing process will help you appreciate the dynamics of team formation, process, and accomplishment. Combining that understanding with managing and leading knowledge and skills presented in Chapters 2 and 3 and this chapter should enable you, as a student or young practitioner, to be an effective team member and, as opportunities arise, form and/or lead a team.

EFFECTIVE PROFESSIONAL MEETING AND CONFERENCE ATTENDANCE

You, your organization, and others will benefit if you continuously meet and effectively interact with colleagues, clients, and competitors. Conferences, workshops, and other gatherings, such as periodic local meetings of professional societies and business groups, are means of networking with people outside your organization. Consistent with the theme of this chapter, that is, developing relationships, you should carefully manage your participation in professional meetings, conferences, workshops, and similar external activities. That way, your involvement will create win-win-win situations. You benefit, your employer benefits, and the organization sponsoring the event benefits. Failure to intelligently network results in wasted resources (time and money) and missed opportunities.

This chapter section offers suggestions for how you can prepare for, participate in, and follow up on these important external events. The remainder of this section is written as though it applies to an entry-level professional desiring to attend a major international, national or regional conference, the most complex of the various kinds of external events. However, many of the ideas offered also apply to local conferences, seminars, workshops, and meetings of professional and business groups. Furthermore, already as an undergraduate or graduate student, you will have opportunities to attend such events and, therefore, some of the ideas presented here may be helpful.

Learning about the Conference

The possibility of participating in a conference or workshop typically begins with you reading a sometimes slick, often multi-colored brochure or website, with an opening photograph of the Golden Gate Bridge, the Eiffel Tower, the Great Wall, or the Sydney Opera House. As long as you can remember, you've wanted to visit San Francisco, Paris, Beijing, or Sydney. You study the brochure. Dozens of sessions and hundreds of papers are described. "Famous" experts will speak. A smorgasbord of tasty technical topics tantalizes you.

But, you hesitate. You haven't been to any major conferences, perhaps because you are relatively young or new to your field or employer. Or, maybe you have been to such conferences but, somehow, you just weren't as satisfied as you expected to be. If any of these situations apply to you, consider the ideas presented here.

Before the Conference

Study the Situation

Scan the program and decide which sessions you would like to attend and what speakers you want to hear. You may have to consider some session jumping, especially when many interesting sessions are scheduled at the same time. Even more important, study the program to identify the people you want to meet. Who is talking about a subject that you are interested in? Who is on a committee you would like to join? Who wrote an article that you found especially stimulating? Who would you like to team with on a project?

Prepare Your Proposal

Get your thoughts together and prepare a proposal to your supervisor, to be delivered in writing or through discussion, depending on how you operate within your office or organization. Explain the benefits to your organization and to you. Tie the discussion to your goals and the goals of your organization. Recognize that you are competing for limited funds whether you are in the private, public, academic, or volunteer sectors.

From a dollar and cents perspective, your supervisor will likely view your attendance at the conference as an investment (the cost of which can be readily measured in dollars and cents) that will produce a return on investment (the magnitude of which cannot be measured in dollars and cents). Your goal: convince the supervisor that the latter will in some way, shape, or form, promise to be greater than the former. Think in terms of return on investment for you and your organization. Plan to be accountable to both.

This is not the time to be altruistic or idealistic such as claiming that "knowledge gained at this conference will enable me to make an even more significant contribution to the breadth and depth of services provided by our organization." Instead, be specific, such as, "there is a woman who is going to speak at the conference about a cost-effective way to map areas of likely concrete delamination on reinforced concrete bridge decks. I've been studying this technology and believe that we could provide that service. I want to go to the conference, hear her paper, talk directly to her, and see what I can learn." Two or three reasons like that ought to do it. Commit to sharing, upon your return to the office, what you learned at the conference. Offer to present a noon "brown bag" or to write a brief memorandum summarizing whom you met, what you learned, and what you will do differently.

Maybe your organization simply can't afford to foot the entire bill for the conference. If the conference is important to you, and some should be, negotiate some sort of partial reimbursement. In fact, you might even be more diligent in what you do at the conference and how you follow-up on it if you have to pay for some or all of it yourself.

If you are even tempted to think that you are owed this conference as some sort of earned "rest and relaxation" (R&R), do a reality check. That kind of thinking, both individually and organizationally, went out somewhere between the demise of the slide rule and the arrival of the personal computer.

At the Conference

Well, you made it. You are in San Francisco, Paris, Beijing, or Sydney. Now you can concentrate on learning, primarily by asking and listening, but also by sharing some of your own ideas and having them evaluated by your peers. How do you do this? Some suggestions follow.

Wear Your Name Tag

Wear your name tag—it is meant to help you and others communicate. Thoughtful conference organizers use large type for the entire name of a person or, more often, the first name. Don't leave the name tag in your room, pocket, or purse. Put it on—preferably on the right side where people can easily see it as they shake hands with you.

Present a Professional Appearance

Dress and behave as though you are the chief public relations officer for your organization. You probably are, at least temporarily. People's perceptions of business, government, academic, and volunteer organizations, are often illogically based on only one contact with a representative of that organization. Chances are, you are your organization's defacto public relations person—unless, of course, your firm or agency is doing so well that many well-groomed, well-dressed, and well-behaved staff members will be attending to offset any bad impressions you make.

Ready Your Business Cards

Take a supply of business cards, not to see how many you can give away, but to make sure that you have them when you need them. They should be on your person at all times, not in your room, out in your car, or over there on the other side of the room in your briefcase. Ideally, you have current, attractive cards (as opposed to dog-eared cards with handwritten information) that include an address, a telephone number, an e-mail address, and a website.

Look for opportunities to ask other people for their cards. Write notes on them. You can even do this in front of the other person—it shows you are interested and intend to follow through. If there is any reason to follow-up on a one-on-one conversation, jot that on the card as one of your action items.

Mix and Mingle

If you should be so lucky or unlucky, depending on the situation, to be attending the conference with one or more people from your organization, generally stay away from them. Don't hang around together and certainly don't sit together at breakfast, lunch, dinner, and in sessions. You already know them, and if the truth would be known, one of the reasons you wanted to come to this conference in the first place was to get away from some of them! Every time you sit next to someone you already know, you deny yourself and someone else an opportunity to get to know each other. Keep moving; don't spend excessive time with one or a few individuals. Recall some of the reasons you asked to come in the first place.

Talk to Strangers

Adopt the philosophy that you are surrounded by opportunities to learn and make contacts and you often have to take the initiative. But, you say, I don't know what to talk about at the morning coffee break or the noon lunch. Mixing and mingling can be a challenge for engineers and other technical persons because we tend to be introverted. Frankly, most people like to talk about themselves and their interests.

Therefore, give them an opportunity to do so by asking questions. If they are half as thoughtful as you are, they will reciprocate and the result will be informative conversations. Even if they don't reciprocate, you'll probably learn something by asking sincere questions. The person who asks questions, directs the conversation and benefits the most from it.

If you can't think of questions, here are some open-ended inquiries that cannot be answered with a "yes" or "no" and, therefore, will encourage conversation:

- "So what do you hope to get out of this meeting?"
- "I notice that you are with Noitall Consulting—what attracted you to the consulting business?"
- "I am working on an artificial intelligence project and need some ideas. Who in your organization does this?"
- "You certainly seem to have a lot of experience. What advice do you have for a young person like me?"
- "I have never been to your country. What should I try to see while I'm here?"

If you simply enjoy a conversation you had with a person you met because you share some good "chemistry," that's a good reason to take action. Your thoughtful follow-up might be as simple as an e-mail to them after the conference saying, "It was a pleasure to meet you. I enjoyed our conversation. I hope to see you in the future."

Listen for Relevance

As you listen reflectively to a speaker relate his or her ideas and information to your current responsibilities, projects, activities, and, quite frankly, your dreams. Think also of the needs of co-workers. At least one valuable idea or piece of information will jump out at you during each presentation. Jot it down immediately in the form of a "to do" or action item. Don't rely on your memory.

Speak Up at Sessions

Most speakers feel disappointed if nobody asks questions or offers comments after their presentation. You can help them and the audience by breaking the ice and asking that first question. Your question will likely lead to more and an exchange of ideas and information will occur. Everyone benefits because of your effort. Furthermore, to the extent your question reveals some of your interests, one or more individuals may approach you after the session to discuss the matter further. As a result, your network of contacts will expand.

Personal

After one speaker concluded his presentation at a conference, I raised my hand, he acknowledged me, and I asked a question. At the conclusion of the session, as I walked out, another person who had been at the session introduced himself and commented on the question I asked. This individual turned out to be the President of an engineering firm. We connected during that brief conversation and occasionally communicated over the next few years. Then his firm retained me for a major consulting project. I share this story to suggest the very real possibility that, by being sincere and proactive at meetings and conferences, you can make connections that will benefit you and your organization.

Communicate with Your Office

Keep in touch with your office and, as appropriate, clients, owners, customers and others whom you serve. After all, people are covering for you while you are away, just as you would do for them. Today's electronic communication tools facilitate such contacts. But, don't over do it. The people back home don't expect you to be there and at the conference at the same time. If you are going to spend hours on your cell phone or at your notebook computer in your room, you might as well have stayed at the office. Selective, effective communication with the office and other locations requires discipline.

See the Sights

Does your attendance mean you have to dutifully and doggedly dash from session to session, focusing totally on speakers while taking copious notes? No, certainly not. Plan to have a good time. There will be plenty of organized, disorganized, and unorganized opportunities to do so. Smart people learn how to mix business with pleasure; life is too short to do otherwise. Remember, one of the reasons you wanted to come to this conference in the first place was to cross the Golden Gate Bridge, climb the Eiffel Tower, walk the Great Wall, or visit the Sydney Opera House.

After the Conference

Well, it is over. It was fun and stimulating; you learned a lot; you made some potentially useful contacts; and you saw the sights. Now back to the reality of office routine. Right?

Act on Action Items

Wrong! The conference isn't over. You need to follow-up on all those "to dos" and action items you identified at the conference in order to realize the return on the investment made by you and your organization. Most of the follow-up tasks will be simple. Write some emails, send some materials, make some calls, and talk to somebody in your office.

Follow-up should start on your trip back home from the conference. Remember, as simple as follow-ups may be, there should be many. You will need to discipline yourself to get them done because not only will you soon be back in your normal, busy living and working environment, but you have to catch up on the things that accumulated while you were gone. Don't let the follow-ups slide; get them done now.

By keeping small promises, you build big relationships. And, don't forget promises that you made to your supervisor or colleagues before you went to the conference, such as presenting a "brown bag" or writing a summary memorandum. Your credibility is on the line.

Say Thank You

One of the wonders of professional and business organizations, as exemplified by their conferences, is the tremendous amount of unselfish volunteer effort contributed by very few people so that very many people can have the opportunity for a stimulating learning experience—and a good time! If you had such an experience, write a brief thank you note to one or more of the conference organizers and hosts.

Looking Ahead

Let's assume you use your drive and imagination, supplemented with some of the advice offered here, to become an effective conference attendee. Now you may want to move up the conference ladder to bigger and better things, such as chairing a session, giving a paper, or serving on or chairing a committee. As you have probably guessed by now, there is a certain amount of art and science to such highly active participation in conferences. For a discussion of active involvement in professional organizations, refer to the section of Chapter 2 titled "Involvement in Professional Organizations: Taking and Giving."

CONCLUDING THOUGHTS ABOUT DEVELOPING RELATIONSHIPS

Your technical knowledge and skill is important, but not sufficient for achieving your potential. You need more and that includes continued expansion of interpersonal knowledge and skills and further development of relationships with many and diverse individuals. "All the technology in the world will not help us," according to communications consultant Dorothy Leeds, "if we are not able, at the core, to communicate with each other and build strong, lasting relationships."Seek to understand the motives and perspectives of others; become an effective delegator; contribute to and later orchestrate meetings; appreciate and work effectively with technologists, technicians, and other team members; select appropriate co-workers and "manage" your "boss;" appreciate peers and others who have high expectations for you and reciprocate; value coaching and being coached; participate in and eventually form and lead teams; and proactively participate in professional meetings and conferences.

I'll tell you what makes a great manager:
a great manager has a knack for making ball players
think they are better than they think they are.
He forces you to have a good opinion of yourself.
He lets you know he believes in you.

He makes you get more out of yourself.
And once you learn how good you really are,
you never settle for playing anything less than your very best.

(Reggie Jackson, Hall of Fame baseball player)

CITED SOURCES

American Society of Civil Engineers, Paraprofessional Task Committee. 2010. *Final Report to the ASCE Board of Direction*. September 3, ASCE: Reston, VA.

Annunzio, S. with J. Liesse. 2001. *eLeadership: Proven Techniques for Creating an Environment of Speed and Flexibility in the Digital Economy*. The Free Press: New York, NY.

Bredin, J. 1988. "Confessions of a Management Mole." *Industry Week*, September 19, p. 32.

Brenner, R. 2007. "Using the Parking Lot." *Point Lookout*, e-newsletter from Chaco Canyon Consulting, September 12.

Brenner, R. 2009. "Letting Go of the Status Quo: The Debate." *Point Lookout*, e-newsletter from Chaco Canyon Consulting, December 30.

Brown, T. L. 1992. "It's One Thing to Reorganize." *Industry Week*, May 4, p. 17.

Covey, S. R. 1990. *The 7 Habits of Highly Effective People*. New York: Simon & Schuster.

Culp, G. and A. Smith, 1997. "Six Steps to Effective Delegation." *Journal of Management in Engineering—ASCE*, January/February, pp. 30–31.

Dychtwald, K. and D. J. Kadlec. 2009. *With Purpose: Going From Success to Significance in Work and Life*. Collins: Toronto, Ontario, Canada.

Finzel, H. 2000. *The Top Ten Mistakes Leaders Make*. Cook Communications: Colorado Springs, CO.

Gabarro, J. J. and J. P. Kotter. 1993. "Managing Your Boss." *Harvard Business Review*, May–June, pp. 150–157.

Gross, R. 1991. *Peak Learning: How to Create Your Own Lifelong Education Program for Personal Employment and Professional Success*. Jeremy P. Tarcher: Los Angeles, CA.

Harris, S. J. 1968. *Winners and Losers*. Argus Communications: Allen, TX.

Hoop, W. J. and M. L. Spearman. 2001. *Factory Physics: Foundations of Manufacturing Management*. Irwin McGraw-Hill: New York, NY.

King, B. R. 2007. "Virtual Teams Should Have Real Meetings." Community Post, e-newsletter of PMI, October 26.

Lillibridge, E. M., 1998. *The People Map: Understanding Your People and Others*. Lilmat Press: Lutz, FL.

Martin, D. D. 1986. "Motivation and the Developing Engineering Manager." *Journal of Management in Engineering—ASCE*, October, pp. 246–252.

Martin, P. and K. Tate. 1999. "Climbing to Performance." *PM Network*, June, p. 14.

McGregor, D. 1960. *The Human Side of Enterprise*. McGraw-Hill: New York, NY.

McQuillen, Jr., J. L. 1986. "Motivating the Civil Engineer." *Journal of Management in Engineering—ASCE*, April, pp. 101–110.

Mole, J. 1988. *Management Mole: Lessons from Office Life*. Bantam Press: London.

Norton, R. 1994. "New Thinking on the Causes—and Costs—of Yes Men (and Women)." *Fortune*, November 28.

Parkinson, C. N. 1957. *Parkinson's Law*. Ballantine Books: New York, NY.

Patterson, K. J. 1991. "Organizations: The Soft and Gushy Side." *The BENT*. Tau Beta Pi, Fall, pp. 19–21.

PM NETWORK. 2006. "Speak Up: People Spot Project Problems—They Just Don't Say Anything." October, pp. 6–8.

Raudsepp, E. 1978. "Why Managers Don't Delegate." *Chemical Engineering*, September 25, pp. 129–132.

Thompson, J. W. 1996. "Engineers Don't Always Make the Best Team Players." *Electronic Engineering Times,* Sep. 30, p. 124.

Tompkins, J. 1998. *Revolution: Take Charge Strategies for Business Success*. Tompkins Press: Raleigh, NC.

Walesh, S. G. 2000a. "Agenda Item: Good News." *Indiana Professional Engineer*. May/June.

Walesh, S. G. 2000b. *Engineering Your Future: The Non-Technical Side of Professional Practice in Engineering and Other Technical Fields-Second Edition*. ASCE Press: Reston, VA.

Walesh, S. G. 2004. *Managing and Leading: 52 Lessons Learned for Engineers*. Lesson 28: "Caring Isn't Coddling," ASCE Press: Reston, VA.

Wikipedia. 2011a. "DISC Assessment." (http://en.wikipedia.org/wiki/DISC_assessment), May 7.

Wikipedia. 2011b. "Myers-Briggs Type Indicator." (http://en.wikipedia.org/wiki/Myers-Briggs_Type_Indicator), May 7.

ANNOTATED BIBLIOGRAPHY

Carnegie, D. 1981. *How to Win Friends and Influence People*. Simon and Schuster: New York, NY. (A classic book, first published in 1937 and subsequently updated, that according to the original author addresses "the fine art of getting along with people in everyday business and social contacts.")

Jansen, J. 2006. *You Want Me To Work With Who?* Penguin Books: New York, NY. (Offers advice for working with difficult people. Bases the advice on one or more of these 11 keys: confidence, curiosity, decisiveness, empathy, flexibility, humor, intelligence, optimism, perseverance, respect, and self-awareness.)

Sandberg, J. 2006. "Office Tormentors May Appear Normal, But Pack a Wallop." *Wall Street Journal*, July 11. (Notes that "workplace humiliation" is being "driven underground" because of liability fear and other factors. As a result, the humiliators are much more sophisticated relying on tactics such as condescending tone, facial expressions, no or slow response to email, and, in meetings, ignoring some attendees. Personal irritants based on a 2006 survey by Randstad USA, a temporary-staffing firm, and listed in order of decreasing dislike: condescending tones, public reprimand, micromanaging, loud talking, and cell phone ringing.)

Walesh, S. G. 2004. *Managing and Leading: 52 Lessons Learned for Engineers.* Lesson 51: "The Two Cultures: Bridging the Gap," ASCE Press: Reston, VA. (In the spirit of developing relationships, this lesson suggests that what have been traditionally called the literary culture and the scientific culture increasingly share common interests. Examples are valuing communication, recognizing the importance of creative work, and appreciating the need to understand each other and work together.)

Whitten, N. 2004. "Yesterday, Today and Tomorrow." *PM Network*, p. 23, August. (Suggests that we daily ask ourselves two questions. What are the top three things we did that "made a difference for the best?" What are the top three things we did, or failed to do, that "made a difference for the worse?" Then learn and change behavior accordingly. These questions can be asked within the context of our goals.)

EXERCISES

4.1 YOUR PERSONALITY PROFILE: The purpose of this exercise is to enable you to gain further understanding of your personality profile. This can be accomplished by using one or more systematic processes examples of which are the DISC Assessment (2011a), the Myers-Briggs Type Indicator (Wikipedia 2011b), and People Mapping (Lillibridge 1998). As noted in this chapter, understanding of personality profiles – ours and others – helps us understand and deal with our feelings and needs and those of the people around us. The results of this exercise should be confidential, unless you want to share them with your instructor or other trusted persons. Suggested tasks are:

A. Arrange for a personality profile assessment. If you are a student, seek assistance on your campus such as may be available at the placement office or the health center. As an employee, ask for help from human resources personnel. Perhaps you have already been profiled. Then consider doing it again, perhaps with a different method.

B. Review the results recognizing that there are no "right" or "wrong" profiles. They are what they are. Contemplate your profile relative to your desired roles and goals, as discussed in the time management section of Chapter 2. What are your strengths and weaknesses or assets and liabilities relative to your roles and goals and what might you do about it? Also recall recent positive and negative interactions with students, faculty, co-workers, supervisors, and others in light of the added personal insight.

4.2 ORCHESTRATING A MEETING: This exercise will provide you with an opportunity to apply the meeting principles discussed in this chapter. Suggested tasks are:

A. Seek an opportunity to orchestrate a meeting. Possibilities include an ad hoc committee formed by a student group or a special project in your office. Facilitate the planning, conduct, and follow-up of the meeting following the applicable tips presented in this chapter.

B. Reflect on what you learned about yourself and how others responded to you.

4.3 TEAMWORK: This exercise's purpose is to help you test the validity of the three teamwork essentials described in this chapter and to reflect on the forming-storming-norming-performing process. Suggested tasks are:

A. Recall a very successful team experience or a disastrous one. Consider a wide variety of teams you served on such as sports teams, academic teams, and project teams.

B. On reflection, identify the principal reasons for the success or failure. To what extent are they aligned with the three team essentials presented in this chapter?

C. Did your team experience the forming-storming-norming-performing process, or something similar? If so, how did it contribute to the success or failure of the team? Does encountering the storming step necessarily doom a team to failure?

CHAPTER 5

PROJECT MANAGEMENT: PLANNING, EXECUTING, AND CLOSING

List everything you're going to do on little slips of paper
which you then organize into proper sequence.
You discover that you organize and then reorganize the sequence
again and again as more and more ideas come to you.
The time spent that way usually more than pays for itself in time saved...
and prevents you from doing fidgety things that create problems later on.

(*Robert M. Pirsig, in Zen and the Art of Motorcycle Maintenance*)

After suggesting a widely applicable definition of project, this chapter stresses the centrality of project management in business, government, academic, and volunteer organizations. Noting the relevance of project management to the engineering student and entry-level practitioner, the chapter describes the process for preparing a project plan. The chapter then discusses project execution, concludes with project closure, and serves as the basis for the next chapter, which continues the project management theme.

PROJECT BROADLY DEFINED

The Project Management Institute (PMI 2008), the global project management professional society, defines a project as "a temporary endeavor undertaken to create a unique product, service, or result. The temporary nature of projects indicates a definite beginning and end. The end is reached when the project's objectives have been achieved or when the project is terminated because its objectives will not or cannot be met, or when the need for the project no longer exists."

This definition suggests the wisdom of focusing on the end result, which is the reason to undertake a project. While the majority of tasks you undertake in a project may not be new, the desired deliverable, that is, the product, service, or result will be unique. In other words, the basic tasks in conducting a project (e.g., collecting data, setting up a computer model, estimating costs, interacting with regulatory personnel) are not new. What's new is the context—client, owner, customer, stakeholders, team members, constraints, and the physical and socio-economic-environmental situation.

Examples of technical projects are preparing a proposal, collecting data, developing a plan, writing a report, preparing plans and specifications, manufacturing a product, constructing a structure or facility, and carrying out a training program. Projects typically involve the cooperative efforts of two or more people. This team effort implies communication and coordination—challenges that can be met with the assistance of material presented in this chapter and in Chapters 3 and 4. Although this chapter discusses ways to more effectively manage engineering and other technical projects, many of the ideas and much of the information presented are applicable to other professions and to projects you may lead or participate in for your community, religious group, club, or other organization.

PROJECT MANAGEMENT DEFINED

PMI defines project management as (PMI 2008) "the application of knowledge, skills, tools, and techniques to meet the project requirements." With its emphasis on meeting project requirements, this definition is aligned with the definition of quality underlying Chapter 7 in this book. That definition of quality is, simply put, meeting all requirements. Accordingly, we could also define project management as providing quality.

A common implicit or explicit feature of project management definitions is simultaneously satisfying deliverable, schedule, and budget expectations. Figure 5.1 illustrates the challenge of being "pulled" in three directions at once. The three expectations are often in conflict. For example:

- Providing all deliverables may delay the project completion date
- Meeting the schedule may require inefficient efforts and "burn up" the budget
- Staying within the budget may tempt the project manager to omit some deliverables and, therefore, fall short on satisfying requirements

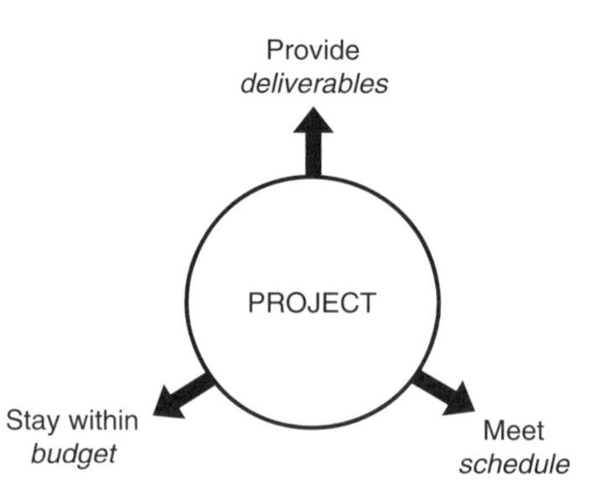

Figure 5.1 Projects are pulled in three usually conflicting directions.

Can all three expectations—deliverables, schedule, and budget—be satisfied? Yes, through careful project management. Project management competence, that is, having the knowledge, skills, and attitudes necessary to consistently provide deliverables on schedule and within budget, is valued in the private, public, academic, and volunteer sectors. Society rewards those who can get things done, make things happen. You, as early as your college days, can begin to build that greatly-appreciated competence by studying and applying the principles and procedures presented in this chapter.

THE CENTRALITY OF PROJECT MANAGEMENT

Project management is the process by which an organization's resources are marshaled to deliver the quality products and services expected by many and varied internal and external clients. As noted and as discussed in Chapter 7, quality is defined as meeting all project requirements. The manner in which an organization manages its projects is the key to developing and retaining clients, owners, and customers. For businesses, project management is the key to profitability and, in the government, academic, and volunteer sectors, project management is necessary to staying within budgets. Essentially everyone in an organization, whether in the private, public, academic, or volunteer sector, works on or is at least indirectly involved in projects. Each person can contribute to the successful completion of projects and derive satisfaction from the team's and organization's achievement.

Whether they are large, small, sophisticated, or basic, all projects require careful project management. Successful project management propagates throughout an organization thus significantly contributing to an organization's success. Conversely, mediocre or ineffective project management has widespread negative and sometimes devastating impacts on an organization. These negatives include, but are not limited to, alienation and/or loss of clients, owners, and customers; litigation; and diminished reputation. Accordingly, project management should be a major, if not the principal, focus of an organization's energies.

In spite of the critical nature of project management, one global survey (PMI 2006) revealed that only 47 percent of "project professionals" have some formal training in project management. Consistent with this, engineers and other technical professionals have traditionally received little or no project management instruction as part of their education.

Some engineering employers meet this challenge by providing project management education and training. Others prepare project management guidelines (e.g., MSA Professional Services 2004, Zipf 1999) and some convene periodic meetings where project managers exchange ideas and information. However, these practices are far from universal in that a large fraction of technical organizations do not have an on-going, systematic approach to preparing and supporting their project managers. For example, studies concluded that only half of U.S. engineering and architectural firms provide project management education and training (Civil Connection 2003) and only 30 percent of corporations worldwide have formal project management career paths (Greengard 2007). If you are using this book and/or similar resources because you are studying project management as part of your engineering or similar education or within your employer's program, you are fortunate.

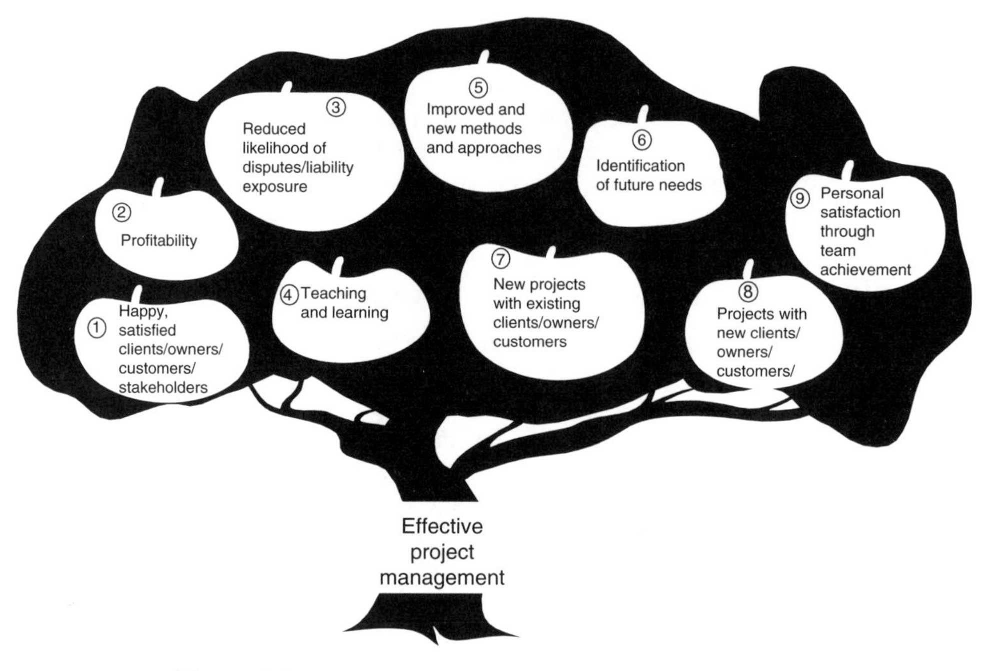

Figure 5.2 Effective project management yields nine fruits.

Back to the centrality of project management in all types of organizations. Figure 5.2 illustrates how effective project management yields nine fruits. Nurturing and harvesting these results is the key to a thriving, as opposed to just surviving or even dying, organization.

Each of the results of effective project management may be briefly described as follows:

1. ***Happy, satisfied clients, owners, customers, and stakeholders:*** Effective project management requires achieving quality, that is, meeting all requirements. Project management is the activity closest to an organization's clients, owners, customers, and stakeholders. They immediately and continuously receive the results, positive or negative, of the way projects are managed.
2. ***Profitability.*** Projects are profit or loss generators in the private sector. Similarly, in the government, academic, and volunteer sectors, the manner in which projects are managed is a major determinant of staying within budgets.
3. ***Reduced likelihood of disputes and liability exposure:*** Refer to Chapter 11, "Legal Framework," for discussion of ways in which individual engineers and their organizations can encounter legal problems and for suggested practices that reduce the likelihood of such problems. Note that all of this is closely connected to managing projects.
4. ***Teaching and learning:*** There is no "make believe" here and, therefore, everyone, especially younger personnel in business, government, academic, and volunteer sectors can be taught and can learn while working on projects.

Individual and corporate technical and non-technical capabilities can be markedly enhanced during a project. Experiences—good and bad—of senior personnel can be shared with others within the project management forum. In contrast, skepticism and even cynicism will thrive in the project management arena, especially among the younger personnel, if they are denied the benefit of learning from senior individuals and if they see inconsistencies between what senior personnel say and do.

5. ***Improved and new methods and approaches:*** Project management often encounters challenges which, in the spirit of "necessity is the mother of invention," leads to improved or new tools and techniques. The resulting advancements can be used on subsequent projects to the benefit of the organization and those it serves.
6. ***Identification of future needs:*** As a result of contributing to projects, which are the interface between the service organization and its clients, owners, customers, and stakeholders, alert project team members become aware of the evolving needs of the organizations they are serving and of internal changes required to satisfy those needs. Following through and meeting those needs enhances a business, government, academic, or volunteer organization's capabilities and reputation.
7. ***New projects with existing clients, customers, customers, and stakeholders:*** Because they are pleased with the results of recent and current projects and because clients, owners, customers, and stakeholders have additional needs, they contract with the serving organization to provide those services. Effective project management earns the trust of the individuals and organizations being served, which is a vital part of earning the privilege of serving them again. Refer to Chapter 14 for a discussion of marketing, noting, in particular, the importance of earning the trust of individuals and organizations as part of the process of having the privilege of serving them.
8. ***Projects with new clients, owners, customers, and stakeholders:*** Satisfied clients, owners, customers, and stakeholders share their successful experiences with their counterparts in other organizations and are often willing to provide formal references on behalf of your organization. Accordingly, your organization has the opportunity to serve other entities. Dissatisfied clients, owners, customers, and stakeholders also share their experiences often to the detriment of service providers who mismanage projects.
9. ***Personal satisfaction through team achievement:*** As discussed in Chapter 4, the most successful teams are those that share and commit to a goal or vision; are composed of individuals having the appropriate, diverse knowledge, skills, and attitudes; and create a trusting communicative atmosphere. A project team with those characteristics is very likely to be successful one result of which is personal satisfaction.

Organizational health is determined principally by its ability to manage projects, because project implementation and delivery of the results affect so many of an organization's vital interests, functions, and relationships. See Walesh (1996) for further discussion of the centrality of project management. The project management tree will flourish and bear much fruit for an organization and its members if they carefully cultivate and care for it. The continuing harvest will yield fruits like those illustrated in Figure 5.2 and described above.

Incidentally, project management can be viewed as an application of the engineering method: diagnose situation, define problem, develop alternatives, select a course of action, and implement it. When project management is presented this way, it may be more likely to engage the interest of engineers and other technical persons who tend to have a technical and analytic bent.

RELEVANCE OF PROJECT MANAGEMENT TO THE STUDENT AND ENTRY-LEVEL TECHNICAL PERSON

You may concur with the need to place a high priority on project management, but as a student or an entry-level engineer or other technical person, you may not see the relevance to you. After all, you will not be managing projects during your first year or so of employment. In fact, your employer may require licensure as an engineer, architect, or other professional as a condition of managing projects. You are partly correct in this assumption. If you are in the very early years of employment with an automobile manufacture, you will not manage the design of the next sports car. But, you may be part of a team responsible for designing a component of the vehicle. If you are in your first year on the staff of a consulting engineering firm, you will not manage the design of an airport, but you may be responsible for a small portion of it. No matter how small your "piece of the pie," it will constitute a project, given the broad definition of project presented at the beginning of this chapter. That is, you are very likely to be subject to deliverable, schedule, and budget expectations.

Manage your mini-project effectively, considering the ideas and information presented in this chapter. As a result, you will increase the likelihood that your part of the overall project will be completed on time, within budget, and in accordance with functional and service expectations. Besides short-term, project-specific benefits, your commitment to smart project management will establish you as someone who makes good things happen. That desirable reputation will lead to more challenging project management assignments and other growth opportunities. Colleagues who notice your success will inquire about and want to emulate your approach to project management. As a result, your efforts will have a positive ripple effect on the organization. The ability to change the behavior of others by example is one aspect of leading.

If you are a student, you also have opportunities to manage projects. Examples are that next big individual or team assignment in one of your classes, the fund raising project being conducted by you and other members of campus organization, and your capstone project.

Is the preceding, optimistic scenario realistic? Yes, because most organizations are project-intensive and because effective project management is one of the highest priority needs in those organizations. You can help to fill that need to your and your organization's benefit by applying the principles and tools described in this chapter. "Project managers fall into three basic categories," according to management consultants Sunny and Kim Baker (1998), "those who watch things happen, those who make things happen, and those who wonder what happened." Be the one who makes things happen on your project, no matter how small.

PLANNING THE PROJECT

All Projects Are Done Twice

In my view, all projects are done twice. The "smart way" is to first think through, then do. Or as World War II air ace Eddie Rickenbacker said, "I can give you a six-word formula for success: Think things through - then follow through." Sounds simple, doesn't it? The "not so smart way," or the "dumb way," of doing projects twice is, first, just start executing. After all, isn't activity progress? Second, rework or redo substantial portions.

Too many projects are done twice the "dumb way." The first time though they, or major portions of them, are done wrong because of poor or no planning. Such waste and the associated frustration, loss of clients and customers or constituents, budget overruns, and, for businesses, low or no profitability, can be largely avoided by the project management philosophy, "plan our work" and "work our plan."

This productive way of doing a project twice is illustrated in Figure 5.3. Think in terms of cumulative resources used, which are represented on the vertical axes of the

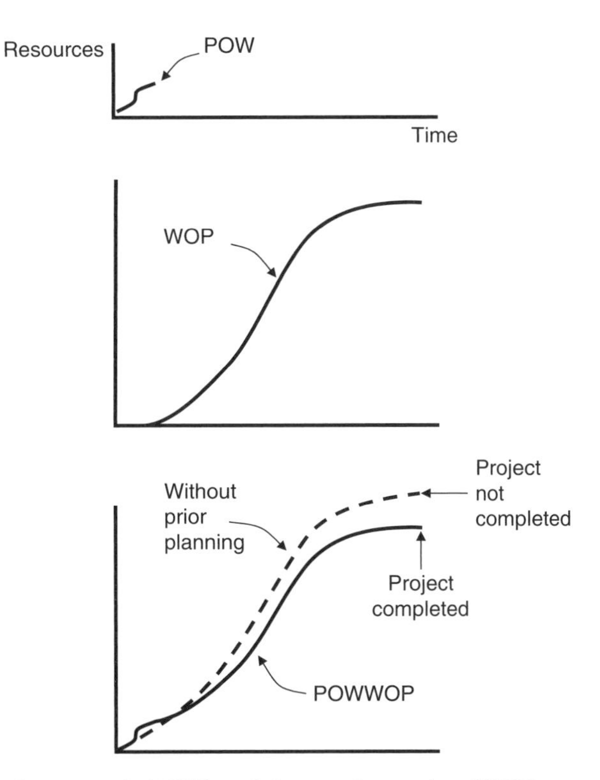

Figure 5.3 The plan or work (POW) and then work our plan (WOP) process is the "smart" way to do a project twice.

three graphs, versus time on the horizontal axes. As illustrated by the solid curve in the first graph, the project manager leads the thinking through effort as the team prepares a project plan (PP). This is the plan our work (POW) part. I'm advocating investing a small amount of time and resources, upfront, to POW.

As shown by the solid curve in the second graph, the resulting PP is used, that is, we work our plan (WOP). Note the typical S-shaped curve of cumulative resource utilization versus time. As suggested by the solid curve in the third graph, which is the sum of the above two solid curves, the project is completed on time and within budget because of the plan our work (POW) and then work our plan (WOP) approach.

In contrast, and as illustrated by the dashed curve in the third graph, absent the "think things through - then follow through" process, as reflected in the PP, the project fails. That is, more resources than necessary are used throughout the project and the project is not completed at the original scheduled completion date. Another way of looking at this: you get a return on investment (ROI) when you plan your project before doing your project, when you apply "the principle that all things are created twice. There is a mental or first creation, and a physical or second creation" (Covey 1990). Simply put, prior proper planning prevents poor performance.

The Project Plan: Introduction

The PP is the mental or first creation. PMI (2008) defines a PP as "A formal, approved document used to guide both project execution and project control. The primary uses... are to document planning assumptions and decisions, to facilitate communication among stakeholders, and to document approved scope, cost, and schedule baselines." The PP is the documented roadmap of the project prepared before the project gets underway. And recall, for purposes of this chapter and book, that project is broadly defined.

Clearly, a PP is not something "in project manager's head." Instead, it is a written, hopefully, consensus product prepared under the leadership of the project manager. A PP is not necessarily a single document. It may be a place or places where project team members and possibly the client, owner, customer, and stakeholders can go, such as a hard file or a project website.

The term PP is used in this book because it is preferred by the PMI, the "gold standard" of project management professional societies. However, other names include project delivery plan, project execution plan, project implementation plan, project management plan, project operations plan, project work plan, task plan, and technical operating plan. What we call it is not important; preparing and using it is.

As an aside, given the apparent value and number and variety of names, you would think PPs, or whatever they are called, would be common. That is not my experience. More broadly, a global study of Information Technology (IT) professionals concluded that only 22 percent of the surveyed professionals reported that their organizations effectively or very effectively used PPs (*PM NETWORK* 2006). Of course, this is for IT operations. Perhaps other types of organizations, such as engineering entities do much better! My experience suggests otherwise.

Consequences of Poor or No Planning

Projects fail largely at the beginning, that is, because of no or poor planning. These failures can take various forms such as:

- Structural or functional failure—because of errors or omissions
- Little or no profit—in the private sector
- Going over budget—in the private, public, academic, or volunteer sectors
- Missing the completion date—in all sectors
- Client, owner, customer, and/or stakeholder alienation in all sectors
- Damaged personal and organizational reputations in all sectors
- Litigation in all sectors

Why go there? Mark Twain seemed to be advocating the POWWOP theme when he said "The secret of getting started is breaking your complex overwhelming tasks into small manageable tasks, and then starting on the first one." Breaking your "complex overwhelming tasks" into "small manageable tasks" is one element of a PP.

Views of Others

"A good beginning is half the work," according to an Irish proverb and Benjamin Franklin, scientist and statesmen, noted that "By failing to prepare, you are preparing to fail." Arthur Bloch, author and television producer, says "If your project doesn't work, look for the part that you didn't think was important." The cartoon character Winnie the Pooh weighs in with "Organize what you do before you do something, so that when you do it, it's not all mixed up." And finally, consider the two millennia-old advice of the Greek philosopher Plato: "The beginning is the most important part of the work."

The Project Plan Avoidance Syndrome

If project planning is so logical and the consequences of not doing it so dire, what are reasons not to do it? Kent Cori (1989), writing tongue-in-cheek, offers these three "reasons" some of us might have for not preparing PPs:

- **Solving problems as they arise is challenging and satisfying:** You, as project manager, may find some satisfaction in this macho "cross that bridge when you come to it" approach. However, your overall project performance will be much better with the more cerebral "cross that bridge in your mind" before you, team members, client, customer, owner, and stakeholder come to it. Identify the bridges, determine how you will cross them, and reflect that in the PP.

- **The necessary time is not in the budget:** If so, why do we always find the time to rework or redo what we did not get right the first time?
- **Avoiding accountability:** This is a selfish "reason" for not preparing a PP because, while the project manager may reduce his or her accountability, the project is more likely to fail to everyone's detriment. PPs are a means to monitor a project manager's effectiveness as well as a means by which the project manager can evaluate the effectiveness of his or her team members.

Daniel Gilbert (2006) tells us that the human being is the only living creature that thinks about the future. Let's use that asset!

Preparing the Project Plan

If you are the project manager, you should take the lead in preparing the PP. Prepare the first draft of the PP by drawing on sources, such as knowledge of the following, which are typically available as a result of the "ramp up" to a project: project type, project location, team members, the agreement/contract, and, of course the client, owner, customer, and stakeholders.

Two possible concerns arise in preparing a PP. One is do you have enough breadth—have you covered all the bases? The other is do you have enough detail—have you drilled deeply enough? Don't let perfect be the enemy of good. There is no such thing as a perfect PP; create a good one and get on with the project.

Involve your project team. An engineer at one of my clients said that, when working on projects, he often does not know the framework and the overall objective. By collaborating with the project team in preparing the PP, your team members will understand the project framework and objectives. Team members are typically an excellent source of ideas on ways to improve your draft PP.

Most business, government, academic, and volunteer organizations have "gold," sometimes not mined, "beneath their feet" in the form of experts. These experienced personnel understand client, owner, customer, and stakeholder requirements or how to determine them. Early participation by these seasoned individuals leverages their knowledge. Use you most experienced personnel in a prospective, preventive mode and not in a retrospective, remedial mode. While these veterans can usually find deficiencies near the end of a project they can also offer advice, at the beginning of a project, to avoid those deficiencies. The cost of prevention will almost always be less than the cost of remediation.

After receiving input from team members and experts, revise the draft PP and consider sharing all or portions of it with your client, owner, or customer. This is a very effective way of confirming that you and he or she are aligned. With his or her concurrence, you may want to share the next draft of the PP, or portions of it, with some stakeholders. For example, your project may eventually need a permit from a state regulatory agency. Consider sharing all or portions of the draft PP with them. This informs regulatory personnel about the likely permit application and provides them with an opportunity to offer early advice that might expedite the later permitting process.

Principal Project Plan Elements

Drilling down further into the PP preparation process, two questions naturally arise:

- What are the elements of a PP?
- What is the order in which the elements are created?

Let's answer the second question first. Clearly, once we have a PP, we WOP by proceeding logically and achieving the objectives. However, we can't WOP until we have a plan to work! Therefore, we must first POW. If you are the project manager, I suggest that you plan the project in reverse, from right to left, as shown in Figure 5.4. Start with your objective and work backward in order to create the needed PP. As succinctly stated by Stephen Covey, "Begin with the end in mind."

Following the backward or reverse idea, start by articulating the project's objective. Then draft, in approximate order, the other nine elements. Let's call these the "Top 10" elements of a PP. Numbers indicate the approximate order in which you and your team, with possible assistance from others, should do them, that is, prepare them. However, expect iterations. For example, assume you draft Element 5, Milestones/Schedule and move through drafting Element 6 and are drafting Element 7, Resources/Budget. Now you find that you can't engage sufficient resources to achieve the milestones/schedule. Therefore, you may need to go back to Element 5, as shown by the dotted line in Figure 5.4, and extend the schedule.

Having mentioned preparing the PP by beginning with the projects objectives and working backwards and having referred to the "Top 10" PP elements, consider three caveats. First, while I believe you must "begin with the end in mind," that is, start the PP by clearly stating and seeking concurrence on the objectives, you do not necessarily then have to prepare the PP elements exactly in the indicated order. Use your judgment. Second, while I identified what I view as the "top 10" PP elements, you will want to tailor PP elements to your project. Again, use your judgment. Thirdly, after

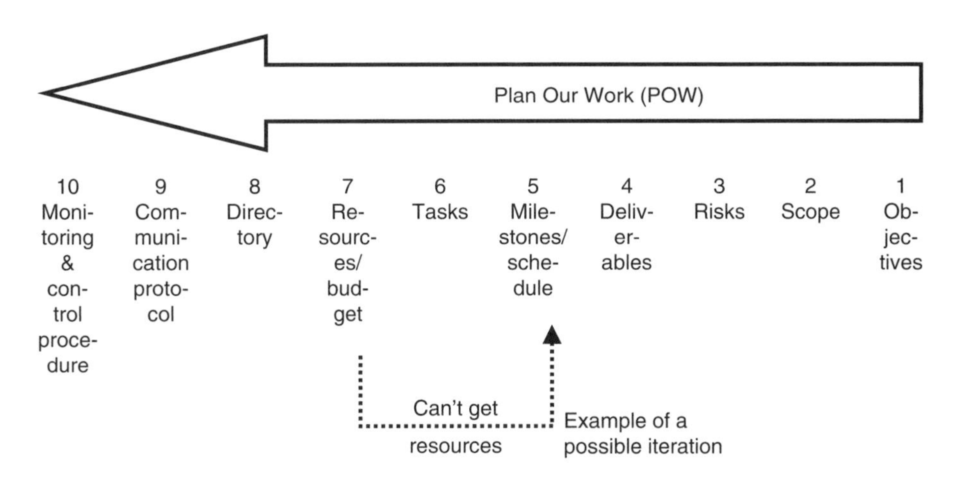

Figure 5.4 Plan the project by beginning with the project objective and working backwards.

describing the "Top 10" PP elements, ten additional elements are named and very briefly described in this chapter. Some of these may be appropriate for your project. Once again, use your judgment.

Use the PP preparation process to get team members "on board." Some of them may be hearing about "your" project for the first time. As a workshop person said to me "There have been projects that I have worked on that I only know the portion I am working on and not the big picture." Avoid this.

Element 1: Objectives – What Do We Want to Accomplish?

Start preparing the PP by writing an objectives element that gets everyone "on the same page," that is, to use PMI language, the product, service, or result your team is going to provide. "Everyone" includes your team members. They will benefit from understanding "how we got here" and "where we are going." "Everyone" could also include the client, owner, customer, and/or stakeholders. Provide project background in the form of history and the current situation. Draw on proposal and intelligence gained from interaction with client, owner, and stakeholders. If one project deliverable is a report, this Objectives element of the PP could become a section in the report. Write it once—use it twice—leverage your effort.

Element 2: Scope – How Are We Going to Do It?

PMI (2008) defines project scope as "the work that must be performed to deliver a product, service, or result with the specified features and functions." In other words, the project scope describes what you and your project team will do, but not in operational detail. The Scope element of a PP typically already exists within the contract or agreement. Now is the time to review the project scope and think about what will be needed to make it operational. That is, begin to think about PP elements that have yet to be prepared such as Risks, Deliverables, Milestones/schedule, Tasks, and Resources/budget.

Once again, remember that at this point in the PP preparation process, some of the most uninformed individuals may be members of your project team. Perhaps you've been pursuing or planning this project for months while others "don't have a clue." Use the PP to include, summarize, or reference the Objectives and Scope elements for them. Get and keep everyone "on the bus" and going in the same direction.

Often, discussions of Scope lead to the possibility of uncompensated scope creep (USC), that is, work outside of the initially agreed-upon tasks that is requested or expected without compensation. An understanding of the project's Scope by project team members will reduce the likelihood of internally and externally-driven USC. Preventing and resolving USC, which is an important aspect of project management, is addressed in detail in the next chapter.

Element 3: Risks – What Could Go Wrong?

Risk is defined by PMI (2008) as "an uncertain event or condition that, if it occurs, has a positive or negative effect on a project's objectives." While unexpected positive effects do occur, our concern is with those that would have negative effects. Consider a simple, effective, four-step process for addressing risk. Other risk analysis approaches are available (e.g., see Hulett and Hillson 2006, Rad 2001, Smith 2003, and Turnbaugh 2005). However, the method described here is quick and effective and, if you are a student, this method can be fruitfully applied to your class and other campus projects.

But first, are you, as the project manager, a worrier or do you have one or more worriers on your team? If so, good! Leadership expert John Maxwell (2003) defines worry like this: "Worry consists of creating mental pictures of what you do not want to happen. Confidence is creating mental pictures of what you want to happen." Worrying, as defined by Maxwell, that is, imagining the things that could go wrong is valuable in PP preparation. If you are a supreme optimist, get a worrier on your team. This helps to cover all the bases. Worriers are especially helpful in identifying risks and in contemplating forecasts of what can go wrong. Their forecasts, or perhaps we should call them "fearcasts" (Gilbert 2006), can help you, as project manager, preclude bad things from happening. Listen to the "fearcasts" and then build preventive measures into the PP. Now consider the four-step risk analysis process.

Step 1 – Identify Risks

Use a simple group technique for risk identification. Begin by assembling your project team, or a subset of team members that represents team functions and areas of expertise. Give each person a pad of "sticky notes" and ask them to individually brainstorm project risks and to write them on the "sticky notes," one risk per note. You might focus on stages of a project such as, for an engineering project, planning, preliminary engineering, design, construction, manufacturing, and operations. Examples of risks that might be listed by individuals are: may have shallow bedrock on the site, owner likes lots of meetings, construction materials may be scarce, and our assembly line has never produced this kind of product.

Views of Others

Merlin Kirschenman (2011) offered the following advice of what could go wrong during construction: "All the construction accidents I reviewed that involved structural failures during construction were, in fact, not accidents but predictable events. The applied loads were far greater than the design loads for the temporary structures. Structural components were missing or taken out. On a continuous pour for a cooling tower, the air temperature was much less than assumed in the design so the concrete did not have the assumed strength and failed. Crane booms were overloaded... All of these so called accidents were predictable events [because] someone did not take care of the details." Again, what could go wrong on your project? Now is the time to identify those risks.

Step 2 – Prioritize Risks

Create two axes, like shown in Figure 5.5, on a whiteboard, newsprint, or other surface easily visible to the team that is conducting the risk analysis. The vertical axis is the

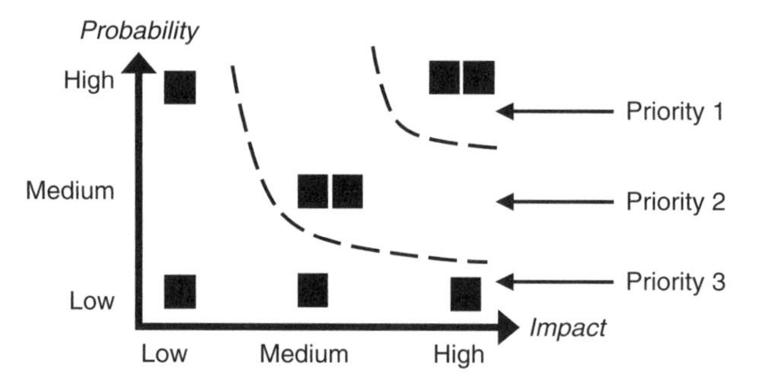

Figure 5.5 The high impact and high probability risks have the highest priority.

probability that a specific risk will occur and the horizontal axis is the impact, if the risk does occur. Take the "sticky notes," one at a time and, using group consensus or group voting, place each one in the agreed-upon portion of the quadrant. Now begin to focus on high probability-high impact risks. These are in the upper right of the quadrant. To do this, you might segment the quadrant, as shown in Figure 5.5, into Priority 1, Priority 2, etc. zones.

Step 3 – Develop Responses to the High-Priority Risks

Consider four different categories of response options for dealing with the high-priority risks (Washington State Department of Transportation 2011). Avoidance is the first category. Develop the PP to eliminate the risk and thus protect the project from its impact. An example is to decline to participate in part of a project, such as the geotechnical investigation in an engineering project.

Transference is the second category. Seek to shift the consequence of a risk to a third party. While this shifts at least some of the ownership and responsibility for the risk's management, it does not eliminate it. An example is to subcontract the geotechnical work.

Mitigation is the third category. Mitigation seeks to reduce the probability and/or impact of a risk. Early action to reduce the probability of an event occurring, or early action to reduce its impact, is more effective than trying to repair the consequences once it has occurred. An example is insisting on more subsurface investigations, such as more soil borings, than originally suggested by the client or owner.

Acceptance is the fourth and last category. In this case, the project team decides not to change the PP to deal with a risk. Active acceptance may include developing a contingency plan should a risk occur. With this category, the project team would do the geotechnical work as requested.

Next take a high-priority risk, like scarcity of a particular construction material as shown in Table 5.1. Work through the four categories for dealing with that risk. The result of the team effort is a list of response options, such as those in the third column. Determine the actual or relative cost of each response option as suggested by the items in the fourth column. This third or "develop response" step ends with a selection of one or more responses for each high-priority risk.

Table 5.1 For each higher priority risk, list possible responses, estimate relative costs, and select one or more responses for inclusion in the project plan.

Risk Priority	Risk	Response Options	Costs	Selected
1	Scarcity of a given construction material	Longer schedule	$$$	
		Alternate material	$	X
		Performance penalty for supplier	$$$$	
		Insurance	$$	
Etc.	Etc.	Etc.	Etc.	Etc.

Step 4 – Integrate Responses into the Project Plan

Having decided how you and your team are going to respond to each risk, build those actions into various PP elements. If, for example, you decided to be prepared to use an alternate material, that decision would be reflected in subsequent PP steps, such as Element 6, Tasks, and Element 7, Resources/budget.

You might be thinking, this four-step process for addressing risk is too much, will take too much time. Experience suggests that a project team can work through the process and generate valuable results in an hour. However, assume you decide not to use the four-step process. Then, early in your project—preferably before you have entered into an agreement—please at least get some diverse potential project participants together, perhaps over coffee or soft drinks, and discuss "what could go wrong on this project?"

Element 4: Deliverables – What Will We Provide to the Client/Owner/Customer?

Achieving the agreed-upon Objectives requires Deliverables, that is, a detailed description of what was referred to earlier as "the product, service, or result" your team is going to provide. A wide variety of deliverables are possible such as a report, plans and specifications, meetings, a manufactured product, and a constructed facility. Now is the time to carefully list the project deliverables. Later cross check Deliverables against PP Element 6, Tasks. Avoid doing unnecessary work and/or missing some essential tasks. For the sake of communicating the essential features of Deliverables, consider obtaining examples or templates drawn from earlier projects and sharing them with members of the project team and possibly the client, owner, or customer, that is, the eventual recipient. These might be typical plan sheets, the table of contents of a report, or a format for cost estimation tables.

Element 5: Milestones/Schedule – When Will We Provide the Deliverables?

The Milestones/Schedule element of a PP is very important to clients, owners, and customers. They want certain deliverables at particular times or milestones. However, notice that while you identify milestones at this point in the project planning process, you don't necessarily get into the detailed schedule until two other PP tasks are done, those being Element 6, Tasks, and Element 7, Resources/Budget.

Construction expert Matt Stevens (NSPE 2007) offers this advice about the relationship between planning and scheduling: "People that really have it figured out know that scheduling is the second thing you do. Planning is the first thing you do. If you're going to be a great scheduler, you have to be a superior planner." In other words, determine what you are going to do before being concerned with when you are going to do it. The essence of planning is identifying the tasks that must be done to provide the deliverables. To reiterate, the first time Element 5, Milestones/Budget, is encountered, it establishes preliminary milestones keyed to Deliverables. Then, after Elements 6 and 7 are completed, return to Element 5, revise the milestones as needed, and develop the detailed schedule.

Beside a simple list of milestones or even tasks, two effective ways of presenting a detailed schedule are the bar or Gantt chart and the network diagram with a critical path. Examples of the list, bar or Gantt chart, and network diagram are presented in the next chapter.

Preparing a schedule is challenging. Achieving it even more so. Consider these scheduling tips:

- **Review time:** Include provisions in agreements to guard against client, owner, or customer delays in reviewing draft materials. For example: "The client, owner, or customer agrees to review draft submittals within 10 working days."
- **Resources:** Confirm the availability of resources, whether they be personnel, equipment, or materials. This is Element 7 in the PP.
- **Client, owner, or customer responsibility:** Examples are arranging for access to private property, providing data, and arranging public or other meetings.
- **Risks:** Earlier we worked through a simple four-step risk analysis process. To reinforce the need for at least a rudimentary risk analysis, consider the observation of Nassim Taleb (2007): "The unexpected almost always pushes in a single direction: higher costs and a longer time to completion." He cites an admittedly extreme example. The Sydney Opera House, which was originally projected to cost AU$7 million, opened ten years late, and ended up costing AU $104 million. On the Boston Artery project in the U.S., final costs were twice the initial estimate (Wikipedia 2011). The point: If you are careless with risk analysis, those risks, when they occur, are much more likely to add to—not reduce—your costs and expand—not contract—your schedule. Perhaps you should build a cushion into your Milestones/Schedule.

Element 6: Tasks – What Tasks Need to be Done and in What Order to Provide the Deliverables?

Thinking through all the project tasks, that is, performing them mentally before doing them physically, is the most challenging element of a PP. The purpose of the Tasks element is to identify all the tasks, and their interrelationships, needed to achieve the project's objectives.

The project manager typically has many sources on which he or she can draw for determining tasks. This is especially true if the project manager was involved in pre-project activities such as meeting with an existing or potential client, owner, customer, or stakeholder and preparing a proposal. Examples of task sources are a request for proposal (RFP), proposal, meeting or discussion notes, scope, preliminary budget, experience with this client-owner-customer-stakeholder, and experience with this type of project.

The most experienced personal should at least assist with Element 6, Tasks. In addition to helping you identify project tasks and their interrelationships, ask them to suggest the best methodologies to be used for some tasks. Examples are the computer model to use for determining traffic volumes or the equipment and labor mix for a construction task. Reason: Reduces likelihood of junior or inexperienced personnel applying unnecessarily elaborate or sophisticated tools or techniques and, as a result, incurring unnecessary cost and elapsed time.

The immediately following discussion of Element 7 includes an example table that could be used to list the tasks. A network diagram, which is discussed in the next chapter, provides a means of presenting the tasks as well as their interrelationships.

An added thought about the tasks you will execute in the project. Many of these tasks, while important, are routine—you've done them many times and others probably do them essentially the same way. In contrast, you will creatively perform some tasks. Your approach and/or the result will be unique. Be sure to protect all of these intellectual assets. You may want to use the contract or agreement to remind your client, owner, or customer of your ownership of memoranda, reports, plans, specifications, and other documents. Somewhere, such as in your general conditions or your contract/agreement, define your intellectual property rights (Cannon 2008).

Element 7: Resources/Budget – How Much Will the Project Cost?

If you are the project manager, the next question is: What resources will be needed to carry out the tasks and what will they cost? Resources include personnel with certain knowledge, skills, and attitudes; data and information; materials; equipment; and facilities (Norton 2008). Estimate the cost of labor and the expenses for each task. Often this process has already been started, in an approximate or generalized manner, possibly for preparation of a proposal that was submitted to the then potential client, owner, customer, or stakeholder. Now detail it for these two reasons:

- Inform/remind project team members of one of the important constraints (money) within which they should plan and carry out their tasks.
- Enable you, the project manager, to monitor the budget and take whatever corrective actions may be needed as the project proceeds.

Try to detail the budget down to the personal level. Then aggregate the budget by discipline, department, office, phase, etc. Table 5.2 shows a way to display and share the project Tasks (Element 6) and the budget breakdown (Element 7). You, as project manager, will find this very useful for later monitoring of project performance. The most important feature of this table is the tasks, the responsible person, and the hours available for the task.

Table 5.2 Example format for the work task and budget breakdown.

		Staffing (Categories and/or specific persons)								
		Project Manager		Construction Engineer		Etc.		Expenses		
Task	Lead person/group	Hours	Labor $	Hours	Labor $	--	--	Type	$	Total Cost $
Totals								--		

Table 5.3 Example format for a project directory.

Organization	Person with title and credentials	Telephone		Email	Fax	Address	Project role(s)
		Office	Cell				
Client/owner/customer							
Noitall Engineers							
Constructor/ manufacturer							
Stakeholders							
Etc.							

Personal

A project team member at one of my clients told me that when he works on a project he often does not know how many hours are available for his tasks. Don't do this to your team members. When you ask them to do a task, your description of the task should include the resources—hours and expenses—that are budgeted for the task.

Element 8: Directory – Who Will Participate?

Projects are ultimately all about people, that is, your project team and those you serve. At this point in the PP process, you know the client, owner, or customer and at least some stakeholders such as environmental groups, suppliers, regulators, business associations, and political entities. Create a directory that includes all project participants. Include names, titles, contact information, and roles. Get this information in writing and be accurate. Table 5.3 is a possible format for the Directory. Possibly supplement the table with a project organizational chart. Get it right—errors propagate! Avoid confusion and embarrassment. Some clients, owners, customers, and stakeholders are very sensitive about spelling, titles, and credentials.

Element 9: Communication Protocol – How Will We Collaborate?

Communication protocol includes identifying principal liaisons within each group participating in a project. Those groups are the entities identified in Element 8, Directory. The protocol may also include guidance for project team members about what should and should not be said to the public and others about project deliverables, schedule, budget, and other features. Communication Protocol is not intended to stifle communication but rather to facilitate its occurrence. Formal, meaning in writing, communication protocol is essential on complex projects.

Early on, we welcome, within reason, the input of clients, owners, customers, and stakeholders because that input may help the project meet requirements, that is, provide quality and be able to do so at an acceptable cost. Later on, as illustrated in Figure 5.6, input from clients, owners, customers, and stakeholders has less influence and costs more to accommodate because of decisions that have been made (CII 1994, CII 1995, NRC 2002). The horizontal axis of Figure 5.6 shows the stages of a construction or manufacturing project. The vertical axis depicts the possible influence and the likely cost of responding to input. Therefore, if you intend to consider input from outside of your project team and, in many cases you should because that input helps the project meet wants and needs, facilitate receiving that input early. Recognize that meaningful interaction with various stakeholders is costly and, therefore, responsibility for that interaction and costs for it should be included in the PP.

Element 10: Monitoring and Control Procedure – How Will We Know How We Are Doing Relative to the Project Plan?

If you are the project manager, or are managing a portion of a project, you must monitor the progress of the entire project, or your portion of it. At minimum, you will

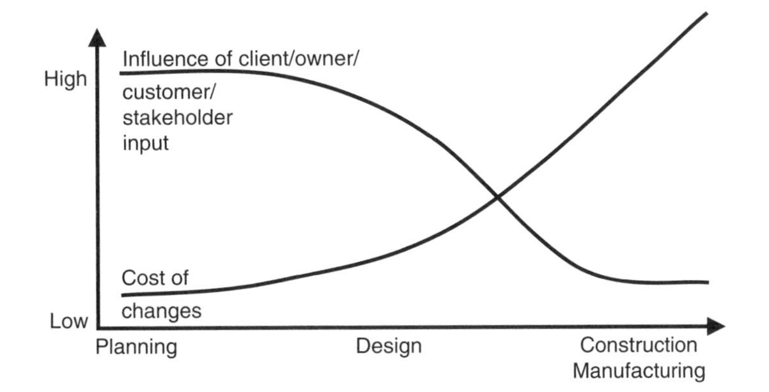

Figure 5.6 The impact of client, owner, customer and/or stakeholder input diminishes and the cost increases as a project progresses.

be concerned with progress on deliverables, schedule, and budget relative to the PP. As you review possible project Monitoring and Control Procedures, consider using the Earned Value Method (EVM). The EVM has other names such as earned value analysis, S-curve method, three-curve method, and integrated budget and schedule method. A detailed treatment of the EVM is beyond the scope of this book. However, useful sources are Anbari (2003), Bettis (2006), Fleming (2010), and PMI (2004).

Begin the Monitoring and Control Procedure PP element by determining the key performance indicators. Besides the usual deliverables, schedule, and budget, those performance indicators might be measures such as strengthening client-owner-customer-stakeholder relationships, receiving federal grants, helping those you serve apply for an award, and developing your and your client, owner, or customer's personnel. Refer back to Element 1, Objectives.

Decide how and when you are going to monitor the performance indicators and then share the indicators and the monitoring system with the project team and possibly the client, owner, customer, and stakeholders. You communicate what is important by what you measure. As noted by Whitten (2003), "inspect what you expect." So, how can you "inspect" and/or "measure?" As project manager, mix and match the approaches, such as the following, to meet your needs: informal one-on-one discussions, periodic accounting reports, project team meetings, and client, owner, customer, and stakeholder input.

Ten Possible Additional Project Plan Elements

The preceding are, in my view, the "Top 10" elements of a PP, subject to the caveats noted earlier in this discussion of project planning. Now consider ten more possible PP elements that might be useful in some project situations. Each is named and briefly described. See Walesh (2010) for details.

- **Assumptions:** If not already part of the agreement or contract, consider including an Assumptions element in the PP.
- **Regulations, Codes, and Standards:** You may want to identify and document, by means of the PP, the local, regional, state, and/or federal regulations, codes, and standards applicable to your project.
- **Style Guide for Written Documents:** If project deliverables include major written documents, prepare a style guide as described in Chapter 3 of this book.
- **Style Guide for Computer-Aided Drafting:** The style guide idea is also applicable to organizational graphics.
- **Quality Control and Quality Assurance Process:** This PP element is likely to be needed if your organization does not provide a quality control/quality assurance (QC/QA) procedure. QC/QA is discussed in Chapter 7.
- **Reference to Written Guidance for Repetitive Tasks:** While each project may be unique, many tasks in projects are typically repetitive, that is, they appear in many projects. Written guidance goes by many other names such as: best practices, bulletins, checklists, guidelines, mini-manuals (Galler 2009), protocols, standard operating procedures (SOPs), and templates (for memos, letters,

reports, data collection, permit applications), and tips. Written guidance, which is a very effective project management and general management tool, is discussed in detail in Chapter 7 of this book.

- **Invoicing Procedure:** As with the QC/QA process, the invoicing procedure may already be fully addressed in your organization.
- **Documentation and Filing Procedure:** This potential element may also be addressed within your organization. If not, describe, in writing, as part of your PP, what is to be documented and where it is to be filed.
- **Subconsultant Management:** If your project will use sub-consultant services, you will want the relationship between you and your "sub(s)" to be as effective as the relationship between you and your client, owner, or customer.
- **Definitions:** John P. Bachner (2008 July) states that definitions are one of the six elements of a good contract and are usually absent. The other five elements, according to him, are project description, scope, general conditions, schedule, and fee.

Project Planning Versus Project Doing

How much time is spent in project planning? There is no "pat" answer to the question because, as they say, "it depends." Consider the following "results based on a survey of 364 project management practitioners at organizations that closed a total of 16,110 projects over 12 months, costing organizations a total of $29.8 billion" (*PM NETWORK* 2006 December):

- Project initiation and project planning accounted for 19 percent of the time (labor) spent on the project. Recognize that, in the private sector, some of the project time spent on project initiation is typically spent prior to the formal start of a project, that is, before a contract or agreement is signed. It is marketing.
- Project execution: 43 percent.
- Project monitoring/controlling: 18 percent.
- Project closing: 20 percent.

These data are not necessarily representative of specific sectors such as engineering. The principal value of the survey results is the suggestion that PP accounts for a significant part of the resources expended on a project.

Review the experience of the Washington State Department of Transportation (2011). This agency addresses the relationship between the level of effort needed to "plan the work" and effort needed to "work the plan." Its experience indicates that it takes about 10 percent of the time and effort to plan the work and 90 percent of the time and effort to work the plan.

Some say they don't have time for all this planning. It is not in the budget. However, consider investing up to ten percent of the overall project effort in the kind of collaborative, broad and deep planning described in the preceding sections of this chapter. As result, you and your team will reduce or eliminate costly omissions or mistakes and the related necessary new work or rework. Your and your team's investment in project planning will yield an attractive return on that investment.

EXECUTING THE PROJECT

Assume that you, as project manager, led the project planning process. You now have the PP, a written, collaboratively-prepared document. You and others, certainly members of your project team and possibly the client, owner, customer, and/or stakeholders, have already obtained value from the PP because of the broad and deep thought that went into preparing it. Now, how do you use the PP? Use it throughout project execution. Don't "leave it on the shelf!" The PP is your project's "road map," "flight plan," or "float plan."

Keep the Project Team on Track

Use the PP to keep the project team "on track." The PP should always be "on the table" at internal meetings and "on the desk" during your discussions with team members. One of the PP uses is holding individuals, groups, and offices accountable. Think of the PP as a "contract" between you and each individual, group, and office. Why? All are involved in preparing the PP, or at least given the opportunity to participate. The PP is typically a dynamic document. Update it as needed and share updates with the project team, and as appropriate, with the client, owner, customer, and/or stakeholders.

As an entry-level engineer, you may receive copies of PPs and be invited to kick-off meetings. If so, you are fortunate to be in an organization that expects and supports sound project management, including the use of PPs. As noted earlier, you should apply sound project management methods to your part of a project, no matter how small. Prepare, using the principles and methods described in this chapter, a written PP for your mini-project. Do so, and you will move from managing mini-projects to managing major projects.

Interact With Client, Owner, or Customer

Use the PP to help you interact even more effectively with the client, owner, or customer. Introduce the PP, or portion of one, at the project kick-off meeting. Use the PP during project progress meetings with those your serve. Sharing the PP with the client, owner, or customer is likely to reduce USC driven by them, that is, being asked to do more than stated in the contract or agreement and not being compensated for it. Reason: Improved communication between the project team and those being served reminds everyone of the scope of the project as set forth in the contract or agreement. As already noted, the important topic of preventing and resolving USC is discussed in the next chapter.

Communicate With Stakeholders

Possibly use the PP, or portions of it, to improve communication with stakeholders such as environmental organizations, community groups, and regulatory agencies. For example, your engineering firm is designing a highway adjacent to a wetland and a local environmental group is concerned about the project's impact. The PP you prepared recognizes the concern and explains how the potential impact will be addressed.

With the client, owner, or customer's approval, share the PP, or a portion of it, with the environmental group.

Monitor Project Progress and Take Appropriate Actions

The project manager has prime responsibility for monitoring the project and taking whatever action may be needed to achieve the project's objectives. Typically, the project manager in a business organization carries out the following monitoring and control tasks, most of which are adaptable to non-business organizations:

1. Tracks the project budget and sub-budgets and corresponding work progress. Notes accidental or other illegitimate labor or expense charges. If illegitimate charges have occurred or if the costs incurred are moving ahead of the products produced, corrective action is needed.
2. Compares tasks completed and milestones achieved to the project schedule, perhaps using periodic updates of a critical path method (CPM) analysis as described in the next chapter. If critical tasks are or soon will be behind schedule, takes corrective actions discussed in that chapter. As noted earlier in this chapter, the project manager may also use the earned value method (EVM) to monitor planned costs, actual costs, and the value of results produced.
3. Remains alert to changes in scope requested by or attributable to the client, owner, or customer, especially those that will increase the cost of doing the project. The preferred remedy to client-driven scope creep is to seek additional compensation commensurate with the additional services.
4. Guards against internally-driven increases in scope. Well-intentioned members of the project team may be tempted to expand the breadth or increase the detail of portions of the project beyond that set forth in the agreement or contract, expected by the client, or in any other way required by the circumstances. Any task can be executed better, but, as discussed in Chapter 7, significantly exceeding requirements does not constitute a quality project while it typically jeopardizes schedules, budgets, and relationships with those being served.
5. Communicates with the key client, owner, or customer representatives with emphasis on empathetic listening. The perception of project progress by those being served should be determined and, if not consistent with reality, corrected. The project manager responds to questions and addresses concerns in a timely fashion. If the project is to be performed by a private organization having one or more marketing personnel who regularly call on clients, owners, or customers, the project manager should ask a marketer, as a third party, to ask the client, owner, or customer representative to share their views on the project's progress.
6. Makes sure all aspects of the project, including meetings and internal and external communications, are being adequately documented. The project manager should take the position that a face-to-face or conference call meeting is not done until documentation is complete. Refer to the "Orchestrating Meetings" section of Chapter 4.

7. Determines the adequacy of internal support services such as drafting, accounting, information technology, and surveying. As appropriate, expresses appreciation for responsive assistance and takes action regarding deficiencies.
8. Stays in touch with sub-consultants and remains informed about their contributions to confirm that the sub-consultants, as part of the project team, are meeting their deliverable, schedule, and budget obligations.
9. Updates the PP as needed and distributes it to members of the project team.
10. Bills the client, owner, or customer, generally in proportion to work completed or in accordance with other provisions in the agreement or contract.
11. Choreographs the project closing as described in the next section.

In carrying out his or her project monitoring and control functions, the project manager is like a juggler who successfully knows the location of and controls many "balls." To do this, the project manager must be organized, assertive, and positive—not careless, passive, or negative.

CLOSING THE PROJECT

Even successful projects could have been done better. Mediocre and failed projects contain the seeds of major future improvements in project management. With respect to mediocre and failed projects, and the good that could come from them, consider the advice of author Napoleon Hill: "Realize, and prove to your own satisfaction, that every adversity, failure, defeat, sorrow, and unpleasant circumstance, whether of your own making or otherwise, carries with it the seed of an equivalent benefit which may be transmuted into a blessing of great proportions." In other words, learn from mistakes.

Accordingly, each project or a sampling of an organization's projects, ranging from the successful to the mediocre or failed projects, should be immediately followed by a post-mortem analysis to determine what can be learned for the benefit of near-future projects. Lessons learned should be documented and widely shared within the organization. Two important components of this post-project review are input from the client, owner or customer and a meeting of the project team. The client, owner, or customer interaction should precede the team meeting so that the team has the benefit of input form those who were served.

Seek External Input

Input from the client, owner, or customer can be obtained in many ways, depending on circumstances. At the informal end of the spectrum, post-project input might be obtained through a private and casual one-on-one conversation. On the other extreme, a formal questionnaire might be sent to the client, owner, or customer. The input method must fit the situation and be based on a sincere desire to view the just-completed project or, more specifically, the services delivered from the client, owner, or customer's perspective as a means of improving the management of subsequent projects.

As an entry-level person, you are not likely to conduct the post-project review with the organization served. But, as a member of a project team, you served internal "clients," that is, members of your organization. Occasionally meet one-on-one with them. Ask for a frank evaluation of the "services" you provided with the goal of doing an even better job the next time. Many individuals are reluctant to volunteer encouragement or criticism, particularly the latter, but will comment on your efforts if you ask.

Conduct Project Team Meeting

After obtaining the external input, the project manager should arrange a meeting of all, or key members, of the project team. The purpose of the meeting is not to determine who was right or wrong, but is to determine what was right or wrong. The group members should resist the natural tendency to dwell on or only discuss negatives. Meeting participants should also celebrate project successes such as developing new approaches; meeting schedules and budgets; and receiving positive client, owner, customer, and stakeholder comments. Possible agenda items include schedule, budget, documentation, internal and external communication, and quality. Members should identify and analyze problems with the idea of avoiding them in future projects. A succinct memorandum or other written result should document the post-mortem meeting, focusing, again, on "what," not "who."

Personal

I assisted an architectural-engineering firm in conducting several lessons learned (LL) workshops. The group documented over 100 LLs. Many were nuances related to their particular organization and clients, not the kind of project management advice found in text and professional practice books. This experience further suggests the value in conducting project post-mortems. Incidentally, the U.S. military calls these After-Action Reviews (AARs). See Whitten (2007) for more ideas.

Leverage the Just-Completed Project

Besides discussing project management lessons learned at the preceding project team meeting, think about the just-completed project in the context of your business, government, academic, or volunteer organization and its existing or potential clients, owners, customers, and stakeholders. Ask questions such as: What additional services might be able to provide to this particular organization, now that you have learned more about them and some of their wants and needs? What other existing or potential clients, owners, or customers might want or need the services provided in the just-completed project? What is the likely life-cycle of the services just provided and what kinds of services are likely to replace them? How can we proudly and professionally tell others about this project?

CLOSURE: COMMON SENSE AND SELF DISCIPLINE

What is essential for successful project management? I interviewed six very experienced project managers (Walesh 1997) and their responses to this question may be summarized as follows: Communicate – communicate – communicate, create a clear and complete mental image before beginning the work, minimize surprises, and maintain intensity. No "silver bullets" here—perhaps just common sense to be converted to common practice through self-discipline.

Why do we never have enough time to do it right,
But always have enough time to do it over?

(Anonymous)

CITED SOURCES

Anbari, F. T. 2003. "Earned Value Project Method and Extensions." *Project Management Journal*, December, pp. 12–23.

Bachner, J. P. 2008. "Six Elements of Good Contracts." *CE NEWS,* July, p. 14.

Baker, S. and Baker K. 1998. *The Complete Idiot's Guide to Project Management*. Alpha Books: New York, NY.

Bettis, N. T. 2006. "Earned Value Management: Manage The Future, Not Just The Past, and Prevent Projects From Trending In A Negative Direction." *CE News*, December, pp. 18–19.

Cannon, H. M. 2008. "Top 10 Items to Include in Your Contracts." *NSPE PEPP Talk-Enewsletter*, September 23.

"Civil Connection." 2003. E-Newsletter from the publishers of *CE News* and *Structural Engineering Magazine*, October 13.

Construction Industry Institute. 1994. *Pre-Project Planning: Beginning a Project the Right Way*. Publication 39–1, The Construction Industry Institute, University of Texas, Austin, TX.

Construction Industry Institute. 1995. *Pre-Project Planning Handbook*. Special Publication 39–2, The Construction Industry Institute, University of Texas, Austin, TX, April.

Cori, K. A. 1989. "Project Work Plan Development." paper presented at the Project Management Institute and Symposium, October, Atlanta, GA.

Covey, S. R. 1990. *The 7 Habits of Highly Effective People*. New York: Simon & Schuster.

Fleming, Q. W. and J. M. Koppelman. 2010. *Earned Value Project Management - Fourth Edition.* Project Management Institute: Newton Square, PA.

Galler, L. 2009. "A Mini-manual Guides Training." e-newsletter, Larry Galler & Associates, January, 25.

Gilbert, D. 2006. *Stumbling on Happiness*. Vintage Books: New York, NY.

Greengard, S. 2007. "A Defined Plan." *CAREERTRACK*, May, pp. 28–29.

Hulett, D. T. and D. Hillson. "Branching Out." *PM Network*, May, pp. 36–40.

Kirschenman, M. 2011. Personal communication. Professor Emeritus, Construction Management and Construction Engineering Department, North Dakota State University, April 19.

Maxwell, J. C. 2003. *Thinking for A Change: 11 Ways Highly Successful People Approach Life and Work*. Warner Business Books: New York, NY.

MSA Professional Services. 2004. *Project Management Guide*. Baraboo, WI.

National Research Council. 2002. *Proceedings of Government/Industry Forum: The Owner's Role in Project Management and Pre-project Planning*. National Academy Press: Washington, D.C.

Norton, J. F. 2008. "Project Patterns: A Discussion of Issues and Opportunities." presented at the Calumet Chapter, Project Management Institute, Merrillville, IN, August 12.

NSPE, 2007. "Want Better Project Schedules? Fix Your Planning Process." *PE*, August/September, pp. 18–19.

Pirsig, R. M. 1981. *Zen and the Art of Motorcycle Maintenance: An Inquiry Into Values*. Bantam Books: New York.

PM NETWORK. 2006. "Governance Gone Awry." *PM NETWORK*, November, pp. 8–10.

Project Management Institute. 2005. *Practice Standard for Earned Value Management*. PMI: Newton Square, PA.

Project Management Institute. 2006. "Deliverables." *PM NETWORK*, December, p. 14.

Project Management Institute. 2008. *A Guide to the Project Management Body of Knowledge – Fourth Edition*. PMI: Newtown Square, PA.

Rad, P. F. 2001. "From the Editor." *Project Management Journal*, June, p 3.

Smith, P. G. 2003. "A Portrait of Risk." *PM Network*, April, pp. 44–48.

Taleb, N. N. 2007. *The Black Swan: The Impact of the Highly Improbable*. Random House: New York, NY.

Turnbaugh, L. 2005 "Risk Management on Large Capital Projects." Forum, *Journal of Professional Issues in Engineering Education and Practice - ASCE*, October, pp. 275–280.

Walesh, S. G. 2010. *Project Plans: Doing Projects Twice the Smart Way*. CreateSpace: Charleston, SC.

Walesh, S. G. 1997. "Project Managers on Project Management." *Indiana Civil Engineer* November, pp. 9–10.

Walesh, S. G. 1996. "It's Project Management, Stupid!" *Journal of Management in Engineering—ASCE*, January/February, pp. 14–17.

Washington State Department of Transportation. 2011. *Project Management: Online Guide*. (www.wsdot.wa.gov/Projects/ProjectMgmt/PMOG.htm). May.

Whitten, N. 2007. "In Hindsight: Post Project Reviews Can Help Companies See What Went Wrong – And Right." *PM NETWORK*, May, p. 21.

Whitten, N. 200. "Leadership Tips for Promoting Project Success." *PM NETWORK*, August, p. 20.

Wikipedia. 2011. "Big Dig." (http://en.wikipedia.org/wiki/Big_Dig), May 9, 2011.

Zipf, P. J. 1999. "The Essential Project Management Manual." *Journal of Management in Engineering—ASCE*, March/April, pp. 34–36.

ANNOTATED BIBLIOGRAPHY

Beckwith, H. 2003. *What Clients Love*. Warner Business Books: New York, NY. (Offers these additional communication suggestions relevant to projects: use your office space to communicate to clients that they belong and you care about them; occasionally erring is acceptable, but never let your communication suggest you don't care; and apply the rule of three—people tend to be able to remember three ideas, topics, etc.)

Delatte, Jr. 2009. *Beyond Failure: Forensic Case Studies for Engineers*. ASCE Press: Reston, VA. (The failure case studies in this book, which are organized by specialty areas—e.g., statics and dynamics, structural analysis, and fluid mechanics and hydraulics—may stimulate a project manager and his or her team to more thoroughly identify what could go wrong on their project.)

Ericson, N. V. 2006. "Risk Management Is Not Just for Principals." *Structural Engineer*, August, p. 14. (Urges creating an organization wide "culture for risk management" recognizing that everyone can play a role. Tactics include sharing lessons learned, showing examples of real claims, explaining costs of claims, providing claim statistics, role playing, and education/training.)

NSPE. 2006. "Claims Happen." *PE*, NSPE, August/September, pp. 32–33. (Offers these risk management guidelines: clear contracts; liability limitation clauses; client, owner, or customer's litigation track record; excellent communication; quality control; requiring subcontractors to carry insurance; and providing education and training.)

Project Management Institute. 2006. *Government Extension to the PMBOK – Third Edition*. PMI, Inc.: Newtown Square, PA. (States that this document "is specifically designed to speak to the distinctive practices found in worldwide public-sector projects." Incorporates these distinguishing aspects of public-sector projects: public accountability, "cradle-to-grave" life cycle, protracted budget cycle, and legislatively-driven financial processes.)

PMI Today. 2009. "You Can Transfer Your Skills to Another Industry." Supplement, August, pp. 1–8. (This article contends, to quote the article, that "project management skills transcend industries.")

Remer, D. S. and M. A. Martin. 2009. "Project and Engineering Management Certification." *Leadership and Management in Engineering – ASCE*, October, 2009, pp. 177–190. ("This paper summarizes the requirements for eight major project management (PM) certificates.")

Tawresey, J. G. 2006. "The Culture of Managing Risk." *Structural Engineer*, April, p. 14. (Argues that "knowledge and skills are not sufficient to limit claims. The engineer must also practice with the right attitudes..." Goes on to argue that "top of the list" attitudes are thoroughness and curiosity.)

EXERCISES

5.1 PLAN FOR YOUR PROJECT: The purpose of this exercise is to give you practice in preparing a PP for a project you are going to do largely on your own. That is, the project is not going to be a team effort. Suggested tasks are:

A. Select a project you are about to begin. Or perhaps you are in the early stages of a personal project and it is not going well; you are "spinning your wheels." Whether you are a student or a young practitioner, you are surrounded by actual and potential projects. Recall that project is broadly defined in this chapter as a temporary activity subject to deliverable, schedule, and budget expectations. Examples from student life are doing a research paper assigned in one of your classes, finding a summer job, and building your own computer.

B. Prepare the PP using, where applicable, the ten plus PP elements described in the chapter.

5.2 PLAN FOR A TEAM PROJECT: This exercise's purpose is to give you and team members experience in preparing a PP for a project you are going to do as a team effort. Proceed in a manner similar to Exercise 5.1 and as you do so, draw on some of the teamwork, meetings, and delegation advice offered in the preceding chapter.

CHAPTER 6

PROJECT MANAGEMENT: CRITICAL PATH METHOD AND SCOPE CREEP

Thought is behavior in rehearsal.

(*Benjamin Franklin, scientist, statesman, and author*)

This chapter begins by discussing the Critical Path Method (CPM). This tool can be used throughout the project planning, execution, and closing process described in the previous chapter. Setting up and then using the CPM reinforces the think through and then do approach stressed in that chapter. Besides explaining the basic CPM, related topics are discussed such as creating a Gantt chart, updating a CPM, and tips for identifying tasks to be included when applying the CPM. The remainder of the chapter addresses scope creep, especially uncompensated scope creep (USC) which, if not prevented before a project begins or resolved once a project is underway, damages budgets and reduces profits. The theme of the scope creep section is that awareness, during project planning, of the possibility of internally and externally-driven USC can greatly reduce its occurrence and negative impacts during the project.

THIS CHAPTER RELATIVE TO THE PRECEDING CHAPTER

This chapter builds on the preceding chapter. That chapter, using this book's broad definition of project, focused on the essential project management process of planning, executing, and closing a project. This chapter supplements that plan our work–work our plan (POWWOP) approach by presenting a practical project management tool, the Critical Path Method (CPM), and by discussing the common uncompensated scope creep (USC) problem and showing how to minimize it. To repeat an observation offered near the beginning of Chapter 5: Effective project management is one of the highest priority needs in most organizations—whether they be in the business,

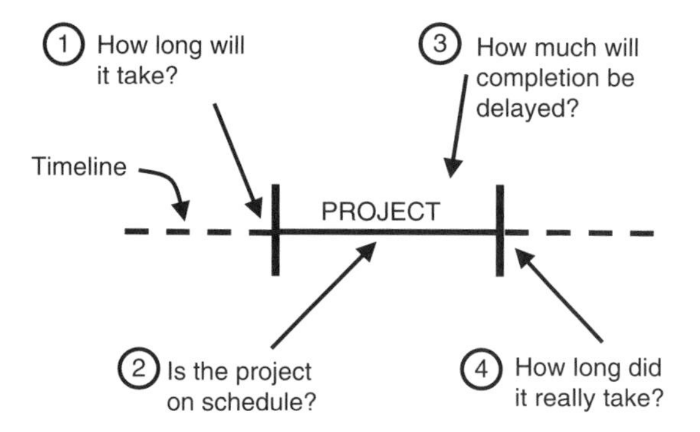

Figure 6.1 The project manager is faced with four questions while planning, executing, and closing a project.

government, academic, or volunteer sectors—and you can help fill that need to yours and your organization's benefit.

THE CRITICAL PATH METHOD

Introduction: The Four Schedule Questions

Project managers typically need answers to the four schedule questions illustrated in Figure 6.1 as they proceed through the POWWOP process. Consider each one.

1. **How long will it take?** That is, when will the project be completed? For example, your supervisor asks you to prepare a conceptual design of a new product to be manufactured by your organization and wants to know how long the design project will take. Many clients, owners, and customers are in a hurry.
2. **Is the project on schedule?** The prudent project manager asks this question often. For example, you and the assistant city engineer are doing a transportation plan for your city. The mayor is preparing for a meeting with the city council and wants to know if the transportation planning project is on schedule.
3. **How much will completion be delayed?** This question arises in situations such as when someone leaves our organization, we can't get expected construction materials or equipment, or a subconsultant fails to provide input on schedule. For example, you are helping manage the construction of a dam designed by your design-build firm. A laborer's strike temporarily stops concrete placement operations for six weeks. The owner wants to know how much the completion of the project will be delayed, if at all.
4. **How long did it really take?** Or more specifically, what factors delayed or accelerated the schedule? What can we learn from the positive and negative events that occurred during this project? Can at least parts of this project serve as a template for some future project? "It's only a mistake if we don't learn from it," according to engineer and author Richard G. Weingardt.

The Critical Path Method (CPM) enables us to answer the four questions. Note that the four questions relate to the Chapter 5 project planning-execution-closing process. That is, first question applies to planning the project, the second and third questions to executing the project, and the fourth question to closing the project.

Alternative Scheduling Methods

To maintain perspective, recognize that you have other tools for the scheduling aspect of projects, that is, for Element 5: Milestones/Schedule introduced in Chapter 5. I am not advocating broad, across-the-board use of the CPM. In some situations, other scheduling tools may be more appropriate. Two common ones are discussed here.

Chronological List

A simple chronological listing of tasks, with the estimated elapsed time for each task, as illustrated in Table 6.1 for the design of a small dam, may be adequate for project scheduling. The most obvious positive aspect of this method is its simplicity.

While potentially useful, a chronological list does not show task overlaps or interdependencies. For example, presumably Task C in Table 6.1, which calls for specification of soil borings, should be done after completion of Task A, which involves site reconnaissance and surveying. However, such interrelationships and overlaps are not shown in the simple chronological listing. Therefore, the elapsed time required to complete a project is not apparent. Nevertheless, sometimes a simple chronological list is all that is needed project scheduling. Frankly, sometimes not even this is done!

Gantt Chart

The Gantt or bar chart method provides a means of showing task overlap, provided that the overlap is known and understood, and provides a "picture" of a project. Four steps are involved, the first two of which are identical to those in the chronological list method: list tasks that comprise the project and estimate the elapsed time for each task. The third step is to estimate the start time for each task, presumably based on the

Table 6.1 The chronological listing of project tasks with elapsed time for each task is the simplest means of project scheduling.

Design of a small dam	
Task	Elapsed time (weeks)
A. Perform site reconnaissance and survey	2.0
B. Draw site map	2.0
C. Specify soil borings	0.5
D. Arrange for and do soil borings	1.5
E. Submit application for preliminary permit	2.0
Etc.	—

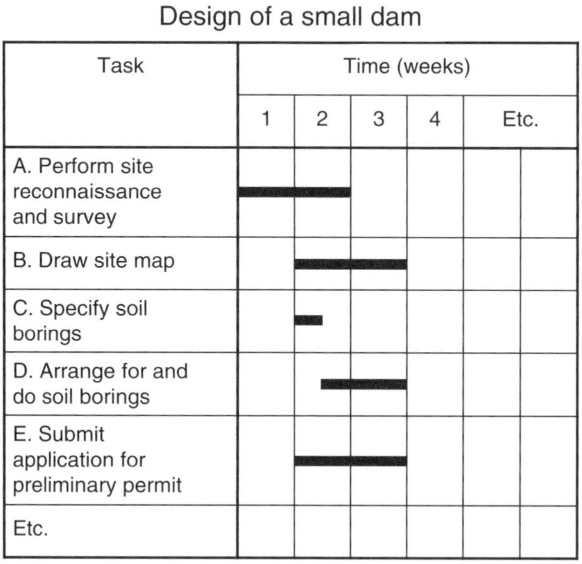

Design of a small dam

Task	Time (weeks)				
	1	2	3	4	Etc.
A. Perform site reconnaissance and survey					
B. Draw site map					
C. Specify soil borings					
D. Arrange for and do soil borings					
E. Submit application for preliminary permit					
Etc.					

Figure 6.2 The Gantt or bar chart shows project tasks and the overlaps among them.

project manager's understanding of the project. The fourth and last step is to draw the bar chart. Figure 6.2 is an example Gantt or bar chart using the first five tasks of the previous example of designing a small dam.

One positive aspect of the Gantt or bar chart method is its graphical nature, which aids the understanding of various users. Both technical and nontechnical personnel generally understand this project scheduling tool. A second advantage of the Gantt chart is that it depicts task overlaps, provided such overlaps are known or can be approximated.

A major disadvantage of the Gantt chart method is that actual interrelationships between individual tasks are not shown. For example, is Task C dependent only on Task B? Second, the Gantt or bar chart fails to identify critical tasks—those tasks that will delay the completion of the entire project if they are off schedule. For example, in Figure 6.2, should Task E, which involves submitting an application for a preliminary dam permit, be started as soon as possible so as not to delay the ultimate completion of the entire project? Or, can it wait? Such interdependency and critical task questions are explicitly addressed by the CPM and, when the CPM is used, one by-product is a Gantt chart as illustrated later in this chapter. Therefore, back to the CPM.

Historic Note

"The first known tool of this type was reportedly developed in 1896 by Karol Adamiecki, who called it a harmonogram. Adamiecki did not publish his chart until 1931, however, and then only in Polish" (Wikipedia 2011a). The chart is named after Henry L. Gantt, an American mechanical engineer and management consultant, who is credited with developing the chart in the period 1910–1915. During the 20th Century, Gantt charts were used on major U.S. projects such as Hoover Dam and Interstate Highway System and it is widely used today (Wikipedia 2011b).

Network Fundamentals

Using the CPM requires understanding network fundamentals. A network is two or more nodes representing interrelated tasks, where the tasks have positive or zero duration, connected by directional branches. Figure 6.3 presents examples of four simple networks. This chapter uses the task-on-a-node format. Another option is the task-on-an-arrow format. Networks are sometimes called "precedence diagrams" because they establish precedence relationships in that "this task must be completed before that task can be completed" or this task "takes precedence, for the time being, over that task."

Network 1 is the simplest network–it means that Task A must be finished before Task B can be started. Network 2 indicates that Tasks C, D, and E must all be completed before Task F can be initiated. In Network 3, Task G must be finished before either Task H or I can be started. Network 4 is presented to note that some tasks have zero duration. For example, Tasks X and Y could be zero duration "Start" and "Stop" tasks. They are sometimes called "dummy tasks"—their purpose is to simply show the beginning and end of the project.

The most important aspect of a network is the connectivity or topology. Relative lengths of branches are irrelevant as is the overall orientation or relative position of tasks in the network. There are, in effect, an infinite number of ways that any network

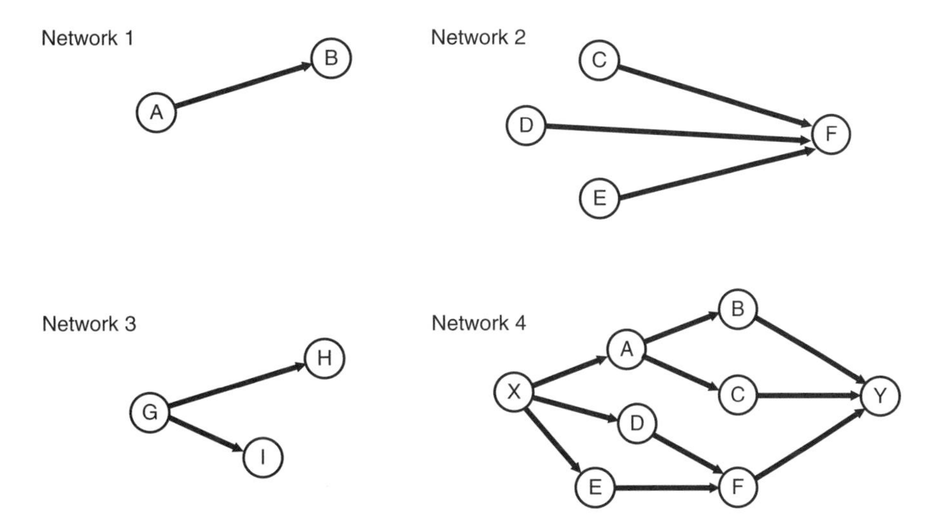

Figure 6.3 Four simple networks illustrate task interdependencies.

can be drawn without changing its connectivity. For example, if Network 2 in Figure 6.3 were rotated 90 degrees counterclockwise it would continue to be the same network.

Critical Path Method Steps

The CPM can be viewed as the following five-step process:

1. ***Determine the tasks:*** This by far the most difficult step. If you've done this particular type of project before, the challenge is lessened. You could go back to the network for an earlier, similar project and edit it to fit the current project. If this is your first time managing a particular type of project, consider getting assistance from someone who has the necessary experience. Determining tasks is not unique to the CPM. I can't imagine any project management tool that does not require defining project tasks. Recall that this is Element 6 in the project plan (PP) preparation process described in Chapter 5 and is also required in the previously-discussed chronological listing and Gantt chart scheduling methods.
2. ***Estimate the duration or elapsed time for each task:*** This step is dependent on availability of resources and outside requirements such as the due date for a grant application, where preparing the grant application is a task or a group of tasks in the project. Again, this step is common with the chronological listing and Gantt chart scheduling methods.
3. ***Identify interdependencies among tasks:*** Step 3 requires an understanding of how projects unfold, that is, how tasks are related. It is also a reality check on Step 1. For example, if a task identified in Step 1 does not "feed into" another task, then the earlier task is not needed. Or if we find that a task could not be started because we would be missing some important information, then we must have omitted a necessary "upstream" task. The first three steps encourage us to carefully think through our project.
4. ***Construct the network:*** This step is mostly mechanical assuming the first three tasks are completed. Software can do Step 4, as explained later in this chapter.
5. ***Determine the project's minimum completion time and the critical path:*** The CP will be defined shortly when the process used to determine it and the minimum completion time are described. Step 5 can also be performed by software.

Historic Note

The CPM was developed in 1956 by the Engineering Control Group, a design and construction unit of E.I. du Pont de Nemours and Company. The method was programmed by the Remington Rand Corporation to run on the UNIVAC computer. The first application of CPM was construction of a $10,000,000 chemical plant in Louisville, KY in 1957. There were 800 tasks in this project

(Dhillon 1987). CPM is now widely used in planning and managing construction projects; however, CPM can also be used for any project consistent with the broad definition of project presented in Chapter 5. A related, but more sophisticated method is the Program Evaluation and Review Technique (PERT) as described by PMI (2008) and Wikipedia (2011c). PERT, which explicitly accommodates uncertainties associated with task durations, is outside the scope of this book.

Example Application of the Critical Path Method

Imagine that you, the project manager, have completed, with the help of others, Steps 1 through 4 for your project. You have determined the eight tasks, A through H shown in Figure 6.4, and estimated task durations in weeks. The network includes zero duration "Start" and "Stop" "dummy" nodes or tasks. You identified, using branches or arrows, interrelationships among tasks. By the way, we are using a very simple network for the instructional purposes. An actual project would have dozens, if not hundreds or even thousands, of tasks.

Purpose

You are using the CPM to, first, determine the minimum elapsed time in which this project can be completed and, second, to identify the critical path (CP). With respect to the first purpose, the minimum time period in which the project can be completed, you should recognize that the initial result may not be acceptable to you, as a service provider, and/or to your client, owner, or customer in that you or they may want the project finished sooner. If that is the case, the CPM can be used to make adjustments so that the project can be completed earlier. More about that later.

The CP included in the second purpose is that sequence of tasks that must not be delayed if the project is to be completed in the minimal amount of time. All tasks

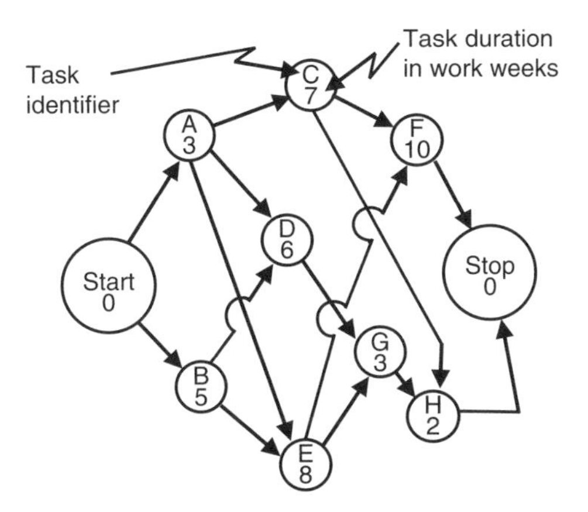

Figure 6.4 Network diagram used for the example application of the Critical Path Method.

included in the network must be done, but all tasks on the CP must be started and completed as soon as possible. To reiterate, if any task in the subset of critical tasks is not started and completed as soon as possible, the project completion time will be delayed.

Determine the Minimum Project Completion Time

The following nomenclature is used so that we can begin Step 5, the last step, in the five-step CPM process:

- EST—Earliest Start Time. The earliest elapsed time, measured from the start of a project, when a particular task can possibly begin.
- EFT—Earliest Finish Time. The earliest elapsed time, measured from the start of a project, when a particular task can possibly be finished.
- EFT = EST + Duration.

Consider the above definition of EST for a task. Let's say that the EST for a particular task is the end of week six. Now note the above definition of EFT for a task and the equation that relates EFT to EST. Back to that task with an EST at the end of week six. If its duration is two weeks, then the EFT of that task is six weeks plus two weeks or the end of the eighth week. Now let's apply these definitions to the example network.

Figure 6.5 shows the EST and EFT for each task. Note the graphic nomenclature, the arrow pointing to the right, which is used when doing the analysis manually. Consider Task A. Its EST is zero because it can be started when the project begins. Its EFT is the end of three weeks because its duration is three weeks. Now focus on Task D. It depends on completion of Tasks A and B. Task A has an EFT of three weeks and Task B has an EFT of five weeks. Therefore, the EFT of five weeks governs and determines the EST of Task D. That is, the earliest Task D can be started is the end of the week five.

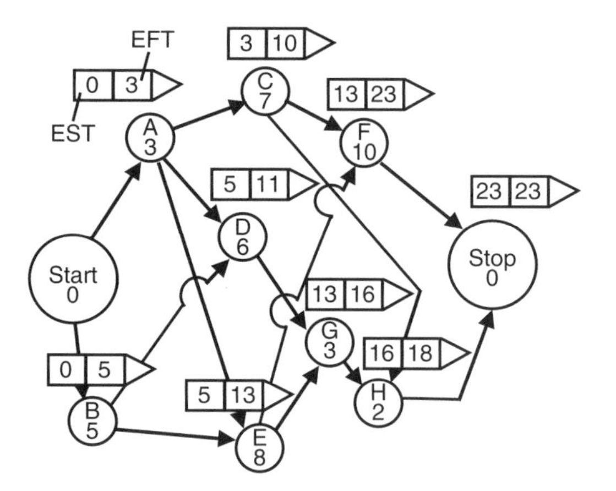

Figure 6.5 The forward pass yields Earliest Start Times and Earliest Finish Times for each project task.

Continuing the above process from Start to Stop and including all tasks yields on EFT for the entire project of 23 weeks. This is called the "forward pass," that is, going from Start to Stop and determining ESTs and EFTs. We have accomplished one of our two purposes: Determining the minimum time in which the project can be completed.

Locate the Critical Path

Now for some more nomenclature:

- LFT—Latest Finish Time. The latest elapsed time, measured from the start of a project, when a particular task can be finished without delaying any other task and, of course, the overall completion of the project.
- LST—Latest Start Time. The latest elapsed time, measured from the start of a project, when a particular task can be started without delaying any other task and, of course, the overall completion of the project.
- LST = LFT − Duration.

Now add the LSTs and LFTs to the network as shown in Figure 6.6. Note the additional graphic nomenclature, that is, the arrows that point to the left. Even for our simple network, we get a complex picture. For an example of how LFTs and LSTs are determined, look at the arrow below Task H. The LFT for Task H is at the end of week 23 because the LST for the Stop task is the end of week 23. Then the LST for Task H is 23 weeks minus two weeks or 21 weeks.

As another example, consider the arrow below Task C. Task C provides input to Tasks F and H. Task F's LST is the end of week 13 while Task H's LST is the end of week 21. Therefore the Task F LST governs because 13 is smaller than 21. Accordingly, Task C's LFT is 13 weeks and Task C's LST is 13 weeks minus Task C's duration of seven weeks or the end of the sixth week.

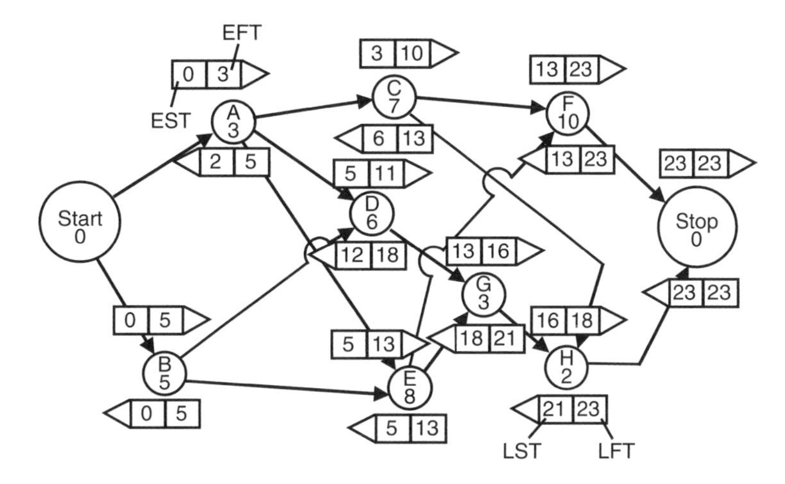

Figure 6.6 The backward pass provides the Latest Start Times and the Latest Finish Times for each task.

This is called the "backward pass," that is, going from Stop to Start and determining all the LFTs and LSTs. Having completed the "forward pass" and "backward pass," we can identify the CP provided that we use this nomenclature:

- Total Float—Amount of time a task can be delayed without delaying the entire project.
- Total Float = LST − EST = LFT − EFT.

The CP is defined by those tasks for which the EST = LST or the EFT = LFT. Thus the CP is defined by Tasks B, E, and F as shown in Figure 6.7. There is no "cushion" for these tasks if the entire project is to be completed at the end of week 23. Each of these tasks must start as soon as they can and be completed as soon as they can, that is, in an elapsed time equal to their duration. Or to use the common CPM term, these tasks have no "float." Other tasks in the example have positive "float." With these tasks, project team members could "relax," or "take their time," because they do not need to be started as soon as possible.

Create the Gantt Chart

Most people, technical and nontechnical, relate well to a Gantt or bar chart, as noted earlier in this chapter. A completed CPM analysis, like that shown in Figure 6.7, can be used to construct the corresponding Gantt chart in the usual left-to-right format as illustrated in Figure 6.8. The horizontal bar for each task is started (its left end) at a time corresponding to its EST. The length of the horizontal bar is depicted as the duration of a task.

The float for a non-critical task is shown as a dashed line beginning at the right end of each bar, that is, beginning at the task's EFT and extending to its LFT. The solid line plus the dashed line for each task thus depicts the "window" within which each non-critical task would need to be started and completed. Each critical task does not

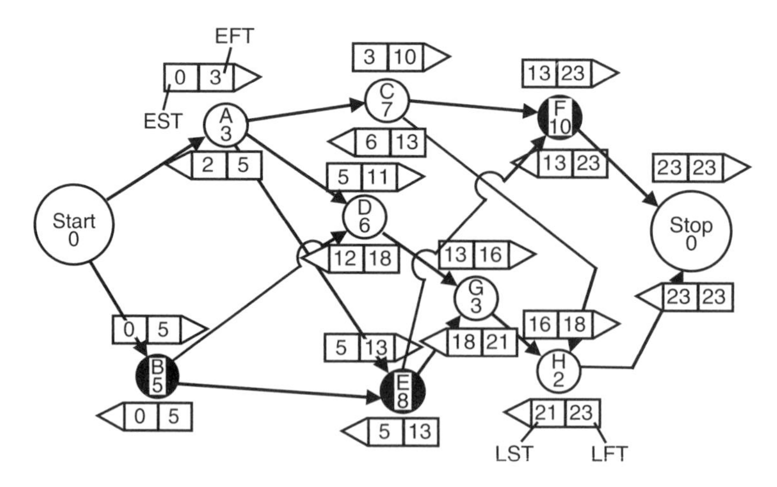

Figure 6.7 The critical path is defined by tasks with zero float.

TASKS	TIME (WEEKS)																						
	1	2	3	4	5	6	7	8	9	10	11	12	13	14	15	16	17	18	19	20	21	22	23
A	█	█	█	--	--																		
B	█	█	█	█	█																		
C				█	█	█	█	█	█	█	--	--	--										
D						█	█	█	█	█	█	--	--	--	--	--	--	--					
E						█	█	█	█	█	█	█	█										
F														█	█	█	█	█	█	█	█	█	█
G														█	█	█	--	--	--	--	--		
H																	█	█	--	--	--	--	--

Figure 6.8 The Gantt chart is developed from the completed Critical Path Method.

have a dashed line segment indicating that its float is zero. Tasks off the CP, like Task C, have positive float. Task C could be completed anytime within the 10 week window—depending on personnel and resources availability. More specifically, the project team could wait as late as the end of the 6th week to start Task C assuming it would then be completed in seven weeks.

Update a Critical Path Analysis

In practice, networks and their CPs frequently are updated during the course of a project, perhaps on a weekly or monthly basis. Updates are required for a variety of reasons including unexpected delays in starting or completing tasks as caused by personnel shortages; material delays; equipment breakdowns; weather; faulty estimates of task durations; missing or unnecessary tasks; and flawed logic, that is, incorrect connectivity. Fortunately, the CPM lends itself to easy updating.

The updating process is simply a matter of showing the new information on the network and then performing a new "forward pass" and "backward pass." As a result of the update, the absolute completion time and/or the location of the CP are likely to change. Let's illustrate the updating process. Assume that you, as project manager, have periodically assessed the status of each task and the project is at the end of the week 10. The status of tasks is shown in Table 6.2.

Consider some of the tasks. Task A is done. Referring back to the completed analysis shown in Figure 6.7, you recognize that Task A should be done in that its LFT is the end of week five. Task C, while underway, will still require five weeks to complete. That is, it will be completed no earlier than the end of the fifteenth week. Referring again to the completed analysis Figure 6.7, you note that the LFT for Task C is the end of the thirteenth week. This is a problem in that Task C, which is on the CP, is going to be completed later than planned. Accordingly, things may be

Table 6.2 Status of project tasks at the end week 10 to be used as basis for updating the Critical Path Method for the project.

Task	Status
A	Done
B	Done
C	Underway with five weeks of work to be completed
D	Underway with one week of work to be completed
E	Underway with two weeks of work to be completed
F	Not started and duration estimate is increased from 10 to 12 weeks
G	Not started and no change in estimated duration
H	Not started and no change in estimated duration
I	A new task, not yet started, that depends only on H and has a two-week duration

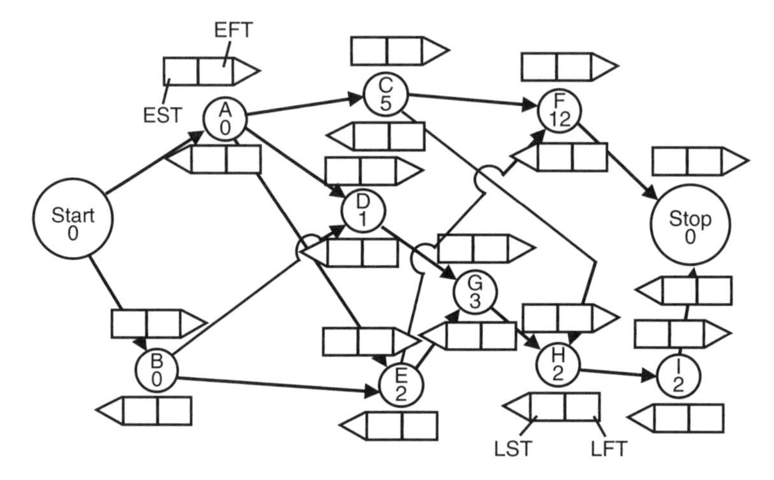

Figure 6.9 The network updated to reflect task conditions at the end of the tenth week.

unraveling in what was to have been a 23 week project. To find out, that is, to define the situation, you decide to update the CPM. Your concerns:

- Has the CP changed?
- Can the project still be completed within the original 23 day schedule?

The update begins by reflecting all the changes in the network diagram as shown in Figure 6.9. New durations and the new tasks are shown keeping in mind that the project is at the end of the tenth week. Perform a "forward pass," as shown in Figure 6.10 to determine the new earliest completion time for the project measured from the end of the 10th week. Then perform a "backward pass" and connect zero float tasks to define the CP.

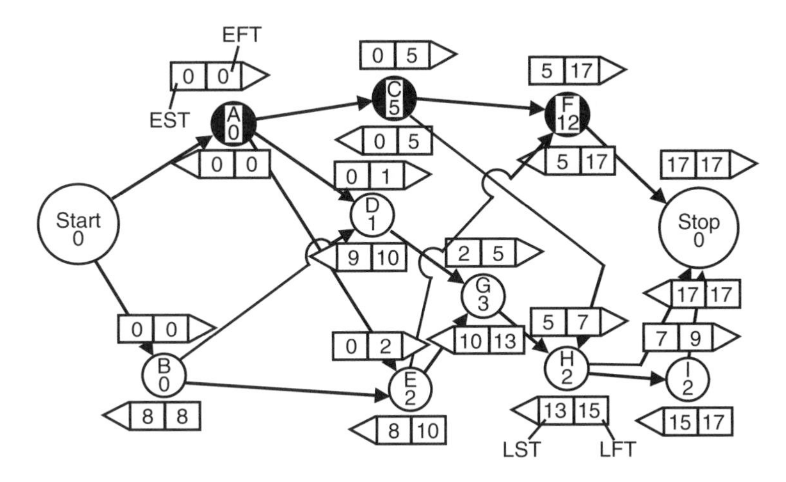

Figure 6.10 The updated earliest completion time, from the end of the tenth week, and the updated critical path.

The updated CPM indicates the following:

- The new completion time is 17 weeks from end of 10 weeks or 27 weeks absolute compared to 23 weeks before.
- The CP is now Tasks A, C, and F, compared to the original CP of Tasks B, E, and F.

Can original completion time of 23 weeks be met? Not unless some changes are made to reduce durations of one or more tasks. Faced with this delay problem, what could you, as project manager, do? A later section of this chapter offers options.

Tips for Determining Tasks

Step 1 in the previously described five-step CPM process is "Determine the Tasks." Tasks—their identification and relationships—are the foundation of the method. Consider some tips for identifying tasks for use in the CPM.

- **Location and timing of tasks:** Think about the location of tasks trying to envision when and how each task is likely to be accomplished. For example, initially surveying for a bridge renovation project might be viewed as a single task. However, on reflection various aspects of survey are likely to be done at different times and/or by different responsible individuals. You, as project manager, will want to get down to the level of individual responsibility so that you can accurately monitor progress. Furthermore, various parts of the survey may, in the context of the network diagram, be prerequisites for different design or other tasks. Therefore, surveying should probably be represented as multiple tasks.
- **Task duration:** Some tasks are long relative to the project's duration. Because of their duration, long tasks are difficult to manage, staff, and monitor. An example is the public information program for a public project or an employee

involvement effort for an internal project. Therefore, consider phasing long tasks. Phase 1 of a public information program might be the responsibility of one person and conclude with an initial public information meeting. Define this as one task and so on for other aspects of the public information effort

- **Individual responsibility:** Try to get task responsibility down to the individual level. This may require breaking some tasks into two or more tasks. One reason to do this is that you, as project manager, will be going to these individuals for task status reports. In general, make tasks small enough so that an individual can be held responsible for each task.
- **When in doubt, go small:** Finally, when in doubt about a task being too large, break it into two or more smaller tasks.
- **Cross check tasks against deliverables:** At this point in the task identification process, you may have many tasks. Avoid doing tasks that are not needed, which adds cost, but not value. Also be careful not to omit necessary tasks. Perform a reality check by studying the evolving network. Begin with the project objective in mind, that is, what the client, owner, or customer expects and work backwards. Confirm that all deliverables are covered, that is, you have all the necessary tasks to produce those deliverables. Also confirm that you do not have unnecessary task.

Some Observations about the Critical Path Method

The generally horizontal format is most common in countries where individuals read from left to right but does not necessarily have to be followed. Assuming an overall directional format, such as horizontal, is selected, arrows may be omitted from the interconnecting branches. Recall again that line lengths have no significance and neither does the orientation of connected nodes. Only connectivity is crucial to the accuracy of a network. All paths must be traversed during the project. CPM is not an exercise in taking the longest path between two points, although the critical path could be found in that fashion.

Crossings of directed branches are acceptable, and, in fact, usually impossible to avoid in actual networks containing many tasks. When crossings occur, they should be clearly indicated using symbols such as the half circles appearing in Figure 6.10.

There must always be at least one critical path. However, there could be two or more critical path segments "in parallel" through all or some of the network. Any delay in a critical activity automatically delays project completion the same amount unless some compensatory action is taken. The critical path is the longest path in terms of time. Time contingencies can be added to each activity or to the overall project.

The CPM example presented in this chapter is small for illustration purposes and to enable rapid manual calculations. Because actual CPM applications usually involve many more tasks and much larger networks which must be frequently updated, manual manipulations are not feasible. Commercial computer programs incorporating the basic algorithms described in this chapter are available for production application of the CPM and other project management tools. Examples of companies offering such software are Microsoft, Oracle, and Softonic. As shown in Figure 6.11,

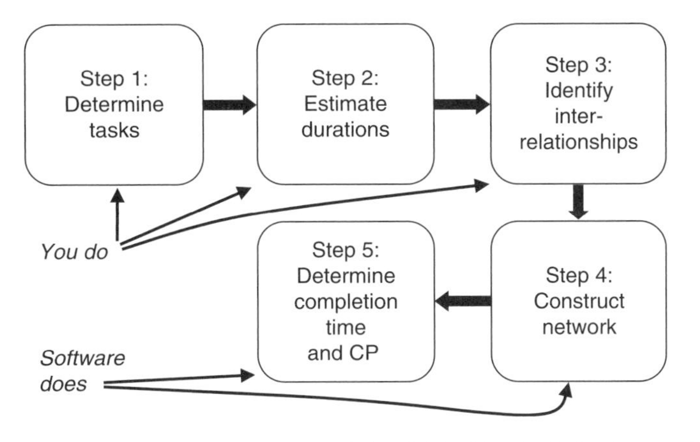

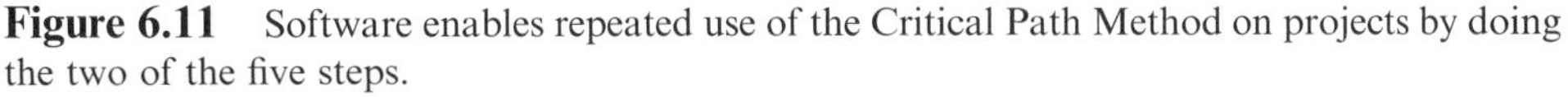

Figure 6.11 Software enables repeated use of the Critical Path Method on projects by doing the two of the five steps.

software does the two mechanical steps, Steps 4 and 5, in the five-step CPM process while the engineer or other technical expert continues to perform Steps 1, 2, and 3.

Earlier, I mentioned that I'm not advocating using the CPM on all projects. So when should it be used? Two project situations come to mind, and clearly there are more. The first is when you undertake a new type of project, that is, you have never managed this kind of project. Using the CPM gently "forces" you and others to think through the project from beginning to end. This might be best done as a small group effort. The second situation in which the CPM could be effective is a project involving multi-departments, disciplines, offices, and/or organizations. Reason: The network diagram explicitly shows interdependencies among tasks and, more importantly, among various entities. This clarification reduces confusion in the "who does what when" arena.

Personal

My initial use of CPM was on the first watershed planning project I had ever managed. The CPM, or more specifically, creating the network forced me to think through the project before starting the project. My first use of the CPM was without the benefit of software—very tedious. Parallel CP segments occurred during the first use which initially caused me to think I had erred.

Review of Earlier Schedule Questions

The introduction to the CPM section of this chapter noted that project managers typically need answers to four schedule questions as they proceed through the planning, execution, and closing phases of a project. The CPM helps to answer those questions as follows:

1. **How long will it take?** The CPM answers this question in that the forward pass provides the EFT for the entire project.

2. **Is the project on schedule?** The CPM provides a clear answer to this question. For example, you could look at the status of each activity relative to its LFT. Or, you could periodically update the CPM analysis and compare the revised absolute completion date to the original absolute completion date.
3. **How much will completion be delayed?** The CPM easily answers this question when the CPM is updated, thus determining if a delay in one or more tasks will delay the absolute completion date of the project. Corrective actions may be needed to reduce the duration of selected CP tasks, such as assigning more personnel or other resources to selected tasks, changing the way some tasks are being performed, removing poorly performing individuals from the project, and working with the client, owner, or customer to change the project schedule.
4. **How long did it really take?** Assume that a significant time under-run or over-run occurred in a project and the project is now being reviewed. If the CPM was used, an examination may reveal flaws in connectivity, task identification, task duration, and assignment of personnel and/or equipment. Assuming that the series of updated CPMs was retained, the type, frequency, seriousness, and types of errors that occurred could be determined and the lessons learned could be used on similar future projects.

Closing Thoughts about the Critical Path Method

The principal value of all three-project scheduling methods—chronological list, Gantt Chart, and CPM—presented in this chapter is that they require that the project be done on paper before it is done in reality, be executed mentally before physically. These tools steer project managers away from thinking that activity is progress. The smart project manager works out the project "map" and then begins the project "journey." The not-so-smart project manager just starts the "journey." Clearly, the CPM is the most powerful of the three methods and much of the value of applying CPM can be realized by carrying it through Step 4, constructing the network.

SCOPE CREEP

Two Types of Scope Creep

As explained in Chapter 5, project scope describes what you and/or your project team will do, although not necessarily in operational detail. That detail is typically revealed in a list of project tasks which are the result of project planning Element 6, Tasks, in the previous chapter. Project scope sometimes includes an explicit list of deliverables. The term scope creep, and its variations, usually refers to a plague on many projects that disrupts schedules, damages budgets, and reduces profit. Scope creep warrants your consideration, beginning as a student.

Uncompensated Scope Creep

Uncompensated scope creep (USC) means doing more than agreed upon and not being paid for it. Doing more usually means being expected to do more tasks than

Figure 6.12 Like a snail, uncompensated scope creep, moves slowly and seems innocuous but gradually adversely impacts budgets and profit.

those explicitly included in a contract or agreement. USC could also mean a suddenly reduced schedule that disrupts the operations of the service provider and adds costs. This is the undesirable kind of scope creep.

A possible alternative definition of USC is "The natural process by which clients discover what they really want" (Helms 2002). This definition has merit and, indirectly, offers advice. You, as a service or product provider, whether in the public, private, academic, volunteer, or other sector, are responsible for helping those you serve "discover what they really want." While that may seem illogical or even unfair to you, it is realty.

USC is like the snail in Figure 6.12. A snail moves slowly, seems innocuous, and yet, given time, goes a long way. Similarly, USC often moves slowly, each incident seems innocuous, and yet, over time, the process goes a long way toward diminishing project profitability in the private sector or destroying budgets in the public, academic, or volunteer sectors.

Compensated Scope Creep

The other kind of scope creep, compensated scope creep (CSC) can be great. For example, a land developer contacts a consulting engineering firm to discuss preparing a conceptual plan for developing her site. The consulting firm and the developer enter into an agreement and the firm prepares the conceptual plan. The consulting firm gets paid, earns a profit, and impresses the land developer. Accordingly, at the request of the pleased developer, the consulting firm contracts with the developer to prepare a detailed plan for the site. The scope has expanded! The firm gets paid, earns a profit, and continues to impress the land developer. The consulting firm then receives a contract to provide detailed design, that is, plans and specifications (more CSC) and then construction management (even more CSC) and so on. CSC, as illustrated here, is great because it:

- Indicates client, owner, customer, and stakeholder appreciation of and confidence in the consulting firm
- Increases revenue and profit for the firm with no or very small marketing costs

The described scenario is realistic and applies, in principle to business, government, academic, and volunteer organizations. It suggests one of the very positive aspects of providing quality products and services, a topic that is discussed in detail in Chapter 7. However, CSC—the positive type—is not the concern in this chapter. The concern here is USC because of the devastating impact in can have on projects.

Consequences of Uncompensated Scope Creep

The three essentials of a project introduced in Chapter 5, that is, deliverables, schedule, and budget, are like simple link and pin truss shown in Figure 6.13 where every side must remain connected. With the link and pin truss, if one member gets longer, then at least one other member must get longer. The same relationship applies to the three project essentials.

For example, if deliverables expand, something's got to give. Often, it's the budget—more costs are incurred, as illustrated in Figure 6.14, or expanding deliverables may increase both the budget and the schedule. Exceptions occur. However, an increase in deliverables usually causes a budget and/or schedule increase. We cannot get something for nothing.

USC can negatively impact the product or service provider in the following ways:

- Consume budget and profit.
- Disrupt schedules.
- Frustrate those conscientious project team members who complete their tasks within the established budget and schedule and then are asked to do more, and often do it without a budget.

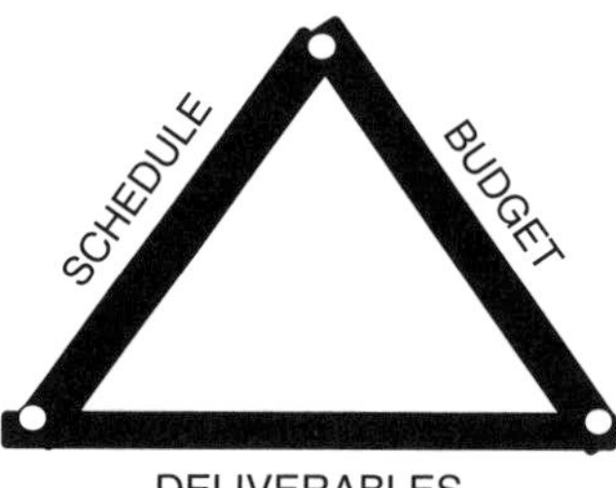

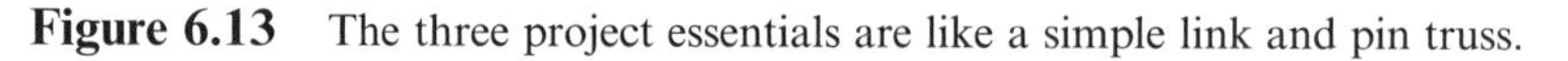

Figure 6.13 The three project essentials are like a simple link and pin truss.

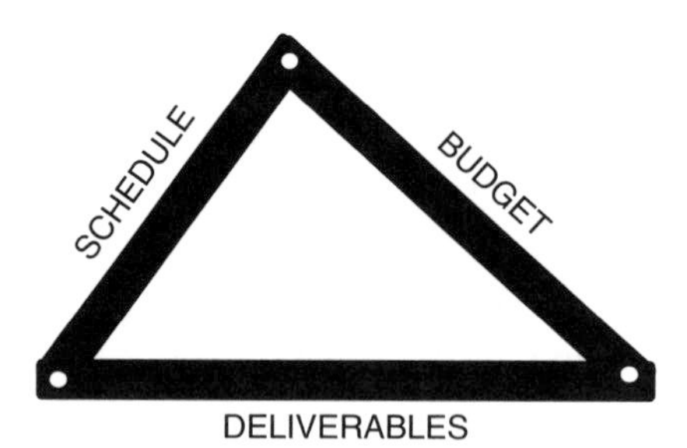

Figure 6.14 If deliverables increase, budget may increase, as shown here, or both budget and schedule may increase.

- Condition the client, owner, or customer to expect even more no-fee products or services. An engineering organization's desire to be helpful coupled with its reluctance to speak up may start and feed a vicious cycle.

Views of Others

Emphasizing the importance of project scope and the service provider's lead responsibility in defining it, consultant Christine C. Brack states that "more misunderstanding arises from the scope of services . . . than any other document" and goes on to say "If the client isn't sure what they want, it's your fault, not theirs" (Brack 2009). Rick Brenner (2002), also a consultant, offers this warning about the consequences of not dealing early with unresolved scope issues: "We'll do almost anything to avoid dealing with conflict directly. We'll even expand the project scope to satisfy all conflicting parties. When we placate conflict, we create a project that nobody can execute."

Drivers of Uncompensated Scope Creep

So why do USC situations arise? Or more bluntly, who is in the driver's seat, as illustrated in Figure 6.15, when USC threatens to crush profit or destroy budgets? Clearly, clients, owners, and customers drive some USC situations. Examples are requesting additional meetings, asking for more alternatives, and adding elements or features during construction or manufacturing.

But are clients, owners, and customers responsible for all scope creep? Certainly not! Might blaming these external entities be a convenient way to hide our own deficiencies? Project team members set up some USC situations. Examples include using a sophisticated technique when a simple one would do; writing a long, formal report when a short memorandum would suffice; and preparing design level cost estimates when planning level estimates are adequate. Like Pogo, the cartoon character, said: "We have met the enemy and he is us!" Be open to this possibility and plan and act accordingly.

External Drivers

We've considered a few examples of how clients, owners, and customers create potential USC situations. Now consider some reasons they do this. Sometimes those you serve simply fail to read, or re-read, the project scope portion of the contract or agreement between them and you and quite innocently ask you or your organization to do out-of-scope tasks. Or perhaps they realize, after entering into an agreement with you, that the original request for proposals (RFP), which you responded to and is now the basis for that agreement, is flawed—it omitted important deliverables. Maybe the organization you are serving wants to capitalize on an unexpected opportunity, such as the appearance of a government grant program, and they need your assistance to prepare the application. Your client, owner, or customer may be responding to

Figure 6.15 Who is in the driver's seat when USC situations arise?

political pressure or maybe they are simply trying to get something for nothing. Frankly, that last cause is, in my view, minor.

You, as project manager, can anticipate some external scope creep drivers during preparation and use of the project plan (PP), which is discussed in the preceding chapter. For example, you saw that the client, owner, or customer's RFP was flawed—it omitted some essential tasks. Therefore, you and they discussed and resolved this issue prior to entering into a service agreement.

Some externally-driven scope creep situations cannot be anticipated. For example, a client, owner, or customer wants to capitalize on a just announced federal grant program related to your already-underway project. The client, owner, or customer asks you to help prepare the grant proposal and to do it at no cost. You and they need to deal with this kind of situation using ideas presented later in this chapter.

Internal Drivers

Now consider some reasons you or members of your team create USC situations. A common reason is a personal preference for using a costly sophisticated method when a less costly, simple one would suffice. Misunderstanding of the desirable but elusive idea of quality, a topic that is treated in detail in the next chapter, is another cause of internally-driven scope creep. For example, some team members see quality as meeting requirements while others see it as striving for perfection and the latter inevitably sets up USC situations. Sometimes tasks budgets are not shared with those asked to do tasks. How can an individual be held accountable for an assigned task if he or she does not know the constraints? Other internal drivers of scope creep could arise such as striving to please a new client, owner, or customer so they provide additional contracts, avoiding internal strife by "giving in" to conflicting team members so that everyone gets to do what they want (Brenner 2002), and avoiding confronting the client, owner, or customer.

You, as a project manager, can account for some internal scope creep drivers as you prepare and use the PP. For example, you know that one member of your project team likes to give clients what that project team member would like, not what the client wants or needs. You reflect this potentially disruptive behavior in the PP and in your guidance of that team member. Or you carefully define, at least for the purpose of the current project, the meaning of quality, using ideas offered in the next chapter, so that all team members are aligned.

Doing Something Extra: The Platinum Rule

Having considered scope creep, especially the adverse consequences of USC and it sometime being internally driven, maybe you are thinking of doing something extra—at no cost—for a client, owner, or customer. This is understandable. At times, most of us want to unilaterally do something extra, for whatever reason.

Caution: This is not the time to practice just the Golden Rule, that is, do unto others as you would have them do unto you. Also apply the Platinum Rule, which is do unto others as they would have done unto them. Make sure you understand what your client, owner, or customer values before you decide to give them something that is beyond the project's scope. Your personal preference is a low priority. As noted by novelist Charles Dudley Warner, "The excellence of a gift lies in its appropriateness rather than its value." And, in mentioning the Platinum Rule, I am emphasizing appropriateness.

Personal

Once while driving to a shopping center during the holiday season, I heard a radio personality offer shopping advice for those of us seeking a gift for a friend or loved one. He asked and answered three questions as follows. Should you buy something you want?—No! Should you buy something the other person should want?—No! Should you buy something the other person does want—Yes! While very difficult, finding a gift that he or she wants or giving extra services that the client, owner, or customer will really value is very satisfying. That's the platinum rule in action.

Relevance to You as a Student

Clearly the preceding introductory scope creep material and the USC prevention and resolution advice presented in the remainder of this chapter are relevant to the young practicing engineer. He or she is likely to be heavily involved in contract projects with budget constraints and possibly profit goals. These young practitioners will want to help control internally and externally-driven USC.

However, an understanding of various aspects of scope creep, especially USC, is also useful for students of engineering or other technical disciplines. For example, when given an individual assignment in one of your classes, strive to understand the scope and meet the related requirements. Frequently going beyond the intended scope of class assignments, because of not understanding what is expected, can add significantly to your overall workload. I am not saying you should not occasionally, based on your interests, delve deeply into some topic. I encourage that. However, do it deliberately. Even more importantly, when working on a team project, encourage all team members to focus on the scope and the tasks needed to fulfill the project requirements. Resist the temptation to drive the project out of its intended scope to the consternation of your team members.

Preventing Uncompensated Scope Creep

If you, as a student or young practitioner, are the manager of a project or a portion of a project, you can, as you prepare the PP and then get the project underway, prevent many USC incidents. Ten suggestions follow that will help prevent USC. Most are simple to understand while requiring self-discipline to consistently apply. These tips are offered, smorgasbord style, for your consideration. Select and apply the tips that will benefit you, your team, and those you serve.

Earn the Trust of Those You Serve

This is the powerful tip as suggested by preacher and professor Harry Emerson Fosdick who said "No virtue is more universally accepted as a test of good character than trustworthiness." Are you and I worthy of being trusted by the representative of the client, owner, or customer? In a capstone course setting, have you, by your words and actions, earned that trust of fellow students, faculty members, and your project sponsor? Considerable time and energy are typically invested in developing mutually-trustful relationships. They cannot be bought; they must be earned.

When we have earned the trust of those we serve, during discussion of a potential project, we can frankly express our concerns about deliverables, schedule, and budget and be confident that those concerns will be respectfully and thoughtfully considered. The Chapter 1 section "Honesty and Integrity" and the Chapter 2 section "Guard Your Reputation" are highly relevant if you value earning trust. For additional discussion of the major effort required to earn trust, plus its foundational role in marketing professional services, refer to the section of Chapter 14 titled "A Simple, Powerful Marketing Model."

Understand Wants and Needs

Our principal purpose, early in the process of pursuing a potential project or as we start to prepare the PP, is to discover what the other person and/or organization really wants—and, eventually, what they need. For a detailed treatment of this important topic, refer to the section of Chapter 7 titled "Strive to Understand Client, Owner, and Customer Wants and Needs."

Prepare and Use a Project Plan

The important process of preparing and using, both internally and externally, a PP is amply discussed in Chapter 5 and need not be repeated here. Not having a PP essentially "guarantees" potential USC situations. If you fail to plan, you plan to fail. Slowly, and surely, like the snail mentioned earlier, scope creep scenarios will arise because everyone is not "on the same page" or you missed necessary tasks and/or deliverables. The only remaining issue is will the cost of the added services or products, or perhaps a reduced schedule, be compensated or a "gift" to the client, owner, or customer? And the question will have to be answered frequently because, in the absence of a PP, the probability of USC rises.

In the discussion of preparing a PP, Chapter 5 suggested the possibility of sharing all or portions of it with the client, owner, customer, and/or stakeholders besides, of course, creating and sharing it with the project team. Sharing the PP with those being served provides another line of communication connecting them with the project team. This added communication should further reduce the likelihood of externally-driven USC.

Personal

Leaders of one of my clients developed and adopted, with my assistance, a policy that required PPs for projects with total fees above a prescribed monetary amount and encouraged project managers to share the PPs with their clients and owners. Initially, the project managers "pushed back," they disliked the policy. They interpreted preparation and use of a PP as a major effort with little or no benefit. After one year of experience with the policy, project managers met to discuss it. Now they favored the policy, including sharing PPs with their clients and owners for two reasons. First, clients and owners indicated to the project managers that they felt they were even more part of the project team. Second, and of particular importance in this USC discussion, the projects experienced less client or owner-driven USC situations probably because sharing of PPs improved communication.

Formalize Client, Owner, and Customer Responsibilities

As project manager, you may be reluctant to raise the issue of client, owner, and customer responsibilities. After all, you are here to be of service—and you and others in your organization like to talk about "full service." However, consider three benefits of defining the responsibilities of those you serve, from the perspective of avoiding USC.

The first benefit is psychological. Most clients, owners, and customers want to be meaningfully involved in certain aspects of their project. For example, their retaining of a consultant does not usually mean that they want a "hands off" approach.

Second, those we serve can perform some tasks best. One example is getting important people to meetings. Assume you would like to have the President of the local

Chamber of Commerce attend the meeting at which you present your firm's recommendations for city-wide street improvements. Your client representative, the community's Director of Public Works, is likely to be in a better position than you in extending a successful invitation. Surveying their community infrastructure, manufacturing facilities, or electric power distribution network is a task in which the client, owner, or customer may be more effective. Sharing information about a project with citizens of a community, the employees of a manufacturing firm, or other stakeholders is another area where the client, owner, or customer "has an edge." Sometimes you or your organization will encounter obstructionist-uncooperative personnel within the client, owner, or owner organization. If you cannot resolve such situations, the organization you are serving probably can.

Finally by discussing and formalizing client, owner, and customer responsibilities at the outset, you are likely to avoid some USC situations later. Examples are permitting, advertising public information meetings, gaining access to private property, and locating utilities.

Watch Your Language: Scope, Tasks, and Deliverables

Chapter 5, in the section "Principal Project Plan Elements," described Scope, Deliverables, and Tasks as, respectively, Elements 2, 4, and 6. The contract or agreement between your organization and those you serve must be carefully written to avoid creating potential USC. But, before offering writing advice, a thought about "scope." This term may have two meanings in a project and those two meanings are nicely captured by the follow two PMI definitions:

- **Project scope:** As noted in Chapter 5, project scope is "the work that must be performed to deliver a product, service, or result with the specified features and functions (PMI 2008)." In other words, the project scope describes what you and your project team will do, although not in operational detail. This is the meaning used in Element 2 in Chapter 5. Project scope is process-oriented.
- **Scope:** PMI (2008) also defines scope as "The sum of the products, services, and results to be provided as a project." While including process, this definition focuses on the deliverables; it is deliverables-oriented.

While your and your project team's understanding of these two "scope" terms is important, your client, owner, and customer's understanding is more important. Some will view scope as being primarily what is to be delivered, that is, consistent with the second definition above. For others, while deliverables are clearly important, they also have specific expectations regarding the actual processes used to develop and provide those deliverables consistent with the first of the above two definitions. Be careful; you and members of your project team should be aware of such preferences. For example, you may provide the expected deliverables but not by the means preferred by the client, owner, or customer—so they are not pleased. Be certain that you and they share the same view of scope.

Personal

I have worked on flood stage profile projects for the U.S. Army Corps of Engineers (USACE). Their definition of scope clearly included delivering profiles for defined flood events. That definition also included, and this is the point, using specified USACE computer programs and procedures in the process of developing those profiles.

Now for some project scope, deliverables, and task writing advice in the interest of avoiding USC and other difficult situations. Avoid words and expressions such as those in the left column of Table 6.3. Such problematic words and expressions might be inadvertently used in proposals, contracts, and agreements to describe scope, deliverables, and tasks. For example, the text might, with good intentions, read "We will be available at the construction site at all times." That is a potential USC problem in that it, if interpreted literally which it could be in contentious or litigious situation, creates an impossible expectation. As suggested by the first entry in Table 6.3, an alternative and achievable version is "We will visit the construction site once per day."

Do not use "red flag" words of promise like insure, ensure, assure, as noted in the second entry of the table. "Periodically," which appears in the table's third entry, is also problematic because it is vague.

The common negative aspect of the last four entries of Table 6.3 is the word "all" in the left column. "All" is almost always bad because it can be interpreted in an absolute manner. For example, in a contentious situation a dissatisfied client, owner, or customer might say "you said you were going to gather all existing information and I found some information that you did not gather." Do a word search on "all" and

Table 6.3 Expressions to use and not to use in contracts and agreements.

Do Not Use	Example Replacements
At all times	Will be done once per...
Insure, ensure, assure	Reasonable effort will be made
Periodically	Every Friday
Supervise, inspect	Observe and report
Certify, warrant, guarantee	Statement as to our judgment based on...
All existing information will be gathered	Readily available information will be reviewed and collected as needed
Prepare summaries of all meetings	Prepare summaries of monthly project status meetings with clients
Close coordination of all stages of the work	Perform interdisciplinary milestone reviews at the 15%, 30%, 60%, and 90% points
Will complete all project services	Will prepare and submit for review and approval normal engineering drawings suitable for construction

(Source: Adapted from Hayden 1987)

make sure it conveys your intended meaning. If not, use replacements similar to those in the right column of Table 6.3.

Recall, as noted in the Chapter 5 project scope discussion, that the project scope in a contract or agreement in not written just for your clients, owners, and customers. It also speaks to members of the project team. Most of them may not have been involved in obtaining the project and may be learning about it for the first time. They need to understand the project scope so that they do not exceed it to the detriment of the project's schedule and budget and so important tasks are not missed.

Consider this project scope language taken from an actual anonymous draft contract prepared by a consulting firm: "Existing and future land use will be reviewed with city staff to reach a consensus on the conditions that will be used to evaluate stormwater impacts." What's the problem with this statement, especially in light of preventing scope creep? It's the word "consensus." Consensus can't be guaranteed and too much time could be expended in trying. As an alternative, drop "to reach a consensus on" and replace with "so that we can determine."

Personal

I once received a letter offering me a dean's position at a university. The letter stated that I would be expected to lead a college by consensus. That was a "flag," an impossible expectation. I accepted the offer but only after the consensus requirement was dropped.

Here's another example of draft contract text. The consulting firm stated that they "will map all storm water facilities." There is the word "all" again. It can drive team members to strive for an elusive goal and it may encourage clients, owners, and customers to expect an elusive end. More specifically, where do you stop in mapping "all" stormwater facilities as you and your team work upstream in the stormwater system? Do you stop when at the upper end of all buried storm drains? Or do you move further upstream and stop at inlets? Or do you go even further upstream to include downspouts? You get the idea. Don't say "all." Instead, be specific, as in "we will map all storm sewers in the system that are 12 inch diameter or larger."

Review this third and last example which, although very short, has two deficiencies each of which could lead to USC situations: "Prepare all information to be submitted to... necessary for regulatory approval... " The first deficiency is use of the hopefully now familiar "all." The second problem is an implied guarantee to obtain approval from a regulatory agency. As committed and careful as a professional services firm may be, it cannot control the decisions of a government regulatory entity. An improved version of the problematic draft statement is: "Provide owner with project information that will support the permit application."

In closing, as they say "the devil is in the details." Careless writing sets you and your organization up for client, owner, and customer and staff-driven USC. Marketing and communications consultant H. Jackson Brown Jr. reminds us "The big print giveth and the small print taketh away."

Include an Additional Services Provision

Consider an Additional Services or Extended Services section in your contract or agreement as a way to raise the issue and to give the client, owner, or customer the benefit of your experience. List potential additional services that might be required in a particular project (e.g., assisting with permit applications, working with vendors) and explain how they would be billed such as according to an attached standard fee schedule or on some other basis. The purpose is to further reduce the likelihood of client, owner, or customer-driven scope creep by flagging, up front, the possible need for additional services. Another approach to reducing the likelihood of client, owner, or customer USC situations is to create an Assumptions section in the contract or agreement. This section could be in addition to or instead of the Additional Services section. Failure to list potential additional services, to state assumptions, or take some other similar action can lead to the client, owner, or customer saying, after problems evolve, "you should have told me I needed to have, or might need to have, you or some other firm perform this service on this project" (Bachner 2008).

Provide a "Cushion" in the Budget

Provide a "cushion" in the budget, especially where the client, owner, or customer is known to expect services beyond the agreed-upon scope. This advice may seem simplistic and unrealistic. However, if you and your organization want to earn a profit or stay within a budget and are knowingly faced with such a client, owner, or customer you have only two responsible choices. One is to provide the "cushion" and the other is to walk away from the project.

If you include a "cushion," where do you put it? One approach, as shown on the left side of Figure 6.16, is to put a small portion of the "cushion" into each task as shown. That is, create a series of "little pots" for task leaders to dip into. I suggest not doing that. Instead, assign the entire "cushion" to the project manager as illustrated on the right side of Figure 6.16. This encourages more focused efforts by task leaders and provides the project manager with a "big pot," a large "cushion" for real crises as determined by the project manager.

Start Preparing and Sharing Deliverables on "Day 1"

Assume, for purposes of discussion, that a report is to be one project deliverable. Then another way to reduce internally and externally-driven USC, and a technique that was introduced in Chapter 3, is to begin the report on "Day 1." Reason: Improves communication—internally and externally—and, therefore, reduces the probability of misunderstanding of the agreed-upon scope, deliverables, and tasks. While I do not literally mean "Day 1," I do mean shortly after the start of the project. If, for example, you are the project manager of a semester capstone project with a report as a deliverable, then begin the report during the first week.

Prepare an outline of the report during the first week, share it with project team members, and ask for their input. This effort will improve communication within the project team. You all will be more aligned regarding the project and thus more likely to avoid internally-driven scope creep. During the second week, give the revised report outline to the professor or professors responsible for the course and/or to the external

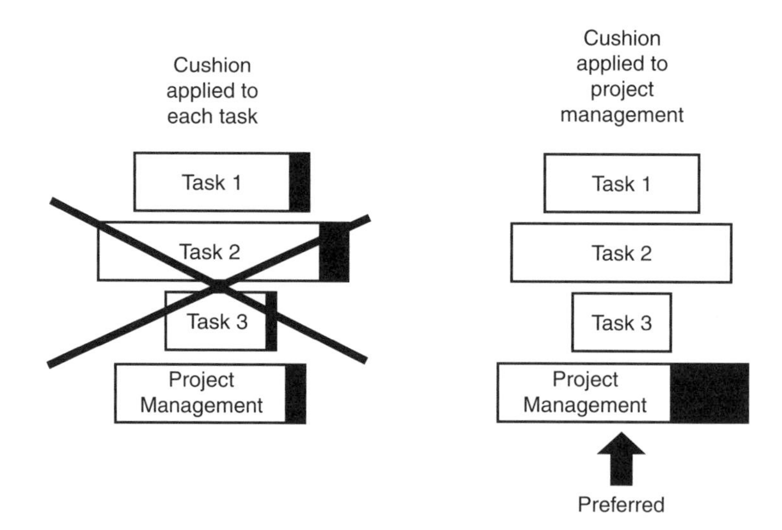

Figure 6.16 When providing a "cushion" in a budget, assign it to the project manager.

sponsor of your project. Indicate that the outline presents the project team's view of the content of the report, a major project deliverable. Ask for input and resolve any differences in the spirit of minimizing the probability of externally-driven scope creep.

During the third week share a draft of Chapter 1 of the report with project team members and ask for input, then share a revised version with the faculty and outside sponsor and continue this pattern. The "frosting on the cake" with the suggested "start on Day 1" approach is that you and your team are much more likely to produce a final report whose quality reflects or matches the quality of the underlying work. Why? Because you avoid "last minute" rushed report preparation and the inevitable errors and omissions. Sadly, such rushing often produces a mediocre document about an excellent project and the document, not the actual work, communicates project mediocrity to the readers. Clearly, the start on "Day 1" advice can be applied to other deliverables such as plans and specifications.

Hold Team Members Accountable

You, I, and others tend to respond to expectations. Use this principle of human nature to reduce the likelihood of internally-driven USC by holding team members accountable. When delegating a task to a team member, describe the context and the expected deliverable. However, and this is critical, explain the resources available to perform the task, such as the number of labor hours, expenses, and the deadline to complete the task. This is one reason for preparing a PP that includes a detailed work task, schedule, and budget breakdown, as discussed in Chapter 5. You, as project manager, can draw on it when you delegate project tasks to members of your team. This scope creep control advice may seem obvious, but junior personnel often tell me that they are assigned project tasks but are not informed about the resources.

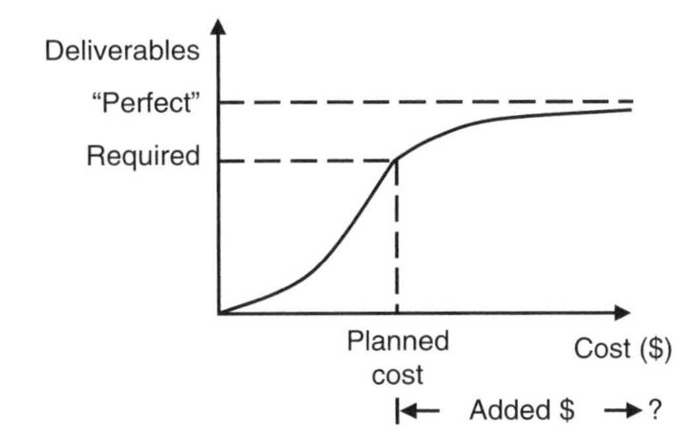

Figure 6.17 Striving for perfection, in the name of quality, will result in costly internally-driven uncompensated scope creep.

Define Quality

Widely-varying understandings, or more accurately, misunderstandings of quality create USC situations, especially the internally-driven variety. Alternative and a recommended definition of quality are described early in the next chapter. Study that material and, as a project manager, obtain an agreed-upon definition of quality, if not for your organization, at least for your project.

If you do not have a common quality definition, you invite internally-driven scope creep. For example, assume you as project manager view quality as meeting all requirements, as recommended in the next chapter. You have determined the tasks, deliverables, and schedule needed to meet those requirements; have estimated the cost; and have entered into a mutually-acceptable agreement with the client, owner, or customer. You are embarking on a potentially successful project—for your organization and for the organization you are serving.

However, some members of your project team define quality as striving for perfection, instead of meeting requirements. They want to shoot for "perfect" deliverables. When striving for perfection becomes the operational definition of quality costs will grow exponentially in the pursuit of an impossible goal, as shown in Figure 6.17. Result: Disaster, at least from a fiscal perspective in that the internally-driven USC situation will incur excessive costs that cannot be recouped.

Resolving Uncompensated Scope Creep

So much for suggested ways to prevent internally and externally-driven USC. No matter how careful you are and how proactive and preventive you try to be, some USC situations may arise as the project progresses. Five suggestions follow for dealing with these incidents. As with the preventive suggestions, the resolution suggestions are simple to understand while requiring self-discipline to consistently apply. Use the tips to assist you, your team, and those you serve.

Recall that the first suggestion for preventing USC was earning the trust of those we serve. That trust enables you as a service provider to negotiate a mutually-beneficial agreement. It also enables you to proactively address potential out-of-scope requests and other issues that inevitably arise once a project is underway.

Personal

I started a six month project with a long-time client. Our agreement called for me to be heavily involved early on in laying the project's foundation and then tapering off. Midpoint in the project, the client started asking me to do many more tasks—tasks way beyond the scope described in our agreement. Because of our mutually-trustful relationship, I immediately reminded the client of his out-of-scope requests. And, again because of our mutually-beneficial relationship, he found a way to compensate me for the extra services.

Stop and Think

When the client/owner/stakeholders requests out-of-scope service, stop and think. If you are not the project manager, don't agree to the extra services without first talking with the project manager. Don't "close the barn door after the horse is gone," that is, don't deliver the scope change and then discuss how you might be compensated.

Instead, first refer to the contract or agreement. Perhaps it is those developed by the Engineers Joint Contract Documents Committee (EJCDC 2011), the American Institute of Architects, "Standard Form of Agreement Between Owner and Architect" (AIA 2011), a similar document from some other organization, or standard language developed by your organization. These documents typically describe circumstances under which a professional services firm is entitled to additional compensation. Examples include the client, owner, or customer changes instructions to the service provider or does not render decisions or provide project input in a timely manner. Another circumstance is revisions to or enactment of codes, laws, or regulations.

If you are the project manager, consider this administrative approach for processing small external scope change requests: Depending on the project and client, owner, or customer, establish a scope change threshold (e.g., hours, dollars) above which you will seek extra compensation and below which you, or selected members of your project team, can perform the out-of-scope service without additional compensation.

Seek Win-Win Solutions

The following win-win options are based in part on PSMJ (1980). Additional compensation tends to be the most attractive option if you are a service or product provider faced with a request for expanded scope. Another possibility is that you and the client, owner, or customer could agree to reducing the scope of your services elsewhere in exchange for new services.

Organizations that we serve are often locked into budget cycles. Perhaps the additional requested services could be performed in and paid for during the client, owners, or customer's next budget cycle. Maybe your service organization has another contract or agreement with this client, owner, or customer under which the requested additional services could be legitimately performed and compensated. Try not to use the previously-mentioned budget "cushion" to fund requested out-of-scope services. Reserve that "cushion" for project situations where you do not have other options.

Suggest Scope Changes

In some situations, you as the service or product provider will be in a position to drive scope change and should do so for the benefit of all concerned. Because of totally unexpected circumstances, you may see the project focus or direction changing before anyone else. Examples are activation of a new grant program and unexpected availability of a new product or manufacturing process. Regardless, you see the project unraveling or you see new, attractive possibilities. Within the context of your mutually-trustful relationship and based on your intimate knowledge of the project, suggest a revised scope, schedule, budget, and means of compensation.

Document Agreed-Upon Solutions

Once a scope change and means of compensation are agreed upon, put all of it in writing. For example, cite it in meeting minutes, use a standard scope change memorandum or letter, or describe the changes and means of compensation in periodic progress reports. Do not rely on anyone's memory—too much is at stake.

Make Deposits in the Goodwill Bank

Assume you decide to "give" your services, that is, for no compensation—for whatever reason. Then document each of those cases in the form of a cumulative list. Possibly give the list to client, owner, or customer either as the "gifts" occur or on an as-needed basis. You, as a service or product provider may get into budget, schedule, or other trouble as a project proceeds. And you are the cause. Now you would like to have help from the client, owner, or customer. You may want to make a "withdrawal" from the "goodwill bank." In order to do that, you need to have made deposits into the bank and have documented them so that you can share that information as needed.

I think that most representatives of clients, owners, and customers would prefer to know early on that USCs are beginning to occur so that they can take action. Client, owner, and customer representatives have "bosses," too, and want to create the best possible outcome. Problems, if not addressed, typically get worse—not better—with time.

Ideas for Clients, Owners, and Customers about Avoiding Uncompensated Scope Creep

Assume your organization retains professional services firms? Perhaps you are a government entity, a manufacturer, an educational institution, or a land developer. Maybe you are a professional services firm that subcontracts some of its services. In contrast with the focus so far of viewing the possibility of USC mostly from the perspective of the provider of services, consider USC as viewed by the user of services.

Scope creep, whether you pay for it or not, is generally not in your best interest. If it is compensated, you incur more costs than you probably planned on. If it is not compensated, additional work may be done grudgingly and/or poorly.

Use Qualifications-Based Selection (QBS) as opposed to Price-Based Selection (PBS). These two methods for selection of professional service firms are discussed in Chapter 14. I know this advice may seem self-serving coming from an independent consultant author. However, experience and studies (e.g., Chinowsky and Kingsley 2010) suggest that the extent of scope creep will be less with QBS than with

Table 6.4 For technically-complex projects, consider selecting an engineering firm with appropriate board-certified personnel on its staff.*

Academy of Coastal, Ocean, Port & Navigation Engineers (ACOPNE)
Academy of Geo-Professionals (AGP)
American Academy of Environmental Engineers (AAEE)
American Academy of Water Resources Engineers (AAWRE)
Structural Engineering Certification Board (SECB)

*Note: This is a representative list of engineering academies and boards. It is not intended to be all inclusive. This is a fluid situation in which the number and variety of academies and boards will increase.

PBS. Furthermore, designs are not likely to be based on lowest life-cycle cost, because of the extra design effort required.

Talk to trusted colleagues. Find out what you can about potential service providers. Ask about scope creep driven by the service provider. Then, if you decide to interview a prospective service provider, use the behavioral or retrospective approach, not the hypothetical or prospective method. The behavioral or retrospective method focuses on actual events and actual behavior. In contrast, the hypothetical or prospective method looks forward in an imaginary manner. Here are examples of essentially the same issue posed in the hypothetical and behavioral modes:

- **Hypothetical:** What would your firm do if you encountered an unexpected field condition that added design or construction costs?
- **Behavioral:** Give an example of when your firm encountered an unexpected field condition that added design or construction costs and indicate how you handled it.

Make sure that the team each firm sends to the interview is the team that, if selected, the firm would assign to your project. Be wary of "bait and switch."

Possibly require those who serve you to prepare PPs, as discussed in Chapter 5, and to share them with you. Ask the potential service provider to provide examples of PPs—without violating confidentiality between that service provider and his or her previous clients, owners, or customers. As part of the agreement or contract, require that both the service provider's responsibilities and your responsibilities will be in writing.

For technically-complex projects, consider requiring that the firm you select has appropriate board-certified professionals on its staff. In engineering, for example, board certification programs are growing as suggested by Table 6.4. The advanced education and extensive experience of board-certified professionals should reduce the likelihood of scope creep on your project.

Personal

When my wife or I need medical services, we insist on a board-certified medical doctor. Not a doctor with just a college degree, a medical degree, and a license. You may be able to do the same when searching for a professional services firm.

Closing Thoughts about Scope Creep

USC, which can be driven internally and externally, consumes budgets, disrupts schedules, frustrates project team members, and conditions some clients, owners, and customers to expect even more services or products while providing no additional compensation. You, as a project manager or member of a project team, beginning as a student, can selectively use the ideas and methods presented in this chapter to greatly reduce the likelihood of scope creep occurring on an imminent project and to proactively deal with the residual scope creep that may arise once even a carefully-planned project is underway. And this scope creep management can be done in a win-win manner, that is, for the mutual benefit of the client-owner-customer and the service or product supplier.

The man who is prepared
has his battle half fought.

(Miguel de Cervantes, Spanish author)

CITED SOURCES

American Institute of Architects. 2011. "Contract Documents." (http://info.aia.org/knowledgebase/2007_Owner-Architect_Agreements.htm), May 9.

Bachner, J. P. 2008. "Six Elements of Good Contracts." *CE News*, July, p. 14.

Brack, C. 2009. "The Scoping Process." *Zweig Letter*, July 20.

Brenner, R. 2002. "Some Causes of Scope Creep." *Point Lookout*, e-newsletter, Chaco Canyon Consulting, September 4.

Chinowsky, P. S. and G. A. Kingsley. 2010. *An Analysis of Issues Pertaining to Qualifications-Based Selection.* American Council of Engineering Companies (ACEC) and American Public Works Association (APWA), Washington, DC and Kansas City, MO.

Dhillon, B. S. 1987. *Engineering Management: Concepts, Procedures, and Models.* Technomic Publishing Company: Lancaster, PA.

Engineers Joint Contract Documents Committee. 2011. (www.ejcdc.org), May 9.

Hayden, Jr., W. M. 1987. *Quality by Design Newsletter.* A/E QMA, Jacksonville, FL, May (Quoted in *Journal of Management in Engineering—ASCE*, October, pp. 284–285).

Helms, H. 2002. "Scope Creep." Published in "A List Apart" (www.alistapart.com/), September 20.

Project Management Institute. 2008. *A Guide to the Project Management Body of Knowledge – Fourth Edition*. PMI: Newtown Square, PA.

PSMJ. 1980. "Getting More Money." *Professional Services Management Journal*, June.

Washington State Department of Transportation. 2005. *Project Management Process.* Fall.

Wikipedia. 2011a. "Gantt Chart." (http://en.wikipedia.org/wiki/Gantt_chart), May 9.

Wikipedia. 2011b. "Henry Gantt." (http://en.wikipedia.org/wiki/Henry_Gantt), May 9.

Wikipedia. 2011c. "Program Evaluation and Review Technique." (http://en.wikipedia.org/wiki/Program_Evaluation_and_Review_Technique), May 9.

ANNOTATED BIBLIOGRAPHY

Bachner, J. P. 2008. "Six Elements of Good Contracts." *CE News*, July, p. 14. (Warns: "Failure to list services the A/E will not perform could result in a client's allegation that 'you should have told me I needed you to perform this service' after problems evolve because the service went overlooked or was dropped from the scope during undocumented discussions." However, isn't the list of services that an A/E will not perform infinite?)

Fisk, E. A. 2000. *Construction Project Administration – Sixth Edition*. Prentice Hall: Upper Saddle River, NJ. (Includes two construction CPM examples in Chapter 13, a university building and a hydroelectric project.)

Halpin, D. W. 2006. *Construction Management – Third Edition*. John Wiley & Sons: Hoboken, NJ. (Provides a 22 task CPM analysis for construction of a gas station.)

Loulakis, M. G. and L. P. McLaughlin. 2005. "The Law: Court Requires CPM Analysis to Prove Shop Drawing Approval Delays." *Civil Engineering–ASCE*, June, p. 88. (The General Service Administration (GSA) Board of Contract Appeals "held that a contractor would have to use CPM schedule analysis to establish merits of its delay claim concerning the GSA's alleged late return of contractor submittals." The contract between the contractor and the GSA required that the contractor "maintain a CPM schedule.")

EXERCISES

6.1 CONSTRUCT A NETWORK: The purpose of this exercise is to help you confirm your understanding of network fundamentals. Suggested tasks are:

A. Consider the following project tasks and their interrelationships: a) A is the initial task; b) E and F can be performed simultaneously and cannot be started before B is completed; c) I depends on F, G, and H; d) K can begin only after E and I are finished; e) L must follow J and K; f) J cannot start until E is completed; g) C must be completed before G can begin; h) B, C, and D cannot begin until A is completed and can be performed simultaneously; i) D must be completed before H can begin; and j) L is the final task.

B. Construct the network.

6.2 APPLY THE CRITICAL PATH METHOD: The exercise provides an opportunity to use the CPM to determine the minimum completion time of a project and its critical path. This lesson will also improve your understanding of project duration and total float. Suggested tasks are:

A. Start with the Exercise 6.1 network and assign the following durations in days for the tasks:

A = 0	E = 6	I = 6
B = 7	F = 23	J = 14
C = 13	G = 25	K = 21
D = 19	H = 10	L = 0

B. Determine the minimum project duration, identify the critical path, and indicate the total float for each task.

6.3 CONSTRUCT A GANTT CHART: The purpose of this exercise is to use a completed CPM analysis to construct a Gantt Chart. Suggested tasks are:

A. Use the results of Exercise 6.2 to develop a Gantt chart for the project and position the "bar" for each task as beginning with its EST and ending with its LFT. Show task duration with a solid line and float with a dashed line. The result is, in effect, a more meaningful Gantt chart.

6.4 APPLY CPM AND CONSTRUCT A GANTT CHART: The purpose of this exercise is to apply the CPM to an actual project. Suggested tasks are:

A. Use, as the actual project, the project you submitted for Exercise 1.2.

B. Estimate the duration of each task.

C. Perform the complete CPM process, including constructing the network, applying the CPM, highlighting the critical path, and developing the Gantt chart showing total floats.

6.5 UPDATING A CRITICAL PATH ANALYSIS: The purpose of this exercise is to gain practice in updating a CPM analysis as would be done once a project is underway. Suggested tasks are:

A. Refer to the solution for Exercise 6.2 and assume the project is underway and, at the end of 20 days, the status of each task is as follows:

Task	Status
B	Done
C	Done
D	Underway with eight days of work to be completed
E	Done
F	Underway with two days of work to be to be completed
G	Underway and on schedule (i.e., 18 days of work to be completed)
H	Not started and duration estimate is increased to a total of 11 days
I	Not started
J	Underway with four days of work to be completed
K	Not started

There are no changes in the "job logic," that is, in the number of tasks and their interrelationships. However, such changes could very easily have arisen at this stage in the project and would have to be reflected in the updated critical path analysis.

B. Determine the following: a) the new critical path; b) the time required, starting now, to complete the project; and c) the projected absolute completion time compared to that projected at the beginning of the project, that is, plus or minus how many days?

6.6 USE OF CPM SOFTWARE: This exercise is intended to provide each student with an introductory, hands-on experience with CPM software. Suggested tasks are:

A. Select and learn how to use, as an individual or as a team, a CPM software package.

B. For personal practice, use the software to redo your Exercise 6.4 results.

6.7 USE OF THE CPM: Completing this exercise will enhance your understanding of how to interpret the status of the CPM being applied to a project. Suggested tasks are:

A. Assume that all tasks in a network diagram are on schedule, except for one that is off the CP and it is delayed, that is, it will take longer to complete than originally estimated. Assuming no corrective action is taken, which one of the following is true: a) project completion will not be delayed, b) project completion may be delayed, or c) project completion will be delayed.

B. Which one or more of the following statements about the CPM, with activities on nodes, are true: a) a CPM network can have two or more parallel paths; b) if any task on the CP is delayed, and corrective action is not taken, then project completion will be delayed; c) lengths of the branches connecting tasks are meaningless other than to show task interrelationships; d) EFT = EST plus duration; e) LST = LFT minus duration; and f) the float of a task = LST−EST.

CHAPTER 7

QUALITY: WHAT IS IT AND HOW DO WE ACHIEVE IT?

Quality is not a naturally occurring event.
It is a result of hard, deliberate work
that . . . never, ever ends.
Achieving quality in project implementation
is not a matter of luck or coincidence;
it is a matter of management.

(*Kenneth H. Rose, consultant and author*)

This chapter addresses quality, that desirable but elusive aspect of professional work. A simple, challenging, but operational definition of quality is offered followed by definitions of quality control and quality assurance. The chapter offers suggestions for developing a quality-seeking culture within a project team, department, or other organizational entity. It then describes and illustrates tools and techniques for stimulating creative and innovative individual or group thinking which help to achieve quality. The chapter concludes with an admonition to commit to quality.

EVERYONE IS FOR IT!

Quality is like apple pie, 4th of July, and long weekends. It's something to cheer about. We want quality in our personal, family, community, and professional lives. And we and our organizations – business, government, academic, and volunteer – like to create quality slogans. I have gradually shifted to the view that if an organization stresses quality slogans, it is in the very infancy of quality. That organization has not yet embedded quality in its culture.

Slogans and the underlying good intentions are fine. However, even with them, client, owner, customer, and stakeholder wants and needs are often overlooked and risks are not identified. Accordingly, the resulting product or service falls far short of what was wanted or needed. John Ruskin, English philosopher, stressed the challenge of providing quality when he said "Quality is never an accident; it is always the result of intelligent efforts."

Yes, everyone is for quality. But, in the context of this book, what is quality and how do we achieve it as we plan and manage projects and carry out our other student and practice activities? How do we make quality operational in our daily work as we provide services and products to our internal colleagues and to our clients, owners, customers, and stakeholders? Those questions are pragmatically answered in this chapter.

Historic Note

The U.S. quality movement over the past few decades was heavily influenced by Dr. W. Edwards Deming. His formal education included a BSEE and a PhD in mathematics and physics. As a consultant to the U.S. War Department, he was sent to Japan immediately after WWII to help the Japanese rebuild their industry. His advice to the shocked Japanese: don't emulate the U.S. companies. More of the advice Deming gave to the Japanese is presented and discussed later in this chapter. Dr. Deming's advice about quality was well-received by the Japanese and they achieved success with it. One indication is the Deming Prize established by Japanese in 1951—the highest business award in Japan, second only to a personal citation from the emperor (Hammer and Stanton 1995, Liker 2004, Rose 2005). A much more obvious reminder of Deming's success with quality in Japan is the large number of Toyotas and Hondas we see on roadways in the U.S. and elsewhere.

QUALITY DEFINED

Clearly, we need a common definition of quality for this book and for your possible use as a student and practitioner, because it is the desired end point. More particularly, what is quality in the consulting business, in government, in academia, and in volunteer organizations?

Quality as Opulence

In the context of professional work or in a general context, the word "quality" may suggest opulence, luxury, "gold-plating," and over-design. Examples of products or results consistent with this opulence concept of quality might be Rolex watches, cashmere sweaters, and safety factors of 5.0. That is, such products and results generally go well beyond what is needed for functional purposes, but not necessarily beyond what may be desired by a few individuals or organizations. In an individual or organizational environment of unlimited or at least great resources, the opulence definition or understanding of quality might be acceptable. The opulent approach to quality, however, is not useful in the vast majority of engineering and business situations.

Quality as Excellence or Superiority

Another approach to quality is the concept of excellence or superiority as suggested by dictionaries which use expressions and words like "degree of excellence" and "superiority" when defining quality. Offering a superior, standard-setting product or service is certainly admirable, but is not likely to be practical for most engineering and business situations. While clients, owners, and customers may want "excellence" and "superiority," they may not want to pay for it. Furthermore, while notions of excellence and superiority may engender positive reactions, they may be too vague, that is, ill-defined. Nevertheless, some technical professionals argue for a superiority approach to quality. For example, Huntington (1989) says, "Quality, however, will not come from automation, but from obsession—a craftsman's obsession with making a thing as good as it can be made." While quality advocates would agree that quality will not necessarily come from automation, many would take issue with the "as good as it can be made" understanding of quality. I do.

Phillip Crosby who wrote *Quality is Free* (1979), one the most influential books about quality, says "The first erroneous assumption is that quality means goodness, or luxury, or shininess, or weight." In the context of professional practice, quality must mean something other than opulence or superiority in kind for the enlightened, progressive, but practical student, young practitioner, or seasoned professional and/or businessperson. This leads to a third definition of quality and one that is used in this book.

Quality as Meeting All Requirements

In his book, Crosby wrote "We must define quality as conformance to requirements." Snyder (1993) offered this definition of quality: "Quality in engineering is a measure of how well engineering services meet the client's needs and conform to governing criteria and current practice standards." Snyder's definition is useful because it elaborates on Crosby's meeting requirements concept. Certainly, the client, owner, or customer has a major role in defining requirements. Learning client needs is part of marketing, which is addressed in Chapter 14. However, for the good of all concerned, the definition of requirements must include and go beyond what the clients, owners, or customers need, or believe they need. For example, the engineer must strive to satisfy government requirements and consistency with the standard of care of the profession, both of which will be discussed shortly.

The requirements of a potential client, owner, or customer may lie outside what you are willing or able to do. In such situations, personal and corporate ethical standards, possibly augmented with liability exposure concerns, may require that you terminate the relationship. Chapter 12 offers engineering students and practitioners an in-depth discussion of ethics with emphasis on application.

As an aside, note the title of Crosby's book. What does he mean by "free?" He explains "free" this way: "Quality is free. It's not a gift, but it is free. What costs are the unquality things—all the actions that involve not doing jobs right the first time."

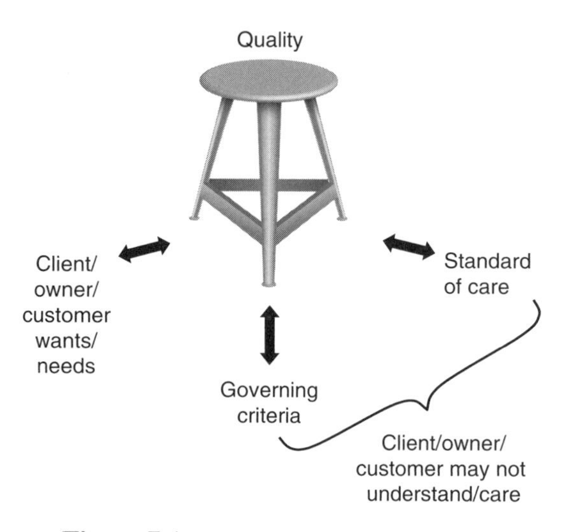

Figure 7.1 Quality stands on three legs.

Consistent with this three-part definition, I see quality as a sturdy stool standing on three legs as shown in Figure 7.1. The first leg is client, owner, or customer wants and/or needs. However, in my view, meeting each and every client, owner, or customer want or need does not constitute a quality project. More is needed. Governing criteria form the second leg. This leg includes local, state, federal, and other requirements as well as design criteria prescribed by others.

The third leg is the standard of care. What does that mean? J. R. Hawkins (2005) offers this short definition: "The level of competence practitioners in their field customarily expect given the circumstances." Standard of care may also be referred to as "duty of care" (Bachner 2007). Regardless of what you decide to use, standard of care or duty of care, please recognize that you are not expected to be perfect. However, you are expected to do your work as well as others doing similar work under the same circumstances. Chapter 11 includes additional discussion of standard of care.

A final thought, as illustrated in Figure 7.1 using the three-legged stool. Your client, owner, or customer may not, at least initially, understand and/or care about two of the three legs—governing criteria or standard of care. Nevertheless, assuming quality is your goal, you must attend to those legs. If you don't build them now, their absence is likely to have negative impacts later.

Figure 7.2 illustrates unfortunate situations in which quality was not achieved even though wants and needs stated by the client, owner, or customer were met. Assume, for example, your firm designed an air pollution control system for a manufacturing plant. It was built and works well – the first quality leg was satisfied. However, your firm neglected to inform the owner about the required permit and now the owner faces a penalty imposed by the state environmental agency. Therefore, the second leg was missing and a quality project was not achieved.

For another example, consider a hypothetical situation. You design the outlet of a stormwater detention basin for a municipality and do not include a safety or debris grate. During a storm, a child is attracted to and goes into the rising water, is carried

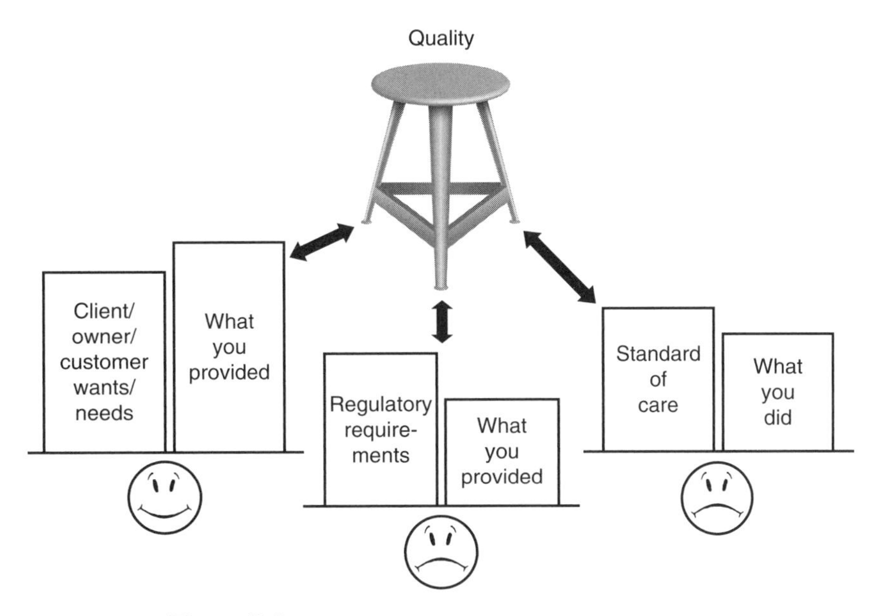

Figure 7.2 Quality requires providing all three legs.

through the outlet and into the downstream underground stormwater system, and drowns. You become a defendant in the subsequent litigation initiated by the municipality. Satisfying the standard of care means that you must be current in your technical specialty, the third quality leg. Three experts called by the plaintiff testify that safety or debris grates are now routinely used. Your design did not reflect the standard of care and, therefore, you were found to be negligent. Clearly not a quality project and clearly a tragedy. Quality requires all three legs!

Personal

Prior to one of my quality presentations, an engineer of one of my clients defined quality this way: "Quality means delivering what the client expects. No more, no less." His is clearly a well-intentioned client, owner, and customer-focused definition of quality. But is that enough? I don't think so. We need to be more specific, particularly with respect to meeting all requirements.

A CAUTION FOR ENGINEERS AND OTHER TECHNICAL PERSONNEL

Frankly, some engineering students and young technical personnel may have difficulty accepting the idea of quality as meeting, but not significantly exceeding, all of the established requirements. Almost any technical activity or project, such as field investigations, laboratory tests, a planning study, and a design culminating in plans and specifications for a manufactured product or constructed facility, can be done better than expected. After all, technical personnel tend to be bright and usually have access to a variety of sophisticated tools and techniques. They know how to "do it better" such as creating a larger spreadsheet, using a computer program instead of performing manual calculations, or writing a report rather than a memorandum. Furthermore, a young person's education may have encouraged him or her to go well beyond what was needed.

However, in the world of practice and business, going well beyond what is needed tends to increase labor and other costs and cause delays, both of which are ultimately disruptive to budgets and profits and to relationships with clients, owners, and customers. Weigh your personal desire to produce a superior or even opulent product or service against the best interests of your employer and those individuals or organizations your employer serves.

But you may be thinking, I want to "delight" my clients-owners-customers-stakeholders by exceeding their expectations, that is, by giving them more than they expected. That's fine, subject to these three conditions:

- You first meet all requirements which is typically a challenge
- They value the "extras" you are going to "give" them
- You can afford to exceed expectations, to provide the "extras," given budget and schedule constraints

Views of Others

Back to the importance of defining quality, the Quality Management Task Force of the Construction Industry Institute says "Quality is conformance to established requirements." Some advice, from industrialist Henry J. Kaiser, about what to do when you have provided quality: "When your work speaks for itself, don't interrupt." English writer John Ruskin said "Quality is never an accident, it is always the result of intelligent efforts." And finally, this more philosophical thought, from minister and writer Ralph Waldo Emerson, that suggests thinking more about quality as we contemplate our daily expectations: "To affect the quality of the day, that is the highest of arts."

QUALITY CONTROL AND QUALITY ASSURANCE

The American Society of Civil Engineers (ASCE 2000) defines quality control (QC) as "the review of services provided and completed work, together with management and documentation practices, that are geared to ensure that project services and work meet contractual requirements." The "what" might be checks, reviews, inspections, tests, and verifications. Each of the "whats" is carried out against a standard or expectation. For example, "All plan sheets will be reviewed by an experienced engineer not directly involved in the project" or "Digital photographs (close up and vicinity view) will be taken of all survey monuments" (Henstridge, 2006).

ASCE (2000) defines quality assurance (QA) as "planned and systematic actions focused on providing the members of the project team with confidence that components are designed and constructed in accordance with applicable standards and as specified by contract." For example, providing initials and dates indicating which experienced engineer, not directly involved with the project, actually reviewed the plan sheets or requiring a sign off when verifying files containing digital photographs of all survey monuments. The American Society for Quality (2011) uses these definitions (quoted):

- **QC:** The observation techniques and activities used to fulfill requirements for quality
- **QA:** The planned and systematic activities implemented in a quality system so that quality requirements for a product or service will be fulfilled

The essence of the preceding ASCE and ASQ definitions is that QC refers to those things that are supposed to be done (e.g., tests) to achieve quality and QA is proof or verification that they were done (e.g., observing a test or reviewing the test result) and appropriate action taken.

Regardless of the phase of a project (e.g., design, construction, manufacturing, operations), QC and QA are typically the responsibility of different individuals or entities within and/or among organizations. For example, the QC/QA process may be applied within an engineering firm during the design process. The work product of designers, carried out within the organization's QC requirements (e.g., double checking calculations) might be reviewed as part of the organization's QA requirements by other designers (e.g., to assure that calculation double checking occurred and/or to make general or detailed parallel calculations). A similar two-part QC/QA process could be used in construction and manufacturing.

SUGGESTIONS FOR DEVELOPING A QUALITY SEEKING CULTURE

Now for some very specific suggestions for developing a quality-seeking culture within your team, department, or other organizational entity. Such a culture shares a common understanding of quality and seeks to provide it to those it serves. In this context, culture means the way things really work around here when serving clients, owners, and customers, especially when faced with problems and challenges. The right kind of quality culture, or underlying system, determines if your team, department, or other organizational entity will thrive, just survive, or die as suggested by growth expert Steven S. Little: "A smart system can work with a little stupidity, but a stupid system can't work with even a lot of smarts."

Before proceeding, consider the International Organization for Standardization. Its short name is ISO which is pronounced "eye so" (ISO 2011). ISO produces standards for just about everything—literally from "nuts and bolts" to, yes, quality management programs. ISO 9000 is a subset of standards—a family of standards—concerned with quality management. Within this section of the chapter, devoted to developing a quality-seeking culture, I will make reference to some of the eight principles underlying

Table 7.1 Eight principles underlie the ISO 9000 Quality Family of Standards.

Principle	Example Benefit
1. Customer Focus	Increased revenue and market share
2. Leadership	Personnel understand where the organization is going
3. Involvement of People	Individuals held accountable for their performance
4. Process Approach	Focused and prioritized improvement
5. System Approach to Management	Ability to focus on key processes
6. Continual Improvement	Improved organizational capabilities
7. Factual Approach to Decision Making	Increased ability to challenge and change opinions and decisions
8. Mutually-beneficial Supplier Relationships	Flexibility and speed of joint responses to changing conditions

(Source: Adapted from International Organization for Standardization, 2011)

ISO 9000. Table 7.1 lists the principles and some example benefits. Why occasionally reference the ISO quality-related principles? Because they reinforce the advice offered here. For example, developing a quality-seeking culture coincides connects with ISO 9000 Principle 2, Leadership.

Strive to Understand Client, Owner, and Customer Wants and Needs

Begin at the beginning. Discover what this person and/or organization really wants and needs. This is ISO 9000 Principle 1, Customer Focus, in action. Study the Request for Qualifications (RFQ) or Request for Proposal (RFP) provided by the organization seeking services. Visit the organization's website. Search for and examine foundational documents such as mission and vision statements and lists of core values.

Personal

I am skeptical—not cynical—but cautious with foundational documents such as missions, visions, and core values statements. At the outset, I accept them at face value—and then look for consistent actions at various levels and places in the organization.

Review documents such as a city's capital improvement plan (CIP), a consulting firm's strategic plan, or a manufacturer's business plan. These are typically more substantive and objective than mission, vision, and core values. Talk to colleagues who have worked with or otherwise know the organization you hope to serve. For example, are personnel in this organization open to new ideas or do they prefer the tried and true?

Ask and Listen

Recognize that RFQ or RFP and the other listed sources are just the "tip of the iceberg" in revealing wants and needs. You need much more data, information, and knowledge in order to understand wants and needs and the way to accomplish that is to ask many and varied questions. Front end your asking and listening, as illustrated in Figure 7.3, in order to determine wants and needs. The horizontal axis depicts the life cycle of a project and the vertical axis shows the amount of asking-listening relative to telling. Focus on asking and listening early on. If you feel compelled to do a lot of telling, to talk much, do that later.

Skillfully mix two kinds of questions with the being closed-ended questions. These are answered with yes, no, or statements of fact. They typically begin with words like how much, where, what, who, is, are, can, will, and should (Parkinson 2010). For example:

- How much is budgeted for the hospital expansion?
- Where will project implementation funds come from?
- Who will be the principal obstacle to project success?
- What data/information are most suspect?

Engineers and other technical personnel tend to ask all or mostly closed-ended questions to get the facts. While this practice is useful, also ask the second type of question, that is, open-ended which often begins with words like why, how, and what. For example: "Why are you adding on to the existing hospital rather than constructing a new one?" Open-ended questions motivate the other person to elaborate. Robert Half, founder of a staffing firm, said: "Asking the right questions takes as much skill as giving the right answers."

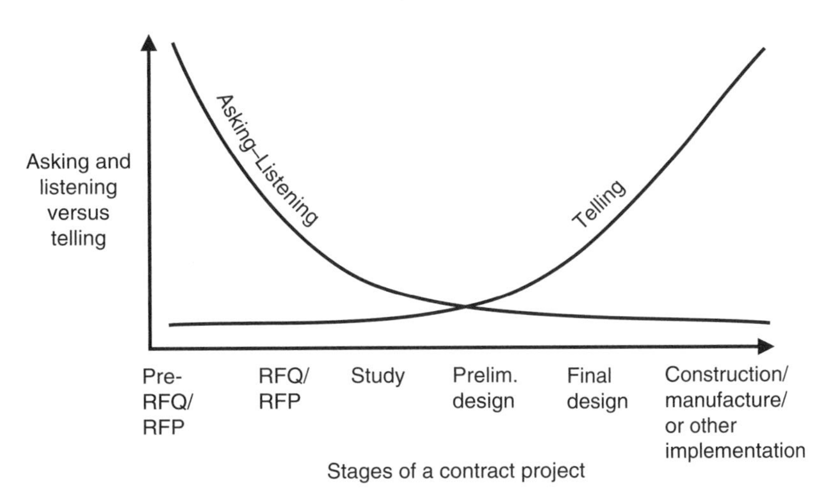

Figure 7.3 Front end asking and listening in order to learn the wants and needs of clients, owners, customers, and stakeholders.

Personal

Years ago, my wife and I decided to repaint the exterior of our house. We had to buy five gallons of different paint mixes and test each on our siding before we found the right color. Getting the right color required patience and skillful mixing of different colors. Similarly, getting the right information needed to understand client, owner, and customer wants and needs requires skillfully mixing closed and open-ended questions.

The "5 Whys" is a specific type of open-ended questioning that helps you and others move beyond symptoms to real concerns, causes, wants, and needs (Liker 2004). Diplomatically and persistently ask "why?" or variations on "why?" up to five times. There is nothing "magic" about five, but you will need at least several "whys."

Refer to Table 7.2 for an example of series of "why?" questions being asked of a hypothetical, potential industrial client who expressed interest in having more entrances/exits for the company parking lot. On hearing the answer to Question 1, you might be tempted to stop asking. However, you don't. Instead you ask Question 2 and learn more but still think that more or modified entrances/exits are needed. You continue to ask. As a result of the answer to Question 3, you begin to think in ways that differ from what you were thinking after the first two"whys." Good, that's why you are using the "5 Whys." So you ask Question 4 and the answer intrigues you if for no other reason than that "always done it that way" often indicates an inability to look at old practices in new ways. There's still more to learn, so you ask Question 5 and, as the potential client thinks about your question and answers it, the client solves the problem.

Table 7.2 The "5 Whys" is an effective way to discover wants and needs.

Question	Answer
1. Why do you need more entrances/exits for your employee parking lot?	Because the lot cannot handle number of vehicles.
2. Why does the lot not accommodate the number of vehicles?	We have major congestion and frustration when workers enter and leave the lot.
3. Why does this happen?	Because day-shift workers leave the lot at 4:00 PM when night-shift workers are arriving and night-shift workers leave at 4:00 AM when day-shift workers are arriving.
4. Why do all day and night shift workers arrive/leave at the same time?	Because we have always done it that way.
5. Why do all workers in a shift need to start and leave at the same time?	They don't—we could stagger start—stop times for subgroups in 15 intervals. That would solve our problems and save us engineering fees and construction costs. I like the way you help us think through a problem.

Speaking of thinking, you may be thinking that as a result of using the "5 Whys," the questioner just talked himself or herself out of a project! That raises another question: What is the service provider's highest priority, that is, is it winning projects or is it gaining or retaining clients, owners, or customers? The answer to that question depends on an organization's marketing philosophy and policies and that topic is addressed in Chapter 14.

English writer Rudyard Kipling offered another method for effective questioning when he said "I had six honest serving men – they taught me all I knew. Their names were Where and What and When and Why and How and Who." Hard to imagine a potential client, owner, or customer and/or a project that does not embody Kipling's six elements. Imagine you are sitting across the desk from a representative of an organization you want to serve. You can recall Kipling's "six" and use them to guide your questions.

Mixing closed and open-ended questions, using the "5 Whys?" technique, applying Kipling's six honest serving men, or any persistent questioning, enables you to "drill down" – to get to the bottom of things – to move past symptoms and get to causes. You go up the learning curve and gain a better understand client, owner, or customer wants and needs. Such understanding is the "secret" of quality.

In closing this asking and listening discussion, consider the following possible consequences of not asking lots of questions and, as a result, not understanding wants and needs:

- You and your team correctly solve the wrong problem, or only part the actual problem, and then you see the wisdom of this comment by humorist Josh Billings: "It ain't so much the things we don't know that gets us into trouble. It's the things we know that just ain't so."
- Then you dissipate a lot of energy justifying what you did. Author Don Miguel Ruiz explains it this way: "Because we are afraid to ask for clarification, we make assumptions, and believe we are right about the assumptions; then we defend our assumptions and try to make someone else wrong."

Distinguish Between Wants and Needs

As you learn about wants and needs, try to distinguish between them. Modifying "wants" to become realistic "needs" is one way you show concern and reveal your expertise. And, sometimes, as a result of your effort, the client, owner, or customer will realize an economic benefit. Consider this example: A community's director of public works says she wants end-of-pipe storage to prevent combined sewer overflows (CSOs). You, as a consultant knowledgeable about solving CSO problems, diplomatically say the community needs control of CSOs. And, you go on to explain that there are other means, besides end-of-pipe storage, to prevent CSOs and they may be less costly.

In contrast, sometimes caring and competent asking leads to a situation in which your organization's judgment indicates that the cost of what is needed to solve a problem exceeds the cost of what the client, owner, or customer wants. Now what? My advice is to perform the trusted advisor role and tell the prospective client, owner, or customer that their "needs" will cost more than their "wants." My advice assumes that your overall goal is quality, that is, conformance to all requirements.

Define the Other Project Requirements

Recall that the first leg on which quality stands, as illustrated in Figure 7.1, is client, owner, or customer wants/needs. Presumably you now understand that leg. Therefore, turn attention to the other two legs. The second leg is governing criteria, that is, local, state, federal, and other requirements as well as design criteria prescribed by others. You or your organization know the governing criteria or can find them. Furthermore, you are "keeping up" and are competent, that is, you know the applicable standard of care, which is the third and last leg supporting quality.

Assess and Manage Risk

Now that client or stakeholder wants and needs and other project requirements are defined, what risks will the project team encounter in meeting those requirements? The likelihood of meeting all requirements is inversely proportional to the team's ability to assess and manage risk. Risk, including an effective process for assessing and then managing it, is discussed in Chapter 5 as one element of a project plan. Assessing and managing risk reflects ISO 9000 Principle 5, System Approach to Management and Principle 7, Factual Approach to Decision Making.

Think Upstream, Not Downstream

W. Edwards Deming, the quality movement leader introduced earlier in this chapter, advocated the upstream approach as illustrated in Figure 7.4. That is, focus on the process used to create the deliverable, not the deliverable itself. He urged studying the production process, identifying the tasks and their interrelationships, determining the effectiveness of each task, and improving each task. The process referred to might be all the tasks completed by a project team in designing a manufacturing process, constructing a bridge, and writing software.

Deming told the Japanese after WWII that the U.S. used the downstream approach, as also shown in Figure 7.4, and he urged them to use the upstream approach. With the upstream approach advocated by Deming, the organization first commits to understanding all the tasks in its production process and their interrelationships. Then the focus shifts to improving each task. This goes beyond doing each task right to continuously striving to do each task better. The Japanese are credited with strengthening Deming's advice by practicing Kaizen, that is, continuous, incremental improvement (Rose 2005).

In contrast, the downstream approach recognizes the production process but does not consistently delve into it. Instead, heavy reliance is placed on inspecting deliverables and, if they do not meet requirements, "fixing" them. Sometimes when an organization experiences quality problems, they apply the downstream approach even more energetically by doing more checking. Or they blame individuals or search for scapegoats. Such reactive approaches are 180 degrees from Deming's upstream

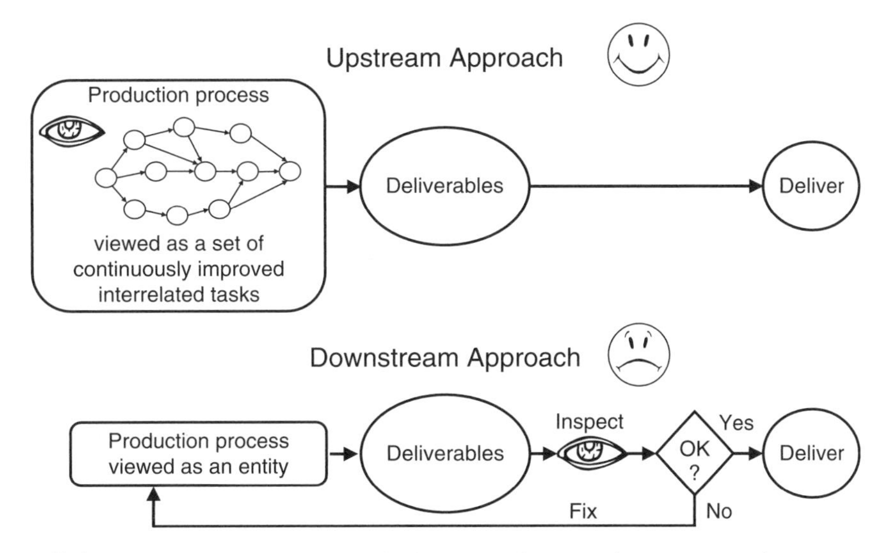

Figure 7.4 The upstream approach, with its emphasis on continuous process improvement, is more likely to achieve quality than the downstream approach with its emphasis on inspection and fixing.

method. Thinking upstream coincides with ISO 9000 Principle 4, Process Approach, and Principle 6, Continual Improvement.

> **Personal**
>
> An engineer wrote this to me: "As long as the work is being reviewed prior to delivery to the client (reports, plans, specs., etc.) then a quality project should result." Maybe, but relying heavily on the "quality by inspection" approach, that is, the downstream approach, can be very costly.

Create, Use, and Continuously Improve Written Guidance for Repetitive Tasks and Processes

While each of your organization's projects, whether internal or for others, is likely to be unique, many of the tasks and processes (series of tasks) included in a project are repetitive. Examples are delineating watersheds, designing a circuit, sizing members of a truss, or preparing next year's budget. Consider documenting how you and your organization currently do each repetitive task or series of task. One reason: Don't "reinvent the wheel" each time. Instead, make existing wheels roll even better (Weiss 2003). My admonition to develop, use, and continuously improve written guidance is supported by ISO 9000 Principle 4, Process Approach, and ISO Principle 6, Continual Improvement.

Written Guidance Explained

What do we mean by written guidance? Written guidance has many names and comes in many forms. Examples of other terms are best practices, bulletins, checklists, guidelines, mini-manuals (Galler 2009), standards, tips, and templates. See the caveat later in this written guidance section about using the term "standards." The label or format aside, the intent of written guidelines is to capture the current cumulative knowledge and experience of an organization's personnel. Write it down and widely share it and then, once understood and applied, continuously improve it.

Personal

Based on my consulting work and management experience, individuals and organizations tend to strongly resist preparing, using, and continuously improving what I refer to as written guidance. Two principal arguments against written guidance typically arise. The first is that preparing the written guidance takes too much time. While an initial significant investment of thoughtful time is necessary, the reward will be a great return on that investment because of the many benefits that flow from written guidance. The second argument is that written guidance stifles creativity. On the contrary, implementing written guidance for routine, repetitive tasks and processes frees up personnel to collaborate, create, and innovate. Other objections to written guidance include concern with maintaining job security, writer's block, and not fully understanding the why and how aspects of a task or process. Nevertheless, my experience as a manager in government, business, and academic organizations convinces me that written guidance yields benefits way in excess of the cost.

Henry Ford experienced opposition to what he called "standardization." He said: "If you think of 'standardization' as the best you know today, but which is to be improved tomorrow, you get somewhere. But if you think of standards as confining, then progress stops."

Benefits of Written Guidance

As suggested by this Project Management Institute (PMI) data presented in Figure 7.5, only a small fraction (24 percent) of the surveyed organizations use written guidance throughout. My experience resonates with these survey results.

And, as suggested by Figure 7.6, organizations that do use written guidance perform better, although their performance is not stellar. If you and/or your student or practitioner team, your department, or your organization do not use written guidance and are not receptive to the idea, is this one study sufficient to cause you to at least experiment with written guidance? Probably not.

Speaking of experiments. In 1927, the Western Electric Company at its Hawthorne plant in Illinois, attempted to discover the effect of incentives on workers' production. At first, to their delight, as incentives were increased production went up. But, somewhat to their dismay, when incentives decreased production still climbed. This

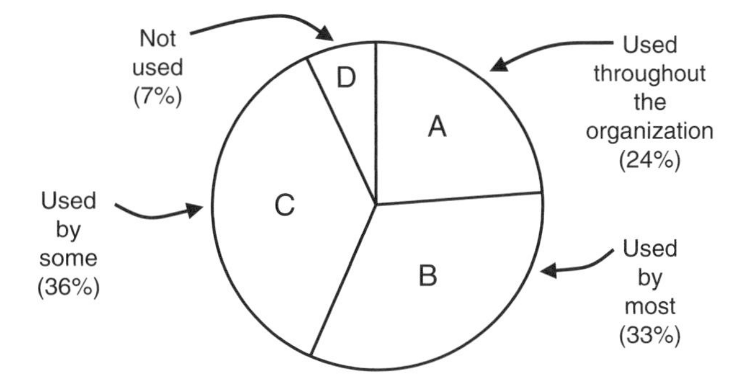

Figure 7.5 Written guidance is typically not used throughout organizations. (Source: Adapted from PMI, 2007)

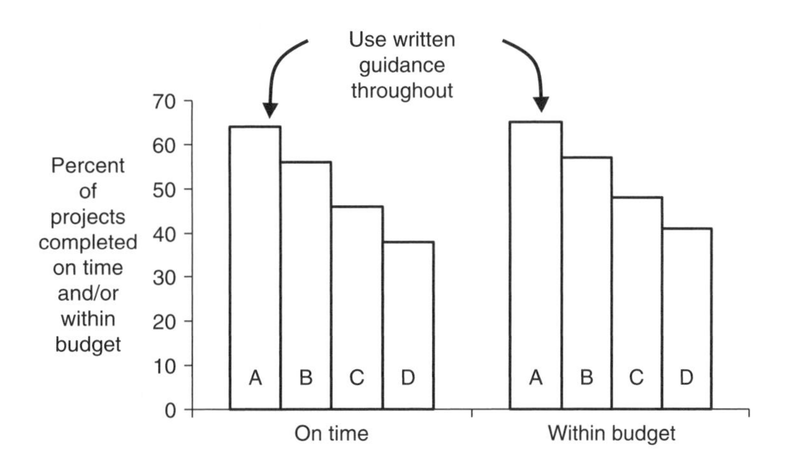

Figure 7.6 Organizations that use written guidance perform better with respect to schedule and budget. (Source: Adapted from PMI, 2007)

later became known as the "Hawthorne effect." Production increased because workers were being made aware of what they were doing and, naturally found better ways to do it. Their point of view was changed; they were not just working but participating in a worthwhile experiment (Nierenberg 1996). Being more aware of what we are doing and then ways to improve, is one benefit of preparing, using, and continuously improving written guidelines.

Perhaps the PMI data and the introduction to the Hawthorne effect and further evidence presented near the end of this written guidance section, even though they may not pertain directly to whatever you and your organization do, will cause you and to give further thought to implementing written guidance for repetitive tasks and processes. It will help you achieve quality, that is, meet all requirements.

As a result of creating, using, and continuously improving written guidance, an organization and its members will derive the following nine long-term benefits:

1. ***Eliminate Valueless or Marginal Activities:*** Unnecessary, redundant, outdated, and other marginal or valueless tasks and steps are likely to be identified and then updated or removed. The process of thinking about, discussing, and then describing, in writing, possibly with graphics, steps to be taken in doing a task or tasks to be completed in a process inevitably identifies valueless or marginal activities. Finding and eliminating tasks reduces expenses and frees up staff for more productive tasks.
2. ***Increase Efficiency:*** More efficient approaches typically arise as a result of thinking through steps that comprise a process. For example, some tasks previously done in series may subsequently be done in parallel, thereby reducing elapsed time. Tasks formerly done by personnel with high hourly rates may subsequently be accomplished just as effectively by personnel with lower hourly rates, thus reducing costs. Delegating tasks is simplified because ready reference can be made to written descriptions. Tasks previously done manually may, as a result of the insight gained by analyzing a process, subsequently be executed with software thereby reducing costs and saving time.
3. ***Avoid "Reinventing the Wheel":*** Knowledge acquired and experience gained by an organization's personnel, assuming it is reflected in the written guidance, can be readily shared with other personnel. As a result, much less time is wasted in redeveloping methods.
4. ***Facilitate Interdiscipline and Interoffice Projects:*** The multidiscipline, multi-office organization should be structured and operated to provide its clients, owners, and customers with the optimum mix of personnel and other resources regardless of the physical and organizational "homes" of personnel. Written guidance that captures best practices applicable across discipline and office lines facilitates the desired corporate team approach. The lack of guidance frustrates interdiscipline and interoffice cooperation even when personnel want to work as a team.
5. ***Train New or Transferred Personnel:*** New personnel and personnel transferred from one office, division, or unit to another can receive the appropriate written guidance as part of their on-the-job education and training. This approach requires less supervisor and colleague time than relying on verbal descriptions and the shared ideas and information are more likely to be understood because they are written.
6. ***Invoke the Novice Effect:*** When new or transferred individuals use written guidance to do a task that is new to them, and are asked to suggest ways to improve those guidelines, the novice effect that was introduced in Chapter 2 is likely to occur. They read the guidance and think of better ways to explain what is to be done or improved ways to do it.
7. ***Reduce Liability:*** Negligence, the principal cause of liability claims in the consulting engineering business, is reduced. Errors and omissions are less likely to occur when work is influenced by tested, written guidance.

8. ***Mitigate Negative Impact of Personnel Turnover:*** Some personnel turnover is inevitable, even in the best-managed and led organizations. Contributions that departed personnel made to an organization are more likely to remain with the organization if some of those contributions were captured and documented in the form of written guidance prepared by the now-departed personnel. In his book about the growing importance of intellectual capital, Stewart (1997) defines such capital as having three components: knowledge embedded in an organization's processes, knowledge emanating from its clients or customers, and knowledge held by the organization's employees. The last portion of intellectual capital can go down the elevator and out the door at any time. Accordingly, proactive organizations try to capture some of the knowledge held by their personnel. Written guidance is one way to do this.
9. ***Support Marketing:*** External and internal clients, owners, and customers are increasingly concerned about the quality of the services and products they receive. The test is: do or will these services or products meet their wants and needs? Using written guidance is one way of demonstrating a unit or organization's commitment to quality and especially a desire to do things right the first time.

If you are still not convinced of the value of written guidance, then consider the results of a study from the medical profession in which checklists (one form of written guidance) were used in surgery (Gawande 2009, Wilson 2010). Under the sponsorship of the World Health Organization (WHO), medical professionals developed a 19-step surgery checklist. The three parts were: before anesthesia, after anesthesia but before incision, and at the end of operation before the team wheels the patient from the operating room. They tested the checklist at eight hospitals around the world and compared surgery results before and after its use. The amazing results: major complications down 36 percent, deaths down 47 percent, and infections decreased by almost half.

Atul Gawande (2009), the author of *The Checklist Manifesto*, the book that describes the surgery checklist study, stated that "The checklist gets the dumb stuff out of the way, the routines your brain shouldn't have to occupy itself with." This prompts me to ask two questions:

- Would you and your colleagues like to get "the dumb stuff out of the way" so that you could use your brains for higher level thinking?
- Do you occasionally, or maybe frequently, experience counterparts of the medical profession's "major complications," "deaths," and "infections" and would you like to markedly reduce them?

If your answer to the two questions is even a tentative yes, give checklists or, more generally, written guidance a try. Gawande also notes that checklists improve outcomes with no increase in skill. That's a thought for those of us who, when an operational problem arises, may immediately focus on the need for more education and training. Maybe, instead we should make more use of checklists or other forms of written guidance. The author shared an anecdote from surgery which indicated members of a surgical team favor checklists because "they improve their outcomes with no increase in skill."

By the way, aircraft pilots routinely use checklists. This includes Captain Chesley B. "Sully" Sullenberger III and First Officer Jeffrey Skiles on January 14, 2009 when they safely landed an Airbus with 155 people on board in New York City's Hudson River. They routinely ran through checklists at LaGuardia Airport before starting the plane's engines. And, after takeoff as soon as the impact with geese stopped the plane's two engines, First Officer Skiles reached for the how to "relight" the engines and "ditching" checklists (Gawande 2009).

Caveat

A word of caution is in order. While written guidance has many benefits, as described above, they could increase liability exposure in litigation. Therefore, be prudent. For example, if you and your organization are using, or will be experimenting with, written guidance, don't call them "standards." Instead, use one of the other terms noted earlier in this section. Reason: Possible confusion, if litigation occurs, with the "standard of care" concept introduced earlier as one of the three legs of quality.

Assume your organization does call the written guidance "standards" and that your organization is the defendant in a liability case and you are testifying. The plaintiff's attorney says, "Are these your standards?" You answer "yes." "Did you use them on this project?" Your answer: "Not entirely, they are guidance." The attorney's assertion: "Oh, so you did not adhere to the standard of care." Written guidance should include appropriate disclaimers and be used prudently. If intelligently applied, the benefits of written guidance greatly offset the disadvantages.

Closing Thoughts About Written Guidance

To conclude this discussion of creating, using, and continuously improving written guidance as a means of achieving quality, consider an observation from the book *If Only We Knew What We Know* (O'Dell and Grayson 1998). The authors write: "You would think... better practices would spread like wildfire to the entire organization." And then they follow their observation with this blunt contrary statement: "They don't."

Why? Because the better practices are not systematically documented, widely shared, and continuously improved. Is your team, group, or other organization making optimum use of its better practices? You know you have them. However, having them for the benefit of a few and leveraging them for the benefit of many are two very different situations.

Expect Each Person to Check His or Her Work

One danger of the QC/QA aspect of a quality program is that it becomes the rationale for not being responsible for one's work. The argument goes like this: "Well, I don't have to be that careful with my calculations, someone else is going to check them anyway." Adopt a quality policy that requires double checking. Push work back to people who make errors, don't correct their work. Personally exemplify responsibility for the accuracy and completeness of your work products. Expecting everyone to check their work is supported by ISO 9000 Principle 3, Involvement of People.

Personal

While serving as a university professor, I sometimes encountered a casual attitude toward work products among some students. I graded homework and these students were careless, apparently thinking I would correct mathematical errors and provide the correct units. Therefore, I established a policy of an "off the top" 50 percent reduction in credit on an assigned problem for a mathematical error or for missing or erroneous units. I held the students accountable for accuracy and units rather than having them hold me, or later on, their colleagues and others, accountable.

Arrange for External Reviews

The idea of external reviews is to leverage knowledge and experience assets within your organization. Build external reviews into your project plan by identifying experts outside of the project team, assigning hours to them, defining when their help will be needed, and indicating what you want them to do. Early input by "high-priced," experienced personnel might result in great savings later on by diminishing rework, additional work, or unnecessary work. Focused participation by experts could help you achieve quality and do it within the budget. Consider an example from the construction field. In an article about constructability reviews, P. D. Folk (2006) concludes that poor constructability "usually results from lack of input by builders or construction managers during the planning, design, and procurement processes" and often leads to liability claims against the engineer. Timely participation by appropriate experts can help you achieve quality.

Reduce Cycle Time

Cycle time is the total elapsed time needed to complete a process, typically a process that includes many repetitive—you've done them before—tasks. Cycle time is measured in hours, days, weeks, and months. It is not the absolute time, for example, person-hours invested in the process. Examples of processes are preparing a proposal, designing a manufacturing process, preparing a capital improvement plan (CIP), obtaining soil borings, and constructing a bridge.

Studies suggest that reducing cycle time can reduce costs and improve quality (Liker 2004, Wilson 2007). One reason costs are reduced is that we experience less wasteful "stop and restart" efforts. However, quality improvement may seem counter-intuitive. For example, you might think that "rushing" and "hurrying" would result in more errors and, therefore, more rework. However, when individuals involved in a process are challenged to do it faster, that is, seek reduced cycle time, they are more inclined to do it right the first time or, better yet, devise an even better way to do it. Remember the Hawthorne effect mentioned earlier. Process diagramming, one tool for reducing cycle time, is described in the next major section of this chapter. Reducing cycle time aligns with ISO 9000 Principle 4, Process Approach and Principle 6, Continual Improvement.

TOOLS AND TECHNIQUES FOR STIMULATING CREATIVE AND INNOVATIVE THINKING

The Need for and Value of Tools and Techniques

To supplement the preceding suggestions for achieving quality, we are fortunate to have many tools and techniques that stimulate a group, such as project team, a committee, or a department to collaboratively think more deeply and widely and more creatively and innovatively about a problem to be solved, an opportunity to be pursued, or an issue to be resolved. This chapter section describes 13 tools and techniques which are a small subset of the many are that available. A principle supporting the use of these methods is that a problem well defined is half solved. You, as a student or young practitioner, can unilaterally use some of these and lead their use by others. That is, all the suggestions do not necessarily require participation of your team, group, or organization.

By using one or more of these tools, an individual or a group is more likely to achieve quality, that is, meet all requirements. Why? Because these methods encourage interaction, synergism, and use of both hemispheres of the participant's minds. (For an introductory discussion of the brain's hemispheres, see the subsequent section of this chapter titled "Why Mind Mapping?"). "We know where most of the creativity, the innovation, the stuff that drives productivity lies," according to former GE Chairman Jack Welch, "in the minds of those closest to the work." Creativity and innovation lie within most of us; we need catalysts to release them.

To reiterate, I am offering these methods to facilitate collaboration, creativity, and innovation within your project team, committee, department, or other group. We don't have to rely only on what Gerard Nierenberg, author of *The Art of Creative Thinking* (1996), calls "accidental creativity." While we certainly welcome the "Aha" moment or "insight out of the blue," we can also stimulate intentional creativity and innovation by using the methods offered here. These tools and techniques are provided because innovation and creativity don't come easy. "All engineers must synthesize, some will innovate, but only a few are able to be truly creative" (Beakley et al. 1986). My hope, in describing an array of methods, is that more engineers will move from synthesis to innovation and from innovation to creativity. These methods clearly serve the quality focus of this chapter. They also are referenced from subsequent chapters, mainly Chapter 8, which discusses design, and Chapter 15, where the emphasis is on effecting change.

Use of these methods to help achieve quality is connected to many ISO 9000 principles, especially Principle 3, Involvement of People. Frankly, we often are so busy doing our work that we fail to collaborate with colleagues. Use of appropriate tools and techniques encourage productive collaboration and increases the likelihood of achieving quality, of meeting all requirements.

Create and Innovate Defined

Having used variations on the words create and innovate, let's define and distinguish between them as follows:

- Create: Originate, make, or cause to come into existence an entirely new concept, principle, outcome, or object. An example of creating is Velcro, the hook-and-loop fastener made of Teflon loops and polyester hooks. It was created in 1948 by George de Mestral, a Swiss electrical engineer. He was returning from a hunting trip with his dog, noticed burdock burrs (seeds) on his clothes and on his dog's fur, examined the burrs under a microscope, and noted many "hooks" that caught on anything. This led him to see the potential for binding two materials reversibly (Wikipedia, 2011b).
- Innovate: Make something new by purposefully combining different existing principles, ideas, and knowledge. Johannes Gutenberg's reusable type printing press, which he first used in about 1439, is an example of innovation. In designing it, he borrowed from woodblock printing, weapon and coin forging, and the screw press process used by winemakers and olive oil producers (Murray 2009, Wikipedia 2011a).

These definitions, which were influenced by similar ones offered by consultant Ned Herrmann (1996), teacher and consultant John Kao (2007), consultant Gerard I. Nierenberg (1982), and by engineer and educator John E. Arnold and engineering educators George C. Beakley, Donovan L. Evans, and John B. Keats (Beakley et al 1986), suggest that innovate and create differ by degree of originality. While innovation is, in effect, "integrative and aspirational" (Kao (2007)) and "grounded in already-invented products or processes" (Herrmann 1996), creativity is "grounded in originality" (Herrmann) and "coming up with something [completely] new" (Nierenberg 1982). We might think of innovate and create as actions that differ by degree of newness where to create is the ultimate.

Brainstorming

This is a great place to start whether trying to determine the cause of a problem, find a solution to one, or explore ways to pursue an opportunity. Invite a wide variety of participants. For example, if the problem at hand is how to reduce the drag on an automobile component invite representative engineers, scientists, technologists, technicians, and administrative personnel – seek representatives of all functions that are involved in any aspect of automobile design and manufacture.

Create a non-threatening environment so that all participants feel free to say what they think. Provide background and pose the problem and then invite, in fact, expect everyone to offer ideas being careful not to evaluate any ideas during the brainstorming session. Great ideas often appear because of:

- Widely varied knowledge, skills, and attitudes represented by the participants
- Some participants are very close to problem causes—but no one ever asked for their views
- Contributed ideas stimulate the thinking and creation of more ideas by others

Go for quantity, not quality (Byrne, 2005) and, remember, absolutely no evaluation of ideas during the brainstorming session. As noted by poet and critic Mark Van Doren, "Bring ideas in and entertain them royally, for one of them may be the king."

Or, as observed by scientist Linus Pauling, "the best way to have a good idea is to have lots of ideas."

Mulitvoting

Possibly use multivoting after brainstorming starting with the ideas generated during the brainstorming session. Participants might be the original brainstorming group or new participants. Again, set the scene, list the available ideas, and give each person multiple votes. As shown in Figure 7.7, ask each participant to anonymously cast their allotted votes any way they wish. A guideline for determining the number of votes per person: Divide the number of ideas by 5. Assume the group has 25 ideas. Then each person receives five votes to anonymously cast any way they want including assigning all five votes to one idea. Rank the results and develop one or more high priority ideas (Rose 2005).

Strengths-Weaknesses-Opportunities-Threats

To apply SWOT, assemble a heterogeneous group—certainly don't clone yourself. Provide background and pose the problem. Assume it is a poorly performing service line in a professional services firm. Create a matrix like Figure 7.8 and start with the first or personnel row. In the Strength column members of the group agree to write

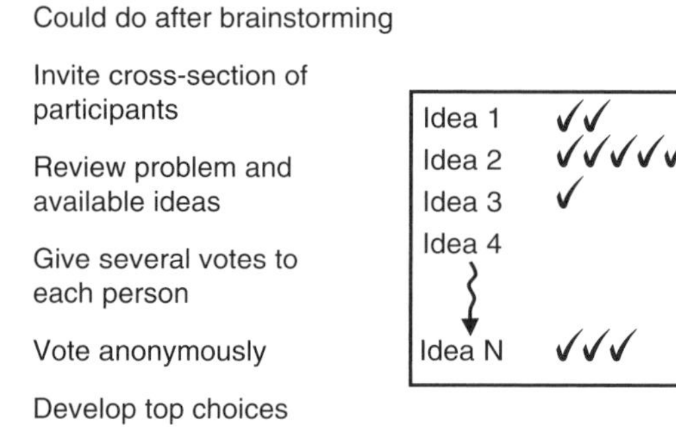

Figure 7.7 Multivoting helps a group prioritize, for further evaluation, a list of options.

Element	S	W	O	T
Our personnel				
Our tools				
Existing clients				
Potential clients				
Regulatory environment				
Financing available				
	Internal		External	

Figure 7.8 A SWOT analysis defines the positive and negative aspects of a challenging situation.

"Bill is an expert" and in the weaknesses column "Bill does not listen to clients or colleagues." The Opportunities column entry might be "Betty contacted us concerning possible employment and she is known to have great client communication skills." The Threats column entry might be "We will lose even more clients if our personnel continue to miscommunicate." Based on what is collectively learned as a result of the SWOT analysis, develop an action plan.

Observation: The SWOT approach is valuable partly because it cannot degenerate into a complaining or whining session. While Weaknesses and Threats are unpleasant or negative, Strengths and Opportunities are uplifting and, therefore, positive. Furthermore, with respect to the group's potential to act and as shown in Figure 7.8, while opportunities and threats are external and largely out of the group's control, strengths and weaknesses are internal and can be addressed by the group. SWOT can be used in a variety of ways. For example, Milosevic (2010) describes use of SWOT to determine how investors and contractors view a particular construction project.

Stakeholder Input

Your or your organization's stakeholders are those internal or external individuals or organizations that have a significant interest in what you and your organization do. Your actions or inactions affect them. As discussed earlier in this chapter, essential to achieving quality is understanding client, owner, and customer wants and needs. The best way to find out how well you and/or your organization are doing in meeting wants and needs is to go to the source—ask your stakeholders. Your organization could do this informally or use formal surveys. As a student or entry-level technical person, you can informally query your stakeholders—fellow students, professors, colleagues, supervisors—to determine how you are doing and how you can improve.

Process Diagramming

Process diagramming, which is also called flow charting or network diagramming, is useful in helping a team or group more fully understand the entire system, the big picture. The premise is that most repeated processes involve individuals from various departments, offices, and specialties. While each knows his or her job, few, if any, may see the big picture. They can't see the "forest," only their "tree," and therefore process improvements are less likely to occur.

Assume selection of a process that involves individuals from various departments, offices, specialties, and/or other characteristics. The process might be preparing a proposal, conducting laboratory experiments, designing and implementing a manufacturing process, or conducting field reconnaissance.

Collaboratively identify the steps or tasks in the process, and their interrelationships, as currently practiced. This may be the first time anyone in the group "sees" and fully understands the process. Assemble the tasks and their interrelationships into a process diagram or flow chart or network diagram as illustrated in the upper part of Figure 7.9. This is the "as is."

Study the result. Look for unnecessary steps/tasks or steps/tasks that could be combined. Search for others that could be done in parallel to reduce the process duration. Then create the new process diagram. As shown in lower part of Figure 7.9,

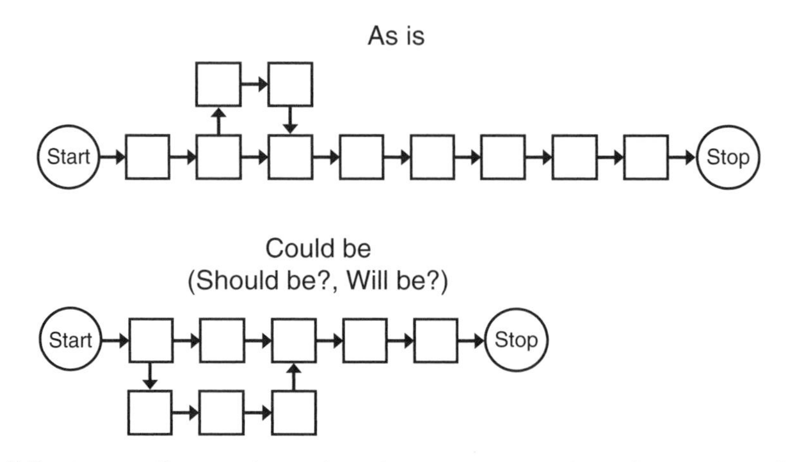

Figure 7.9 Process diagramming seeks to improve a process by actions such as eliminating useless tasks and doing more tasks in parallel.

this is the "could be," "should be," or "will be." Note that in the hypothetical case that the original ten tasks are reduced to eight tasks and the elapsed time is shortened. Use the could be-should be-will be to perform further analysis and/or to develop an action plan to implement the newly-created process.

Fishbone Diagramming

This method, which is also called a cause and effect analysis or the Ishikawa analysis (Black and Kohser 2008, Hensey 1993, Rose 2005), provides a systematic way to thoroughly identify widely-varying possible causes of a problem. Once again, assemble a diverse group, provide background, pose the problem, and construct the fishbone diagram.

Assume that an engineering project team is studying a recently constructed stormwater detention facility that seems to have failed to protect downstream properties during a large rainfall event. Begin constructing the fishbone diagram shown is Figure 7.10 by drawing the "head." Collaboratively identify possible "bones," that is, categories of possible causes of failures such as causative storm, maintenance, the original design, and other. Then creatively detail each "bone," make judgments as to the likely cause or causes of failure, and develop an action plan.

Pareto Analysis

Recall the "vital few-trivial many" rule, "20/80" rule, or Pareto's Law discussed in Chapter 2. The approach introduced here is called Pareto Analysis after Vilfredo Pareto, an Italian sociologist and economist, who is credited as the source (Kuprenas et al 1999, Rose 2005). When agreed-upon quality is not being achieved, Pareto Analysis can be used to identify the most influential causes of substandard performance so these causes can be addressed first.

Assume that a consulting engineering firm analyzed design projects, over a five-year period, that failed to be profitable. Failed profitability was defined, from a quality

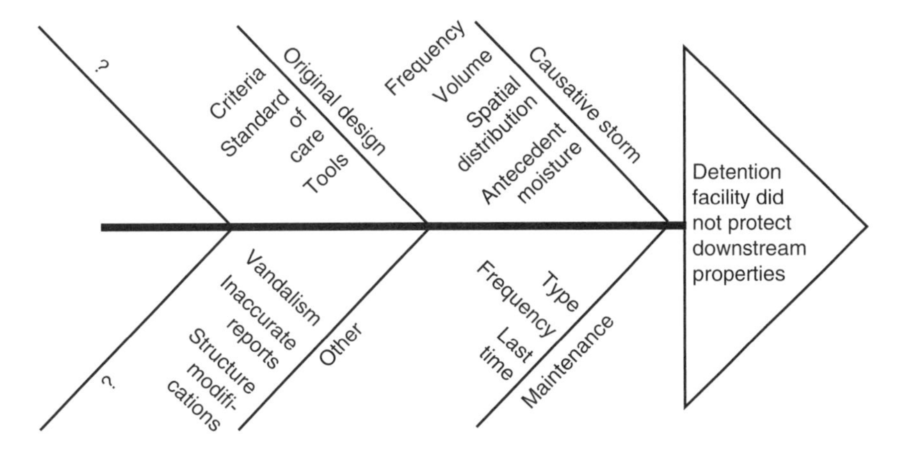

Figure 7.10 Fishbone diagramming used to find possible causes for the failure of a storm-water detention facility to prevent downstream flooding.

Table 7.3 Possible causes of poor project performance and the number of surveyed personnel selecting each cause as being the most important.

Possible cause of poor profitability	Number of surveyed personnel reporting the cause as being the most important
1. Technical errors	3
2. Non-technical errors	4
3. Client-driven uncompensated scope creep	17
4. Internally-driven uncompensated scope creep	85
5. Poorly-defined scope	9
6. Unresponsive client-owner-customer	8
7. Change in project manager and/or key personnel	6
8. Absence of a project plan	135
9. Unexpected conditions and/or events	5
10. Insufficient project monitoring	5
Total:	277

perspective, as earning less than half of the project's planned profit. Table 7.3 presents the results of a staff survey in which they were asked to select, among ten options, what they viewed as the most dominant cause of not achieving quality.

In Figure 7.11, the possible causes are arranged on the horizontal axis in descending order of number of surveyed personnel selecting the causes as most influential. This display, which is typical of Pareto Analysis results, emphasizes that two causes (8. Absence of project plans and 4. Internally-driven uncompensated scope creep) account for about 80 percent of poor project profitability. Clearly, remedial measures should focus on these two causes. Kuprenas et al (1999) present a case study in which

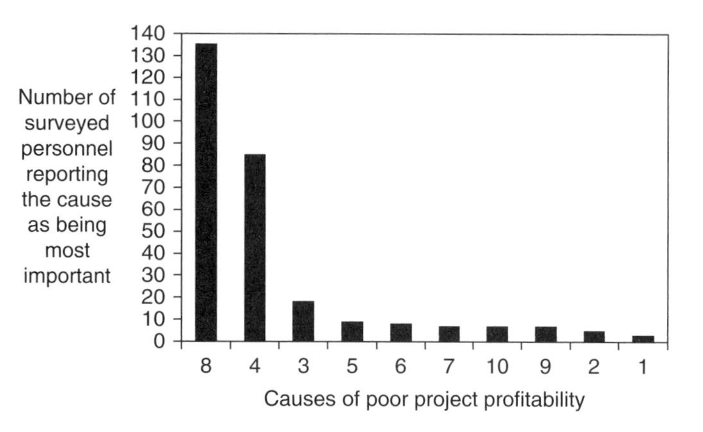

Figure 7.11 Pareto Analysis displays the relative importance of causes of poor project profitability performance so that they can be prioritized for possible correction.

Pareto Analysis was used to reduce defects in the manufacturing of spacecraft electronic components. Examples like these suggest many possible applications of Pareto Analysis.

Problems-First Meetings

Project and other management meetings are often devoted primarily to reporting on progress. Instead, report progress in writing and start meetings with and focus on problems (Liker 2004). Go around the room and ask each person to share a problem they are facing and then expect others to offer initial ideas to solve the problem. The expectation to offer ideas coupled with the group's diversity is likely to lead to synergistic interaction and generation of creative and innovative ideas. Such short, intense collaborative efforts also plant problems and the need to solve them in the subconscious minds of participants. Some of this "planting" may lead to later unexpected "growth" of creative or innovative ideas.

Mind Mapping

When faced with the need, either individually or as a group, to "get started," to generate ideas consider mind mapping, which is also called clustering (Arciszewski 2009, Gelb 2004, Gross 1991, Rico 2000). It offers a simple, quick, and powerful means for assembling a set of connected ideas.

For example, I wanted to write text and prepare visuals for a presentation about the marketing topic "cold calls." Such calls are narrowly defined in the marketing field as meeting with a stranger with the goal of eventually making a sale of services or products. Therefore, I prepared a mind map over a four-day period devoting no more than a total of 15 minutes over the four days to the effort. I spread the effort over four days because by planting an issue in our subconscious mind, it will "work on it" and, the next time we consciously consider the issue, new ideas will appear.

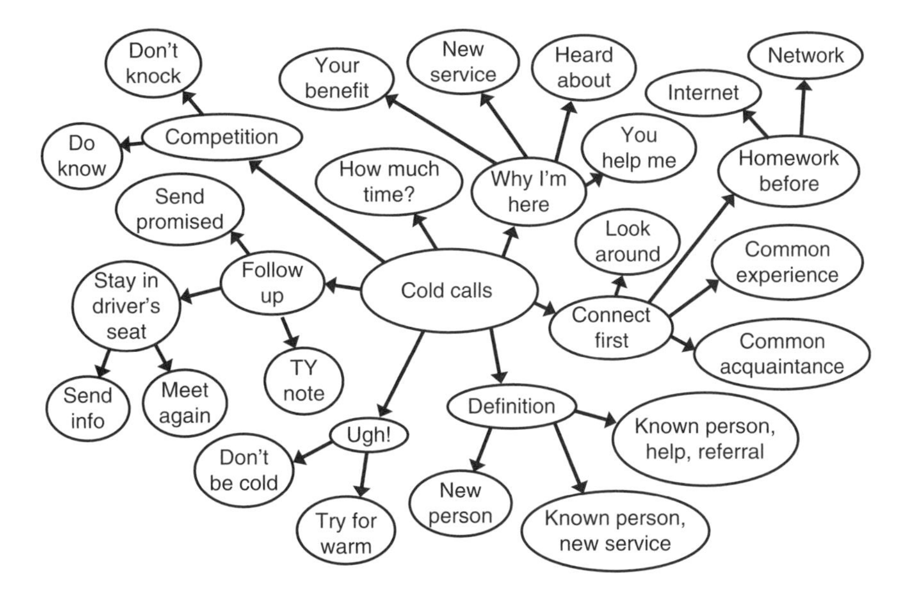

Figure 7.12 An example mind map that suggests the manner in which this process can generate ideas.

The Process and Results

For the purpose of describing mind mapping in this chapter, the subject of the mind map – cold calls – is not important. The important aspects are the results and the process used to obtain them. I took the hand-written version of the mind map and used it to make the neater version that appears as Figure 7.12 Much more time was needed to prepare the figure than was used to prepare the original mind map. To reiterate, the original mind map was completed with no more than 15 minutes of effort over four days.

The mind map started with the center oval labeled "Cold calls." If you or your team were to do this, the initial oval sets the scene. Then do as I did and ask "what does this make me think of?" or "what comes to mind?" "Ugh!" was the first thing that popped into my mind so I added it to the mind map. After that initial negative thought, I became more positive and added the "Don't be cold" and "Try for warm" ovals. This led me to the "Definition" branch and thinking about who might we be calling on and under what circumstances? Then the diagram quickly appeared. The mind map provided a wealth of ideas which I then used to prepare the "cold calls" presentation.

By the way, draw by hand, don't use software. Hand drawing is uninhibited, will enhance spontaneity, and engage both hemispheres of participant's brains (Arciszewski, 2009). Also, while creating the mind map, don't be overly judgmental. If an idea "appears," add it to the mind map. While I did this mind map by myself, as you could do, mind maps can also be developed by a group.

So what do you do with the mind map, that is, assume your group generated lots of ideas, now what? One possibility is to convert the mind map into a draft outline or text and then provide it to the group that performed the mind map or to others for discussion, further study, refinement, and action. View the mind map as an initial individual or group brain dump that provides content for subsequent, in-depth thinking.

Why Mind Mapping?

Mind mapping is an effective means for generating ideas, whether performed individually or by a group, because:

1. It can be done quickly in real time by simply drawing on your knowledge and experience, or the combined knowledge and experience of members of a group.
2. No preparation is required other than to select the topic and perhaps the participants in that they should be diverse in terms of characteristics such as knowledge, skills, attitudes, and experiences.
3. Once the highly-visual process starts, ideas flow in that one idea leads to another. The process is all about generating ideas for later consideration.
4. The process is non-linear, that is, it does not require one item to logically follow another in step-by-step fashion, and, as a result, many and highly-varied ideas are generated. Stated differently, the process, whether done by one individual or a group, engages both hemispheres of participant's brains.

With respect to the last item, most of us typically rely on left-brain thinking which is verbal, analytic, symbolic, abstract, temporal, rational, digital, logical, and linear (Edwards 1999, Walesh 2011a). For example, the left side of our brain likes to think Step 1, Step 2, Step 3, etc. Left-brain thinking is powerful. However, we have another kind of mental ability and it is draws on the brain's right hemisphere.

Mind mapping is a tool to engage more right-brain thinking which is nonverbal, synthetic, actual, analogical, non-temporal, non-rational, spatial, intuitive, and holistic. For example, the right side of our brain tends to see the big picture and new possibilities. We should strive to supplement our valuable left-brain abilities, the development of which is typically stressed in our formal education, with equally valuable right-brain abilities. As a result of mind mapping, you or a group are better equipped to take more creative, innovative approaches to identifying and solving problems, seeing and pursuing opportunities, and addressing issues. A half brain is good, a whole brain is better! (Edwards 1999).

Gelb (2004) suggests trying stream of consciousness writing, which is similar to mind mapping in that is encourages spontaneous, unstructured thinking. Select an issue, problem, or opportunity and begin to write whatever occurs to you. Gelb says "the secret of effective stream of consciousness writing is to keep your pen moving; don't lift it away from the paper or stop to correct your spelling and grammar – write continuously."

Ohno Circle

Named after Taiichi Ohno, an early innovator of the Toyota Production System, the technique is "used to make deep observations of a process or scene with the goal of improving what you see" (Wilson, 2011). This method differs from the others in that it is done by an individual, not a group or team. However, it is intended to enable an individual to find ways to improve group or team efforts to effectively and efficiently meet requirements. "The story is that Mr. Ohno would have his engineers and managers draw a circle about two feet in diameter on the floor in the area of a process in a factory... and ask them to stand in the circle and observe what was going on for up to eight hours at a time" (Wilson 2011). Yes, that does seem extreme.

But how about taking 30 minutes, finding a quiet, unobtrusive place in one of your classes, meetings, offices, or laboratories, or in the field at a project site, and simply observe. More specifically, in the spirit of continuous improvement, look for wasteful or otherwise undesirable situations like these.

- Underutilized resources – Capital, equipment, or personnel that are not being fully deployed or utilized
- Excess motion – Unnecessary movements
- Defects – Production that does not meet standards and needs fixing
- Excess inventory – Retention and storage of production which is not needed now
- Non-value added processing – Performing work for which clients/owners aren't willing to pay
- Transportation – Using excessive resources to move materials, equipment, or personnel
- Waiting – Not having the materials, information, or resources at the right place or time
- Safety and health hazards

Then document what you saw and possibly a few improvement ideas, share with your team, discuss, decide what to do, and act. By the way, while you are watching, also listen. Because is causes us to focus, the Ohno Circle method enables us to see and hear what we would probably otherwise have missed.

Metrics

Metrics, the art and science of measuring and using the results to guide future action, helps us understand how things are going in an absolute sense and in terms of historic trends as we strive to achieve quality, and stimulate us to think more creatively and innovatively about issues, problems, and opportunities illuminated by the data. Metrics are useful in any project management or other management effort committed to achieving quality. The idea is to determine the set of requirements that define quality, select those that can be quantified, and then implement a metrics program to track them and, of course, act on what is learned. Metrics is based on recognizing an element of human nature – what gets measured gets done.

Consider, for example, a manufacturing organization that has defined and is striving to achieve quality. The organization might use the following metrics, maybe on a quarterly or annual basis:

- Time required to respond to a request for a price quote
- Time required to design and test a new product or a major modification to an existing product

- Number of products manufactured per shift
- Percent of each component rejected per shift
- Cost per manufactured product
- Warranty claims on each product per year
- Market share

Similarly, a consulting engineering business committed to quality might use the following metrics, perhaps on an annual basis:

- Percent of projects completed early, on-time, and late
- Percent of projects completed under budget, on budget, and over budget
- Number and percent of new clients served compared to total clients served
- Number and percent of contracts signed by state, province, or country of origin of the client; monetary size of contract could also be used
- Number and/or dollar amount of liability claims and statistics and how they were resolved
- Chargeable hours as a percent of total hours worked, expense ratio, and multiplier as discussed in Chapter 10

A variety of metrics are available to municipalities to help them achieve their quality objectives. Examples include number of flooding complaints; number and percent of construction projects completed early, on-time, and late; and number and percent of employees who receive education and training. As part of their quality program, colleges and universities might track metrics such as alumni with advanced degrees, graduates placed in employment or graduate school, Scholastic Aptitude Test scores of incoming first-year students, teaching load credits per faculty member, and funds raised from external sources.

Freehand Drawing

Freehand drawing is another means of assisting individuals and groups to more fully use their mental resources and thus be more creative and innovative. Drawing, which might be defined as converting "a mental image into a visually-recognizable form" (Beakley et al 1986), has been used for over two millennia by the predecessors of what are now engineers, architects, and other similar technical professionals.

Drawing on the History of Drawing

Almost everyone has seen some of Leonardo da Vinci's 16th Century freehand drawings which, according to Gelb (2004), exemplify whole-brain thinking. "Beginning about 1820, engineers in [the U.S.] began to be taught projection or mechanical drawing, based on the French system that was first developed by Gaspard Monge" (Beakley et al 1986). This systematic manual, but not freehand drawing, used tools like straight edges, triangles, and circle guides. This is a primarily left-brain process.

Near the end of the last century, the drawing component of engineering education and practice changed drastically in that manual drawing was gradually eclipsed by computerized graphic tools. While computer-aided drawing and design tools are more sophisticated than manual drawing, they share one characteristic—they are primarily left-brained. "By moving to computerized graphic tools, engineering design has become constrained in subtle ways by the primitives and processes intrinsic to those tools" (Arciszewski et al 2009). "Computers make it easy to draw the wrong thing," according to consultant Dan Roan (2008). Or we might say, easy to draw the same old things. "The spontaneity of freehand design . . . permitted direct expression by parts of the brain that are not engaged by computer-aided drafting tools" (Arciszewski et al 2009). Today's computerized drawing, like its predecessor manual projection drawing, "draws heavily on interaction with the left hemisphere of the brain."

On the other hand, freehand drawing or sketching, being free of technical processes and symbols, is dominated by the right hemisphere of the brain (Beakley et al 1986). However, freehand drawing instruction is rare in U.S. engineering education. Therefore, while advances in drawing, from freehand to the more disciplined projection drawing and into today's computerized drawing, have been largely beneficial they probably have had the negative effective of removing some right-brain stimulus potential from engineering education and, as a result, from engineering practice.

Personal

I am an amateur artist, working in graphite and colored pencil. In 2008, after an almost six decade lapse that began after the third grade, I returned, on a whim, to art by taking a pencil drawing class, loving it, and doing a variety of drawings. I soon discovered that I would draw for two or more hours and be oblivious to the passage of time. In returning to art, I initially envisioned no connection to engineering education or practice. This was simply a pleasant diversion.

While thinking about and doing the preceding, talking to my art instructors and other students, and doing some reading, I began to see possible connections between visual arts and improving engineering education and, ultimately, practice. I was referred to and read Betty Edward's book *Drawing on the Right Side of the Brain* (1999). That led to creativity and innovation research, interacting with colleagues, presenting and writing (e.g., Walesh 2011a, 2011b, 2012), designing a workshop, presenting part of a university course, starting a book, and writing the freehand drawing section of this chapter. One thing leads to another!

What if you, whether in school or in practice, were to learn more about freehand drawing and then use it more? What if those you studied, researched, or worked with were to do the same. What benefits might we enjoy? Three likely benefits are offered (Walesh 2011a, 2011b, 2012).

Seeing, Not Just Looking

A principle guiding freehand drawing is to draw what we see contrasted with drawing something the way we think it should look. For example, before taking pencil drawing lessons, if I were asked to draw a boat, tree, dog, or other object, I would have been thinking mainly about what such an object should look like and try to draw it in that preconceived manner. Now, having benefited from drawing lessons, I draw what I see, that is, composition, shapes, and values. Artists first carefully examine the object or thing to be drawn and then, and only then, draw what they see. While each of us has his or her own style of converting what we see into pencil strokes on paper, the process is driven by careful observation.

And I suspect most amateur and professional artists tend to more closely observe everything as a result of the habit we acquired, or are acquiring, as a result of drawing or practicing some other visual art. Artists see more than they did in their pre-artist days. When I look consciously at any object, even though I have no interest in drawing it, I see more especially in terms of shapes and shadows then I used to.

So, what has enhanced seeing and not just looking got to do with engineering, with being more creative and innovative? Improved seeing, whether literally as described here or possibly, by extension, figuratively, further enables an engineering student or practitioner to more completely and accurately define an issue to be resolved, a problem to be solved, or an opportunity to be pursued. To paraphrase and expand the common expression "a problem well defined is half solved," an issue, problem, or opportunity more completely and accurately seen, both physically and figuratively, is half resolved, solved, or pursued. Engineering students and practitioners are likely to gain valuable enhanced actual and figurative vision as a result of participating in freehand drawing or other visual arts.

Increased Awareness of the Right Hemisphere's Powerful Functions

As a result of learning and applying freehand drawing principles, or studying and doing other visual arts, an engineering student, researcher, or practitioner is likely to become more aware of the different and valuable functions of the brain's right hemisphere relative to the left hemisphere. This enhanced awareness may be implicit as in increasingly viewing issues, problems, and opportunities more intuitively and holistically.

Or, as in my case for example, enhanced right-brain appreciation may result from studying literature that connects art and education (e.g., Arciszewski 2009, Beakley et al 1986, and Edwards 1999). This, again based on my experience, may lead to imagining how fuller use of right hemisphere functions by engineers could enhance their individual and group effectiveness.

This curiosity may, in turn, lead to discovery, study, and application of many tools and techniques, like those described in this chapter, available to assist individuals and groups further engage the brain's two hemispheres. Expanded right-mode thinking, coupled with typically strong left-mode thinking, will enable individuals and groups to more creatively and innovatively address issues, solve problems, and pursue opportunities.

A Tool for Group Collaboration

Taking improved seeing further, the enhanced seeing that can enable an engineer, or an engineering team, to more fully define an issue, problem, or opportunity can also, through continued enhanced visualization, help him or her resolve the issue, solve the problem, or pursue the opportunity. Dan Roan (2008) refers to this process as visual thinking which he defines as "an extraordinarily powerful way to solve problems" and explains it as consisting of four steps, mainly, look, see, imagine, and show. In application, each step in visual thinking involves drawing.

To elaborate, simple freehand drawing enhances a person's and/or group's ability to perform the first two steps, that is, to look and then see, really see at least the physical aspects of issues, problems, and opportunities. Then more "art," in the form of simple shapes, lines, arrows, stick people, and things, facilitates the remaining imagining and showing steps. The idea is to engage both of the brain's hemispheres because, as noted earlier, a whole brain is better.

Take a Break

You, as a students or practitioner and whether working individually or within a group, sometimes get bogged down when intensely studying, researching, analyzing, designing, writing, or engaging in some other challenging endeavor. You hit a wall, draw a blank, or experience writer's block. Take a break! Changing from one kind of activity to another stimulates creativity. This is especially true when transitioning from work to leisure, exercise, or a hobby. Perhaps by resting our conscious mind we engage or release our subconscious mind.

Rollo May (1976) tells the story of Albert Einstein asking a friend "Why is it I get my best ideas in the morning while shaving." The friend's answer was "often the mind needs the relaxation of inner controls—needs to be freed in reverie or daydreaming—for the unaccustomed ideas to emerge." Expanding on the take a break suggestion, May recognizes that, what he refers to as "the insight [that] comes at a moment of transition between work and relaxation," is common. He goes on to suggest that these vivid breakthroughs of the subconscious mind to the conscious mind are preceded by "intense, conscious work." Recognize that while creative-innovative breakthroughs occur during relaxation or exercise, the insight instance is preceded by intense work. Our conscious mind must make an intense effort before resting and hopefully engaging the subconscious mind. Apparently we cannot relax our way into creativity and innovation.

Personal

One morning, first thing, I began work on a book proposal for submission to a publisher. After about two hours, I had a good start, including an outline and some text. However, I began to bog down and also get hungry. I biked to a nearby restaurant and, while enjoying a light breakfast, three proposal-related ideas "popped" into my head. I wrote about them on the backside of paper placemats and tucked the results into my pocket. I then began a ten-mile bike ride, during which I stopped three times to briefly jot down more ideas that appeared "out of the blue." These specific situations are typical of many similar creative experiences I've enjoyed over the years that were stimulated by intensely working on a project and then "changing gears." By the way, you are reading that book.

CLOSURE: COMMIT TO QUALITY

Quality means meeting all requirements of those we serve, whether within or outside of our organization. Defining and then satisfying those requirements is challenging. You can rise to that challenge if you:

- Are aware of and respond to individuals and organizations—internal and external stakeholders—who have an interest in the products or services you produce
- Ask, listen, study, and work to provide those individuals with quality products and services
- Continuously improve the tasks and processes you use, or contribute to, as you do your work
- Apply appropriate tools and techniques, both individually and in groups, to stimulate creativity and innovation
- Expect and enable, to the extent of your ability, everyone to contribute to your organization's efforts to achieve quality

---

Care and quality are internal and external aspects
of the same thing.
A person who sees quality and feels it as he works
is a person who cares.
A person who cares about what he sees and does
is a person who's bound to have some
characteristics of quality.

(*Robert M. Pirsig, in Zen and the Art of Motorcycle Maintenance*)

CITED SOURCES

American Society of Quality, 2011. "Basic Concepts—Quality Assurance and Quality Control." (http://asq.org/learn-about-quality/quality-assurance-quality-control/overview/overview.html), May 10.

American Society of Civil Engineers. 2000. *Quality in the Constructed Project: A Guide for Owners, Designers, and Constructors – Second Edition*. Manuals and Reports on Engineering Practice No. 73, ASCE: Reston, VA.

Arciszewski, T. 2009. *Successful Education: How to Educate Creative Engineers*. Successful Education LLC: Fairfax, VA.

Arciszewski, T., E. Grabska, and C. Harrison. 2009. "Visual Thinking in Inventive Design: Three Perspectives." *Soft Computing in Civil and Structural Engineering*, Topping, B. H.V. and Y. Tsompanakis, (Editors), Saxe-Coburg Publications: UK, Chapter 6, pp. 179–202.

Bachner, J. P. 2007. "Duty of Care." Risky Business, *CE News*, November, p. 18.

Beakley, G. C., D. L. Evans, and J. B. Keats. 1986. *Engineering: An Introduction to a Creative Profession*. Macmillan Publishing Company: New York, NY.

Black, J. T. and R. A. Kohser. 2008. *DeGarmo's Materials and Processes in Manufacturing – Tenth Edition.* John Wiley & Sons: Hoboken, NJ.

Byrne, J. 2005. "Brainstorming Helps Engineers Generate New Ideas."*Engineering Times*, December, p. 6.

Crosby, P. B. 1979. *Quality is Free: The Art of Making Quality Certain*. Mentor Books: New York.

Edwards, B. 1999. *Drawing on the Right Side of the Brain*. Jeremy P. Tarcher/Putnam: New York, NY.

Folk, P. D. 2006. "Constructability Reviews Enhance the Quality of Construction Documents."*PE*, October, pp. 22–23.

Galler, L. 2009. "A Mini-manual Guides Training." e-newsletter, Larry Galler &Associates, January 25.

Gawande, A. 2009. *The Checklist Manifesto: How to Get Things Done Right*. Metropolitan Books: New York, NY.

Gelb, M. J. 2004. *How To Think Like Leonardo da Vinci: Seven Steps to Genius Every Day*. Bantam Dell: New York, NY.

Gross, F. 1991. *Peak Learning*. Jeremy P. Tarcher, Inc.: Los Angeles, CA.

Hammer, M. and S. A. Stanton. 1995. *The Reengineering Revolution: A Handbook*. Harper Business: New York, NY.

Hawkins, J. R. 2005. "The Standard of Care for Design Professionals – Part 1." *Structural Engineer*, June, p. 16.

Henstridge, F. 2006. "Quality Assurance in Surveying and Mapping, Part 2." *The American Surveyor*, July/August. (Offers specific QC/QA procedures.)

Herrmann, N. 1996. *The Whole Brain Business Book: Unlocking the Power of Whole Brain Thinking in Individuals and Organizations*. McGraw-Hill: New York, NY.

Hensey, M. 1993. "Essential Tools of Total Quality Management." *Journal of Management in Engineering – ASCE*, October, pp. 329–339.

Huntington, C. G. 1989. "A Craftsman's Obsession."*Civil Engineering—ASCE*, February, p. 6.

International Organization for Standardization. 2011. "Quality Management Principles." (www.iso.org/iso/home.htm). May 23.

Kao, J. 2007. *Innovation Nation: How America Is Losing Its Innovation Edge, Why It Matters, and What We Can Do To Get It Back*. The Free Press: New York, NY.

Kuprenas, J. A., R. L. Kendall, and F. Madjidi. 1999. "A Quality Management Case Study: Defects in Spacecraft Electronics Components."*Project Management Journal*, Project Management Institute, June, pp. 14–21.

Liker, J. K. 2004. *The Toyota Way: 14 Management Principles From the World's Greatest Manufacturer*. McGraw-Hill: New York, NY.

May, R. 1975. *The Courage to Create*. Bantam Books: New York: NY.

Milosevic, I. N. 2010. "Practical Application of SWOT Analysis in the Management of a Construction Project."*Leadership and Management in Engineering – ASCE*, April, pp. 78–86.

Murray, D. K. 2009. *Borrowing Brilliance: The Six Steps to Business Innovation by Building on the Ideas of Others*. Gotham Books: New York, NY.

Nierenberg, G. I. 1996. *The Art of Creative Thinking*. Barnes & Noble Books: New York, NY.

O'Dell, C. and C. Jackson Grayson, Jr. 1998. *If Only We Knew What We Know: The Transfer of Internal Knowledge and Best Practice*. The Free Press: New York, NY.

Parkinson, J. R. 2010. "Building a Question Can Prove as Important as the Answer."*Herald-Tribune*, Sarasota, FL. December 4.

Pirsig, R. M. 1981. *Zen and the Art of Motorcycle Maintenance*. Bantam Books: New York, NY.

PMI, 2007. "Standard Issue."*PMI Network*, June, p. 18.

Rico, G. 2000. *Writing the Natural Way: Using Right-Brain Techniques to Release Your Expressive Powers*. Jeremy P. Tarcher/Putnam: New York, NY.

Roan, D. 2008. *The Back of the Napkin: Solving Problems and Selling Ideas with Pictures*. Penguin Group: New York, NY.

Rose, K. H. 2005. *Project Quality Management: Why, What, and How*. J. Ross Publishing: Boca Raton, FL.

Snyder, J. 1993. *Marketing Strategies for Engineers*. ASCE Press: New York, NY.

Stewart, T. A. 1997. *Intellectual Capital: The New Wealth of Organizations*. Doubleday Currency: New York, NY.

Walesh, S. G. 2011a. "Enhancing Engineers' Creativity and Innovation: Why and How." Presented at the 2nd Reunion Conference on Environmental Engineering, University of Wisconsin-Madison, August 3–5, 2011.

Walesh, S. G. 2011b. "Enhancing Engineers'Creativity and Innovation: A Whole Brain Approach." Kirlin Lecture, University of Maryland, October 12.

Walesh, S. G. 2012. "Art for Engineers: Encouraging More Right Mode Thinking."*Leadership and Management in Engineering-ASCE*. January 2012

Weiss, A. 2003. *Great Consulting Challenges and How to Surmount Them*. Jossey-Bass/Pfeiffer: San Francisco, CA.

Wikipedia. 2011a. "Johannes Gutenberg." (http://en.wikipedia.org/wiki/Johannes_Gutenberg), October 7.

Wikipedia. 2011b. "Velcro." (http://en.wikipedia.org/wiki/Velcro), July 8.

Wilson, R. W. 2007. "Getting Better: Another Angle on Speed."*Indiana Professional Engineer*, July/August, pp. 4–8.

Wilson, R. W. 2010. "Value of Simple Checklist."*Indiana Professional Engineer*, March/April, pp. 4–5

Wilson, R. W. 2011. "Getting Better: The Ohno Circle."*Indiana Professional Engineer*, January/February, p. 4.

ANNOTATED BIBLIOGRAPHY

American Council of Engineering Companies. 2003. *Quality Management Guidelines*. ACEC: Washington, DC. (ACEC indicates that the Guidelines help "you and those in your organization identify and address the issues most critical to customer satisfaction and retention, effective and efficient operations, and a healthy bottom line.")

Bachner, J. P. 2008. "A Process for Documentation."*CE NEWS*, January, p. 19. (Urges that an organization's documentation program, in order to reduce risks, should include "written guidance that tells people what should be written," and "how long various types of documentation should be kept." These recommendations are in keeping with this chapter's advice to develop, use, and continuously improve written guidance on all repetitive tasks and processes.)

Cachadinha, N. M. 2009. "Implementing Quality Management Systems in Small and Medium Construction Companies: A Contribution to a Road Map for Success."*Leadership and Management in Engineering – ASCE*, pp. 32–39. (Describes a case study in which a quality management system was implemented in a small Portuguese construction company.)

Cooley, K. J. 2004. "QC/QA—Turning a Potential Problem Into An Opportunity." Forum, *Leadership and Management in Engineering – ASCE*, October, pp. 123–124. (Argues that consistently achieving quality in the professional service business requires a formal QC/QA program. Urges a "systemic, firm-wide approach" that "includes a prescribed process to properly perform QC reviews on all projects before products are released" and an "aggressive compliance monitoring and corrective action program.")

Fisher, A. 2004. "Get Employees to Brainstorm On Line."*FORTUNE*, November 29. (Cites the online brainstorming experience of the chemical-manufacturing division of the W. R. Grace company. Since 2001, they ran 34 online brainstorming campaigns that yielded 2,685 ideas, 76 new products, and 67 improvements in how things get done.)

Los Angeles Department of Public Works, Bureau of Engineering. 2007. "Developing the QA/QC Plan." in Chapter 9 of the *Project Delivery Manual*, November, 10 pages. (Describes the procedure for developing a QA/QC plan.)

Rad, P. F. 2001. "From the editor."*Project Management Journal*, June, p 3. (Defines risk management as "the systematic process of identifying, analyzing and responding to a project's unplanned events" and goes on to describe these three categories of risks: 1) out of control of project manager, 2) statistical treatment of project duration, and 3) flaws in physical deliverables.)

Schwinger, C. W. 2009. "Quality Assurance for Structural Engineering Firms."*STRUCTURE*, A joint publication of NCSEA/CASE/SEI, June. (Describes many ways to strengthen an organization's QC/QA program.)

Shiramizu, S. and A. Singh. 2007. "Leadership to Improve Quality Within An Organization."*Leadership and Management in Engineering – ASCE*, October, pp. 129–140. (Stresses first defining quality and then notes these three broad steps in improving quality: commitment, investment, and maintenance.)

White, J. B. 2009. "How Detroit's Automakers Went From Kings of the Road to Roadkill."*Imprimis*, A Publication of Hillsdale College: Hillsdale, MI, February. (Contends that Detroit automakers made many mistakes including underestimating the competition, squandering money on ill-conceived diversification schemes, failing to anticipate changes in consumer expectation, and, especially relevant to this chapter, failing to make quality "job one.")

EXERCISES

7.1 APPLY CREATIVTIY AND INNOVATION TOOLS: This exercise will enable you and your team to collaboratively apply and evaluate some of the of the creativity and innovation tools presented in this chapter. Suggested tasks are:

A. Select, preferably in a team mode, a real issue to be resolved, problem to be solved, or opportunity to be pursued by your group. The issue, problem, or opportunity does not have to be an engineering or technical problem. It does have to be real and worth addressing.

B. Define the issue, problem, or opportunity using one or more of the tools and techniques presented in this chapter for stimulating creative and innovative thinking. Include a definition of outcome requirements, that is, define what would constitute a quality result.

C. Develop alternatives, options, etc. again using one of more of this chapter's tools and techniques.

D. Prepare an implementation program, that is, who will do what and when will they do it and what resources are needed?

E. If feasible, implement or start to implement the program. However, this task is not necessary for the successful completion of this exercise.

F. Prepare a memorandum that addresses all of the preceding tasks. Include a section that discusses the extent to which the tools and techniques enhanced your group's creativity and innovation.

CHAPTER 8

DESIGN: TO ENGINEER IS TO CREATE

The glory of the adaption of science
to human needs is that of engineering.

(*Hardy Cross, engineering professor and author*)

The chapter begins with a discussion of design as the essence or root of engineering noting that the design process is omnipresent in all engineering specialties and in related disciplines. Design is then presented in the context of related functions and the interactions among them all of which results in meeting a need, solving a problem, or pursuing an opportunity. The disproportionate impact of the design process on cost is explored followed by design viewed in terms of implementation deliverables such as drawings, technical specifications, and non-technical provisions. The chapter then moves to the idea of design as risky business and design as a creative, personally-satisfying, and people-serving experience and concludes with the linguistic link between the words "create" and "engineer."

THE ROOT OF ENGINEERING

Design, whether used as a verb to represent a process or interpreted as a noun to refer to the result of the process, is omnipresent in engineering and related disciplines. Design as process pervades all of these disciplines and is their essence in all sectors of the economy. Broadly speaking, the design process—the root of engineering—begins with defining the requirements, that is, defining a need or describing a problem or opportunity, followed by logical thinking, applying scientific principles, developing alternatives, considering socio-economic-environmental effects, deciding on a course of action, and communicating the results in a manner that enables implementation. While the process typically relies heavily on traditional means and methods, it may include innovative and creative approaches. The goal of design is quality, that is, meeting all requirements as discussed in the previous chapter.

Informed by this broad definition of design, recognize that various engineering disciplines offer definitions tailored to their specialties. For example, civil engineer Choi (2004) defines design as "a multidisciplinary process involving detailed analysis, judgment, and experience aimed at producing construction drawings, technical specifications, and bid schedules required to allow contractors to bid and construct physical projects." Mechanical engineers Ulrich and Eppinger (2008) describe "design for manufacturing" (DFM) as consisting of estimating the manufacturing cost, reducing the cost of components, reducing the cost of assembly, reducing the cost of supporting production, and considering the impact of the decisions on other factors such as overall development time and the life-cycle cost of the designed item.

The ultimate result of the design process – the fruit that grows from the root – is a useful structure, facility, system, product, or process. Aeronautical engineers design aircraft and spacecraft, civil engineers design high-speed rail systems, chemical engineers design processes to convert raw materials into finished products, and mechanical engineers design hybrid automobiles. As a result of their design orientation, all engineering disciplines deliver functional results some of which are stunning and widely acknowledged while others are unnoticed or taken for granted. Essentially all engineering designs contribute to the quality of life for untold users.

Mathematics, natural sciences, humanities, and social sciences are the foundation of engineering, as explicitly described, for example, in the U.S. civil engineering body of knowledge (ASCE 2008). While being students and appreciators of that foundation, engineers go beyond, as a result of the design process, to develop plans for structures, facilities, systems, products, and processes useful to and sometimes aesthetically pleasing to society (Billington 1986). These plans are the root of the engineering process and the fruit is that which is ultimately constructed, manufactured, or otherwise implemented.

Views of Others

"Scientists define what is," according to aeronautical engineer Theodore von Karman, "engineers create what never has been." Civil engineering professor David P. Billington (1986) continues contrasting science and engineering as follows: "Science is discovery, engineering is design. Scientists study the natural, engineers create the artificial. Scientists create general theories out of observed data; engineers make things, often using only approximate theories." And this final thought about design in the broad sense from engineer, industrialist, and philanthropist Eugene C. Grace: "Thousands of engineers can design bridges, calculate strains and stresses, and draw up specifications . . . , but the great engineer . . . can tell whether the bridge should be built, where it should be built, and when."

THIS CHAPTER'S APPROACH

So how do we approach this vast, pan-engineering topic of design in one chapter? First, the treatment of design in this chapter is necessarily broad in scope; it is introductory.

By the time a student uses this book in his or her formal education, he or she will have taken an in-depth design course in his or her chosen discipline, or will soon do so. Either way, for students, this chapter will provide context for future, current, or past design courses that have or will rely on books (e.g., Choi 2004, Ulrich and Eppinger 2008) and other resources that present discipline-specific descriptions of design. The young practitioner reader of this book will already have completed many design courses and hopefully have begun to experience design. For him or her, this chapter will also provide context as well as an opportunity to reflect, to revisit the all-important design function.

Second, design is presented in this chapter from several very different perspectives. Included are design as the essence of all engineering disciplines; design as one of several engineering functions; the disproportionate impact of the design on cost; the implementation deliverables generated during the design process; design as risky business; and design as a personally-satisfying and people-serving experience. The natural connection between the words "engineer" and "create" end the chapter.

DESIGN IN THE CONTEXT OF MAJOR ENGINEERING FUNCTIONS

Four Engineering Functions

Regardless of the engineering discipline, the process of design may be viewed as the second of four steps as shown in Figure 8.1. More specifically, engineers are involved in all four steps as follows:

- Participate, if not lead, in defining a need to be met, a problem to be solved, or an opportunity to be pursued
- Lead and manage the design process which ends with documentation, often plans and specifications, sufficient to bring to reality that which was designed
- Lead, manage, monitor, or otherwise participates in the necessary construction, manufacture, or other implementation of that which was designed
- Assist with the fruitful use, operation, and maintenance of that which was implemented

Certainly, one can argue for identifying and including other functional areas within engineering such as research and marketing. A counter argument, particularly for the sake of simplicity, is that such functions are included in or part of the four fundamental functions presented in Figure 8.1. For example, marketing efforts often identify the need, problem, or opportunity included in the first of the four steps and research is often included in the second step, that is, design.

Interaction

Note how the four functional areas presented in Figure 8.1 interact with each other to result in sound decisions and optimum structures, facilities, systems, products, or processes. Consider, for example, the iteration between design and construction as illustrated by the hotel walkway collapse described in Chapter 11. In that case, the

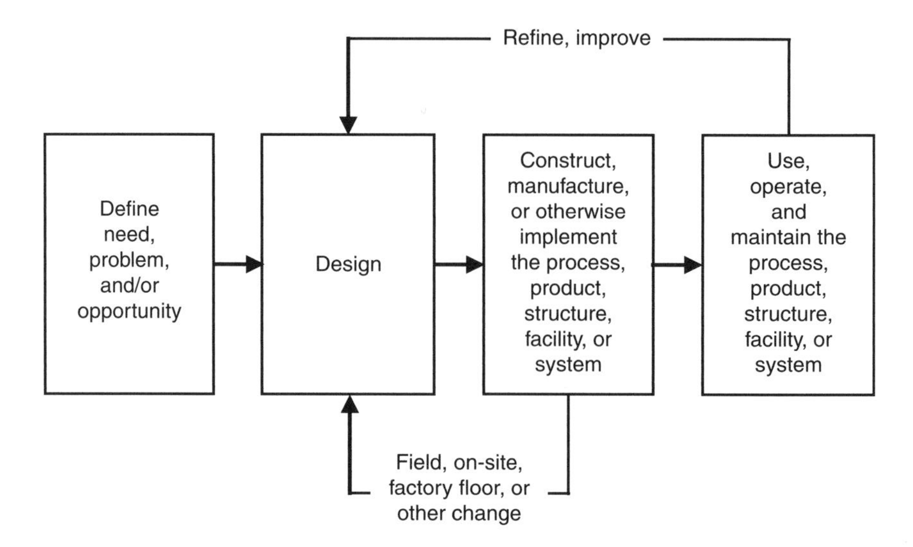

Figure 8.1 Engineers participate in these four major functions to meet needs, solve problems, and pursue opportunities.

apparent failure during the design function to consider constructability led to a design change during construction, which, because of inadequate review, had catastrophic results. More thought should have been given to construction during design—the two functions are inextricably connected.

Within a longer time scale, interaction occurs between the second and fourth functions shown in Figure 8.1. Thoughtful engineers observe the use and operation of their designs; discover ways to improve those systems, facilities, structures, products, or processes; and integrate those improvements into their future design efforts.

At the micro level, interaction, more commonly referred to as iteration or trial-and-error, is common within the design function. All but the most trivial designs typically involve numerous trial and error loops during which ideas are formulated, tested, analyzed, and changed or refined. As an engineering student or a recent graduate used to assignments and problems requiring analysis of existing entities and being asked to obtain unique and "correct" answers, you may find design to be somewhat unsettling. Rarely in practice is there a best or correct solution. Rather, the technical professional strives to arrive at a design that is within that "best" subset of all possible solutions.

"Back-of-the-Envelope" Sketches and Calculations

Conceptual, preliminary, or "back-of-the-envelope" sketches and calculations, based on an engineer's experience, are useful in the early stages of the design process (e.g., see Peck 1996 and Petroski 1991). Roughing out some alternative approaches, layouts, or configurations is likely to be more productive in the typically inevitable trial-and-error design process than seizing, at the outset, one possible approach and carrying its design forward in great detail. While quick "back-of-the-envelope" sketches and calculations by experienced engineers may be the creative impetus for the ultimate design, bringing that design to completion usually requires a major effort by a multi-disciplined team. "The sketch that fits on the back of an envelope," according to Petroski "can turn into box cars of plans and specifications and environmental impact statements and . . . years of hearings, debate, and deliberation."

Design Phases

What has been referred to so far in this chapter as the design process is often an effort involving two or more phases. For example, the preliminary design of the processes and controls needed to manufacture a product, documented in a report, is likely, after approval and with requested refinements, to be followed by a final design documented in the form of detailed plans or drawings and specifications suitable for project implementation. Similarly, the preliminary design of a watershed-wide system of flood control facilities, if generally acceptable to the client, owner, and stakeholders, is likely to be followed by the detailed design of individual facilities. When the end point is the manufacture of a product, Black and Kohser (2008) describe the conceptual, functional and production phases or stages of the design process.

As designs progress through phases, the cost estimates become more accurate. Using the watershed flood control system as an example, cost estimates in the preliminary design might have an accuracy of plus or minus 25 percent while cost estimates prepared for the final design of each individual flood control facility would be accurate to plus or minus one to five percent.

Each phase in the design process has primary emphasis or purpose. The preliminary design report is likely to focus on what to do to meet a need, solve a problem, or pursue an opportunity and explain why it should be done. In contrast, the final design plans or drawings and specifications typically show and describe, in great detail, how to do what has been recommended as a result of the preliminary design. What, why, and how thinking drive the design process, hopefully in a creative and innovative manner stimulated by tools and techniques like those described in the previous chapter.

Personal

I managed the preliminary design of a flood mitigation project that involved 0.7 miles of channel modifications, dikes, floodwalls, and bridge removals and replacements. The preliminary design report discussed and illustrated the channel's peak flow capacity, alignment, grade, trapezoidal cross-section, thickness, and special features such as transitions at existing bridges and superelevation of the channel sidewall at a curve. Upon approval, the project entered the second or detailed design phase which was to be managed by another engineer in our firm. Although I thought our preliminary engineering phase had addressed all of the important issues, I vividly recall, at the hand off meeting, how the team that was to prepare the final design saw many, challenging new issues to be addressed.

Hard and Soft Results

In contemplating the ultimate results of the four engineering functions shown in Figure 8.1 you, as an engineering student or young practitioner, may think in terms of structures, facilities, systems, and products composed of metal, concrete, synthetics, and other substantive materials. However, the illustrated process can produce "soft" results such as a computer model, a project management process, or a way to organize a technical organization to improve utilization of human and other resources. You may also be tempted to confine the four functions to your engineering work. Try not to be myopic in your view of the scope and results of the four-functions depicted in Figure 8.1. That process is widely applicable both within and outside of engineering and other technical fields. It can be used to meet needs, solve problems, and pursue opportunities in various aspects of your professional and personal life. Think of the four functions as a process to use in making things happen.

THE DISPROPORTIONATE IMPACT OF THE DESIGN FUNCTION

One reason the design function is so important, among the four engineering functions illustrated in Figure 8.1, is that while it accounts for a small fraction of the total project cost it is the primary determinant of the total project cost. That total cost is the sum of the cost of design; construction, manufacturing, and other implementation costs; and the subsequent and operation and maintenance (O&M) costs for a structure, facility, system, product, or process. The disproportionate or leveraging effect of the design process on total project costs and on the overall quality of the result is generally applicable across technical fields.

Stating that "detail design decisions can have substantial impact on product quality and cost," Ulrich and Eppinger (2008) describe the previously-mentioned DFM methodology. Applied by a team, the DFM methodology seeks to reduce manufacturing costs, subject to meeting requirements, by systematically reducing the costs of components, assembly, and overhead. The last item includes items such as materials handling, QC/QA, purchasing, shipping, receiving, facilities, and maintenance.

Design, or more specifically, the engineers and other technical professionals who do it, are the principal determinants of cost and quality. The importance of selecting those who will do a design, because of their impact on total costs, is addressed further in Chapter 13, "Role and Selection of Consultants."

DESIGN IN TERMS OF DELIVERABLES

The design process, as shown in Figure 8.1, typically results in the production of drawings and other written and visual information specific enough to be used by other individuals or organizations to construct, manufacture, or otherwise implement a structure, facility, system, product, or process. As discussed in Chapter 1 and illustrated in Figure 1.1, and as reinforced by Figure 8.1, the individual or organizational entity responsible for design is often not the same individual or organizational entity responsible for constructing, manufacturing, or otherwise implementing that which was designed. Conveying the essence of the "designer's" creation to the "builder" in sufficient detail and with adequate understanding so that the latter can produce what the former intended is a monumental communication challenge. Accordingly, the full range of communication techniques discussed in Chapter 3, tailored as needed to design process, can be drawn on.

The design process, especially in civil engineering and similar disciplines, often results in the production of deliverables called bidding documents, which later become contract documents. Bidding documents typically consist of a package containing the following three components, each of which is discussed in the next sections:

- Drawings
- Technical specifications
- Non-technical provisions

Drawings

Drawings, which may also be called plans, graphically portray the type and arrangement of components that comprise the desired structure, facility, system, product, or process. That is, a set of drawings shows what is to be built, executed, or established and where. Given the typical complexity of design process results, a visual representation is essential. Many of us need to see something so that we can understand it, including physicist Albert Einstein who said "If I cannot picture it, I cannot understand it." Drawings could include a few sheets to up to hundreds or thousands of sheets, depending on the size and complexity of the intended result. Drawings are sometimes produced manually, but are usually generated with computer-aided drafting (CAD) software.

As described by Choi (2004), "the following drawings would fall into the category of civil drawings:

- Plan of existing conditions, showing topography and surrounding features and structures
- Plan of survey control, showing baselines, existing and new benchmarks, and coordinate grids
- General plan of new structures, showing location and final grading around the new structures
- Sections, profiles, and details of new structures showing existing ground surface, limits of excavation, locations of backfill, and final grades"

Often, when a construction project is completed, the drawings used to guide the construction are updated to show changes made during construction. Such drawings, which are called record or as-built drawings, are subsequently useful to the owner when the structure, facility, or system is used, operated, maintained, and modified.

"The graphical quality of design drawings has improved with the use of CAD software," according to Choi (2004), "but attractive drawings alone are inadequate for successful construction." In the interest of producing adequate drawings, Choi provides a detailed, useful discussion of the roles and responsibilities of the design engineer and the CAD technician or technologist. Consistent with the leading and managing theme of this book, he goes on to say that problems caused by misuse of CAD "are caused by the management responsible for staffing and managing some design projects."

Technical Specifications

Within civil engineering and related disciplines, "technical specifications [or just specifications] are written instructions and requirements that accompany construction drawings . . . In general, specifications contain all the necessary information that is not shown on the drawings." That is, drawings and specifications should not include duplicate information and, if they do, "the specifications take precedence" (Choi 2004). Technical specifications, which ideally should be prepared by engineers involved in the project's design, could consist of a few pages or run on to hundreds or thousands of pages.

Specifications are typically developed by combining proven "boiler plate" text extracted and edited, as applicable and appropriate, from preceding or parallel projects or other standard sources, and carefully written original text peculiar to the design at hand. As noted by Choi (2004), specifications are legal documents that should be written in a "simple and brief style" following a "say it once and say it right" and "when in doubt, spell it out" approach. Instructions for the contractor on a specific project are written in the imperative mode as in "The contractor shall place concrete in lifts not exceeding 24 inches and compact each lift with mechanical vibrator equipment." Each word is important and the specification writer must combine technical understanding with great writing skill because, as they say, "the devil is in the details."

Specifications typically address a wide variety of technical and nontechnical topics. Some examples are: material requirements, testing requirements, installation or placement instructions, lists of materials or equipment, submittal and schedule requirements, safety and environmental protection needs, permits to be obtained, and coordination with other contractors. In North America, specifications often follow the format of the Construction Specifications Institute (CSI). The institute's mission is "to advance building information management and education of project teams to improve facility performance" which they achieve by "improving the communication of construction information" (CSI 2011). For facility construction, construction products or methods are organized under the divisions listed in Table 8.1. "Each division is divided

Table 8.1 The Construction Specifications Institute organizes its facility construction specifications into 14 divisions.

01	General	08	Openings
02	Existing Conditions	09	Finishes
03	Concrete	10	Specialties
04	Masonry	11	Equipment
05	Metals	12	Furnishings
06	Wood, Plastics, and Composites	13	Special Construction
07	Thermal and Moisture Protection	14	Conveying Equipment

(Source: Adapted from CSI, 2011)

into sections. CSI has specific guidelines and recommendations for the subject matter, numbering system, and format for the sections" (Choi 2004).

Specifications, along with drawings and contract documents, are used by contractors in preparing bids, and if successful, in constructing the structure, facility, system, or product. The ultimate owner is typically interested in the specifications and drawings because they describe the end point in terms of what the owner will eventually use, operate, and maintain. Finally, engineers who will be involved in construction or manufacturing, as described in the next chapter, use the specifications and other documents (Choi 2004).

This discussion pertains mostly to specifications in the context of civil engineering and related disciplines. The term specifications can have a very different meaning in some other engineering disciplines. For example, mechanical or industrial engineers may define "product specifications" as "the precise description of what the product has to do." Other terms used by these disciplines are technical specifications and just specifications (Ulrich and Eppinger 2008). As such, this type of specification is written before the product is designed, not after as in the case of civil engineering. In the context of the preceding chapter, specifications as used in mechanical and industrial engineering is, in effect, a big part of the quality definition for the desired product.

Non-Technical Provisions

The third and last portion of the deliverables typically produced by the design process may include one or more of the following: agreement between the client/owner and contractor, general conditions, supplemental conditions, bid schedule and forms, instructions to bidders, and other items such as supplements to bid forms, agreement forms, bonds and certificates, addenda, and modifications. As with the drawings and specifications, these are legal documents. These largely non-technical provisions are introduced here drawing on Choi (2004) and Clough et al (2005). Refer to these and similar sources for more detail.

Non-technical provisions also include the engineer's construction cost estimate which is "a designer's prediction regarding the probable cost of a construction project" (Choi 2004). Because of liability concerns, some design professionals use the expression "opinion of probable cost." This cost estimate, or whatever it is called, should remain confidential until after the bid opening.

General conditions define the rights, duties, and responsibilities of three parties discussed in Chapter 1 and illustrated in Figure 1.1. These conditions describe procedures generally accepted in engineering or other technical services. Examples of items typically included under the umbrella of the general conditions are payment and completion procedures, scope change provisions, insurance and bonds, schedule, and means of settling disputes.

Supplemental conditions, which are also called special provisions, are extensions of the general conditions and address site-specific requirements and other idiosyncrasies of a project. Examples include special times when work may proceed, specific insurance and bonding requirements, daily damages for delays, permits that will be needed, hourly wages to be paid, temporary facilities to be provided, and the need for security personnel.

The client, owner, or customer may require that bids be submitted in a specific format or fashion. This leads to the engineer developing bid forms that will be completed by bidders. Similarly, special instructions to bidders may be prepared to explain steps in the bidding process such as how to obtain a set of bidding documents, place and time to submit a proposal, withdrawal of a submitted bid, and conditions under which proposals could be rejected.

DESIGN AS RISKY BUSINESS

The design process can also be viewed as "risky business" (Delatte 2009, Florman 1987, Petroski 1985). When an individual or an organization undertakes design, they are aware of the possibility, however remote, that a quality design may not result. The resulting structure, facility, system, product, or process may fail to meet all requirements as discussed in Chapter 7 and, therefore, not achieve quality.

Engineers and other technical professionals, as well as other innovators and creators such as writers, composers, painters, and poets, share an apprehension or fear that they won't be able to do the task at hand. Their innovative-creative process is sometimes stymied by "writer's block." This fear is probably best surmounted by recognizing and acknowledging it, reflecting on one's depth and breadth of understanding of the problem at hand, drawing on one's understanding of science and engineering fundamentals, conferring with colleagues, being open to creative and innovative approaches, and working hard and persistently. Most design efforts will be energized by applying some of the tools and techniques for stimulating creative and innovative thinking as described in the preceding chapter.

Failure of a structure, facility, system, product, or process can have dire consequences in terms of loss of life or great economic cost, as illustrated by descriptions of failures, a few of which are presented in Chapter 11 and many of which are described by (Delatte 2009). Because each non-trivial design is new and unique, there cannot be a 100 percent guarantee of success. That which is designed is only as safe as its weakest element. Each design is an untested hypothesis. The test is the structure, facility, system, product, or process itself and how it functions. Failures can, in a cold academic sense, be explained as disproved hypotheses.

Author Gay Talese (Fredrich 1989) provides this insight into the consequences of disastrous failures for engineers: "Every time there is a bridge disaster, engineers who are unaffiliated with its construction flock to the site of the bridge and try to determine the reason for the failure. Then, quietly they return to their own plans, armed with the knowledge of the disaster, and patch up their own bridges, hoping to prevent the same thing... When a bridge fails, the engineer who designed it is as good as dead."

The designer strives to reduce the probability of disastrous failure by conducting risk analyses, as discussed in Chapter 5; providing redundancies and safety factors; and studying failures. Having mentioned studying failures, I heartily recommend *Beyond Failure: Forensic Case Studies for Civil Engineers* by Norbert J. Delatte, Jr. (2009). Speaking to you, the student, and I might add you, the young practitioner, the author says: "I would like to instill a sense of failure literacy in you. Poets and authors are expected to have intimate familiarity with the work that has gone on before:

Shakespeare's sonnets, Hemingway's short stories, and so forth. In the same way, engineers analyzing and designing structures and systems need to know how similar facilities have performed in the past and when and how they failed."

Forty case studies are presented with the typical elements of each including, but not limited to, technical, procedural, and ethical lessons learned. Much can be learned from failures, as illustrated by Delatte's book and the few examples presented in Chapter 11. Analyzing failures for learning and adjudication purposes is called forensic engineering. Forensic engineering is to engineering what Monday morning quarterbacking is to football—it's much easier to analyze disasters than to prevent them (Petroski 1985).

DESIGN AS A PERSONALLY-SATISFYING AND PEOPLE-SERVING PROCESS

More Than Applied Science

Another way to understand and appreciate design is to see it as part of an often creative-innovative process that culminates in a tangible, personally-satisfying, and people-serving result. Recall from the previous chapter that to create means to originate, make, or cause to come into existence an entirely new concept, principle, outcome, or object. Similarly, to innovate means to make something new by purposefully combining different existing principles, ideas, and knowledge. Recall, also from the preceding chapter, that essentially all engineers synthesize, some innovate, and a few create. In the context of this design chapter, this section seeks to move more engineers into the innovative and creative functions because that is the key to the ultimate in service and personal satisfaction.

Although technical professionals use science in design, design is much more than rote application of science. A designer's work is much like that of the writer, composer, painter, sculptor, and poet in that bits of what is known or has been experienced are re-combined, typically via a trial and error, iterative process, in a unique and new fashion. Author and engineering professor Henry Petroski (1985) says it this way: "It is the process of design, in which diverse parts of the given-world of the scientist and the made-world of the engineer are reformed and assembled into something the likes of which Nature had not dreamed, that divorces engineering from science and marries it to art."

Engineer and author Samuel Florman (1987) argues that the creativity and innovation necessary and prevalent in the best design can be emotionally fulfilling. The fear of personal failure is more than offset by the deep and lasting satisfaction associated with the design of a structure, facility, system, product, or process that serves the user and society. The possibility of that satisfaction is a magnet that pulls many young people to the study of engineering and other technical fields.

Petroski (1985) reinforces the preceding thoughts about the anxiety and satisfaction found in design by noting that the image of the writer staring at a blank page with a wastebasket full of false starts is analogous to the technical professional starting a design. Likewise, the image of the writer learning of a reader's enjoyment and enlightenment resulting from his or her writing or the image of the painter seeing the enjoyment of people viewing his or her work is very similar to the image of the engineer or other technical professional witnessing the aesthetic impact and effective functioning of his or her creation. While recognizing the importance of efficiency and economy in design, Billington (1986) asserts that achieving an aesthetic result requires something else and "that something is imagination – a talent for putting things together in unique ways that work, that are beautiful, personal, and permanent." Using bridges as examples of function and beauty, he cites Joseph Strauss' Golden Gate Bridge and the Roebling's Brooklyn Bridge in the U.S. and Robert Maillart's bridges in Switzerland.

Creators and innovators derive great personal satisfaction from the fruits of their efforts partly because of the uniqueness of the result. However, engineers and other technical professionals often experience an even higher level of satisfaction because the creative or innovative result is useful to society. Recall, in Chapter 1, the words of President Herbert Hoover, who practiced engineering internationally prior to beginning public service, describing the satisfaction of involvement in inherently useful creations. He referred to the combination of creating, innovating, and serving as "the engineer's high privilege."

Aspiring to Creativity and Innovation

Each of us has access to the satisfying and productive creativity and innovation inherent in the design process. That creativity and innovation can be accidental as mentioned in the preceding chapter. A better option is to practice intentional creativity and innovation using the tools and techniques to stimulate creativity and innovation that are described in that chapter. While those methods are presented as a means of achieving quality, they can also be viewed as means of stimulating individual and group creativity and innovation during the design process. To emphasize the applicability of the Chapter 7 tools and techniques to design their names are listed in Table 8.2 with the hope that they will be viewed as integral to the design process.

THE WORDS "ENGINEER" AND "CREATE"

Not only is creativity, as exemplified by design, one of the principal functions of engineering, the words "create" and "engineer" are closely intertwined linguistically. Petroski (1985) and Florman (1987) both explore the origins of the word "engineer."

Table 8.2 These tools and techniques stimulate creativity and innovation during the design process.*

Brainstorming	Problems first meetings
Multivoting	Mind mapping
Strengths-Weaknesses-Opportunities-Threats	Ohno circle
Stakeholder input	Metrics
Process diagramming	Freehand drawing
Fishbone diagramming	Take a break
Pareto analysis	

*Note: For detailed descriptions of the listed tools and techniques see the section of Chapter 7 titled. "Tools and Techniques for Stimulating Creative and Innovative Thinking."

Although they follow different routes and arrive at slightly different conclusions, both agree that "engineer" has its roots in creativity.

Petroski states that "engineer" originally meant "one (a person) who contrives, designs, or invents." That is, "engineer" was synonymous with creator. This use preceded by a century the idea of an engineer as one who manages an engine. According to Petroski, the association between engineer and engine began in the mid-1800s with the emergence of the railroad as the metaphor of the industrial revolution. Petroski concludes his exploration of the origins of the word "engineer" by noting that even today there is a "confusion of the contriver and the driver of the vehicle."

Florman traces "engineer" back to the Latin word "ingenium," which meant a clever thought or invention and was applied in about 200 A.D. to a military battering ram. That is, "engineer" was synonymous with that which was created. Later, in medieval times and during the Renaissance, the French, Italian, and Spanish words, respectively, "ingenieur," "ingeniere," and "ingeniero" came into use originally referring to those who designed and built military machines such as catapults and battering rams. In English, the word progressed from the fourth through seventeenth centuries as "engynour," "yngynore," "ingener," "inginer," "enginer," and, finally, "engineer."

Thus, Petroski and Florman agree that "engineer" has deep roots in creativity. Petroski claims that the first emphasis was on the creative person and Florman believes it was on what was created. However, both agree that "engineer" has its roots in contriving, inventing, designing, and creating. Or, to reiterate the subtitle of this chapter, to engineer is to create.

CLOSING THOUGHTS ABOUT DESIGN

Design is the essence or root of engineering because this often personally-satisfying and people-serving effort results in useful structures, facilities, systems, products, and processes. These fruits of design meet needs, solve problems, and realize opportunities. Creating, an essential element of some design, is historically and linguistically linked to engineering. Aspirationally, to engineer is to create.

I would propose a simplified two-part definition of design:
figure out everything that can possibly go wrong.
Make sure it doesn't happen.

(*Norbert J. Delatte Jr., engineering professor and consultant*)

CITED SOURCES

American Society of Civil Engineers. 2008. *Civil Engineering Body of Knowledge for the 21st Century: Preparing the Civil Engineer for the Future-Second Edition*. ASCE Press: Reston, VA.

Billington, D. P. 1986. "In Defense of Engineers." *The Bridge*. National Academy of Engineering: Washington, DC. Summer, pp. 4–7.

Black, J. T. and R. A. Kohser. 2008. *DeGarmo's Materials and Processes in Manufacturing–Tenth Edition.* John Wiley & Sons: Hoboken, NJ.

Choi, Y. 2004. *Principles of Applied Civil Engineering Design*. ASCE Press: Reston, VA.
Clough, R. H., G. H. Sears, and S. K. Sears. 2005. *Construction Contracting - Seventh Edition*. John Wiley and Sons: New York, NY.
Construction Specifications Institute. 2011. "MasterFormat." (www.csinet.org). May 11.
Delatte Jr., N. J. 2009. *Beyond Failure: Forensic Case Studies for Civil Engineers*. ASCE Press: Reston, VA.
Florman, S. C. 1987. *The Civilized Engineer*. St. Martin's Press: New York, NY.
Fredrich, A. J. (Editor). 1989. *Sons of Martha: Civil Engineering Readings in Modern Literature*. American Society of Civil Engineers: Reston, VA.
Peck, R. B. 1996. "Contributors to Engineering Judgment." *The BENT of Tau Beta Pi*, Spring, p. 9.
Petroski, H. 1985. To *Engineer is Human: The Role of Failure in Successful Design*. St. Martin's Press: New York, NY.
Petroski, H. 1991. "On the Backs of Envelopes." *The Bridge*, Fall, pp. 18–22.
Ulrich, K. T. and S. D. Eppinger. 2008. *Product Design and Development – Fourth Edition*. McGraw-Hill Higher Education: New York, NY.

ANNOTATED BIBLIOGRAPHY

Billington, D. P. 1996. *The Innovators: The Engineering Pioneers Who Made America Modern*. John Wiley and Sons: New York, NY. (" . . . examines what the great engineers actually did, the political and economic conditions in which they worked, and the impact of these designers and their work on the nation.")
Kao, J. 2007. *Innovation Nation: How America is Losing Its Innovation Edge, Why It Matters, and What We Can Do to Get It Back*. Free Press: New York, NY. (According to the author, "Innovation has become the new currency of global competition as one country after another races toward a high ground where the capacity of innovation is viewed as the hallmark of national success." By extrapolation, engineers need to be bring more creativity and innovation to their designs.)

EXERCISES

8.1 BOOK REVIEW: The purpose of this exercise is to provide you with an opportunity to study, in depth, one book of your choice, subject to instructor approval, about design, the root of engineering. In so doing, you will be further introduced to the many facets of this sometimes creative and innovative process. Suggested tasks are:

A. Select one "design" book. It could be recent or it might be old. Some sources are books listed in the Cited Sources (e.g., Petroski 1985) and the Annotated Bibliography (e.g., Billington 1996) sections of this chapter, books recommended by others, and books you find by searching under "design" or similar terms.

B. Request approval of the book from your instructor.

C. Read the book and prepare a review in which you do the following: a) cite your book (e.g., name, author, publisher, date), b) describe some of the key ideas and/or theses presented in the book, c) identify the evidence in support of the ideas/theses, and d) indicate whether or not you agree with the key ideas/theses. Refer to Chapter 3 of this book for writing guidance.

CHAPTER 9

BUILDING: CONSTRUCTING AND MANUFACTURING

When your work speaks for itself,
don't interrupt.

(*Henry J. Kaiser, industrialist*)

The chapter begins by portraying the engineer as a person who finds challenge and satisfaction as a builder of people-serving structures, facilities, systems, products, and processes. After noting that constructing and manufacturing are two ways in which engineers build, the chapter introduces construction by describing its importance, explaining what gets constructed and why, outlining the roles of engineers, and identifying trends in constructing. In parallel fashion, those same topics are presented for manufacturing. Having stressed similarities between constructing and manufacturing, the chapter discusses differences between these two major categories of building and the offers some closing thoughts.

THE ENGINEER AS BUILDER

Most engineers and related technically-intensive professionals, while valuing science, are driven by the desire to do something with science—to build something to improve the quality of life. This compulsion may be the primary source of their professional satisfaction and the pleasure of building may have begun during childhood. On learning of a scientific discovery, engineers are likely to say "congratulations!" to the scientists and then immediately ask, primarily themselves, "so what's it good for; what's the application, or what structure, facility, system, product, or process can we now construct or manufacture more effectively or efficiently?" Engineers appreciate science while deriving their principal satisfaction from its people-serving applications.

Engineer and author, Samuel C. Florman (1976) argues that talents and impulses deep within us underlie what we engineer and then build. Our main goal, according to him, is to "understand the stuff of the universe, to consider problems based on human needs, to propose solutions . . . and to follow through to a finished product." Creating useful things for society's welfare gives us "existential delight." The late engineering professor and author Hardy Cross (1952) offered this "bottom line" description of the building, people-serving essence of engineering: "It is not very important whether engineering is called a craft, a profession, or an art; under any name this study of man's needs and of God's gifts that may be brought together is broad enough for a lifetime."

This chapter addresses building, or more specifically, the basics of engineered constructing and manufacturing in more depth. Recall Figure 8.1, which shows the four functions performed by engineers. While the second of those functions, design, is the root of engineering as indicated in the previous chapter, it typically leads to the fruit of engineering, which is the third function, that is, constructing, manufacturing, or some other form of useful implementation.

As noted in the "The Engineer as Builder" section of Chapter 1, the creation of communities in river valleys, which first occurred about five thousand years ago, prompted the building process, that is, constructing and manufacturing of infrastructure to provide shelter, food, and other goods and services to the concentrated and growing populations. Looking over the span of history, most engineers and other technical professionals, and many non-technical people, marvel at the function and beauty of what our ancestors have built and what we are building. A few representative, highly-varied examples of the fruits of constructing and manufacturing, spanning almost five thousand years and listed in approximate chronological order, are (based, in part, on Fredrich 1989, Hopp and Spearman 2001):

- The Egyptian pyramids
- Tools developed during the Iron Age
- The Great Wall of China
- Athens' Parthenon
- The Roman Pont du Gard in what is now France
- China's Grand Canal
- Printing press
- Steam engine
- The telegraph system
- Transcontinental railroad in the U.S.
- Powered flight
- Mass production of affordable automobiles and untold other consumer and producer products
- The Panama Canal
- San Francisco's Golden Gate Bridge
- Digital computer
- Commercial jet aircraft

- The Eurotunnel connecting England and France
- Many electronic devices interconnected via the internet
- Robotic prosthetics
- The International Space Station

As with the treatment of design in the previous chapter, the presentation of constructing and manufacturing in this chapter is necessarily introductory. For students, this chapter will provide context for future, current, or past constructing and/or manufacturing courses. Perhaps the introductory material presented here will help you, as a first year engineering student, select a major or stimulate you to learn more about studying manufacturing and/or constructing in contrast with, for example, design. The young practitioner reader may have studied either constructing or manufacturing as part of his or her formal education and now, as a result of their work responsibilities, want to learn more about the other form of building.

Views of Others

"The difficult we do immediately, the impossible takes a little longer" is how the U.S. Navy Seabees, the World War II construction battalions led by civil engineers, proudly expressed what is today the "can do" attitude of many constructors and manufacturers (Florman 1976). On building versus destroying, literally and metaphorically, Speaker of the U.S. House of Representatives Sam Rayburn said "Any Jackass can kick down a barn, but it takes a craftsman to build one." In the spirit of learning as we construct each project and manufacture each product, this thought from the French writer Alexandre Dumas: "One's work may be finished some day, but one's education never." And finally, this idea from Jim Rohn, business philosopher: "Whatever good things we build end up building us."

CONSTRUCTING

Importance of Constructing

A nation's quality of life is heavily dependent on its infrastructure which includes, but is not limited to, its water supply and wastewater systems, transportation systems (e.g., highways, bridges, tunnels, railroads, airports, ports), stormwater management and flood control systems, communication systems, power generation and distribution systems, and other structures, facilities, and systems supportive of daily life. The infrastructure is built and maintained by the construction industry. Annual U.S. construction "is on the order of a trillion (1,000 billion) dollars" making it the largest product-based, contrasted with service-based, industry in the country (Halpin 2006).

Speaking from the construction industry perspective, Clifford Schexnayder (2011), one of the authors of a construction management book (Knutson et al 2009), offers this view of construction: "In its simplest terms, construction is about getting work, doing work, and keeping score." The resulting construction project can be a very complex undertaking. Accordingly, engineers committed to succeeding in constructing need technical competence supplemented with the many non-technical knowledge and skill areas treated in this book such as, but not limited to marketing, communication, project management, quality, business accounting, law, and ethics.

Consider the economic impact of construction in the U.S. It accounts for about eight percent of the Gross National Product (GNP) and six percent of the private, non-farm employment. Taking into account the production, transportation, and distribution of construction equipment and materials, constructing provides about 12 percent of the U.S. employment (Clough et al 2005).

What Gets Constructed and How?

Given the diversity of what gets constructed and how, the four categories described by Knutson et al (2009) and Clough et al (2005) are useful in gaining insight into the construction industry. These categories are:

- **Residential construction:** This category includes structures built for habitation such as individual homes, that is, tract or custom homes, condominiums, apartments, and assisted living facilities. The need for residential construction is determined largely by the private sector and the structures are typically designed by architects. Small firms dominate residential construction which annually accounts for 40 to 45 percent of construction in the U.S.
- **Commercial construction:** Included here are a wide range of structures, in terms of size and cost, such as office buildings, stores, banks, schools, universities, hospitals, libraries, theaters, sports complexes, and automobile dealerships. The need for commercial construction, which comprises 25 to 30 percent of annual U.S. construction, is determined by both private and government entities. As with residential construction, commercial facilities are typically designed by architects. However, engineers often provide a support role in areas such as structural; constructability; lighting, security, and other electrical aspects; fire protection; and heating, ventilating, and air conditioning (HVAC).
- **Industrial construction:** Examples of structures and facilities in this category are heavy manufacturing plants, steel mills, refineries, nuclear power plants, pipelines, electric power-generating facilities, ore-handling installations, and other highly-technical projects. Industrial construction, which accounts for five to ten percent of annual U.S. construction, is typically initiated and financed by the private sector, engineering firms perform the design, and specialized contractors normally bid for and perform the construction. Sometimes the client or owner may retain a single firm that both designs and constructs the industrial project.
- **Heavy engineered construction:** Included here are projects where the owner is typically a government entity and the publicly-financed structures, facilities, or systems are major public infrastructure elements. Examples are highways, bridges, tunnels, airports, ports, harbors, dams, flood control works, water and wastewater systems, and storm water management systems. This construction

category represents 20 to 25 percent of the annual U.S. construction. Licensed engineers, who may be government employees or members of consulting engineering firms, direct the designs. Then bidding is used to select a contractor to build the project. This is the common design-bid-build process. Sometimes, because of financial constraints, time limitations, or complexity challenges, a design-build or design-construct approach is used in which the client or owner contracts with one firm that fulfills both the design and construction functions.

Inherent in the preceding discussion of the four types of construction is the previously-mentioned process illustrated in Figure 8.1. That is, regardless of the construction category, in each case, as shown in Figure 8.1:

- Some individual or entity defines a need, problem, or opportunity
- Others perform design
- An organization constructs that which is designed, and
- The resulting structure, facility, or system is used, operated, and maintained

In addition to the process perspective, construction can also be viewed in terms of principal participants. Fisk (2000) suggests that regardless of what is being constructed, a successful project requires the knowledge and skills of three principal participants, namely, an owner or client, a designer, and a constructor. This essential triangle is a subset of the interaction process illustrated in Figure 1.1.

Roles of Engineers in Constructing

Project Manager

One of the many diverse roles played by engineers in construction is project management and each organization participating in a construction project will typically have a project manager. Well before construction begins, the firm that is to design the project will designate someone to manage the design process in accordance with the contract between the client or owner and the design firm. This effort will continue to at least delivery of plans, specifications, and other documents, including construction cost estimates, all of which are described in an introductory manner in the preceding chapter. The project manager may also assist the client or owner with selection of the lowest cost, responsible contractor through a publicly-advertised competitive bidding process.

This design firm's project manager, or someone designated by the design firm, may also extend his or her duties into the construction phase in the form of monitoring the effort on behalf of the client or owner, relative to the design requirements. That individual may be required to certify that the project has been substantially completed in accordance with the intent of the design firm's plans and specifications prior to the client or owner accepting and operating the constructed project.

The contractor's project manager, who is sometimes referred to as superintendent, may be an engineer. His or her responsibilities typically include the construction means and methods, schedule, budget, and safety. More specifically, the contractor's project manager is usually responsible for (Clough et al 2005, Fisk 2000):

- Procuring, marshalling, and allocating labor, equipment, and materials
- Planning the type and sequence of construction activities
- Coordinating the work of subcontractors
- Construction site safety and security
- Informing the client or owner, designer, and the public about a project's status
- Coordinating with local, state, federal, and other regulatory agencies

The owner or client is likely to have a project manager who may be an employee of the owner or client organization or may be on the staff of a professional services firm that specializes in managing construction projects. This project manager's duties may include assisting with selection of an engineering or architecture firm to design the project, providing advice about selection of a contractor, administering the construction contract, monitoring quality control and quality assurance, arranging for review of shop drawings by the design engineer or architect, and working with the design professionals in responding to design changes proposed during construction (Fisk 2000).

Resident Project Representative

The term Resident Project Representative (RPR) is suggested by Fisk (2000) to refer to "an on-site full-time project representative to whom has been delegated the authority and responsibility of administering the field operations of a construction project as the representative of the owner or the design firm." The RPR, often an engineer, may also be referred to by terms such as resident engineer, resident inspector, resident technician, resident manager, project representative, or construction observer. He or she may supervise one or more staff-level, on-site personnel. Fisk (2000) notes that construction inspection, which is one of the RPR's duties, "requires a highly-qualified person with, good working knowledge of construction practices, construction materials, specifications, and construction contract provisions." Other possible RPR duties include survey, layout, documentation of completed work for payment, and materials inspection and testing to confirm the acceptability of the materials being incorporated into the project. The RPR and/or his or her supervisees do not direct the construction but watch for departures from plans and specifications and, if observed, advise the contractor, owner, and designer and facilitate a resolution.

Personal

During summer employment, while studying engineering, I worked for a small engineering consulting firm. One assignment was observing construction of a sewer project. As a result of spending a few days in the field on that project, I still have two vivid memories. The first is the high degree of accuracy with which earth moving machines can be operated. The second impression was the many ways construction could deviate from plans and specifications, especially underground construction in which case the deviations may be quickly buried and out of sight. For these and other reasons such as the need to design for constructability, if you are studying civil or similar engineering, I urge you to obtain some construction-related experience, regardless of your ultimate career objective.

Project Engineer: An engineer may also be employed by a contractor as project engineer or assistant project manager, especially if the contractor undertakes large and/or complex projects. In that role, he or she may report to one or more of the contractor's project managers or superintendents, or to an executive of the construction company. This technical specialist may be responsible for "tracking requests for information, managing shop drawings, maintaining daily records of the project, calculating pay estimates, updating project schedules, and resolving errors in plans and specifications" (Knutson et al 2009). More broadly, the project engineer may be asked to resolve technical problems that arise on one or more projects and determine ways to optimize use of the company's labor, equipment, and materials

Trends in Constructing

If constructing interests you, then you may want to learn about some trends in this industry such as the following:

- **Addressing the shortage of specialty trades personnel:** Specialty trades, such as carpenters, electricians, masons, equipment operators, glaziers, painters, plumbers, sheet metal workers, and iron workers are critical, they make up about two-thirds of the construction industry employees. Because of insufficient interest in trades by high school students and because of the high cost of the necessary hands-on training, the construction industry is facing a shortage of trained labor. "Labor and management need to work together to solve this problem." (Knutson et al 2009).
- **Shift to design-build:** Design-bid-build, the traditional project delivery system, is slowly being replaced by the design-build project delivery system (Knutson et al 2009).
- **Computer-aided earthmoving equipment:** Contractors are increasingly requesting digital files that were prepared by the designer to facilitate grading and other earthwork.
- **Web-based project management:** Another trend is document sharing via a common project website. This enables timely sharing of documents such as project plans, meeting agendas and minutes, schedule status, cost summaries, requests for information (RFIs), and change orders with the status of approvals.
- **Lean construction:** According to the Lean Construction Institute (2011), lean construction is "a new way to design and build capital facilities" based on success with lean production management in other engineering disciplines which "caused a revolution in manufacturing design, supply, and assembly."

The objectives of lean construction are to "maximize value and minimize waste." This is accomplished, in part, by simultaneously designing a structure, facility, or system and its delivery process and by redefining control "from monitoring results to making things happen."

- **Improved management:** "On the whole, construction contractors have been slow in applying proven management methods to the conduct of their businesses," according to Clough et al (2005). "Specialists have characterized management in the construction industry as being weak, inefficient, nebulous, backward, and slow to react to changing conditions. This does not mean that all construction companies are poorly managed. On the contrary, some of America's best-managed businesses are construction firms, and it may be noted with satisfaction that the list of profitable construction companies is a long one. Nevertheless, in the overall picture the construction industry is at or near the top in the annual rate of business failures and resulting liabilities." Explanation for the management problems in the construction industry include the uniqueness of projects, which complicates standardization, and the small size of most construction companies which means that decisions on a wide variety of complex topics are made by one or a few people. Another explanation is the ease at which an individual or group can enter the construction business in that "nearly anyone with a pickup truck and a cell phone can get into the business" (Knutson et al 2009). Moving forward, the construction industry must improve its management practices.

Couple the thrill of constructing with the preceding challenges and construction could be in your career plan. If so, learn more about the construction industry by referring to the sources listed at the end of this chapter, talking to people in the field, taking construction courses, seeking construction-related summer or cooperative education assignments, and requesting construction assignments.

MANUFACTURING

Importance of Manufacturing

Manufacturing can be defined as "organized activities that convert raw materials into salable goods" with those goods being consumer goods or producer goods. Consumer goods are those purchased by the general public whereas producer goods are items purchased by other companies to, in turn, use to manufacture producer or consumer goods. Representative consumer goods are electronic devices, clothes, automobiles, furniture, cosmetics, beverages, and books. Examples of producer goods are machine tools lathes, punch presses, drill presses, milling machines, and printing presses. Manufacturing is a value-adding process during which the conversion of materials into goods adds value to the original materials and enables the effort (Black and Kohser 2007).

Our lives, from meeting our most basic needs to our most uplifting moments, are impacted by manufactured goods. Essentially omnipresent in developed countries, these goods are often taken for granted. "Every day we come in contact with hundreds of manufactured items, from the bedroom to the kitchen, to the workplace, we use appliances, phones, cars, trains, and planes, TV's, VCRs, DVDs, furniture, clothing, and so on," according to Black and Kohser. They go on to note that "These goods are manufactured in factories all over the world using manufacturing processes." And some of those manufacturing processes, and the manufacturing systems of which they are a part, are engineering marvels.

The selling price of a manufactured product is the sum of engineering, administrative, sales, marketing, and manufacturing costs plus profit. Because the manufacturing component accounts for about 40 percent of the total cost, profitability is most dependent on the manner in which goods are manufactured (Black and Kohser).

A country's standard of living is determined largely by the goods and services available to its citizens. For example, in the U.S., manufacturing accounts for about 20 percent of the GNP, employs approximately 18 percent of the workforce, and provides 40 percent of the exports (Black and Kohser).

Historic Note

The idea of interchangeable parts, that is, parts that are identical for practical purposes, is credited to Eli Whitney. Interchangeability was one element of his invention, patented in 1794, of the cotton gin ("gin" for engine) which separated seeds from cotton. Interchangeable parts, which are fundamental to manufacturing, permit easy assembly of consumer and producer goods as well as easy repair of those goods (Black and Kosher 2008).

What Gets Manufactured and How?

The question of what gets manufactured and how is answered by the interaction of the manufacturing organization's marketing, design, and manufacturing personnel who function as an interdisciplinary project team (Ulrich and Eppinger 2008). That project team may be staffed solely by personnel of the manufacturing firm or it may consist of them plus external participants representing business partners, consulting firms, and suppliers and, of course, many team members will be engineers. This marketing-design-manufacturing sequence can be viewed as a specific case of the first three functions shown in Figure 8.1, the graphic in the preceding chapter that depicts major functions performed, wholly or partly, by engineers. That is, and as explained in part by Ulrich and Eppinger (2008) who are quoted here, the process is as follows:

- **Marketing:** The marketing activity is part of the "Define need, problem, and/or opportunity" function in Figure 8.1. Marketing personnel facilitate "identification of product opportunities, the definition of market segments, and the identification of customer needs." Chapter 14 of this book is devoted to marketing and, while that chapter focuses on marketing professional services, some of the ideas and information presented there are applicable to the marketing of manufactured goods.

- **Design:** Design of the product is aligned with the "Design" function in Figure 8.1. This activity "includes engineering design (mechanical, electrical, software, etc.) and industrial design (aesthetics, ergonomics, user interfaces)."
- **Manufacturing:** The manufacturing activity is aligned with both the "Design" function and the "Construct, manufacture, or otherwise implement the process, product, structure, facility, or system" function in Figure 8.1. Design is part of the manufacturing activity in that the manufacturing system must be designed, that is, the set of processes and operations that will result in the designed and desired end product. Then the system must be operated to actually produce the product. This manufacturing activity "also often includes purchasing, distribution, and installation."

One of the reasons for relating the preceding marketing-design-manufacturing sequence to Figure 8.1 is that the process shown there was also referenced when discussing construction, the other broad form of building introduced in this chapter. That is, the answer to the question "what gets manufactured and how?" is broadly similar to the answer to the question "what gets constructed and how?" In both cases, needs, problems, or opportunities are identified; designs are prepared; and something or things are built and used with engineers being involved throughout the process. And, parallel to construction, a successful manufacturing project requires the knowledge and skills of three principal participants, namely, customers for consumer or producer projects, a manufacturer, and a designer. This essential triangle is a subset of the interaction process illustrated in Figure 1.1.

Roles of Engineers in Manufacturing

As with constructing, engineers fulfill many and varied roles in manufacturing as members of project teams who execute the marketing-design-manufacturing sequence. As explained by Black and Kohser (2008), design engineers design the product and manufacturing or industrial engineers design the manufacturing system and manage its use. Materials engineers focus on developing new and improved materials. Some engineers assist with the initial marketing activity. Other engineers serve the broader project management role. They manage the design, manufacturing, and other functions by attending to the typical competing demands discussed in Chapter 5, namely deliverables, schedule, and budget.

In fulfilling these roles as members of project teams, engineers work closely with other experts representing diverse areas such as sales, accounting, finance, and law. Engineers involved in manufacturing also work with technologists, technicians, and other team members such as tool and die makers, machine operators, and computer control programmers. Accordingly, as stressed in Chapters 2, 3, and 4, your effectiveness as an engineer in manufacturing will be greatly enhanced first by getting and keeping "your personal house in order" and then by enhancing your communication knowledge and skills and striving to develop relationships with others.

Personal

Years ago, engineering students, business students, a business professor, and I, an engineering professor, visited a U.S. automobile manufacturing plant. As we followed the production line over a period of four hours we stopped at many of the stations. Something was added to the evolving vehicle at each station—e.g., a headliner, a dashboard, or an engine. Each station was staffed by a team and the team was surrounded by tables and graphs. They continuously documented and sought to improve their assigned task, their part of the manufacturing process. The team was publicly accountable for their task and for improving it consistent with Deming's upstream approach as described in Chapter 7. As each car rolled off the end of the production line, with the left front door open and the key in the ignition, a worker slid into the car, started the engine, and drove away. I thought the cars were driven to a downstream inspection station. No, they were going to have their wheels and headlights aligned after which they would be shipped. The desired quality had been built into the vehicle. If you have never seen a modern production line, a manufacturing marvel, I encourage you to do so.

Trends in Manufacturing

As a student or young person early in your career who is interested in manufacturing, you may wonder what the future holds. Black and Kohser (2008) offer the following ideas about global trends in manufacturing:

- **Increased globalization:** Manufacturing will continue to be an even more global activity as companies try to optimize availability of low-cost labor, access to suppliers and materials, and location of customers.
- **Continuous improvement:** Factories are being designed or redesigned to more effectively provide quality while functioning faster and cheaper. Efforts to reduce time-to-market for new products will continue. Continuous improvement will be increasingly stressed to meet higher quality expectations of customers.
- **Increasingly tailored products:** The number and variety of products will increase while production quantities will decrease. Existing manufacturing processes must be more flexible and new processes developed.
- **Reduced time-to-market:** Manufacturers will gain a competitive edge by reducing the time needed to design and manufacture products. Tactics include designing products so that they are easier to manufacture and creating even more flexible manufacturing systems.

These challenging trends, coupled with the thrill of building useful products, suggest that manufacturing is an attractive career choice. If manufacturing interests you, explore this specialty further beginning with the resources cited at the end of this chapter. Then confer with professionals who are knowledgeable about the manufacturing industry, enroll in manufacturing courses, and/or seek manufacturing assignments.

DIFFERENCES BETWEEN CONSTRUCTING AND MANUFACTURING

Up to now, this chapter has stressed the commonalities between constructing and manufacturing, For example, both involve building to meet human needs and, to the extent they do so, both impact the quality of life. Constructing and manufacturing both have a history of several thousand years and both are heavily dependent on engineering and, as a result, offer many opportunities for aspiring engineers, whether they be students or entry-level practitioners. Constructing and manufacturing also exhibit two differences which may be of interest to students or young practitioners who are contemplating their careers.

As explained by Halpin (2006), the first difference begins with understanding that manufactured products are typically designed and produced without a designated purchaser. "The product is produced on the speculation that a purchaser will be found for the item produced... Design and production are done prior to sale." The manufacturer is at risk of not being able to recover the funds invested in design, production, and marketing.

In contrast, as Halpin goes on to explain, with constructing the purchase begins with the client, owner, or customer who has need for a structure, facility, or system and, as a result, often retains an engineering firm to provide design services and later contracts for construction. The risks in this process include the possibility that the one-of-a-kind resulting structure, facility, or system will not function as required.

In anticipation of Chapter 14, "Marketing – A Mutually-Beneficial Process," note that both manufacturing and constructing require marketing and that some engineers may elect to participate in that process. This leads to the second difference between manufacturing and constructing which is the focus of the related marketing efforts. In the manufacturing arena, marketing is conducted by the manufacturer and targets potential buyers of consumer or producer goods. On the constructing side, marketing is typically carried out by engineering firms who seek to provide design services to organizations that need structures, facilities, or systems and by construction companies desiring to build those items.

CLOSING THOUGHTS ABOUT CONSTRUCTING AND MANUFACTURING

While design, as discussed in the previous chapter, is the root of engineering, constructing and manufacturing are the fruit of engineering. These two forms of building share some common elements, the most important of which is their people-serving essence. If either constructing or manufacturing intrigue you, explore one or the other further. Depending on whether you are a student or young practitioner, take an exploratory constructing or manufacturing course, arrange summer employment or a cooperative education assignment with a constructor or manufacturer, converse with construction or manufacturing practitioners, visit construction sites and manufacturing plants, and seek constructing or manufacturing assignments.

It is not a paradox to say that in our most theoretical moods
we may be nearest to our most practical applications.

(*Alfred North Whitehead, English mathematician and philosopher*)

CITED SOURCES

Berra, Y. 1998. *The Yogi Book*. Workman Publishing: New York, NY.

Black, J. T. and R. A. Kohser. 2008. *DeGarmo's Materials and Processes in Manufacturing – Tenth Edition*. John Wiley & Sons: Hoboken, NJ.

Clough, R. H., G. H. Sears, and S. K. Sears. 2005. *Construction Contracting - Seventh Edition*. John Wiley and Sons: New York, NY.

Cross, H. 1952. *Engineers and Ivory Towers*. Edited by R. C. Goodpasture. McGraw-Hill: New York, NY.

Fisk, E. R. 2000. *Construction Management Administration - Sixth Edition*. Prentice Hall: Upper Saddle River, NJ.

Florman, S. C. 1976. *The Existential Pleasures of Engineering*. St. Martin's Press: New York, NY.

Fredrich, A. J. (Editor). 1989. *Sons of Martha: Civil Engineering Readings in Modern Literature*. American Society of Civil Engineers: New York, NY.

Halpin, D. W. 2006. *Construction Management - Third Edition*. John Wiley & Sons: Hoboken, NJ.

Hopp, W. J. and M. L. Spearman. 2001. *Factory Physics - Second Edition*. Irwin McGraw-Hill: New York, NY.

Knutson, K.; C. J. Schexnayder; C. M. Fiori; and R. E. Mayo. 2009. *Construction Management Fundamentals - Second Edition*. McGraw-Hill Higher Education: New York, NY.

Lean Construction Institute. 2011. (www.leanconstruction.org/index.htm) May 13.

Schexnayder, C. 2011. Personal communication, Professor and author, Arizona State University, January 29.

Ulrich, K. T. and S. D. Eppinger. 2008. *Product Design and Development - Fourth Edition*. McGraw-Hill Higher Education: New York, NY.

ANNOTATED BIBLIOGRAPHY

Armstrong, S. C. 2005. *Engineering and Product Development Management: A Holistic Approach*. Cambridge University Press: Cambridge, UK. (The author "takes the disciplines of integrated product development, project management, systems engineering, product data management, and organizational change management and integrates them into a holistic approach for managing engineering and product development. He treats the most important constituent of a program – the people and the organizational culture.")

Billington, D. P. 1996. *The Innovators: The Engineering Pioneers Who Made America Modern*. John Wiley and Sons: New York, NY. ("...examines what the great engineers actually did, the political and economic conditions in which they worked, and the impact of these designers and their work on the nation.")

Iacocca, L. and W. Novak. 1984. *Iacocca: An Autobiography*. Bantam Books: New York, NY. (Describes the life of Lee Iacocca, who was educated as a mechanical engineer and led the creation of the Mustang in the mid-sixties while at the Ford Motor Company and is credited with saving the Chrysler Corporation.)

McCullough, D. 1977. *The Path Between the Seas: The Creation of the Panama Canal 1870–1914*. Simon & Schuster Paperbacks: New York, NY. (Tells the personal and engineering story of the construction project which, at that time, was "the largest, most costly effort ever before mounted anywhere on the earth . . . Great reputations were created and destroyed.")

Walesh, S. G. 1990. "Water Science and Technology: Global Origins – Keynote Address," Proceedings of the Engineering Foundation Conference: Urban Stormwater Quality Enhancement-Source Control, Retrofitting, and Combined Sewer Technology, Davos Platz, Switzerland, October. (Explains how "the roots of the science and technology used today in the water resource field encircle the globe and reach back to the beginnings of recorded history.")

EXERCISES

9.1 BOOK REVIEW: The purpose of this exercise is to provide you with an opportunity to study, in depth, one book of your choice, subject to instructor approval, about constructing or manufacturing, the fruit of engineering. As a result of doing the book review, you will learn more about one of these vital processes each of which contributes to our quality of life. Suggested tasks are:

A. Select a "constructing" or "manufacturing" book. It might describe current practices or offer a historic perspective. Possible sources are books listed in the Cited Sources (e.g., Hopp and Spearman 2001) and the Annotated Bibliography (e.g., McCullough 1977) sections of this chapter, books recommended by others, and books you find by searching under "constructing," "manufacturing," or similar terms.

B. Request approval of the book from your instructor.

C. Read the book and prepare a review in which you do the following: a) cite your book (e.g., name, author, publisher, date), b) describe some of the key ideas and/or theses presented in the book c) identify the evidence in support of the ideas/theses and d) indicate whether or not you agree with the key ideas/theses. Refer to Chapter 3 of this book for writing guidance.

9.2 APPLY THE OHNO CIRCLE METHOD: This exercise will enable you to try the Ohno Circle Method, which is one of the creativity and innovation tools described in Chapter 7, and also provide an opportunity for you to critique a construction project or manufacturing process. Suggested tasks are:

A. Arrange to visit an active construction site, that is, construction will be occurring while you are there, or a manufacturing process that will be operating when you arrive. Find a safe place from which you can, for an extended period, observe all or most aspects of the construction or manufacturing.

B. If you selected a construction site, imagine that you are the newly-appointed construction project manager, that is, on the staff of the general contractor. Or, if you chose a manufacturing process, pretend you are the just-appointed new manager of it. Either way, you want to improve effectiveness and efficiency, that is, do the right things and do them right.

C. Apply the Ohno Circle method, modified so that you stay at your observation point for only one hour, not the up to eight hours that Taiichi Ohno required! The idea is to be there long enough to "see everything." More specifically, look for underutilized resources (e.g., personnel, equipment, materials); excess motion of personnel; unnecessary movement of parts or materials; excessive parts or materials; defects (e.g., production or constructed elements that do not seem to meet requirements); waiting because materials, information, or resources are not available where and when needed; and safety and health hazards. While you are watching, also listen carefully as suggested in Chapter 7, because it will cause you to be even more focused.

D. Recognize that, unless you have constructing or manufacturing experience, you are not qualified to do what you are being asked to do. However, recall the novice effect introduced in Chapter 2. It may serve you well and, as Hall of Fame baseball player Yogi Berra (1998) said, "You can observe a lot by watching."

E. Prepare a memorandum that addresses all of the preceding tasks with emphasis on problems you observed and your ideas for resolving them.

CHAPTER 10

BASIC ACCOUNTING: TRACKING THE PAST AND PLANNING THE FUTURE

Some know the price of everything
and the value of nothing.

(*Anonymous*)

This chapter provides an introduction to basic accounting terminology and concepts. After discussing the relevance of accounting to the engineer, two financial statements—the balance sheet and the income statement—are discussed, as is the relationship between them. The chapter then uses some accounting basics to suggest ways a student or young practicing engineer can plan for the third phase of their life, that is, the returning or retirement phase. Time utilization rate, expense ratio, and multiplier are introduced as accounting-related performance indicators for consulting engineering firms and similar professional service organizations. The income statement is examined further, this time as part of the business plan for a professional services firm, followed by a brief treatment of the project overruns and the implications. A theme of this chapter is the relevance of accounting basics to professional and personal life.

RELEVANCE OF ACCOUNTING TO THE ENGINEER

Accounting is the process of recording, summarizing, analyzing, verifying, and reporting in monetary terms the transactions of a business or other organization. Aspiring and already practicing engineers and other technical personnel should understand the basic terminology and concepts of accounting so they can effectively function within the triangular "playing field" illustrated by Figure 1.1 in Chapter 1. The three types of organizations represented by the triangle's vertices all utilize accounting, with the private entities placing emphasis on their need to be profitable while the public organizations focus on staying within budgets.

The general, overriding reason for accounting is to determine where an organization has been in financial terms and, by analysis, extrapolation, and planning, to determine where the organization is likely to be going, again in financial terms. Some of an organization's accounting is done for internal reasons and some of it is done to satisfy external needs.

Some internal reasons for doing accounting include "score keeping," that is, determining how closely an organization is adhering to its business plan; providing the basis for pricing products or setting charge-out rates for professional services; and guiding internal resource allocation and optimum marginal investing. External reasons for accounting include reporting to stockholders; providing data to lenders in support of loan applications; satisfying the requirements of federal, state, and other tax laws; and giving information to insurance companies as part of a process of securing professional liability insurance. Liability insurance premiums are based partly on annual billings and similar financial measures.

Clearly, the accounting basics that follow will, besides enhancing your effectiveness as an employee in the business, government, academic, or volunteer sectors, be of practical value if you decide to start your own business. Finally, professional interests aside, basic accounting knowledge and skills will help you manage your financial resources and plan for the future. Accordingly, after accounting basics are introduced, personal and professional applications are discussed.

THE BALANCE SHEET: HOW MUCH IS IT WORTH?

The balance sheet, which is one of two financial statements that define the financial condition of an organization, is prepared at regular intervals such as monthly, quarterly, and annually and meets internal and external needs as just discussed. The other financial statement is the income statement which is discussed in the next major section of this chapter.

The balance sheet is referred to by other names such as financial statement, statement of financial condition, statement of worth, and statement of assets and liabilities (Clough et al 2005). The term balance sheet is used in this book because the term seems appropriately descriptive and because it is widely used in engineering literature (e.g., Blank and Tarquin 2005, Clough et al 2005).

The balance sheet shows, at a given time, assets (A), liabilities (L), and net worth (NW) or equity (E) in monetary terms. The basic equation for the balance sheet is:

$$A - L = NW \text{ (or E)}$$

To reiterate, a balance sheet is a snapshot of the financial condition of a person or organization at a specific point in time. While it shows the financial status at that instant, it does not indicate what financial transactions lead to the current situation.

Two example balance sheets are used to illustrate the features and usefulness of balance sheets. The first is a hypothetical balance sheet for a young person's finances and the second is a hypothetical balance sheet for a construction company. The personal and business examples are intended to suggest the broad applicability of balance sheets. The applicability of basic accounting terminology and concepts to both professional and personal matters is stressed in this chapter.

Personal Balance Sheet

Table 10.1 presents the assets and liabilities of a hypothetical young engineer or other technical person within several years after graduation from college. Two observations are in order:

- This balance sheet, and balance sheets in general, are not as accurate as they may appear to be, particularly when line items are entered to the nearest cent. For consistency, accountants usually show all items to the nearest cent. Some line items such as the current balance of a checking account or the amount owed on an automobile loan, can be determined and stated, on any day, to the nearest cent. However, in general, the overall accuracy of a balance sheet is less than that because some values are estimates, such as the current market value of a condominium or other personal property. This inconsistency in a balance sheet is further complicated by the difficulty of getting all values to be simultaneous or coincident.
- While absolute values of line items, assets, and liabilities at any time are important, changes and trends, such as a gradual increase in net worth for an individual, a couple, or a company are even more important. A balance sheet is a snapshot of net worth at a point in time. A series of balance sheets provides a moving picture of a changing situation. For example, a young professional

Table 10.1 This hypothetical end-of-the-calendar-year balance sheet provides an estimate of a young person's net worth.

Assets	
Condominium	$165,000.00
Personal property (e.g., furniture)	14,000.00
XYZ stock	5,283.68
Car	16,000.00
Retirement (vested)	7,500.21
Cash/checking	893.76
Insurance (cash value)	1,012.16
Total	**$209,689.81 (A)**
Liabilities	
Mortgage on condo	$143,293.16
Car loan	12,151.98
Credit cards	2,542.12
College loan	2,016.16
Total	**$160,003.42 (L)**
Net worth (or equity)	
Total	**$49,686.39 (NW)**

might exhibit a negative net worth shortly after graduation from college, but as a result of sound personal financial management, quickly improve the situation so that in a few years assets exceed liabilities and the difference grows.

If you owe or own anything, you have the basis for a personal balance sheet and should consider developing a balance sheet for your personal finances. Use spreadsheet software to construct your balance sheet and update it at least annually, at the end of the calendar year, on your birthday, or on some other day of significance to you, by adding a column for the end of the most recent year. By preparing and maintaining a personal balance sheet your will realize two benefits. First, you will have an on-going measure of how well you are managing your finances as measured, in part, by the upward, static, or downward trend in your net worth. Second, you will have ready access to the kinds of asset and liability data typically required by banks and other lending institutions in support of applications for home mortgages, automobile loans, and other common financial transactions.

Business Balance Sheet

The second example balance sheet, which is shown in Table 10.2, applies to a hypothetical construction company and represents the last day of a calendar year. This balance sheet was constructed and the explanation is based on, in part, on balance sheets prepared and explained by others (e.g., Blank and Tarquin 2005, and Clough et al 2005).

Assets

Current assets are short-lived working capital, that is, assets that are cash or could be converted to cash typically within a year. Current assets in the balance sheet are as explained as follows:

- **Cash:** The construction company needs to have ready access to cash for miscellaneous uses.
- **Notes receivable, current:** Perhaps they have loaned money to someone or a business and it is due now.
- **Accounts receivable:** These are funds due to the construction company for work completed and billed. It may include retainage which is a portion of the total contract amount (typically 10 percent) retained by owner until the contractor finishes all work. It is in addition to performance and payment bonds.
- **Deposits:** An example is a security deposit paid to an engineering/architectural firm for a set of plans to use in bidding.
- **Inventory:** Examples are items and materials to be used for construction such as pumps, motors, pipe, and aggregate.
- **Prepaid expenses:** This might be the prepaid purchase price of pipe not yet delivered or dues to a contractor organization. These items are assets because the contractor is entitled to receive something of value.

Table 10.2 This hypothetical end-of-the-fiscal-year construction company balance sheet provides a net worth estimate.

Assets	
Current	
Cash	$439,215.63
Notes receivable, current	18,624.48
Accounts receivable	1,405,698.10
Deposits	1,400.00
Inventory	31,040.12
Prepaid expenses	9,597.81
Total	**$1,905,576.14**
Fixed	
Land	$26,345.98
Buildings	84,083.19
Construction equipment	453,478.15
Vehicles	86,378.92
Subtotal	**$650,286.24**
Less accumulated depreciation	**489,178.45**
Total fixed assets	**$161,107.79**
Total assets	**$2,066,683.93**
Liabilities	
Current	
Accounts payable	$403,182.15
Due to subcontractors	893,436.03
Provision for income taxes	108,937.69
Equipment contracts	3,456.96
Long-term notes payable	18,915.89
Total liabilities	**$1,427,928.72**
Net worth (or equity)	
Common stock, 3860 shares	386,000.00
Retained earnings	252,755.21
Total	**$638,755.21**

In contrast with current assets, fixed assets usually require more than a year to convert to cash and then only after a major organizational reorientation. Fixed assets in Table 10.2 are explained as follows:

- **Land:** Occupied or not yet used land owned by the construction company.
- **Buildings:** This includes buildings actively used in the construction business and other buildings owned by the company.

- **Construction equipment:** Examples are specialty items such as back hoes, front-end loaders, pavers, and construction cranes.
- **Vehicles:** This would typically be company cars and trucks.
- **Less accumulated depreciation:** Accounts for the gradual loss in value of various fixed assets. This amount is subtracted from the fixed assets subtotal.

Liabilities

The construction company's financial obligations, that is, its liabilities, are explained as follows:

- **Accounts payable:** Funds owed for miscellaneous products, materials, and services.
- **Due to subcontractors:** Based on invoices received.
- **Provision for income taxes:** A projection of funds that will be needed.
- **Equipment contracts:** The company leases some of the equipment used on its projects.
- **Long-term notes payable:** Maybe the company borrowed money to acquire another company.

Net Worth

The company's net worth, which is its assets $2,066,683.93 minus its liabilities $1,427,928.72, that is, $638,755.21, consists of these two components:

- **Common stock:** Investment in company by its stockholders – 3860 shares at $100.00 each.
- **Retained earnings:** These funds, also called earned surplus, are available for use within company.

The book value of the construction company = $638,755.21/3860 = $165.48 per share. The market value of each share of stock, for example, what a buyer of the company would be willing to pay, is unknown and is very likely to be different than the common stock value or book value.

Balance sheets are routinely prepared for a wide variety of businesses, such as manufacturing organizations and consulting firms, and not-for-profit entities. Accounts receivable often dominate the assets of consulting engineering firms, that is, accounts receivable are very large compared to tangible items, and may approximate a firm's net worth (James 1998).

THE INCOME STATEMENT: INFLOW AND OUTFLOW

The income statement is another important type of business financial statement. As does the balance sheet, the income statement also has different names in that it is sometimes referred to as the profit and loss statement, statement of earnings, statement of loss and gain, income sheet, summary of income and expense, profit and loss summary, statement of operations, and operating statement (Clough et al 2005). The term income statement is used here for simplicity and seems to be a common term.

The income statement presents the type and amount of income (I) and expense (E), along with the difference over a specified period of time, and shows the net income (NI). Any time period is possible, but typically income statements are prepared on a monthly, quarterly, or annual basis. The basic equation for the income statement is:

$$I - E = NI$$

As is the case with the balance sheet, two example income statements are presented—the first applies to personal income and expenses and the second applies to a business. Later in this chapter, a second business income statement is used to explain the importance of the income statement for a professional services business.

Personal Income Statement

Refer to Table 10.3, which might apply to a young professional a few years after competing his or her formal education. This might be the income statement for the individual whose balance sheet is presented in Table 10.1. Note, again, that the income statement shows income received and expenses incurred over a period of time, that is, one year. Contrast this with the previously-discussed balance sheet, which shows assets and liabilities at a point in time.

Table 10.3 This hypothetical annual income statement shows a young person's income, expenses, and net income.

Income	
Salary (gross)	$68,052.29
Interest/dividends	596.81
Sale of stock	1,108.23
Total	**$69,757.33 (I)**
Expenses	
Mortgage (interest and principal)	$21,890.67
Utilities and electronic services	3,845.61
Food including restaurants	5,100.00
Clothing	3,600.00
Car payments	7,611.89
Insurances	3,621.09
Taxes	11,500.00
Entertainment/travel	3,500.00
Miscellaneous	2,100.00
Total	**$62,769.26 (E)**
Net Income	**$6,988.07 (N)**

A personal income statement like that shown in Table 10.3 could be used in a postmortem mode to review income and expenses during the past year. In addition, a personal income statement could also be used in a prospective mode to plan income and its use in the near future. Retrospective and prospective uses of income statements are routine in the business environment.

The earlier observation about the variation in accuracy of line items in the balance sheet also applies to the income statement. For example, while items such as salary and interest and dividends earned can be listed to the nearest penny in a retrospective income statement, expense items such as entertainment/travel and clothing would be estimates unless unusually meticulous records were kept. Variation of accuracy within the income statement does not in any significant way detract from its usefulness as an analysis and planning tool.

Finally, note that there is no obvious quantitative connection between Table 10.3, the income statement, and Table 10.1, the balance sheet, even though they could be for the same hypothetical young person. The first applies to a point in time and the second to an interval of time. However, both are needed to understand the young person's financial situation.

Business Income Statement

Table 10.4 follows the same general format as the example of personal income statement but applies to a hypothetical construction company. The hypothetical business for which the income statement was developed is for the same business in the calendar year for which Table 10.2, the balance sheet, was developed. One indication of the relationship between the balance sheet and the income statement is retained earnings of $252,755.21 in Table 10.2 and the identical retained earnings balance at the end of the year in Table 10.4. However, there are a few obvious connections between the construction company's balance sheet (Table 10.2) and its income statement (Table 10.4).

Note also that the accrual method of accounting is being used, not the cash method or basis. With the accrual approach, income is recognized and entered into the books as it is earned (not when the revenue is actually received as when the cash basis is used). Expenses are recognized and entered when they are incurred (not when payments are actually made as when the cash basis is used). The annual income statement for the hypothetical construction company was developed and the explanations that follow are based on, in part, income statements prepared and explained by others (e.g., Blank and Tarquin 2005 and Clough et al 2005):

Income

- **Contracting income:** Done on completed project basis, this is the sum of all income for all projects completed during the calendar year.
- **Discounts earned:** This may be a credit or cash payment for volume purchases.
- **Equipment rental:** Perhaps this is income from rental of some of the company's equipment to other contractors.

Table 10.4 This hypothetical annual income statement shows income, expenses, and net income for a construction company.

Income	
Contracting income	$10,365,195.68
Discounts earned	28,095.62
Equipment rental	26,789.45
Interest revenue	3,456.91
Miscellaneous	14,556.78
Total	**$10,438,094.44**
Expenses	
Project costs	$9,818.665.78
Office overhead	256,678.90
Marketing	111,234.67
Miscellaneous	2,305.67
Total	**$10,188,885.02**
Net income	
Net income before income taxes	$249,209.42
Federal and state taxes on income	104,667.78
Net income after taxes on income	**$144,541.64**
Retained earnings	
Balance on January 1	221,568.45
Dividends paid	113,354.88
Total retained earnings	108,213.57
Balance on December 31	$252,755.21
Earnings per share on net income (net income after taxes divided by the number of shares)	$37.45

- **Interest revenue:** This may be earnings on company checking and/or savings accounts.
- **Miscellaneous:** An example would be sale of building materials to other contractors.

Expenses

- **Project costs:** This may be raw labor, fringe benefits, equipment rental, materials, subcontracts, and a portion of general overhead allocated to completed projects.
- **Office overhead:** This is the portion of overhead not allocated to projects.
- **Marketing:** The company tracks personnel costs and expenses.
- **Miscellaneous:** An example is legal fees.

Net Income

- **Net income before income taxes:** The income minus expenses. Note how small this is relative to the total income, that is, net income before taxes is only 2.39 percent of the total income. This suggests risky profitability.
- **Federal and state taxes on income:** This is 42.0 percent of the net income before taxes.
- **Net income after taxes on income:** The difference of the preceding two items. This is discretionary income to be used for purposes such as paying dividends to stockholders, raising salaries, putting back into the business, and holding onto as a cushion.
- **Retained earnings:** What follows is an accounting of the changes in retained earnings over the year.
  - **Balance on January 1:** The retained earnings at the beginning of the year.
  - **Dividends paid:** Payments to stockholders during the year. $113,354.88 of the retained earnings carried over from the beginning of the year was paid to stockholders during the year.
  - **Total retained earnings:** The difference between the preceding two items.
  - **Balance on December 31:** This is the preceding item plus the difference between the net income after taxes and the dividends paid. The company may be in the process of building its retained earnings to help finance a new venture.
- **Earnings per share on net income:** Recall from Table 10.2, that there are 3860 shares so that the earnings per share are $144,541.64/3860 = $37.45 which is an index of company performance.

RELATIONSHIP BETWEEN THE BALANCE SHEET AND THE INCOME STATEMENT

The balance sheet and income statement for an individual or for an organization are inextricably linked. Consider a reservoir, with its continuously-varying inflows, outflows, contents, and water level, as an analogy. The balance sheet for the end of a reporting period is analogous to the current contents of the reservoir. The income statement is analogous to an accounting of how much water came into the reservoir and from where, and how much water went out of the reservoir and where it went during the time period. Another common analogy to the balance statement and income statement is a checking account. The end-of-month (or end-of-reporting-period) balance is like the balance sheet. The listing of deposits, checks written, fees charged, interest earned, and other transactions for the month, is like an income statement.

Recall the balance sheet (Table 10.2) and the income statement (Table 10.4) for the hypothetical construction company. The one explicit connection between the two financial statements was the retained earnings at the end of the year. The income statement showed the basis for the retained earnings on the balance sheet. While not immediately apparent, the balance sheet and income statement for an organization or individual are completely linked. The linkage may be presented as follows:

- Balance sheet at end of period
- Income statement for next period
- Balance sheet at end of period
- Income statement for next period
- Etc.

In a business or other organization, or even in one's personal financial affairs, neither the balance sheet nor the income statement is sufficient for a full understanding of organizational or individual financial condition. Using the reservoir analogy again, even if all inflows to and outflows from the reservoir were known for a year (income statement), the reservoir contents at the end of the year (balance sheet) could not be determined without having the reservoir contents (balance sheet) for the end of the preceding year.

There may be some exceptions to the general statement that both a balance sheet and income statement are desired. For example, a modest sole proprietorship consulting business will need an income statement. However, it may not require a balance sheet because it operates on a cash basis with no significant liabilities and has little property or other assets.

ACCOUNTING FOR YOUR FUTURE

A person's life can be viewed, in a very simple way, as consisting of three phases, learning, earning, and returning (Maxwell 2003). The three phases are interrelated. For example, the learning phase, which may be viewed as ending with an individual's formal education, hopefully prepares that person for the earning phase. This is the long period—about four decades—during which an individual adds value and, in exchange, earns which is hopefully a comfortable income.

The returning phase starts with formal retirement from work. During this third and last phase, which could last for two or more decades, an individual could participate in various combinations of activities such as relaxation, travel, volunteer work, compensated part-time work, and possibly even a new career. However, experience suggests that for many, regardless of their chosen combination of "retirement" activities, this final phase includes some volunteer or giving back activities and, therefore, the label "returning," as in returning something to society, is appropriate.

Estimating the Necessary Net Worth at the End of Your Earning Phase

If you connect with this learn-earn-return life model, then you might be receptive to some thoughts about how you, as a student or young technical person, could use the "earning" phase to be able to afford a financially comfortable "returning" phase. That is, to use an expression introduced in this chapter, what net worth will you need to earn during your earning phase to enjoy a vibrant returning phase and how can you earn it? "You can be young without money," according to playwright Tennessee Williams, "but you can't be old without it." As an illustration of one way to answer the "how much will you need?" question, consider the ten step process presented in Table 10.5.

Table 10.5 This process provides an estimate of net worth needed at the beginning of an individual's returning phase.

Item	Scenario 1: Conservative which means that selected financial factors are not favorable. (Assumes a high inflation, a gradually enhanced life style, no Social Security, and a low return on investment during retirement)	Scenario 2: Optimistic which means that selected financial factors are favorable. (Assumes a low inflation, no change in life style, some Social Security, a high return on investment during retirement)	Scenario 3: Your situation: Reflect your views of the future financial situation and your aspirations.
1. Starting annual income ($). Same for Scenarios 1 and 2.	60,000	60,000	
2. Years in the earning phase. Same for Scenarios 1 and 2.	45	45	
3. Annual rate of inflation (%).	4.0	3.0	
4. Annual income required at retirement adjusted for inflation ($). Calculated as: (Item 1) * (1.0 + Item 3/100) ^ (Item 2)	350,471	226,896	
5. Life-style change factor, that is, desired ratio of actual buying power at the beginning of retirement to buying power at beginning employment.	1.25	0.9	
6. Annual income required at retirement adjusted for life-style change ($). Calculated as: (Item 4) * (Item 5)	438,089	204,206	
7. Annual income at retirement from Social Security ($).	0.0	50,000	
8. Annual income required at retirement adjusted for Social Security ($). Calculated as: (Item 6) – (Item 7)	438,089	154,206	
9. Annual yield on net worth during retirement (%).	5	9	
10. Net worth required when retirement begins ($). Calculated as: (Item 8)/(Item 9)	8,761,775	1,713,404	

An explanation of each item in Scenarios 1 and 2 and, if you participate with your assumptions, Scenario 3, follows:

1. Young engineer or other technical person's annual gross income after earning a master's degree and as he or she begins full-time employment.
2. Assume the individual begins full-time employment on his or her 22rd birthday and retires from full-time employment on his or her 67th birthday.

3. Based on the historic average annual inflation rate in the U.S., financial planners assume somewhere between three and four percent (Hoover, 2011).
4. Assumes that if the individual could support his or her initial life-style at the income earned at the beginning of the earning phase then that same life-style could be continued into retirement, assuming the income is adjusted upward for inflation.
5. However, as some people move through their earning phase, net life-style enhancement is inevitable and many of these enhancements will be desired after the earning phase ends and the returning phase begins. Examples of these enhancements, most of which mean higher annual costs extending into retirement are bigger and/or second homes, nicer and/or more automobiles, expanded wardrobes, international travel, and increased giving to various causes. In contrast, some individuals may anticipate less annual expenses upon retirement and, accordingly, apply a life-style change factor that is less than one. Therefore, the annual income received at retirement, which has already been adjusted for inflation, should be adjusted further upward or downward, using a life-style change factor. This factor is 1.25 for Scenario 1 and 0.9 for Scenario 2.
6. This is simply the product of the Step 4 and Step 5 results.
7. Looking ahead, an individual may want to and be able to roughly estimate annual income, in future dollars, from Social Security. This projected income, if it materializes, significantly reduces the net worth needed at the beginning of an individual's returning phase. Given the uncertainty of the Social Security system at the time this book was written, Social Security is assumed to be zero in Scenario 1 and a nominal value of $50,000, in future dollars, is assumed in Scenario 2.
8. This is the Step 6 result minus the Step 7 estimate and begins to illustrate the large differences between and huge implications of the financially-conservative Scenario 1 and, at the other end of the spectrum, the financially optimistic Scenario 2.
9. Assume that, as the individual enters the returning phase, he or she wants to live solely off earnings from his or her net worth plus, in the case of Scenario 2, Social Security. This approach reflects the individual's uncertainty about his or her life span, the desire to leave a large estate for others, and/or an intent to leave a legacy for special causes. This approach requires an estimate of the annual yield of the net worth at retirement. A conservative value of five percent is used in Scenario 1 and an aggressive value of nine percent is used in Scenario 2 (Hoover 2011).
10. The net worth required when the returning phase, that is, retirement begins is the quotient of the required annual income required and the annual yield.
11. The required Scenario 1 net worth of $8,761,775 may initially seem astounding. Recognize that it is a function of the conservative, that is, "safe" set of financial assumptions. With the optimistic Scenario 2, the net worth required at retirement is dramatically reduced to $1,713,404. However, it is still almost two million dollars.

The approach used in Table 10.5 to estimate the net worth needed at the beginning of retirement uses annual income adjusted for inflation, life-style change, and Social Security. Another approach is for you to prepare an annual retirement budget, at any point in time; adjust it upward annually or every few years for inflation and other factors, to your desired retirement year; and convert it to the net worth required at retirement. This may appeal to an individual or couple that plans a dramatic change in life style upon retirement.

Speaking of couples, the net worth analysis presented above and the means of accumulating that required net worth presented next can obviously be performed for a couple. While, contrary to the common statement, two cannot live as cheaply as one, many economies do occur and can be readily reflected in the analyses.

Estimating net worth required at retirement, when projected over 45 years as in Table 10.5, is at best very difficult. The intent of this section is to encourage you, as an engineering student or entry-level technical person, to begin thinking about your life's finances. Prepare an estimate of your required net worth at retirement so that you have a target, update it occasionally, and track your actual net worth against the target. Use the blank column in Table 10.5 to get started or develop your own process. But do something because you will need to be a multi-millionaire by the time you retire.

Accumulating the Necessary Net Worth by the End of Your Earning Phase

Assume you generally concur with the preceding analysis, expect to be somewhere within the Table 10.5 spectrum, and want to be financially prepared for the returning phase of your life. How might you accumulate the required net worth by the end of your earning phase? The short answer is to first commit to investing at least ten percent of all of your income beginning with the first pay check you receive when you begin your first full-time employment as an engineer or other technical professional.

The suggested minimum ten percent of income might be invested in a variety of ways such as employer-sponsored retirement programs, which are often matched by the employer; self-funded retirement programs such as when you have your own business; stocks and bonds; and real estate, all of which could gradually grow your net worth. This habit, and it must become a habit, when combined with the power of compounding, will assure accumulation of substantial assets—millions—during your earning phase. "You can create far more wealth by how you use the money you already earn," according to financial advisor Charles Givens, "then you can by earning more."

As an example, assume that a student or young professional, perhaps you, completes Scenario 3 in Table 10.5. Because a moderate approach was taken, that is, somewhere between conservative Scenario 1 and optimistic Scenario 2, the resulting net worth required at retirement is $5,000,000. The forward-looking professional then decides to create some "ten percent" scenarios with the goal of achieving all or most of the necessary $5,000,000 net worth at the end of his or her earning phase. Table 10.6 presents results of three "ten percent" investment scenarios.

At the end of the earning phase in conservative Scenario A, the individual would have invested $381,685 and, because of compounding, have accumulated $1,205,279. This is only 24 percent of the previously projected $5,000,000 net worth required when

Table 10.6 The scenarios suggest that a moderate to aggressive ten percent, career-long investment program can accumulate substantial funds by the end of one's earning phase.

Item	Scenario A: Conservative	Scenario B: Moderate	Scenario C: Aggressive
Starting annual income ($) (Same for all scenarios)	60,000	60,000	60,000
Years in the earning phase (Same for all scenarios)	45	45	45
Annual increase in income (%)	1.5	3	4.5
Portion of income invested annually (%) (Same for all scenarios)	10	10	10
Annual return on investments (ROI) (%)	5	7	9
Amount invested by end of earning phase ($)	381,685	556,319	833,101
Value of investments at end of the earning phase ($)	1,205,279	2,583,128	5,477,205

retirement begins. The balance might be partly made up by various means, such as inheritance, proceeds from a reverse mortgage on a home, royalties, and Social Security. Furthermore the individual may elect to work part-time and/or may have income from royalties. Given the shortfall, the person may also use some of the net worth's principal each year.

Scenario B, the moderate approach in Table 10.6, doubles the annual percent increase in income and increases the annual return on investment from five to seven percent. Given these changes, the amount invested by the end of the earning phase increases to $556,319 while the value of the investments increases to $2,583,128, which is 52 percent of the net worth required at the beginning of retirement. Once again, other resources would be needed during retirement.

Scenario C, the last and most aggressive scenario in Table 10.6, increases the annual increase in income from 3.0 to 4.5 percent and the annual return on investment from seven to nine percent. With these optimistic conditions, the amount invested at the end of the earning phase would increase to $833,101 and the value of the investment would increase to $5,477,205 which is 110 percent of the $5,000,000 needed at the beginning of the returning phase.

Recognize, as mentioned earlier, that a person would probably have additional resources upon entering retirement such as inheritance, proceeds from a reverse mortgage on a home, royalties, Social Security, income from part-time employment, and income based on drawing on the net worth's principal. And, of course, a person could invest more than ten percent of their income. In summary, the optimum ten percent, career-long investment program for the hypothetical student or young professional would probably lie between Scenarios B and C.

You can achieve your desirable multi-million dollar net worth by combining the ten percent, or more, process with at least moderate annual increases in income and by prudent investing. "A part of all you earn is yours to keep," according to author George S. Clason, "it should not be less than a tenth no matter how little you earn. It can be as much more as you can afford. Pay yourself first." An exercise at the end of this chapter provides you with an opportunity to perform an investment analysis tailored to your present situation and your aspirations. Therefore, the details of the spreadsheet used to generate the numbers discussed in this section are not provided.

Is This Overkill?

Before closing this discussion of using accounting basics to encourage you to begin to think about how you will fund the retirement or returning phase of your life, you may have been thinking the treatment presented here is overkill. That is, certainly everyone knows that they need to project the resources needed for retirement and then start to accumulate them. Unfortunately, logical as that approach may seem, it is not common. For example, the 2010 Retirement Confidence Survey conducted by the Employee Benefit Research Institute (2011), a non-profit retirement think tank, concluded that less than half of U.S. workers have estimated how much they will need for retirement and, even more surprising, only 60 percent were saving for retirement!

THE IMPACT OF TIME UTILIZATION RATE AND EXPENSE RATIO ON PROFITABILITY IN THE CONSULTING BUSINESS

Although providing service is the primary reason for their existence, consulting firms must generate a profit. This profit-oriented discussion is targeted primarily at aspiring and entry-level engineers and other technical personnel who want to be or are on the staff of consulting engineering, architecture, and similar firms. The ideas and information that follow are applicable to businesses that generate most of their revenue by selling time as opposed to selling products—offering services rather than producing goods. The factors determining profitability should be of interest to consulting firm employees, as well as government personnel, private sector personnel, and others who retain consultants so they are in a better position to understand how consulting firms operate.

Utilization Rate and Expense Ratio

Consulting firm profitability may be stated as Profitability = f (U, R, other factors) where, for a specified time period (Norris 1987):

- **U = Utilization Rate (or Chargeable Rate or Billable Rate):** Consider all the hours worked by salaried and hourly personnel at all levels for a specific time period such as a week, month, or year. Recognize that some of the hours are devoted to contract projects and, therefore, are charged to and eventually paid for by clients. Then, for that time period, U = (charged time in hours)/(total time in hours). U is always less than 1.0 for an organization because not everyone can be working directly on client projects all the time. At any time, various employees will be doing non-project administrative work, completing marketing tasks, participating in education and training, and enjoying a vacation or holiday.
- **R = Expense Ratio:** Let S = non-salary costs of a business that are not billed directly to clients (e.g., Social Security, professional liability insurance, unemployment insurance, rent, utilities, and entertainment). Let P = total payroll cost,

that is, the dollars paid to employees, excluding benefits, for all the hours they worked as described under the description of U. Then R = Expense Ratio = S/P.

The preceding terms will be defined further including why they are critical to the profitability of a consulting firm and who in the organization has primary control over U and R. Common sense tells us that raising U and decreasing R should increase profitability.

Analysis of a Consulting Firm's Income Statement

As a further introduction to the importance of the time utilization rate and the expense ratio, consider Table 10.7, which is a year-to-date income statement for a hypothetical consulting firm.

Note that the table uses the accrual method of accounting, as described earlier in this chapter. Review of each line item:

- **Total revenues:** Money paid to or due to the consulting firm.
- **Less reimbursable expenses:** Expenses (e.g., travel, copying) billed to clients and paid to or due to the consulting firm (excludes "markup," overhead, etc.). This revenue "comes in," but is "immediately" used by the firm to reimburse expenses—no gain here for the firm. Therefore, reimbursable expenses are subtracted from the total revenues.
- **Less outside consultants:** Similar to the preceding item. The consulting firm may mark up invoices from outside consultants partly to cover the consultants costs of administering the process. The mark up amount is not included in outside consultants because it is income.
- **Net revenues:** Revenues generated by in-house labor. As indicated below, this revenue has to cover many items.

Table 10.7 Hypothetical year-to-date income statement for a consulting firm shows how various indicators are calculated.

Total revenues	$1,800,000
Less reimbursable expenses	100,000
Less outside consultants	500,000
Net revenues	1,200,000
Less direct labor (P – P′)	400,000
Less non-reimbursables	10,000
Gross income	790,000
Less overhead (S + P′)	600,000
Net income before taxes	$190,000

(Source: Adapted from Birnberg 1985)
Pre-tax profit as a percent of net revenues = 190,000/1,200,000 = 15.8%
Overhead ratio = O = (S + P′)/(P – P′) = 600,000/400,000 = 1.5
Multiplier = M = 1,200,000/400,000 = 3.0

- **Less direct Labor:** Raw labor cost—money paid to or due to employees while they worked on projects—excludes fringe benefits. This amount is subtracted from net revenues.
- **Less non-reimbursables:** Expenses incurred as a result of projects, but not billable to client (e.g., unexpected lab test needed and not covered in contract or acceptable to client as a contract change). Also includes invoices that were not paid and are written off. Non-reimbursables are subtracted from net revenues.
- **Gross Income:** Income after project expenses are accounted for, but before overhead. That is, all of the preceding are income and expenses directly related to projects.
- **Less overhead:** While projects were underway, many and various expenses were being incurred within the firm which were not directly billed to the clients. Overhead $= S + P'$ where P' is the sum of salaries and hourly pay not billable to client such as vacation, illness, holidays, bonuses, office staff, and business development. Recall that S is non-salary costs not billed to client and that P is total payroll. Then $P - P'$ = payroll paid by project income. The overhead ratio is $O = (\text{overhead})/(\text{direct labor cost}) = (S + P')/(P - P')$. The overhead ratio, which is usually greater than 1.0, is "burden" on direct labor—direct labor has to be "marked up" to recoup overhead and to generate profit. For the hypothetical income statement shown in Table 10.7, $O = 1.5$. Overhead is subtracted from gross income.
- **Net income before taxes:** Note how overhead takes a "big cut" in that gross income is reduced from $790,000 to $190,000, because of overhead, to become net income or pre-tax profit. As noted in Table 10.7, net income or profit before taxes is 15.8 percent of net revenues.

Sensitivity of Profit to Time Utilization and Expense Ratio

Profitability is sensitive to overhead in that an increase in overhead goes directly to the "bottom line," that is, pre-tax profit is decreased by the same amount. For example, assume for the hypothetical consulting firm presented in Table 10.7, that overhead increases 10 percent from $600,000 to $660,000. Then pre-tax profit drops $60,000 to $130,000, or 31.6 percent.

Therefore, we can gain further insight into the operation of consulting firms by looking further at the components of O, the overhead ratio (Norris 1987):

- From before: $O = (S + P')/(P - P')$
- Divide numerator and denominator by P: $O = (S/P + P'/P)/(P/P - P'/P)$
- Add and subtract 1 in numerator where 1 is (P/P): $O = (S/P + P'/P + 1 - P/P)/(P/P - P'/P)$
- Rearrange: $O = (1 + S/P - (P/P - P'/P))/(P/P - P'/P)$
- Note that $S/P = R$, the expense ratio, based on earlier definition and it is controlled largely by principals and upper management.

- P/P – P′/P is an approximation of what was defined earlier as U, the utilization rate.
- Therefore, O = (1 + R – U)/(U)

Refer to Figure 10.1 for a graph of O = f(U, R) and consider the sensitivity of O to U and R. Focus, for example, within the zone in which consulting engineering firms tend to operate centered around U = 0.600, R = 0.500, and O = 1.500. These three typical values exactly satisfy the previously derived equation.

Assume R remains constant, but U drops one percentage point from 0.60 to 0.59 (a 1.7 percent drop). Then O = (1.00 + 0.50 – 0.59)/(0.59) = 1.542. A small (one percentage point or 1.7 percent) decrease in U causes a large (over four percentage points or 2.8 percent) increase in O. Note how this affects the net year-to-date profit of the firm whose income statement is shown in Table 10.7.

- Assume O increases from 1.500 to 1.542, (2.8 percent)
- Then overhead increases from $600,000 to $616,800 or by $16,800 (2.8 percent)
- Therefore, pre-tax profit drops by $16,800 from $190,000 to $173,200—a 8.8 percent decrease
- Thus, a 1.7 percent decrease (one percentage point) in U causes an 8.8 percent decrease in pre-tax profit—about five to one

Assume U remains constant, but R increases one percentage point from 50 percent to 51 percent (a two percent increase) for the firm whose income statement is shown in

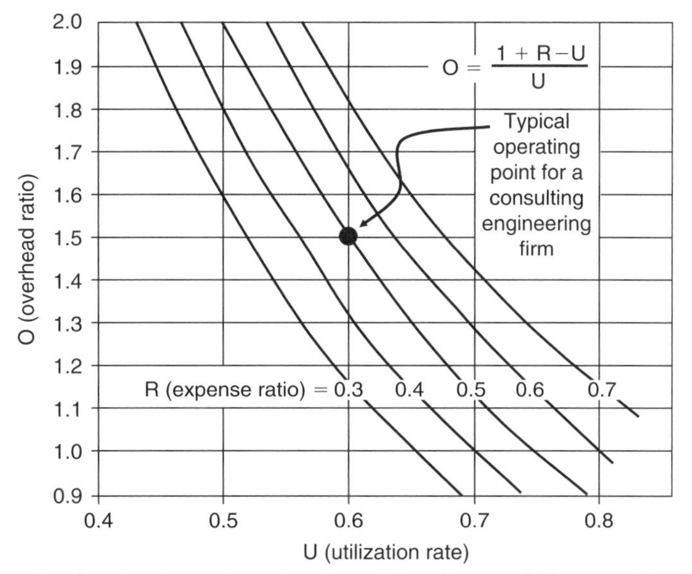

Figure 10.1 Overhead ratio can be reduced, and therefore profit increased, by increasing the utilization rate and decreasing the expense ratio. (Source: Adapted with permission of ASCE from Norris 1987)

Table 10.7. Then $O = (1.00 + 0.51 - 0.60)/(0.60) = 1.517$. Therefore, a two percent (one percentage point) increase in R causes a 1.1 percent increase in O. See how this affects the year-to-date pre-tax profit for the consulting firm whose income statement is shown in Table 10.7.

- If O increases from 1.500 to 1.517, overhead increases from $600,000 to $606,800
- As a result, pre-tax profit drops by $6,800 from $190,000 to $183,200—a 3.50 percent decrease
- Thus, a one percentage point (two percent) increase in R causes an almost four percent decrease in profit.

What are the personal, project, and organizational management implications of the impact of time utilization rate and expense ratio on profit? Top managers will watch R, the expense ratio, very closely. They control most of it—refer again to its components. They should be very aware that a one percentage point increase in R will cause a roughly four percent drop in profit.

Top managers and all staff control U. Time is typically accounted for (e.g., logged into each employee's time sheet) to at least the nearest 0.25 hours. Time utilization is usually tracked on a weekly basis for the entire organization to the nearest 0.1 percent. All time legitimately worked on projects must be charged to clients—provided "the budget can take it."

Something as harmless looking as everyone "knocking off" an hour early on Friday to clean up the office (not chargeable to clients) could have a disastrous effect on profitability. Assume a typical project team member normally charges 24 hours/week to projects ($U = 0.60$). Then, if $R = 0.5$, $O = (1.00 + 0.50 - 0.60)/(0.60) = 1.50$. A one-hour decrease drops this to $U = 0.575$ for a 4.2 percent decrease. Then, if $R = 0.5$, $O = (1.00 + 0.50 - 0.575)/(0.575)$. Therefore, O increases 7.3 percent. If everyone in the previously discussed hypothetical consulting firm "dropped" (failed to charge) one hour per week, overhead would increase by approximately 7 percent and profit would decrease by roughly 23 percent. Stated differently, the approximately four percent reduction in billable time as a result of "knocking off" early to clean up the office would diminish profit by approximately 23 percent.

In summary, consulting firms must be profitable. The income statement shows profit and factors leading to it. Overhead goes to bottom line where it impacts profit. The absolute value of overhead is determined by the overhead ratio, which is a function of the utilization rate and the expense ratio. A one percent increase in U or a one percent decrease in R, will typically cause a several-fold percent increase in profit with profit being more sensitive to U than R.

While tracking time utilization is critical, we need to understand what it is and what it isn't. It does measure the percent of hours that are billable while it does not indicate if those hours are being used effectively and efficiently. Delegation, which is discussed in detail in Chapter 4, is one way to encourage effective and efficient use of everyone's time. Absent enlightened delegation, an engineering firm can have a high U while incurring unnecessarily high labor costs on projects, low profit, and demoralized personnel because too many of them are doing work that should and could be done by less costly personnel.

THE MULTIPLIER

Multiplier, a common term in the consulting business, is one measure of a firm's efficiency. Refer again to Table 10.7, which is a hypothetical year-to-date income statement for a consulting engineering firm. The multiplier (M), a dimensionless parameter, is defined as net revenues divided by the direct cost of labor used to produce the revenues. In other words, the hours of labor that cost the firm $400,000 must generate total net revenues of $1,200,000 so that $M = 3.0$. Therefore, the multiplier is a factor that the salary chargeable to projects must be "marked up" to cover the raw salary itself, non-reimbursables, overhead, taxes, and profit.

Another way of viewing the situation is that the employer (the consulting firm) buys labor wholesale at the raw labor rate and sells it to clients, owners, and/or customers retail, at the marked up or multiplied rate. Consider a young engineer employed by the hypothetical consulting firm and receiving an annual salary of $60,000. The young engineer's raw labor rate is $28.85 per hour—$60,000 divided by 52 weeks and 40 hours per week. For each hour the young engineer works on a contract project, the client or owner will be billed at a charge-out rate determined by the product of the multiplier and the engineer's hourly rate or 3.0 * $28.85 = $86.54.

The Multiplier as an Indicator of Cost Competitiveness?

The multiplier is one possible measure of cost competitiveness between consulting firms. Assuming a particular project requires a fixed number of different kinds of personnel and that raw salaries are similar between firms, the smaller the multiplier of a given firm the less charge there will be to the client and the more cost-competitive the firm will be. Accordingly, potential clients often ask consulting firms for their multiplier and most consulting firms try to keep their multiplier as low as possible.

However, consider further the assumptions that must be met before a multiplier comparison is useful. Each firm is assumed to require about the same number of hours of various types of professional personnel for a given project and compensation rates for various levels of personnel are essentially the same among the consulting firms. Situations may readily occur which invalidate these assumptions and, therefore, eliminate or greatly reduce the value of comparing multipliers. For example:

- A consulting firm may have a personnel strategy under which premium levels of compensation, including benefits, are provided to existing employees and offered to prospective employees to attract and retain the best personnel. Furthermore, an unusually high level of education and training are provided, resulting in above-average costs and below-average time utilization. On the surface, this personnel strategy might produce an above-average multiplier. On the other hand, this firm might be able to do any given project with significantly fewer hours of high-priced labor than other firms because of superior personnel and other practices such as effective delegation. Even though compensation levels and education and training costs, and, therefore, the multiplier are higher than average, the total fees paid by clients for particular projects might be less than the fees charged by most firms.

- A consulting firm may emphasize productivity-enhancing technology. The resulting capital, maintenance, and education and training costs would increase the firm's multiplier to well above the industry average. However, because of dramatically increased productivity, the total fees charged to those they serve could be much less than industry norms.

Reducing the Multiplier

One way a consulting firm can reduce its multiplier is to reduce its overhead. As already discussed, a firm can reduce overhead by increasing U (the utilization rate) or by decreasing R (the expense ratio). Another way to reduce the multiplier, as suggested by focusing on the bottom line of the income statement shown in Table 10.7, is to reduce the profit expectation. The reduction of expected profit permits a reduction in expected net revenue, which, in turn, tends to reduce the multiplier.

Consider an example in which an increase in utilization rate results in a decrease in the multiplier. For the base line situation, assume the income statement presented in Table 10.7. Assume further that the hypothetical firm is able to increase its utilization rate from 0.60 to 0.61, a 1.7 percent increase. Based on the Figure 10.1 relationship showing the overhead ratio as a function of expense ratio and utilization rate, the stated increase in U would reduce the overhead ratio by about 2.8 percent or the overhead by $16,800—from $600,000 to $583,200. If pre-tax profit is held at $190,000, the firm can reduce annual total revenues and, therefore, net revenues by $16,800 to $1,783,200 and $1,183,200 respectively. Therefore, the revised multiplier is equal to $1,183,200 divided by $400,000 = 2.96, which is a 1.33 percent decrease from the base line value of 3.00. In summary, a 1.7 percent increase in time utilization rate for the situation used in the example yields a 1.3 percent decrease in the multiplier with no reduction in pre-tax profit.

Note also, and consistent with the previous discussion of time utilization, if the utilization rate is increased 1.7 percent from 0.60 to 0.61, the multiplier is kept at 3.0, and revenues remain the same, the profit increases 16,800 or 8.8 percent from $190,000 to $206,800. To reiterate, U, the utilization rate, is one of the most important indicators in consulting firm accounting.

Caveat about Cost and Consultant Selection

The preceding discussion of cost competitiveness is not meant to encourage use of price (fee plus expenses) as a means of selecting consulting firms. The discussion simply recognizes that price-based selection (PBS) does occur and, when it does, is fraught with flaws such as erroneous assumptions about the meaning of the multiplier. Prudent selection of consultants, with emphasis on qualifications, that is qualifications-based selection (QBS), is discussed in Chapter 13.

THE INCOME STATEMENT AS PART OF THE BUSINESS PLAN FOR A CONSULTING FIRM

The preceding discussion of business income statements emphasizes their use to document and analyze what happened over the last year or recent years in a consulting firm, construction firm, or other business. Income statements can also be one part of the coming year's business plan for those organizations, whether they are continuing operations or are just starting up. That is, income statements, in addition to being used retrospectively, can also be used prospectively. Some other elements of a business plan are start-up financing, legal form of the organization, assessment of competition, marketing strategy and tactics, and means of finance. Exercises at the end of this chapter and Chapters 11 and 14 provide opportunities to think through some elements of a business plan for a start-up consulting engineering firm.

Table 10.8 illustrates the use of an income statement for a contemplated new business. The example income statement scenario applies to the planned first year's operation of a small consulting firm that would employ two licensed engineers, two engineer interns, four technicians, and an administrative assistant with some personnel working part-time during the first year of operation. The firm's founders may be investigating various possible personnel mixes and the resulting implications, including the necessary charge-out rates.

The income statement was created on a spreadsheet to facilitate running many scenarios. Careful review of the Table 10.8 scenario illustrates the application of various topics covered in this chapter, including raw labor rate; utilization rate; overhead as a sum of non-billable direct labor cost and non-billable, non-salary costs; expense ratio; overhead ratio; and charge-out rates. The calculated values of parameters such as utilization rate (U), expense ratio (R), overhead ratio (O), and multiplier (M), when compared to values common in the consulting industry, provide a check on the reasonableness of any scenario. Unreasonable values of parameters would be cause for exploring additional scenarios until a workable income statement can be developed.

PROJECT OVERRUNS: IMPLICATIONS FOR PROFITABILITY AND PERSONNEL

Budgets for planning, design, and other projects typically performed by consulting firms, for construction projects carried out by contractors, and for design and production projects performed by manufacturing organizations are usually prepared as part of the process of negotiation between the firm and the client, owner, or customer. The contract or agreement between the consulting firm, contractor, or manufacturer and the client, owner, or customer typically "locks in" a total fee. An exception to this is successful use of the formal change of scope provision usually included in the contract or agreement.

Consider what happens if the budget is exceeded. An examination of a typical annual income statement (e.g., Table 10.7) for a consulting engineering firm indicates that the net revenue will not increase as a result of the troubled project (unless a change of scope is negotiated with an additional fee). All expenses and labor costs incurred as a result of continued effort on the project may not be billed to the client and, therefore, will go into the company's overhead, with the appropriate overhead factor applied to the raw labor costs. Increases in overhead as a result of the project will go directly to and subtract from the company's bottom line.

Table 10.8 Annual income statement scenario for a new consulting firm.

Key:	
us	User supplied
c	Calculated
(1), (2), (3), (4), (5)	Calculation sequence

		Personnel Category			
Item		Professional engineer	Engineer intern	Technician	Administrative assistant
1.	Number in category. (us):	2	2	4	1
2.	Annual raw salary, assuming full time, in $. (us):	85,000	55,000	30,000	40,000
3.	Raw labor rate, in $/hr. (c):	40.87	26.44	14.42	19.23
4.	Portion of full-time work, in percent. (us):	100	100	50	75
5.	Total hours worked of all in category. (c):	4160	4160	4160	1560
6.	Annual raw salary of all in category, in $. (c):	170000	110000	60000	30000
7.	Utilization rate, U, in percent. (us):	55	70	80	30
8.	Billable hours of all in category. (c):	2288	2912	3328	468

Summary of Preceding:

Total personnel:	9
Total full-time equivalent personnel. (c):	6.75
Total direct labor cost (P), in $. (c):	370000
(Direct labor cost means raw salary, that is, excludes benefits)	
Total billable direct labor cost (P – P′), in $. (c):	227500
Total non-billable direct labor cost (P′), in $. (c):	142500
Utilization rate (U) based on hours, in percent. (c):	64
Utilization rate (U) estimated based on (P – P′)/P, in percent. (c):	61

Overhead:

Total non-billable direct labor cost (P′), in $. (c):		142,500
Non-billable, non-salary costs (S), in $.		
	Office space (us):	28,125
	(9 persons) * (125 ft ^ 2/person)* $25/ft ^ 2-year)	
	Vehicles (us):	30000
	(50,000 miles)*($0.60/mile)	
	Computer HW, SW, supplies (us):	11000
	Supplies, copying, etc. (us):	4500
	Marketing expense. (us):	12000
	Employee benefits. (us):	131,000
	Miscellaneous. (us):	9000
	Subtotal (S), in $. (c):	225,625
Total overhead (P′ + S), in $. (c):		368,125

Key:					
us	User supplied				
c	Calculated				
(1), (2), (3), (4), (5)	Calculation sequence				

Item	Personnel Category				
	Professional engineer	Engineer intern	Technician	Administrative assistant	
Ratios:					
Expense ratio = R = S/P. (c):				0.61	
Overhead ratio = O = (P′ + S)/(P − P′). (c):				1.62	
Income Statement:					
Annual total revenue, in \$. (c):				679,625	(5)
Less reimbursables, in \$. (us):				30000	
Less outside consultants, in \$. (us):				10000	
Net revenue, in \$. (c):				639,625	(4)
Less billable direct labor (P − P′), in \$. (c):				227,500	(3)
Less non-reimburseables, in \$. (us):				4000	
Gross income, in \$. (c):				408,125	(2)
Less overhead, in \$ (P′ + S), in \$. (c):				368,125	
Net income before tax, in \$. (us):				40000	(1)
Multiplier:					
M = (Net revenue)/(Billable direct labor) (c):				2.81	
Charge Out Rates:					
Personnel category	Raw labor rate, in \$/hr. (c)	Charge out rate, in \$/hr. (c)			
Professional engineer	40.87	114.89			
Engineer intern	26.44	74.34			
Technician	14.42	40.55			
Administrative assistant	19.23	54.07			

Personal

While employed in the private sector, I managed many projects and mismanaged some. Inevitably, regardless of how carefully projects are planned and executed using the principles and procedures discussed in Chapters 5, 6, and 7, budget problems will arise. Accordingly, in order to minimize the impact on the bottom line, especially within consulting firms, salaried personnel will sometimes be expected to work on the projects on their own time. You should not consider entering the consulting field if you are not willing to occasionally do this, even when "it's not your fault." While your self-discipline, communication ability, development of relationships with others, and proactive project management can significantly reduce the incidence of project over runs, they will occur and must be dealt with in a profit-conscious manner.

CONCLUDING THOUGHTS ABOUT YOU AND ACCOUNTING

You, as a student or young practitioner, should understand accounting basics. Accounting knowledge and skills will enable you to be a more effective employee. Those basics will also help you plan your personal financial future and you could draw on them if you establish your own business.

Profit is like health. You need it, and the more the better.
But it is not why you exist.

(*Thomas. J. Peters and Robert H. Waterman, Jr., management consultants and authors*)

CITED SOURCES

Birnberg, H. G. 1985. "Communicating the Company's Operating Performance Data." *Journal of Management in Engineering—ASCE*, January, pp. 12–19. (Discussed by M. D. Hensey in the *Journal*, July 1985, p. 175.)

Blank, L. and A. Tarquin. 2005. *Engineering Economy*. McGraw-Hill Higher Education: New York, NY.

Clough, R. H., G. H. Sears, and S. K. Sears. 2005. *Construction Contracting - Seventh Edition*. John Wiley and Sons: New York, NY.

Employee Benefits Research Institute. 2011. "The 2010 Retirement Confidence Survey: Confidence Stabilizing, But Preparations Continue to Erode." Executive Summary of EBRI Issue Brief # 340, January 27.

Hoover, D. J. 2011. Personal communication, Certified Financial Planner and Chartered Financial Consultant, February 14.

James, L. R. 1998. "Ten Ways to Stay Out of Financial/Business Trouble." *Compendium of Conference Education Materials*. American Consulting Engineers Council Fall Conference, The Greenbrier, WV, November 5–7.

Maxwell, J. C. 2003. *Thinking For a Change: 11 Ways Highly Successful People Approach Life and Work*. Warner Books: New York, NY.

Norris, W. E. 1987. "Coping with the Marketplace–A Management Balancing Act." *Journal of Management in Engineering—ASCE*, July, pp. 194–200.

ANNOTATED BIBLIOGRAPHY

Clason, G. S. 1955. *The Richest Man in Babylon*. New American Library: New York, NY. (Although first published in the 1920s, this classic book uses a set of parables to offer ageless advice about financial planning, thrift, generosity, personal improvement, and creating personal wealth.)

Levitt, S. D. and S. J. Dubner. 2005. *Freakonomics: A Rogue Economist Explores the Hidden Side of Everything*. HarperCollins: New York, NY. (One theme woven through this book is need to be skeptical, not cynical, coupled with the opportunity to ask lots of questions in money and other matters.)

Zofnass, P. J. 2009. "Making Your Firm's Balance Sheet Work for You." *CE NEWS*, December, pp. 23–24. (The article "presents a step-by-step approach to creating and using a financial model to ensure your firm's future success.")

EXERCISES

Note to instructor: Exercises 10.2 and 10.3 should be assigned together because, as students do them, they are likely to iterate between the two exercises. Furthermore, students will learn more and produce more thoughtful and thorough products if the exercises are carried out by a team.

10.1 YOUR FINANCIAL FUTURE: This exercise is intended to get you started, or help you move further forward, on personal financial future. The results of this exercise should be confidential, unless you want to share them with your instructor or other trusted persons. Suggested tasks are:

A. Prepare your personal balance sheet following the general structure of Table 10.1. As suggested in this chapter, consider keeping this balance sheet and update it at least annually, even beginning in college. Contemplate how your balance sheet might look five years after you complete your formal education. This will require that you "rough out" income and expenses, including major purchases, over that period.

B. Estimate the net worth needed at the beginning of your returning phase, that is, at retirement. Be able to produce results in the format of Table 10.5.

C. Develop one or more scenarios describing your career-long investment program or programs that will enable you to accumulate the financial resources needed as you enter your returning phase. Begin with the process used in Table 10.6. State your assumptions. For example, is the annual investment made at the beginning of each year, so that it earns during the year, or does the investment occur at some other time during the year?

10.2 PROJECTED INCOME STATEMENT FOR FIRST YEAR OF A NEW CONSULTING BUSINESS: The purpose of this exercise is to help you, and others if carried out as a team effort, understand and see the value of income statements prepared in the prospective mode. Successful completion of this exercise will probably require iteration with Exercise 10.3. Suggested tasks are:

A. Assume you completed your master's degree in engineering six years ago and have since worked three years as assistant engineer for a municipality; worked three years for an engineering firm as a doer/seller, that is managing projects and helping to obtain new projects; and have earned your PE license. You recently attended a national engineering conference and talked at length with two college classmates that you have stayed in touch with. Based on the conference discussion and many subsequent communications, you and your two friends have decided to start an engineering consulting firm. The current date is October 1 and your new firm will kickoff with the three of you as full-time personnel in three months, that is, in January 1. You will employ other personnel as determined by you three principals.

B. Develop, using a spreadsheet, an annual income statement, similar in structure and function to Table 10.8.

C. Perform at least five possible trials for the first year of business. For each trial, show or state all assumptions. Check each trial against "industry" patterns or typical values.

D. Prepare an executive summary memorandum, with spreadsheets attached, that describes the analysis and sets forth recommendations. This document is part of your business plan. The start up financing part of the business plan is addressed in the next exercise, the legal organization in an exercise at the end Chapter 11, and the marketing portion of the business plan is a Chapter 14 exercise.

10.3 FINANCING THE START UP OF A NEW CONSULTING BUSINESS: This exercise will help you, or you and others if a team exercise, think through the process of raising the funds needed to start and sustain, for the first year, a new consulting business. Suggested tasks are:

A. View this is a continuation of the scenario described in Task A of the preceding exercise.

B. Estimate the funds you will need, in addition to your projected income, to start and run your business during its first year of operation. Possible cost items are developed in the preceding exercise.

C. Determine how you will raise the necessary funds. For example, from the three principals, a loan, and/or investors?

10.4 EFFECT OF TECHNOLOGY ON MULTIPLIERS AND FEES BILLED TO CLIENTS: The intent of this exercise to suggest to the student that the multiplier is not a reliable indicator of consulting firm cost competitiveness in light of technological advances. Suggested tasks are:

A. Use your team's income statement spreadsheet from Exercise 10.2 to illustrate how use of productivity enhancing technology could significantly increase your firm's multiplier. Start with your final simulation from the earlier assignment. To do this exercise, you must reflect the added annual cost (expenses, no labor) to your firm of purchasing and maintaining a new generation of productivity-enhancing tools. Do this by increasing the "computer" expense by a large factor. Also recognize the additional training and equipment maintenance time (labor) needed by engineers and other technical personnel in learning how to use the new equipment. How can you factor this into your simulation? What effect will your selected approach have on the multiplier?

B. Run the simulation, obtain your firm's new "high tech" multiplier, and compare it to your earlier "low tech" multiplier. How much has it increased?

C. Discuss how your clients, owners, and customers could benefit (e.g., possible reduced consulting prices, improved results) even though the multiplier increases. What does this say about using only the multiplier in evaluating the effectiveness of a consulting firm? What benefits might the "high tech" approach provide to the firm?

D. Prepare a memorandum describing what you did, including all assumptions used in Task A; why you made those assumptions; and the answers to all questions.

CHAPTER 11

LEGAL FRAMEWORK

As always happens in these cases,
the fault was attributed to me, the engineer,
as though I had not taken all precautions
to ensure the success of the work.
What could I have done better?

(Nonius Datus, the Roman engineer responsible for the design and construction of a water supply tunnel through a mountain, in what is now Algeria, upon visiting the construction site in 152 A.D. He learned that the two segments of a tunnel being excavated from both ends were out of alignment and had passed each other. (Source: de Camp 1963))

After citing examples of circumstances in which the entry-level professional should know some legal fundamentals, the chapter explains selected legal terms and notes the increased tendency to initiate legal actions. Three ways in which liability is incurred are explained followed by examples of failures and lessons learned from them. The chapter discusses ways to minimize liability including liability insurance, organizational preventive practices, and personal preventive practices. An admonition to keep liability minimization in perspective follows and the chapter concludes with an introduction to the legal forms of business.

WHY LAW FOR ENGINEERS?

"Law constitutes the rules under which civilized individuals and communities live and maintain their relationships with one another. It includes all legislative enactments and established controls of human action" (Bockrath and Plotnick 2011). Engineers should not practice law just as lawyers should not practice engineering. However, just as knowing the basics of business accounting, as described in Chapter 10, helps young engineers practice engineering, knowing the basics of law also helps them practice engineering.

Many of your actions, or inactions, could have legal implications for your and for your business, government, academic, or volunteer organization. You should know enough about the legal aspects of engineering practice to recognize when you need to take certain actions, when to do nothing, and when to ask if legal counsel might be prudent. This chapter will provide you, as a student or entry-level engineer or other technical professional, with an understanding of some legal principles and related behavioral guidelines.

Bockrath (1986) describes situations in which the entry-level technical professional may need to know and apply legal principles. Supplementing these with additional circumstances leads to the seven categories of tasks presented here. Consider using legal principles presented in this chapter, beginning as a student, to help guide your actions when involved in tasks such as these:

1. ***Helping prepare contracts for professional services:*** You may be asked to help prepare a contract or agreement for services, goods or other outcomes between various entities like those illustrated in Figure 1.1, such as a consulting engineering firm and a client or owner, a municipality and a constructor, or a manufacturer and a customer. The desired end-point may be a set of plans and specifications; a constructed structure, facility, or system; or a manufactured product.
2. ***Interpreting contracts once a project is underway:*** Even a well-crafted, mutually-acceptable contract or agreement requires numerous interpretations during planning, design, construction, manufacturing, or operations. For example, the customer's representative may call the design engineer at a manufacturing firm, after reviewing a draft design from the manufacturer, and request that more alternatives be developed and examined. The design engineer must decide if the request is reasonable, that is, within or beyond the scope of the contract.

   Or assume that the agreement between the consulting firm and a client or owner indicates that the latter will contract for geotechnical services with a third party when such services are needed. However, once the overall project is underway, someone on the client or owner's staff contacts a young engineer at the consulting firm and says "Why don't you retain the geotechnical firm as sub-contractor to your firm—you are more familiar with geotechnical firms anyway?" The young engineer ought to determine if this request is as logical as it sounds or if prudence indicates that the consulting firm should decline the suggestion.
3. ***Managing to minimize personal and organizational liability:*** As young engineers or other technical professionals go about their work, especially when doing what may appear to be relatively mundane tasks, they should be aware of ways in which personal liability and the liability of their organization can be minimized. Suggestions, many of which are both simple and potentially powerful, are presented in this chapter.
4. ***Anticipating and/or preparing for expert witness testimony:*** This situation is similar to the preceding, but much more focused in that it assumes the engineer or other technical person is going to be involved as a consultant or expert witness in litigation process. You, as a young engineer, are not likely to serve as

consultant or expert witness in such situations. However, as your gain experience and assume more responsibility the likelihood of being asked or required to do so will rise. So begin to think about that possibility. Consulting and expert witness roles as part of litigation are not discussed in detail in this chapter. However, if these topics interest you, refer to Babitsky and Mangraviti (2002, 2005) for an in-depth treatment. When serving as consultants and/or expert witnesses, engineers are subject to strict ethics expectations in accordance with the engineering society codes of ethics discussed in the next chapter.

5. ***Understanding project-related requirements in local, state, and federal laws and rules:*** The practice of engineering and some other technical professions in the public and private sector is typically heavily influenced and constrained by the requirements of local, regional, state, and federal laws and administrative rules. You are strongly urged to learn about those requirements in the early stages of a project.

   Consider, for example, a consulting engineering firm responsible for planning or designing the storm water management system for a new residential development on the periphery of a growing U.S. city. The community is likely to have zoning, subdivision, drainage, and other codes and requirements that must be satisfied. County and regional rules and regulations may also apply, particularly if the development will initially be wholly or partly outside the corporate limits of the city. State water laws may also be applicable. For example, if the storm water system includes a detention facility, the outlet control structure may qualify as a dam under state law and require a state permit. Federal regulations and codes may also apply. Perhaps a stream passing through the area has a 100-year flood plain delineated under the flood insurance program administered by the Federal Emergency Management Agency (FEMA). A storm water discharge permit may be required under the National Pollution Discharge Elimination System (NPDES).

   Note that the preceding examples apply only to the water resources portion of the project. A similar set of city, county, regional, state, and federal legal requirements may also exist for other aspects of the new development such as its waste water system, its water supply system, and its streets and highways. Unfortunately, too many technical professionals move well into a planning or design project before determining the applicable rules and regulations and meeting with the regulators.

6. ***Being aware of the ways in which state, federal, and other programs may provide funding for client's projects:*** While local through federal regulations and legislation may sometimes be reviewed as a problem because of the many requirements that must be met, some legislation, particularly at the state and federal level, also includes a "carrot" in the form of partial funding. The professional service firm that strives to be of full service will commit the necessary resources to be aware of existing legislation that could be useful to their clients, owners, and customers and you, by being alert, can help with that effort.

   For example, some states in the U.S. have legislation which enables local communities to implement storm water management utilities that provide, through user and other fees, a means of equitably generating revenue to finance the planning, design, construction, operation, and maintenance of storm water systems. Consulting firms and local government entities should know about such legislation and how to help a given community establish a utility.

7. ***Recognizing relevant pending or recent legislation and possible impacts on projects:*** In addition to knowing about already-enacted legislation, you or somebody in your organization should be tracking pending legislation. Given the multi-year span of some planning-design-construction projects and manufacturing projects, legislation being debated at the beginning of the project might be enacted and available to benefit a client, owner, or customer before the end of the project.

LEGAL TERMINOLOGY

Like the engineering, medical, and other professions, the legal profession has its special terminology. Understanding some of that terminology is essential for the purposes of this chapter. Consider the following terms and their definitions:

- **Liability:** Obligated, according to law or equity, to make good on acts or omissions that cause loss or damage to others. The cause could be breach, fraud, or negligence or various combinations. To be found liable means that an individual or organization has to pay or do something.
- **Breach:** "Violation of a right, of a duty, or of a law, either by act of commission or by non-fulfillment of an obligation. Thus, breach of contract is the unexcused failure to satisfy one's contractual undertaking." (Bockrath and Plotnick 2011). The focus is on results in that breach has little to do with intention.

Personal

I once served as a consultant to a law firm and was subsequently declared an expert witness. The law firm was representing an engineering firm. The engineering firm designed a facility and was then retained by the owner to provide construction management services. The engineering firm, perhaps under pressure to minimize their costs, decided not to perform some testing specified in their contract with the owner. The facility failed. Regardless of whether or not the failure to perform testing contributed to the facility failure, the failure to test was a breach of contract and, therefore, was a major negative factor in the lawsuit and in the engineering firm incurring large costs. DWYSYWD: Do what you said you would do.

- **Fraud:** "An intentionally deceitful practice aimed at depriving another person of his or her rights or doing injury to him or her in some respect." (Bockrath and Plotnick 2011). Fraud has everything to do with intention.

- **Negligence:** "...consists of a failure to follow such a pattern of behavior as, under the circumstances, a reasonable person would have pursued, or, contrarily, of doing what a reasonable person would not have done. Thus negligence is conduct that is abnormally likely to cause harm to others, or that is, all circumstances considered, unreasonably dangerous, though not intentionally so." Part of the test is the stated standard of care, which was introduced in Chapter 7 and is defined later in this list. Negligence is the claim of choice for most plaintiffs. However, a plaintiff must prove, by preponderance of evidence, all of the following: a duty of care; breach of that duty; causation (connection between the negligence and the harm or injury); and loss or damages (Bockrath and Plotnick 2011). Therefore, an engineer might be negligent but not legally liable. Incidentally, being honest and well-intentioned, noble as they are, are not enough to protect you from negligence.
- **Tort:** "A wrongful behavior... for which a civil action will lie; the unprivileged commission (or omission) of an act whereby another person incurs loss or injury; such breach of duty as results in damage to plaintiff" (Bockrath and Plotnick 2011).
- **Standard of care:** "The level of competence practitioners in their field customarily expect given the circumstances" (Hawkins 2005). (Recall that this definition was used in Chapter 7 to help explain that quality means meeting all requirements and standard of care is one of those quality requirements). This similar definition is offered by the Online Ethics Center (2011): "The degree of care that a reasonably prudent person would exercise in some particular circumstances."

Breach, fraud, and negligence are the three ways, individually or in combinations, in which an engineer or other technical professional and/or his or her organization can incur liability.

Historic Note

The concept of liability and its incorporation into laws goes back to ancient times. For example, 4000 years ago the Babylonians developed the Code of Hammurabi which clearly stated the importance of individual responsibility. As quoted in Petroski (1985), the Code included these "eye-for-an-eye" house-building provisions: "If a builder build a house for a man and do not make its construction firm, and the house which he has built collapse and cause the death of the owner of the house that builder shall be put to death. If it cause the death of the son of the owner of the house, they shall put to death a son of that builder. If it cause the death of a slave of the owner of the house, he shall give to the owner of the house a slave of equal value. If it destroy property, he shall restore whatever it destroyed, and because he did not make the house which he built firm and it collapsed, he shall rebuild the house which collapsed from his own property. If a builder build a house for a man and do not make its construction meet the requirements and a wall fall in, that builder shall strengthen the wall at his own expense." And then we have this rigorous dam maintenance requirement, as quoted in Biswas (1970), which was part of the Code: "If anyone be too lazy to keep his dam in proper condition, and does not keep it so; if then the dam breaks and all the fields are flooded, then shall he whose dam the break occurred be sold for money and the money shall replace the corn which he has caused to be ruined."

CHANGING ATTITUDES: FOREWARNED IS FOREARMED

Decades ago, people were more inclined to take risks and accept the consequences, whether the results were favorable or unfavorable. However, today, individuals seem more inclined to blame others for the negative consequences of their actions and, in some situations, start legal proceedings. A new concept of social injustice has evolved. As in the past, accidents and failures are recognized as being expensive as a result of factors such as medical costs, additional materials, and schedule delays. However, it is increasingly common to expect someone else to pay the costs, or at least part of them. As noted by Robert Half, staffing firm founder, "The search for someone to blame is always successful." However, while finding someone to blame may always be successful, successfully blaming them is another matter. Smart individuals and their organizations manage their affairs so as not to become scapegoats.

More specifically, the tendency is to look for "deep pockets." This may be a government agency, a contractor, a consulting engineer, a manufacturing firm, or an individual engineer. Forewarned is truly forearmed in minimizing exposure to litigation. Fear of litigation may stifle creativity and innovation at the very time society is faced with increasingly complex issues, problems, and opportunities that require new approaches.

LIABILITY: INCURRING IT

Unfortunately, there are many ways in which today's individual engineer or other technical professional and his or her organization can incur liability. Liability-incurring opportunities abound in the practice of engineering and related technical professions.

One way to illustrate the potential liability exposure of an engineering or similar organization and its members is to mention some of the services typically provided by or within such an organization and then think through some of the possible related liabilities. This approach was used to develop the scenarios presented in Table 11.1. Each scenario notes a service typically provided by an engineering organization and then gives an example of a potential liability in terms of fraud, breach, and/or negligence.

To reiterate, and as illustrated by Table 11.1, liability-incurring opportunities abound in the practice of engineering and related technical professions. As noted in the earlier definitions of legal terms, breach has little or nothing to do with intention, but nevertheless consists of violating a right, a duty, or a law. Simply failing to deliver

Table 11.1 Essentially all of the services provided by an engineer or similar professional can expose the firm to liability.

Service	Potential liability for engineer or other professional	Comment
Facilitating meetings	Breach: Contract promised X meetings and only Y were conducted	—
Interpreting use of the site	Negligence: Failure to allow for building set backs	—
Investigating underground site conditions	Negligence: Failure to find shallow bedrock or utilities such as a water main or gas line	Unknown site conditions, especially subsurface, can be a major source of liability. Use contract clauses that allocate or shift the risk associated with the site conditions.
Helping to secure project financing	Fraud: Engineer/architect "steers" client, owner, or customer to a specific lender in return for a kick-back from the lender	—
Assisting in obtaining project approval from regulatory agencies	Fraud: Falsely claim that an agency granted approval	—
Preparing drawings and specifications for electrical work	Negligence: Inadequate insulation leads to fire	—
Preparing cost estimates	Negligence: Numerical error	—
Reviewing contractor bids	Fraud: Altering a bid to favor a contractor	—
Examining contractor's shop drawings	Negligence: Failure to note an unsatisfactory change	Use means and methods clauses stating that shop drawing review is for general conformance with the contract documents and that means and methods are the contractor's responsibility
Maintaining accurate books and records	Breach: Failure to do and submit when required	—
Creating as-built drawings to show final construction including location of utilities	Negligence: Incorrectly locating buried electric line leading later to a disastrous excavation accident	—

Source: Most of the scenarios, excluding the liabilities, are adapted from Bockrath and Plotnick (2011)

plans and specifications on time as specified in a contract or agreement could constitute breach. Fraud, which is explicitly intentional and deceitful action, consists of actions such as billing a client for products not delivered or falsely stating that a necessary government permit had been secured.

Negligence, the most common of the three ways in which technical professionals and their organizations may incur liability, includes failing to exercise care and provide expertise in accordance with the profession's standard of care. For example, calculation error could lead to a negligence determination. To reiterate a point made earlier, you must recognize that being honest and well-intentioned are simply not enough to avoid negligence. You and your organization must be disciplined in the manner in which you provide services if negligence and possible resulting liability are to be avoided. And to restate a point made as part of the earlier definition of negligence, while negligence is always a slippery slope, it does not necessarily equate to liability.

Views of Others

"...those who hire architects and engineers are not justified in expecting infallibility; they purchase service, not insurance," according to professors and authors Joseph T. Bockrath and Fredric L. Plotnick (2011). They go on to say, however, that "In the absence of a special agreement, the prevailing view is that the designer's undertaking is to exercise ordinary professional skill and diligence and to conform to accepted professional standards." In other words, think about and respect the standard of care in whatever work you do. Judge John McClellan Marshall offers this thought about the relevance of the legal framework to engineers: "Engineers and architects are, by their training and experience, characteristically practical people whose professional lives are defined by mathematical and physical realities. As a result, the notion that something as fluid as the law should be of importance to that world is something that escapes many engineers and architects" (Bockrath and Plotnick 2011). Hopefully, partly as a result of this chapter, that importance won't escape you. And, finally, this thought from an anonymous source: "The engineer, when faced with a problem, asks how can we fix it? The attorney, when faced with a problem, asks who can we blame?"

LIABILITY: FAILURES AND LEARNING FROM THEM

A review of some actual failures and the resulting death, disruption, economic loss, and other consequences provides insight into the causes of the failures and the related liabilities. Although liability can be incurred as a result of breach, fraud, and negligence, negligence is by far the dominant cause.

The purpose of this section is to suggest that study of engineered structures, facilities, systems, products, and processes that fail provides preventive lessons. Aspiring and young engineering professionals should become students of the history of their profession. By doing so, they will learn to better appreciate the legacy left by engineers and other technical professionals around the globe and also gain insight into failures and how to minimize them. As noted by Harry S Truman, 33rd U.S. President, "The only thing new in the world is the history you don't know."

This section's purpose is accomplished by discussing one failure and mentioning a few others with the hope that the reader will continue to be a student of failures. An excellent resource is Norbert J. Delatte, Jr's book *Beyond Failure: Forensic Case Studies for Civil Engineers* (2009), which uses a case study approach to draw technical, procedural, and ethical lessons from many unfortunate failures. The over 40 case studies are organized by specialty areas such as statics and dynamics; structural analysis; and fluid mechanics and hydraulics; and management, ethics, and professional issues.

The focus in this section is on relatively dramatic structural failures. Lack of strength is typically the cause and the resulting failure is usually instantaneous and catastrophic. There can also be what might be referred to as serviceability failures. Examples of this much less dramatic, but more common type of failure are roof deterioration, excessive floor deflection, and vibration of structural components.

Collapse of Hotel Walkway

In 1981, two walkways suspended one above the other in the atrium of the relatively new Hyatt Hotel in Kansas City, MO, suddenly collapsed, killing 114 people and injuring almost 200. This description of the disaster and subsequent actions is based on data and information provided by Delatte (2009), Goodman (1990), and Petroski (1985).

The cause of the collapse is illustrated in Figure 11.1, which shows original and as-built support details. Subsequent laboratory simulation confirmed the nature of

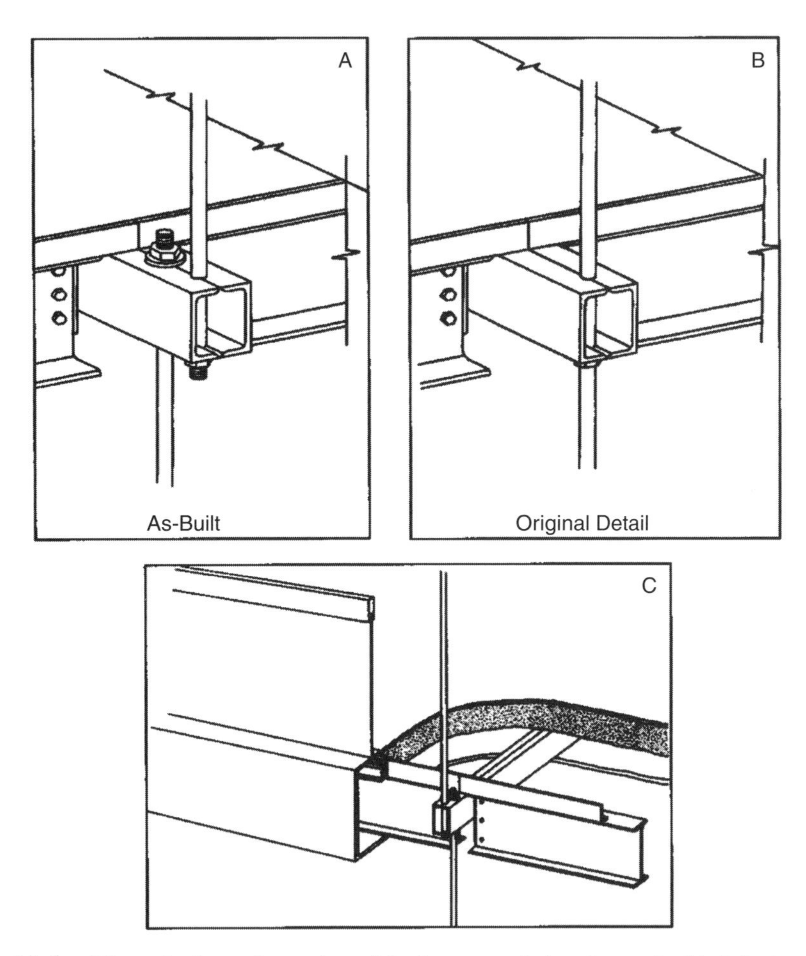

Figure 11.1 The as-built configuration of the hanger rods-box beam doubled the load on the connection and contributed to the walkway collapse. (Source: National Bureau of Standards)

failure of the connection between the hanger rod and the box beam of the type that supported the walkway.

Two technical errors occurred and, unfortunately, were additive. First, the originally designed system for supporting the walkways was under-designed in that it had only 60 percent of the strength specified by the Kansas City building code. However, the failure might not have occurred because of the safety factor explicitly incorporated in the code. This negligence was attributed to the design professionals.

The second technical error occurred when a change was made in the field. Instead of suspending the box beams and the walkway on single rods, which would have been very difficult from a construction perspective, shorter rods were used. At each suspension point on the upper walkway, one rod was connected to the ceiling and terminated just below the box beam supporting the highest walkway. A second rod was connected to that box beam and extended down to and just below the box beam supporting the lower walkway. The net effect of this change, which was never actually designed, was an excessive load at the point where the upper box beams were supported by nuts on the upper rods. The change, in effect, doubled the connecting forces at this location.

Events and circumstances leading to the walkway failure include fast track project delivery, that is, overlapping of design and construction; key members of the design team leaving the design firm during the project; and miscommunication during proposed construction changes. Delatte (2009) summarizes the lessons learned as a result of the Kansas City Hyatt Hotel walkway disaster as:

- Every connection in a structure must be designed and verified as such.
- Technical questions, such as the strength of members, should be answered by referring to project documents, not by trusting memories.
- Personnel changes must be carefully managed so that critical information is transferred.
- Each engineer should have primary responsibility for his or her work.

Notice, from a common sense, personal, and organizational perspective, the ease of understanding and presumably applying the lessons learned. I say presumably because most of the legal and other problems encountered by engineers can be traced to failure to apply common sense practices like those described above. The issue is typically insufficient self and organizational discipline with the result that common sense does not lead to common practice and, as a result, failures occur.

Other Failures

To reiterate, many technical and other lessons can be learned by studying failures. Consider the 1940 collapse of the Tacoma Narrows Bridge in Washington State that led to a better understanding of wind-induced oscillations or vortex shedding (Delatte 2009, Petroski 1985). Then there is the 1986 explosion of the Challenger rocket in which seven astronauts died as the result of a faulty seal (Delatte 2009, Fledderman 1999, Martin and Schinzinger 2005, Whitbeck 1998). Some believe that this disaster could have been prevented had engineers acted more responsibly (e.g., Florman 1987).

In his book, *To Engineer Is Human: The Role of Failure in Successful Design*, Petroski (1985) reminds design professionals that each design is not a fully tested hypothesis. Some mistakes and, unfortunately, a few disasters are inevitable. They offer an opportunity to learn at the individual level and at the professional level and, as a result, advance the state-of-the-art.

LIABILITY: MINIMIZING IT

As noted earlier in this chapter, society is becoming more litigious, more likely to take legal action, to "sue everybody in the hope of collecting from somebody" (Bockrath and Plotnick 2011). Most claims are meritless. For example, a study of New York construction-related professional firms concluded that about 83 percent of the claims had no merit. That is, the "design professionals were determined to have had no responsibility for damage or injury as measured by having no indemnity payment by the insurer on behalf of the design professional." Regardless of the lack of merit in these cases, "these firms were forced to diminish their productivity by expending their time and money to defend themselves" (NSPE 2011a). Therefore, engineers and other technical professionals must be even more diligent in trying to minimize risks, including obtaining liability insurance and adopting preventive practices.

Insurance: Financial Protection

While purchasing liability insurance won't prevent lawsuits, insurance will provide some financial protection if a liability action is initiated. The insurer, in exchange for regular premium payments, agrees to make liability payments and defend suits arising out of negligence or alleged negligence in the provision of professional services by the insured. Not all consulting firms purchase liability insurance in that some "go bare." Small firms are much more likely to go without liability insurance than large firms. Firms without liability insurance are, in effect, self-insured (Parsons 1999).

As with most insurance policies, there are exclusions—actions and activities that are not covered by liability insurance. Examples are the failure of the insured to complete services on time, intentional fraudulent and other acts of the insured, and the insured providing services outside the organization's area of expertise. As is also the case with some other forms of insurance, professional liability insurance typically has deductible provisions—an initial amount of loss that is not covered by the insurance (Bockrath and Plotnick 2011).

Liability insurance is expensive when annual premiums are quantified, for example, as a percent of annual billings for a consulting engineering firm. Liability insurance premiums approximate the after-tax profit of many smaller consulting engineering firms. Premiums tend to be higher for greater risk areas of service such as structural design. Incidentally, insurance premiums are part of overhead and, therefore, go right to the bottom line on the engineering organization's income statement.

Organizational Preventive Practices

Some consulting firms and other organizations may consider liability insurance optional, yet all such organizations should aggressively and systematically establish programs to minimize liability, particularly that which could be incurred as a result of negligence. An organization's upper management can take many corporate-wide actions to reduce liability exposure. While these are largely beyond the area of responsibility of entry-level professionals, some possible preventive actions are briefly listed as follows to provide context for the next section which applies primarily to the entry-level professional:

- Incorporating a private practice and incorporating high-risk services separately
- Limiting practice to "safer" disciplines and avoiding higher litigation-potential service areas such as poorly-financed developers, hazardous waste projects, and geotechnical studies
- Understanding what constitutes quality on each project as discussed in Chapter 7
- Avoiding financial interest in projects because if the organization has financial interest it may be liable for problems even if there is no claim of professional negligence
- Using standard contract forms such as those available from The Engineers Joint Contract Document Committee (NSPE 2011b) and placing standard of care provisions in agreements (Loulakis et al 2011)
- Placing liability-limiting provisions in contracts and agreements
- Creating, using, and continuously improving written guidance as advocated and discussed in detail in Chapter 7
- Establishing software policies and procedures on the premise that the organization has primary responsibility for the computer programs that it uses
- Hiring insured subconsultants
- Limiting comments about on-going projects to knowledgeable personnel to reduce the likelihood of transmitting erroneous information to the client-owner-customer or making promises that cannot be kept
- Using peer review

Personal Preventive Practices

You, by working smart, can reduce your and your organization's liability exposure. The six suggestions offered here for your consideration are largely repeats of topics discussed in detail in other chapters. However, they are repeated here in summary form to emphasize their role in helping you avoid legal problems for you and your organization. These suggestions are largely within the range of responsibility and authority of entry-level technical professionals. If you practice as suggested you will make a significant contribution to reducing your and your organization's liability exposure and, as you progress in your career, you will become an even more valuable practitioner.

Guard Your Reputation

This suggestion repeats, for emphasis, the message of the identically-titled section in Chapter 2. That section said: "Tell the truth. Keep your word. Give credit for ideas and information. Do your share. Don't blame others. Accept responsibility for your errors and, to the extent feasible, correct them." If you follow this "guard your reputation" advice, you are much less likely to be involved in legal complications and, if you are, you are much more likely to be treated with respect.

Maintain Competence

You need a proactive program of formal and informal activities to maintain individual competence and contribute to corporate competence. Technical services that fall below the standard of care may be deemed negligent and result in personal and organizational liability. Your and your employer's best interests require that you remain current and competent in your areas of technical and non-technical specialization. And that responsibility lies primarily with you. Refer to the Chapter 2 section titled "Managing Personal Professional Assets: Building Individual Equity."

Watch Your Language

You may be asked to assist with drafting or reviewing contracts or agreements. Be careful with the words that are used so that you and your organization minimize disagreements with those you serve or work with. I'm not suggesting watching our language in contracts and agreements as a way to "put one over" on a client, mislead an owner, or "slip one by" a customer. I am suggesting watching our language in order to communicate effectively. Consider the three categories of problematic words, and examples of each, presented in Table 11.2.

The kind of problems created by improper use of Category 1 words, absolutes/superlatives, is illustrated by stating in a proposal that your engineering or architectural firm will provide the "best" project team. Would you be able to substantiate that "best" claim, or want to be asked to substantiate that claim, in a contentious or litigious situation? Using "best" in this situation invites having your organization held, in litigation, to a standard above the standard of care.

Consider the Category 2, words of promise. An example of a potential related problem is you "guarantee," as you draft a section of a scope of services, that your firm will obtain the necessary permits. While you can assist with the permit application or even prepare it, you cannot assure that the government entity will grant the permit.

Table 11.2 Carefully use the words in these three categories of problematic words.

Category	Examples
1. Absolutes/superlatives	All, always, any, best, every, highest, maximum, minimum, never, none, only
2. Promises	Approve, assure, ensure, examine, certify, guarantee, insure, investigate, reach consensus, supervise, test, warrant
3. Multiple meanings, interpretations	Complete, defend, equal, essential, estimate, expert, final, full, furnish, install, necessary, periodic, required, safe, specialist, thorough

(Source: Adapted from Dixon 2007)

Table 11.3 Problematic words can be readily replaced with words that more accurately convey the intent.

Problematic words	Category	Possible replacement
All existing information will be gathered	1	Readily available information will be reviewed and collected as needed
At all times	1	Will be done once per...
Insure, ensure, assure	2	Reasonable effort will be made
Periodically	3	Every Friday
Prepare summaries of all meetings	1	Prepare summaries of monthly project status meetings with client
Supervise, inspect	3	Observe and report
Will complete all project services	1	Will prepare and submit for review and approval normal engineering drawings suitable for construction

(Source: Adapted from Hayden 1987)

Finally, we have words with multiple meanings/interpretations, that is, Category 3. Writing in the scope of services that you will make "periodic" visits to the construction site sounds good, but will you visit the construction site daily, weekly, or monthly?

At the draft stage in order to reduce legal and other problems later on, search for problematic words in the above categories in contracts and agreements, and also in emails; memoranda; letters; statements of qualifications (SOQs); and promotional materials such as advertising, brochures, and websites. To conclude this watch your language section, refer to Table 11.3 which shows how to avoid problematic words.

Practice Empathy

Related to the topic of watching our language, is how we, as service providers, listen to and talk to owners, clients, and customers because it apparently impacts the probability of litigation. Bachner (2009) cites a study in which the researcher recorded hundreds of physician-patient conversations. "Half of the physicians had never been sued; the other half had been sued at least twice . . . " Why was that? "Those who had never been sued spent 20 percent more time with their patients and were far more likely to exhibit active-listening skills. Importantly, those who had never been sued didn't give their patients more or better information; the difference was not in what they said, it was in how they said it."

Might how engineers and other technical professionals communicate be the cause of some of our interpersonal and legal problems? Bachner goes on "As for the quality of your work being a substitute for the quality of your relationships, wake up!" He concludes by saying, "When errors occur, as they inevitably will, having a relationship can mean the difference between resolving them amicably and moving on, or having to defend a lawsuit."

Communicate-Communicate-Communicate

This advice applies to the interaction between you and your supervisors, colleagues, paraprofessionals, support personnel, client-owner-customer representatives, and other project stakeholders. No more need be said here in that many and varied communication principles and tips are presented in Chapters 3, 4, and elsewhere in this book. You are less likely to make project errors if you understand the expectations of others and if they understand your expectations.

Document Everything

Everything means essentially everything, including, but not limited to, meetings, telephone calls, e-mails, field reconnaissance, and conversations. The "Orchestrating Meetings" section of Chapter 4 stresses the importance of documenting meetings and the Chapter 14 section titled "Selectively Share Data, Information, and Knowledge" emphasizes the need to document marketing efforts. These are just examples of what should be your habit, that is, documentation. As you do your engineering work, record what you did, when you did it, and why. And, once again, watch your language—assume that anything you write will someday be viewed by your peers or, worse yet, by opponents in litigation.

Your organization may have a uniform documentation system consisting of components such as filing procedures, special forms, and a project management system. For example, Bachner (2008) urges that an organization have a documentation program that includes "written guidance that tells people what should be written," and "how long various types of documentation should be kept." If your employer doesn't have a documentation system, perhaps you should offer to assist in its development. If such a suggestion is not well-received, develop your own system so at least your work is carefully documented.

Personal

One of my most interesting consulting and expert witness assignments was the result of poor or no documentation of meetings during a project, especially one meeting at which design criteria were discussed. After that meeting, the engineering firm designed a flood control project. Soon after it was constructed, a large rainfall occurred and resulted in flooding. The owner brought a suit against the engineer claiming negligence. The engineer replied by indicating that, based on the undocumented design criteria meeting, the facilities were to be designed for a "moderate" storm, not a "big" storm like the one that occurred. The owner contended that, based on the undocumented meeting, the facility was to be designed for a "big" storm. The case was eventually settled at a cost of several hundred thousand dollars to the engineering firm and their liability insurer—all for lack of documentation of a one meeting.

Please note that once a litigation process begins, counsel for plaintiffs and defendants will request, during the discovery process, minutes and summaries of meetings, emails, and notes about telephone conversations. "Telephone conversations, in general, are not discoverable during litigation." However, telephone conversation logs are discoverable and can demonstrate your and your team's efforts to communicate

(Matsuoka 2009). Claims as to "who said what and when they said it" that are backed by written documentation carry much more weight than undocumented recollections.

Back to meetings. A possible explanation for not preparing meeting summaries is that "we each took our own notes so we don't need official group notes" (Domalik 2010). The fallacy here is that the individual notes are likely to have many differences on important decisions and action items.

An article in *NSPE's Professional Engineers in Professional Practice (PEPP)* newsletter (NSPE, 2010), written by a liability insurer, offered the following thoughts which stress the value of documentation in the event of litigation: "To have to 'put it in writing' will cause you to stop and think. In the event you are drawn into a law suit, those writings are likely to be admissible evidence as proof of what was said and done, even when those who participated cannot remember, seek to misrepresent facts, cannot be found, or have died." The quote also highlights another benefit of documentation. That is, when we "put it in writing," we "stop and think." We often don't know what we really think about something until we write about it. The article also says: "In suits alleging negligence in the performance of professional services, well-drafted documents and well-kept, comprehensive project records are the strongest defensive weapons available to professionals."

MAINTAINING PERSPECTIVE ON LIABILITY MINIMIZATION

"The greatest mistake you can make in life," according to writer and editor Elbert Hubbard, is to be continually fearing you will make one." In keeping with that advice, your firm or other organizations should guard against letting the "tail wag the dog," that is, becoming overly fearful if not paranoid about liability.

Recognize that much of what engineering and similar organizations do to minimize liability exposure is also being done for one or more other reasons, some of which may even be more important—this is just good managing and leading. For example, peer review, which is one way to minimize liability, is also likely to yield a more cost-effective design as measured by minimizing life-cycle costs. Documentation during a project, another liability minimization device, is also helpful in coordinating a project, writing a report on a project, as a resource for "surprise" meetings, and as a guide for future, similar projects. While timely response to client, owner, and customer requests will surely minimize liability exposure, it is also a mark of good service. Use of standard, tested contract and agreement language is another liability-minimization measure and also a time-saving device. Finally, written technical and other guidance will certainly minimize the probability of errors or omissions within an engineering or other technical organization. Such guidance will also, in the long run, greatly reduce the time required to complete tasks and processes and serve as a very effective orientation and training tool for new personnel.

LEGAL FORMS OF BUSINESS OWNERSHIP

Businesses such as consulting firms, manufacturing companies, and constructors, usually use, from a legal perspective, one of three basic forms of business ownership.

These forms are the sole proprietorship, partnership, and corporation. As an aspiring engineer or as a young professional, you should understand the elements of the three forms for at least two reasons. First, doing so will help you understand part of the legal context of your employment. For example, why you may or not be able to invest in your company. Second, some day you may, individually or with others, start your own consulting, manufacturing, construction, or other business and will need, among many other decisions, to select a basic form of business ownership.

While the introduction presented here will certainly not prepare you for either of the stated applications, it could form the foundation on which you can gradually, by study and observation, build a superstructure of understanding how businesses are organized from a legal perspective. Recognize that variations exist on the three basic forms of form of business ownership. The following descriptions of the sole proprietorship, partnership, and corporation are based, in part, on Clough et al (2005), Halpin (2006), and Walesh (2000).

Sole Proprietorship

With the sole proprietorship or, as it is sometimes called, the individual proprietorship or sole ownership, an individual owns and operates the business. One advantage of the sole proprietorship is that it is the simplest and least expensive to establish and operate. An example of simplicity is that U.S. income tax reporting consists of completing one tax schedule (Schedule C, Profit or Loss from Business) for inclusion with the proprietor's personal tax return. Furthermore, there is maximum freedom of action in that the owner is the "boss" and can act unilaterally in all decisions. Being a sole proprietor is enabled by the internet and its communication and research resources.

The most significant negative aspect of the sole proprietorship is that the owner is personally liable, including personal, non-business assets, for all debts and obligations. This disadvantage may be partially offset by purchasing liability insurance. Another disadvantage of the sole proprietorship is that the size of the business and the ability to expand are limited by the sole proprietor's resources, including his or her knowledge and skill and ability to obtain financing. The knowledge and skill limitation can be somewhat offset by hiring employees or by developing arrangements with other sole proprietors that offer needed complementary services.

Personal

I have functioned as a sole proprietor for over a decade. Being a one person business has not hampered me in the development of services and in securing desirable clients. Instead of hiring personnel for necessary expertise, I have contracted with or otherwise used the services of other mostly sole proprietors and small businesses to provide expertise in areas such as information technology, website design and maintenance, graphics, word processing, and government grant programs.

Partnership

In a general partnership, which is an unincorporated entity, two or more persons own and operate the business although ownership, decision-making, debts, losses, and profits are not necessarily equally shared. Partnerships are formed and operated under state or other government entity partnership laws. A partnership is recognized as a legal entity entitled to own property, hire employees, and sue or be sued. The laws generally recognize the partnership as an entity separate from the individual partners. For example, the partners pay income taxes, not the partnership. Any partner can act on behalf of the partnership provided the action is in keeping with the scope of the business.

The appeal of a partnership is that it combines the financial assets, facilities, equipment, and talents of two or more individuals who are interested in engaging in the same type of business. On the negative side, particularly when compared to a sole proprietorship, individual partners are restricted in their business actions although they typically have more latitude than owners of a corporation. For example, a partner cannot sell or mortgage his or her share of the partnership's assets without permission of the other partners. Furthermore, in a fashion similar to the sole proprietorship, each partner is financially responsible for the acts of all the partners to the full extent of his or her personal assets.

Corporation

Although various kinds of corporations exist, such as private and public and profit nonprofit, this discussion is limited to private, for-profit corporations. Such a corporation is an entity created under state or similar entity incorporation laws, consisting of one or more individuals, owned by one or more stockholders, and considered separate from the employees or the owners. A board of directors, elected by the owners, provides general control. The corporation can buy and sell real estate, enter into agreements, and sue and be sued. A corporation can be dissolved by surrender or expiration of its state or other charter with its business obligations settled in accordance with those laws. Owners pay taxes only on dividends received after the corporation pays state and federal taxes. A subchapter S Corporation is a special U.S. form of corporation available to businesses that have 35 or fewer stockholders and meet other requirements. The advantage of this form of corporation is that it has less tax liability than a standard corporation.

A significant advantage of a corporation, contrasted with a sole proprietorship or partnership, is the limited liability of the owners. Stockholder liability is limited, with a few exceptions, to the amount of their investment in the corporation. Examples of exceptions—situations in which one or more individuals within a corporation risk personal liability—are when fraud is committed or when a corporation is underfunded or underinsured. Other advantages of corporations are the ability to raise large amounts of capital and the corporation's perpetual organizational life, that is, continuation of the organization is not dependent on particular employees or owners because the corporation is an entity separate from the employees and owners. Another advantage of a corporation is that it provides ease of multiple ownership.

A possible disadvantage of a corporation is that the stockholders are not in any way agents of the corporation. The board of directors controls the corporation and the officers of the board act as agents for the corporation; although each stockholder owns a part of the organization, he or she cannot act unilaterally on behalf of the organization.

Closure

From a legal perspective, businesses such as consulting firms, manufacturing companies, and constructors use one of three basic forms of business ownership: the sole proprietorship, the partnership, and the corporation. As you move through your career in the private sector, when you are an employee, you may work for all three types of organizations. If you decide to start you own business, various business ownership avenues and scenarios are available. For example, you may begin as a sole proprietorship, then join with others in a partnership, and finally convert to a corporation.

CONCLUDING COMMENTS ABOUT THE LEGAL FRAMEWORK

This chapter argues that, while engineers should not practice law, they should know enough about law to avoid having their actions or inactions impact negatively on them or their organizations. You as an entry-level practitioner, and you as a student who will soon be one, should think about reducing your organization's liability exposure by applying the knowledge, skills, and attitudes (KSAs) explained and advocated in this chapter. In addition, by acquiring these KSAs, you will enhance your career opportunities as an employee and/or a business owner.

I was to learn later in life that
we tend to meet any situation by reorganizing;
a wonderful method it can be for
creating the illusion of progress
while producing confusion, inefficiency and
demoralization.

(Petronius, Roman writer)

CITED SOURCES

Babitsky, S. and J. J. Mangraviti, Jr. 2002. *Writing and Defending Your Expert Report: The Step-by-Step Guide with Models*. SEAK: Falmouth, MA.

Babitsky, S. and J. J. Mangraviti, Jr. 2005. *How to Become a Dangerous Expert Witness: Advanced Techniques and Strategies*. SEAK: Falmouth, MA.

Bachner, J. P. 2008. "A Process for Documentation." *CE News*, January, p. 19.

Bachner, J. P. 2009. "Be Nice," *CE NEWS*. November, p. 18.

Biswas, A. K. 1970. *History of Hydrology*. North Holland Publishing Company: Amsterdam, The Netherlands.

Bockrath, J. T. 1986. *Contracts, Specifications, and Law for Engineers - Fourth Edition*. McGraw-Hill: New York, NY.

Bockrath, J. T. and F. L. Plotnick. 2011. *Contracts and the Legal Environment for Engineers and Architects-Seventh Edition*. McGraw-Hill Book Company: New York, NY.

Clough, R. H., G. H. Sears, and S. K. Sears. 2005. *Construction Contracting - Seventh Edition*. John Wiley and Sons: New York, NY.

de Camp, L. S. 1963. *The Ancient Engineers*. Ballantine: New York, NY.

Delatte, N. J. 2009. *Beyond Failure: Forensic Case Studies for Civil Engineers*. ASCE Press: Reston, VA.

Domalik, D. 2010. Personal communication, Vice President, East Region Quality Director, HDR, Pittsburg, PA, February 22.

Dixon, P. A. 2007. "*Choosing Your Words Carefully*." Holmes Murphy & Associates: Peoria, IL, October 23.

Fleddermann, C. B. 1999. *Engineering Ethics*. Prentice Hall: Upper Saddle River, NJ.

Florman, S. C. 1987. *The Civilized Engineer*. St. Martin's Press: New York, NY.

Goodman, L. J. 1990. "Revisiting the Hyatt Regency Walkway Collapse." *Civil Engineering News*, March 20.

Halpin, D. W. 2006. *Construction Management - Third Edition*. John Wiley & Sons: Hoboken, NJ.

Hayden, Jr., W. M. 1987. "Quality by Design Newsletter." May, A/E QMA, Jacksonville, FL as quoted in *Journal of Management in Engineering—ASCE*, October, pp. 284–85.

Hawkins, J. R. 2005. "The Standard of Care for Design Professionals – Part 1." *Structural Engineer*, June, p. 16.

Loulakis, M. C., L. P. McLaughlin, and B. McLaughlin. 2011. "Engineers Standard of Care Cannot be Expanded by Opposing Expert's Testimony." The Law, *Civil Engineering*, American Society of Civil Engineers, March, p. 88.

Martin, M. W. and R. Schinzinger. 2005. *Ethics in Engineering - Fourth Edition*. McGraw Hill Higher Education: New York, NY.

Matsuoka, C. M. 2009. "Manage Risk with Communication Tools." *Structural Engineer*, December, p. 22.

National Society of Professional Engineers. 2011a. "Claims Study: Study of Exposure Times and Meritless Claims Against Design Professionals in New York." By Victor O. Schinner & Company and in Professional Engineers in Professional Practice, February.

National Society of Professional Engineers. 2011b. "Standard Contract Documents." (www.nspe.org/ejcdc). May 13.

NSPE. 2010. "Professional Liability/Risk Management Brief: Project Documentation During the Planning and Design Phase." *PEPP Talk*, February.

Online Ethics Center for Engineering and Research. 2011. "Glossary." National Academy of Engineering. (www.onlineethics.org/Default.aspx?id=2960) May 13.

Parsons, J. 1999. "Beware of These Six Traps in the Liability Insurance Trail." *American Consulting Engineer*, September/October, pp. 23–25.

Petroski, H. 1985. *To Engineer is Human The Role of Failure in Successful Design*. St. Martin's Press: New York, NY.

Walesh, S. G. 2000. *Flying Solo: How to Start an Individual Practitioner Consulting Business*. Hannah Publishing: Valparaiso, IN.

Whitbeck, C. 1998. *Ethics in Engineering Practice and Research*. Cambridge University Press: Cambridge, UK.

ANNOTATED BIBLIOGRAPHY

Brack, C. 2009. "The Scoping Process." *Zweig Letter*, July 20. (Argues that, when all project documents are considered, "more misunderstanding arises from the scope of services . . . than any other document." Stresses the importance of a service provider's scoping role by stating "If the client isn't sure about what they want, it's your fault, not theirs.")

Ericson, N. V. 2006. "Risk Management Is Not Just for Principals." *Structural Engineer*, August, p. 14. (Urges creating an organization wide "culture for risk management" recognizing that everyone can play a role. Tactics include sharing lessons learned, showing examples of real claims, explaining costs of claims, providing claim statistics, role playing, and education/training.)

NSPE. 2010. "Professional Liability/Risk Management Brief: Practice Management — How Design Professionals Can Avoid Liability Caused in In-House Processes." *PEPP Talk*, July 22. (This article, authored by the liability insurer, Victor O. Schinnerer & Company, Inc., identifies the following "areas within the design process which most frequently give rise to claims:" 1) Failure to supervise and review the work of new employees, 2) Inadequate project coordination and in-house communication, 3) Failure to communicate between prime professionals and consultants, 4) Lack of quality control on design changes, and 5) Poorly worded contracts.)

Quinn, B. and L. Willard. 2009. "Reducing Your Risk in Using Structural Software: Selecting a Champion." *Structural Engineer*, October, p. 24. (Suggests, as part of a organization's risk reduction effort, a Software Error-Reduction Plan. The SERP includes: selecting a "champion" for each program, providing education and training, creating and using written software use guidelines, and performing internal reviews of software models. While the article addresses structural software, the SERP idea is clearly applicable to software used in other engineering specialty areas.)

Walesh, S. G. 2008. "Avoiding Digital Disaster." The Softside of Engineering column, *Indiana Professional Engineer*, Indiana Society of Professional Engineers, November/December. ("Regardless of our age, but especially, if we are young, let's think twice about sending that email, leaving that voice message, or creating or forwarding that video. The digital record is likely to always be out there and subject to retrieval by friends who want to help us and, more importantly, by "enemies" who don't. Let's not in a moment of anger, carelessness, or frivolity, set ourselves up for a digital disaster years, if not decades, from now.")

EXERCISES

11.1 DESIGN FIRM'S RESPONSE TO A FAILURE: This exercise seeks to increase your and your team's awareness of the need for a timely and thoughtful response to the failure of an engineered structure, facility, system, product, or process. Suggested tasks are:

A. Select a failure - and it does not have to be a dramatic structural failure - of a constructed or manufactured structure, facility, system, product, or process designed by engineers. Some sources of ideas are comprehensive books by Delatte (2009) and Petroski (1985) and a search of the literature. Don't use the Hyatt Hotel failure because it has already been discussed in this chapter.

B. Learn about design and the failure. Understand the design criteria and the participants in the project. Go back to the time of the failure and determine the actual or possible cause of the failure in terms of what was known at that time.

C. Assume your group represents the engineering firm or other entity that performed the design and you just heard about the failure. Your group includes the project manager. The chief executive of your organization is not present. Further assume that your group quickly met to discuss the failure and your organization's response.

D. Conduct the meeting, evaluate the failure relative to your organization's design and possible other involvement such as construction management, develop action options, select a preferred course of action, and write a memorandum to your chief executive summarizing your deliberations. Compare your findings and recommendations to what actually happened.

11.2 CONSTRUCTOR OR MANUFACTURER'S RESPONSE TO A FAILURE: The purpose of this exercise is similar to that of the previous exercise. Suggested tasks are the same as Exercise 11.1 except the group represents the constructor or manufacturer.

11.3 OWNER'S RESPONSE TO A FAILURE: Once again, the purpose is similar to that of Exercises 11.1 and 11.2. Suggested tasks are the same as Exercise 11.1 and 11.2 except the group represents the owner of the structure, facility, system, product, or process.

11.4 FAILURE CASE STUDY: The exercise, to be done by you, will increase your sensitivity to and awareness of management, leadership, and technical actions and methods that help to reduce the likelihood of failure of a structure, facility, system, product, or process. Suggested tasks are:

A. Study the failure of an engineered structure, facility, system, product, or process. Given the Hyatt Hotel failure is discussed in this chapter, don't select it.

B. Prepare a memorandum that: a) describes the failure, including the consequences in terms of human loss, monetary loss, delays, ruined careers, and failure of organizations; b) identifies the design criteria and the explains the design process and the resulting construction, manufacturing, or other implementation; c) discusses the cause or causes of the failure; and d) summarizes the lesson(s) learned.

11.5 CONSULTANT AND/OR EXPERT IN LITIGATION: Given that you may someday be a consultant or expert in litigation, the purpose of this exercise is to introduce you to these potential roles. Suggested tasks are:

A. Study the roles, responsibilities, and ethical expectations of consultants and designated experts in litigation involving engineered structures, facilities, systems, products, and processes. Possible initial sources include Babitsky and Mangraviti, Jr. (2002, 2005), Bockrath and Plotnick (2011), Martin and

Schinzinger (2005) and the next chapter in this book. Try to confer with a law student, law professor, and/or an attorney. Read relevant sections of the next chapter.

B. Document your findings in the form of recommendations to a young engineer, perhaps you someday, who is about to participate for the first time as a consultant or expert. Recall the writing guidance included in Chapter 3 of this book.

11.6 LEGAL FORM OF OWNERSHIP FOR A NEW CONSULTING BUSINESS: This exercise will help you, or you and others if a team exercise, think more deeply about the three forms of business ownership described in this chapter. Suggested tasks are:

A. Review the scenario described in Chapter 10, Exercise 10.2, Task A. Exercise 11.6 builds on that income statement exercise and the Chapter 10 finance exercise (Exercise 10.3) as part of the overall effort to develop a business plan for your new consulting business. The business plan can be completed by preparing the marketing element as an exercise at the end of Chapter 14.

B. Select a legal form of ownership for your new business and explain the rationale for the selection.

CHAPTER 12

ETHICS: DEALING WITH DILEMMAS

What you are speaks so loud
I cannot hear what you say.

(*Ralph Waldo Emerson, minister, lecturer, writer*)

The chapter begins with the inevitability of ethical decisions and sometimes major dilemmas, starting during your formal education and extending throughout your career. After defining ethics, the chapter turns to the challenge of teaching and learning ethical processes in the academic and practice environments. The legal and ethical domain is discussed to relate ethical issues to legal issues. A discussion of the need for ethics codes and the limitations of codes is followed by an introduction to the codes developed and administered by engineering societies, other professions, businesses, government, and academia. The chapter offers suggestions on how individuals and groups can make decisions in difficult ethical situations. After presenting an instructive case study the chapter concludes with thoughts about each individual's responsibility and opportunity to exemplify ethical behavior.

INEVITABLE ETHICAL DILEMMAS AND DECISIONS

Television, the internet, radio, newspapers, and magazines often report on ethical issues or, more precisely, alleged or actual unethical behavior. A wide range of society's institutions and organizations are typically involved, including government, business, academia, and religious groups as members of those groups struggle with ethical issues.

As you study for and then begin to practice your profession, you will increasingly face a wide spectrum of ethical issues. At one extreme you will encounter ethical situations requiring decisions that you will be able to make with ease while very difficult dilemmas will confront you at the other end of the ethical spectrum. For example, during your studies or in the early years of employment, you may be tempted to:

- Plagiarize, that is, write a research paper assigned in class using ideas, data, and/or information developed by others without properly crediting them.
- Cheat on a take-home examination.
- Steal from someone else's examination during an in-class test.
- Copy another person's laboratory report.
- Embellish your resume.
- Shift blame for your errors to others.
- Fail to fulfill agreed-upon responsibilities within a team setting.
- Share sensitive information about a client, owner, or customer with a third party outside of your employer.
- Claim expertise you do not possess.
- Fail to express concern, as a member of a project team, about a team decision that you believe would have an adverse impact on the environment.
- Log more time to a project than you actually worked.
- Provide negative information about a competing consulting firm, manufacturing company, university, or other organization.
- Hide, during construction or manufacturing, life-threatening errors discovered in plans and specifications.
- Accept a gift offered by a vendor even though doing so conflicts with your employer's policy.
- Ignore unfair treatment of another employee.
- Participate, as a team member, in an interview with a potential client or owner even though you know that "bait and switch" is in play. That is, if your firm is selected the interview team would not provide the promised services; a less capable group would be assigned to the project.

The preceding merely suggests the varied ethical situations frequently faced – most likely daily – by student and entry-level engineers. The list should sensitize you to the certainty of confronting ethical decisions and dilemmas. This chapter provides you with ideas and information to help you navigate those often choppy and sometimes turbulent ethical waters. Regardless of your technical and non-technical expertise, unethical behavior can sink your career. Accordingly, consider developing now a sense of what you will and will not stand for and use this chapter to help refine and implement that commitment, that is, chart and follow an ethical course.

Personal

When we, as students, cheat on homework, lab reports, and other teaching-learning assignments, who is cheating who? The student who cheats probably thinks he or she is cheating the system—getting something for nothing. However, based on my experience, the student is actually cheating himself or herself by missing out on growth opportunities such as understanding a principle, learning how to use an analysis or design technique, or enhancing research and writing ability all of which are likely to be valuable during the remainder of one's formal education and during practice. While I make no claim to have been the perfect student, I learned so much by doing the homework, preparing the lab reports, doing the assigned reading, and reviewing for examinations. And, this is my point, throughout my career, I have repeatedly drawn on what I've learned and/or the underlying principles. Furthermore, as nicely stated by Russell (2004), "Cheating ultimately affects the cheater negatively, because by cutting corners or sweeping things under the rug, we deprive ourselves of the opportunity to learn from our mistakes and build open and honest relationships with others and with ourselves." And, I might add, good news about you travels fast; bad news even faster. If you were known by students and faculty as a cheater in college, you will probably be remembered by them as a cheater for the rest of your career. You think your professional reputation does not start in college? Think otherwise!

DEFINING ETHICS

What is ethics? Is it the same as law? Do ethics and law overlap or are they different? Are unethical and illegal acts synonymous? Is ethics something you have or something you do? Is ethical behavior the same as moral behavior? Chapter 11, which describes the legal framework within which technical professionals function, discusses breach, fraud, and negligence. What is left, if anything, to consider within the context of ethics?

Definitions

To begin to answer these questions, consider some definitions of ethics as means of extracting the essence of ethics for use in this chapter. The following definitions are intentionally taken from a wide variety of sources (the last definition is one of mine) and begin with those that specifically mention engineering and then transition to those that don't:

- "Engineering ethics is the study of decisions, policies, and values that are morally desirable in engineering practice and research" (Martin and Schinzinger 2005).
- "Engineering ethics is a body of philosophy indicating ways that engineers should conduct themselves in their professional careers" (Fledderman 1999).
- "Ethics is the study of systematic methodologies which, when guided by individual moral values, can be useful in making value-laden decisions" (Vesilind 1988).
- Ethics is "the standards of conduct that indicate how one should behave and act. The standards are derived from the community's values, norms, and principles" (Valparaiso 2011).

- "Ethical conduct is often defined as that behavior desired by society which is above and beyond the minimum standards established by law" (Onsrud 1987).
- Ethics is what you do when no one is looking. Your reputation is what others think of you; your ethics is what you really are and do.

Distilling the Definitions

These definitions suggest a benchmark concept with behavior below the benchmark being considered unacceptable, in that it is punishable by legal means. In contrast, behavior above the benchmark is desired, but within the purview of law. The definitions also suggest that ethics is the process each person uses to make what is referred to in one of the definitions as "value-laden decisions." Accordingly, let's define ethics for use in this book, as the process used to make value-laden decisions beyond the law in professional matters.

This process or action-oriented perspective reinforces the idea of values, which is largely a personal matter. As suggested by Table 12.1, which draws on Covey (1990), McCuen and Gilroy (2010), and McCuen and Wallace (1987), an individual's values profile may be composed of many individual values, each having a different relative importance. Just as each snowflake is unique, so each person has a unique values mosaic. For example, one person may rate honesty as one of the most important values and, as a result, strive for absolute honesty in all matters. In contrast, another person might also regard honesty as an important value, but trust may be valued even higher. Each person's values profile or mosaic determines how he or she works through the process of making value-laden decisions. Because the mosaics may differ profoundly, an action one individual considers ethical could be considered unethical by another individual.

In summary:

- Ethics is related to, but different than and above, laws. Ethical behavior is referenced to, but more than legal behavior.
- Ethics is mostly action (what you do) not knowledge (what you know). It is the personalized way you use your values profile to make value-laden decisions.
- For purposes of this book, ethics is defined as the process used to make value-laden decisions beyond the law in professional matters.

TEACHING AND LEARNING ETHICS

Ethics is difficult to teach in a university or within a place of employment in the sense that static mechanics, machine design, circuits, and project management can be taught, learned, tried, and tested. Learning ethics fundamentals and then applying them within the study and practice of engineering is strongly influenced by individual values. This teaching-learning-application challenge is heightened because of the skeptical, but fortunately not cynical, perspective of many of today's young people.

By the time you began your began college studies, you had learned much about the unethical behavior of many private and public individuals and organizations partly as

Table 12.1 Each person's values profile is composed of many values having relative importance peculiar to the individual.

Value	Definition
Accountability[a]	Being answerable for obligations.
Confidentiality[a]	Assurance that important information will not be disclosed.
Diligence[b]	Long, steady application to one's occupation or studies; persistent effort, attentive care.
Efficiency[b]	The quality or property of acting or producing effectively with a minimum of waste, expense, and unnecessary effort.
Equality[b]	The state or instance of being equal; especially, the state of enjoying equal rights, such as political, economic, and social rights.
Equity[b]	The state, ideal, or quality of being just, impartial, and fair.
Excellence[a]	The condition of providing superior service.
Fairness[a]	Selection of an action that would not unduly emphasize self-interest or show lack of objectivity in making [a] decision.
Freedom[b]	The condition of being free of restraints; the power to act, speak, or think without the imposition of restraint.
Honesty[c]	Telling the truth—in other words, conforming our words to reality. (Note: Honesty is retrospective, it is what you say about what you've done.)
Honor[b]	Esteem, respect, reverence, reputation, applicable to both the feeling and the expression of these characteristics.
Integrity[c]	Informing reality to our words—in other words, keeping promises and fulfilling expectations. (Note: Integrity is prospective, it is what you do about what you said.)
Knowledge[b]	Familiarity, awareness, or understanding gained through experience or study; cognitive or intellectual mental components acquired and retained through study and experience – empirical, material, and that derived by inference and interpretation.
Loyalty[b]	Feelings of devoted attachment; the condition of being faithful; the unfailing fulfillment of one's duties and obligations in a close and voluntary relationship.
Persistence[d]	The act or fact of persisting; the quality or state of being persistent, especially perseverance.
Pleasure[b]	An enjoyable sensation or emotion; satisfaction; sometimes, though not invariably, suggests superficial and transitory emotion resulting from the conscious pursuit of happiness.
Prudency[a]	Exercising good judgment.
Reliability[a]	Dependability in meeting duties and obligations.
Respect[d]	An act of giving particular attention; the quality or state of being esteemed.
Safety[b]	Freedom from danger, risk, or injury.
Security[b]	Freedom from doubt; reliability and stability concerning knowledge of the future.
Sensitivity[d]	Awareness of the needs and emotions of others.
Thoroughness[d]	[Carrying] through to completion; [care] about detail.
Tolerance[d]	Sympathy or indulgence for beliefs and practices differing from or conflicting with one's own.
Trust[b]	Firm reliance on integrity, ability, or character of a person or thing; implies depth and assurance of such feeling, which may not always be supported by truth.

a) Values and definitions quoted from McCuen and Gilroy (2010).
b) Values and definitions quoted from McCuen and Wallace (1987).
c) Values and definitions quoted from Covey (1990) with parenthetic comments added.
d) Definitions quoted from Merriam-Webster (2011).

a result of the news media's probing into and reporting on the personal lives of people holding, or wanting to hold, high positions in government, business, academic, and religious organizations. Even though that behavior may be atypical, it gets great attention and could influence your thinking.

As a result, when your professors and university administrators make pronouncements about ethics, you may be skeptical and give little credence to their statements. If you mistrust them, you may even, by extension, doubt what they tell you about technical matters. Similarly, as you begin your professional career, you may view your supervisors and others in your organization, including the executives, in the same skeptical manner. This is a positive perspective provided that healthy skepticism does not degrade into cynicism. The skeptic has doubts, thinks critically, hopes for the best, and wants to be shown. In contrast, the cynic is contemptuously distrustful of human nature and the motives of others. Assuming you are skeptical and not cynical, recognize that the healthy skepticism you direct to faculty, fellow students, managers, colleagues, and others is likely to be reflected back to you—at least until you all become well acquainted and prove yourselves to them.

While this chapter cannot cause you to be ethical, it can alert you to the need to have or develop a personalized ethical framework. Books, professors, supervisors, and friends cannot tell you what your ethics should be, but they can help you learn the value of consciously formulating and refining an ethical framework to use as guidance in your student and professional life and beyond.

Personal

Based on my experiences in academia and the private and public sectors, I know that some academics believe the university is the last bastion of ethical behavior in society, that it is the place where the highest ethics are practiced or, if not actually practiced, at least advocated. Universities certainly have the corner on ethics rhetoric, but they do not have a monopoly on ethical conduct. Accordingly, you and your student peers may get the erroneous idea that ethics—either personal or organizational—are irrelevant or devalued off the campus, in the business and professional world. You may believe that "it is downhill from here" in that ethical expectations will be less "out there in the real world" than they are in the college and university setting. You may be conditioned to think that professional ethics is an oxymoron. To the contrary, ethical conduct is an everyday issue in the practice of the profession and you will encounter some ethics exemplars.

As a student and later a professional, you do not have a choice as to whether or not you will be confronted with ethical decisions and dilemmas. What will your rules of personal conduct be? Will you strengthen or detract from the ethical climate of your school or employer? What values will you hold most dearly? What will you stand for and not stand for? What will you use as your ethical framework? What decision process will you follow to make value-laden decisions? You cannot escape the domain of ethics. Your true values will be gradually revealed by many situations that arise during the normal course of your student and work days and your response to them.

LEGAL AND ETHICAL DOMAIN

The connection between legal and ethical behavior was suggested in the earlier definitions of ethics. Figure 12.1 (McCuen and Wallace 1987, Onsrud 1987) is a useful model of the legal and ethical domain. In this model, the position of the vertical line, which separates legal from illegal actions, is set largely by statute and common law. Accordingly, the definitive separation between legal and illegal acts is shown by a solid vertical line. The position of the horizontal line, which separates ethical and unethical actions, is much less definitive because it is based primarily on personal values informed by various codes of ethics when professional matters are involved. As a result, the less-definitive separation between ethical and unethical lines is shown by a dashed horizontal line. Whereas most engineers agree on the legality of an act, they might not agree about whether or not some aspect of the act is ethical or unethical.

Quadrant sizes have no meaning. The rectangular axes are simply intended to define the "space" within which all possible legal-ethical transactions occur in the business, government, academic, and volunteer sectors. Clearly, as suggested conceptually in Figure 12.2, most such transactions occur in Quadrant 1 the legal-ethical quadrant. However, given human nature, transactions will also occur in the other three quadrants but with much less frequency as also suggested by Figure 12.2.

Therefore, engineers and similar professionals are in Quadrant 1 almost all the time. Figure 12.3 presents some hypothetical examples of professional activities and their placement in quadrants, beginning with Quadrant 1.

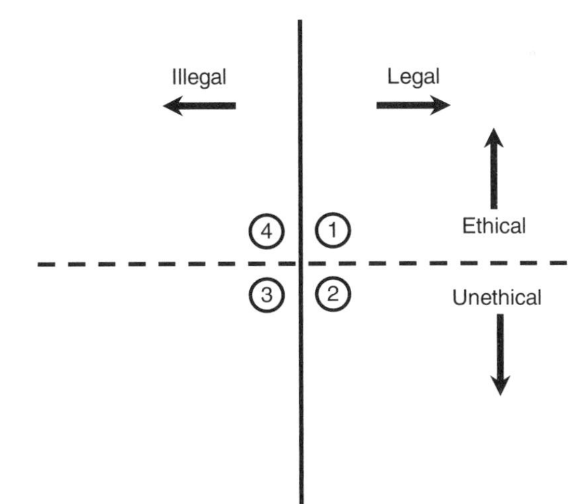

Figure 12.1 The legal-ethical domain helps understand and then resolve legal and ethical issues.

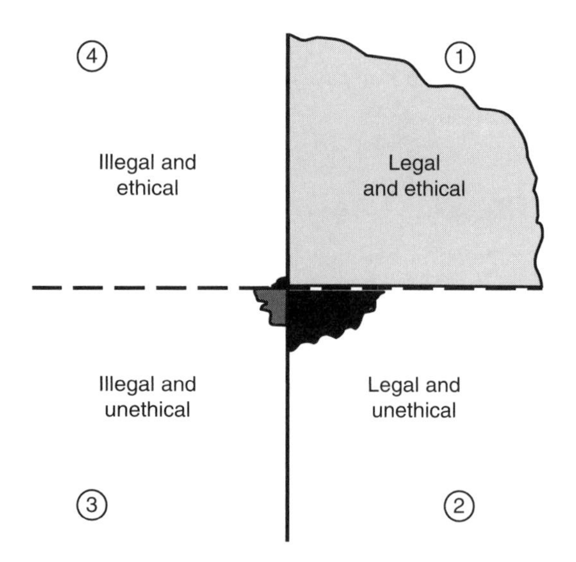

Figure 12.2 The four legal-ethical quadrants exhibit widely varying relative occurrence in engineering and similar professions.

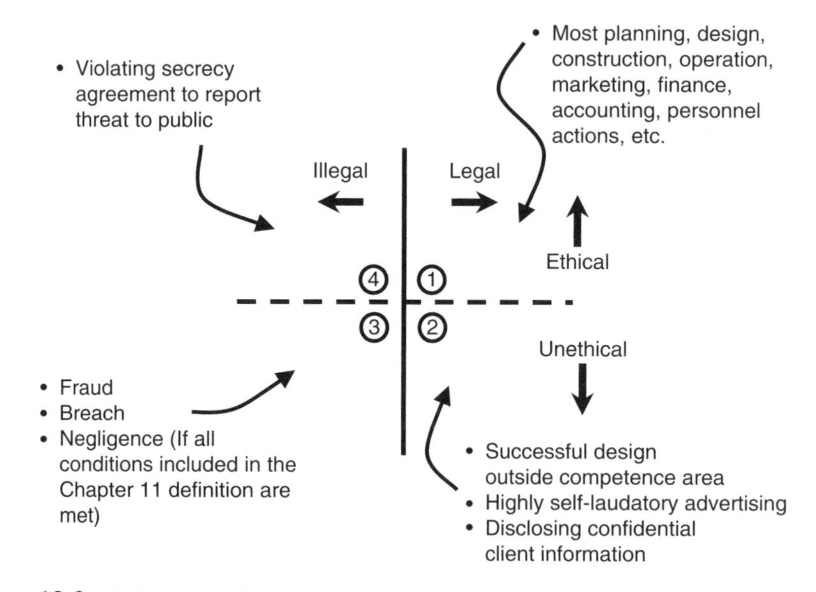

Figure 12.3 Selected professional activities can be placed in each of the four quadrants.

- **Quadrant 1—Legal and Ethical:** Most planning, design, construction, and operation activities fall within this quadrant along with most support activities such as marketing, finance, accounting, and personnel matters.
- **Quadrant 2—Legal and Unethical:** An engineer successfully designs a structure or facility outside his or her area of competence. A consulting firm advertises its

services in a highly self-laudatory manner. An individual engineer discloses confidential information developed for or with a client or former client. As explained later in this chapter, these hypothetical actions are usually legal, but may be considered unethical, because they conflict with accepted codes of ethics established by engineering organizations.

- **Quadrant 3—Illegal and Unethical:** An example is an act of fraud (intentionally deceitful practice), such as bid rigging or collusion with others to secure a contract for professional services. Another example is breach (nonfulfillment of an obligation) such as missing the contractual delivery date for a set of plans and specifications. In general, illegal acts such as fraud and breach of contract, are unethical, but there can be exceptions, as indicated in the following explanation of Quadrant 4. Negligence is, in effect, illegal as explained in Chapter 11 in the discussion of the definition of negligence, if certain criteria are met.
- **Quadrant 4—Illegal and Ethical:** This somewhat problematic quadrant is best introduced by using a non-engineering or non-business example such as a concerned citizen stopping at the scene of an automobile accident, putting an injured child in his or her vehicle, and exceeding the speed limit to get the child to the emergency room of a local hospital. While the speeding was illegal, most would probably agree that the citizen's overall actions were ethical. A professional example would be an engineer violating a signed secrecy agreement with an employer to "go public" and report on a situation that he or she believes is hazardous to the public at large.

Clearly, you are urged to be involved in activities and transactions that almost always fall in Quadrant 1. Occasionally, you may find yourself in Quadrant 4, the resolution of which may pose some serious threats to your professional career. Just as clearly, you should avoid Quadrant 2 partly because, to do otherwise, you risk losing the hard-earned trust of your colleagues. Quadrant 3 is extremely dangerous ground.

Personal

My hope for you, as an engineering student and/or an entry-level professional, is that you will interact with at least a few ethically-influential professors and practitioners. By ethically-influential professors and practitioners I mean individuals who consistently offer you lessons, by words and, more important, by actions in how to process ethical questions, especially ethical dilemmas. These individuals, based my grateful experiences with some of them, are concerned about the public, the environment, the profession, and the individuals they interact with. Learning from them will keep you almost always in Quadrant 1. To the extent you learn and practice what they teach, you will be just that much better prepared for a career characterized by success and significance. That is, because of your solid reputation and your ability to deal with difficult situations, you will do well materially and you will leave a legacy.

CODES OF ETHICS

Introduction to Codes: What They Are

Essentially all engineering societies and some business, government, university, volunteer, and other organizations have developed, adopted, and refined codes of ethics. The overall purpose of such codes of ethics is to "express the rights, duties, and obligations" of organization members and provide "a framework for ethical judgment" (Fledderman 1999). The framework metaphor is used because a code, no matter how carefully crafted, cannot anticipate all of the ethical situations and dilemmas a student or practitioner may encounter. A code attempts to reach and document consensus, or at least the majority opinion, among members of a group and do so in a manner that recognizes the already-discussed natural variation in the values profiles or mosaics of individuals.

What Codes Aren't

A code is not a legal document in that failing to follow it is illegal, although such failure may result in expulsion of an individual from an organization. Furthermore, a code typically does not "create new moral or ethical principles" because "these principles are well established in [a] society, and foundations of our ethical and moral principles go back many centuries" (Fledderman 1999). Instead, a code provides guidance for applying those principles to professional activities. Reference to principles established in a "society" reminds us that codes of ethics are likely to vary from culture to culture. The codes of ethics discussed here and, more broadly, the overall approach to ethics in this chapter, reflects what one ethics author refers to as applicable in "technologically developed democracies" (Whitbeck 1998).

Limitations of Codes

Fledderman (1999) and Martin and Schinzinger (2005) describe limitations of ethics codes such as:

- The already-mentioned inability of codes to anticipate all of the ethical decisions and dilemmas a student or practitioner may encounter
- Lack of prioritization of competing demands such as maintaining client-owner-customer confidentiality versus addressing environmental concerns
- Existence of many codes within a particular sector, such as across engineering disciplines as discussed later in this chapter, may suggest to an individual engineer or other technical professional that ethical conduct is linked more to a discipline than to the overarching profession
- Limited power partly because, unfortunately, too few engineers are members of professional societies and, therefore, those engineers may not feel bound by professional society codes

Nevertheless, as you may already begin to imagine and as will be emphasized in the remainder of this chapter, codes of ethics are very useful in the business, government, academic, volunteer, and other sectors. Ethics codes help conscientious individuals work though value-laden situations and dilemmas.

Engineering Society Codes of Ethics

This section begins by listing a cross-section of engineering societies having ethics codes and the website addresses for access to the codes. The primary purpose of the representative list is to encourage you to read a code, or a few codes, for an engineering discipline or disciplines that interest you. By so doing, you will more fully understand this discussion of ethics codes and, from a practical perspective, be better prepared to use one or more codes beginning as an engineering student. The codes and their websites follow:

- American Council of Engineering Companies (ACEC) www.acec.org/about/ethics.cfm
- American Institute of Chemical Engineers (AIChE) www.aiche.org/About/Code.aspx
- American Society of Civil Engineers (ASCE) www.asce.org/Leadership-and-Management/Ethics/Code-of-Ethics
- American Society of Mechanical Engineers (ASME) www.asme.org/groups/educational-resources/engineers-solve-problems/code-of-ethics-of-engineers
- Institute of Electrical and Electronic Engineers (IEEE) www.ieee.org/portal/cms_docs/about/CoE_poster.pdf
- National Society of Professional Engineers (NSPE) www.nspe.org/Ethics/CodeofEthics/index.html

Some Critical Similarities

The preceding codes have many common elements, the most important of which is protection of public safety, health, and welfare. For example, Canon 1 in the NSPE code, after the words "Engineers, in fulfillment of their professional duties, shall:" lists, as Fundamental Canon 1, "Hold paramount the safety, health, and welfare of the public." All of the other engineering society codes presented above include responsibility for "safety, health, and welfare of the public" in their first listed provision. Holding paramount protection of public safety, health, and welfare is the first consideration in the vast majority of engineering codes of ethics.

To carry this safety-health-welfare theme beyond a canon or accepted principle and into expected actions, note that the NSPE code, in the Rules of Practice section, states: "If engineers' judgment is overruled under circumstances that endanger life or property, they shall notify their employer or client and such other authority as may be appropriate." Similarly, the Guidelines to Practice section of the ASCE code says "Engineers whose professional judgment is overruled under circumstances where the safety, health and welfare of the public are endangered, or the principles of sustainable development ignored, shall inform their clients or employers of the possible consequences" and goes on to say "Engineers who have knowledge or reason to believe that another person or firm may be in violation of any of the provisions of Canon 1 [the safety-health-welfare canon] shall present such information to the proper authority in writing and shall cooperate with the proper authority in furnishing such further information or assistance as may be required."

One of the most well-known U.S. cases of "whistle-blowing" is the action of engineer Roger Boisjoly, then employed by the Morton Thiokol company, in connection with the January 28, 1986 shuttle Challenger disaster that resulted in the death of seven astronauts. Boisjoly raised concerns about a joint and its O-rings before the fatal launch and shared company confidential documents with investigators after the launch (Delatte 2009, Fledderman 1999, Martin and Schinzinger 2005, Whitbeck 1998).

As another example of the high bar set by codes of engineering societies, consider this competency provision in the IEEE code which indicates that individuals are to "maintain and improve [their] technical competence and to undertake technological tasks for others only if qualified by training or experience, or after full disclosure of pertinent limitations." In similar fashion, the ASME code includes a Fundamental Canon which states "Engineers shall continue their professional development throughout their careers and shall provide opportunities for the professional and ethical development of those engineers under their supervision."

Codes and You

Demanding provisions, like the preceding, which are small parts of the referenced codes, can put great demands on you as you progress through your career. Recognize that adherence to the applicable code of ethics is a condition of membership in engineering and other technical organizations. Unfortunately, many professionals of all ages and levels of experience join such organizations without a detailed review of the provisions of the code. This, in turn, undoubtedly leads to some of the ethical problems that arise. You should study an engineering organization's code of ethics before joining. If you cannot embrace the code, don't become a member.

Engineers and other technical professionals who are members of professional societies with codes of ethics are not relieved of ethical responsibilities, as defined within those codes, because they happen to work for organizations in which the management or culture does not support the codes (McCuen and Wallace (1987). The young professional seeking his or her first employment is often already a member of one or more professional societies through the student chapter structure. During the employment interview process, you should ask questions about the code or codes of ethics that apply within the business, government, academic, volunteer, or other organization you are considering joining. As noted later, some employers have codes of ethics that supplement or complement those of professional engineering and other organizations. If you cannot accept a potential employer's code, or attitude toward codes, move on to other opportunities.

Codes Evolve

Engineering society codes, as well as codes used by other organizations, are likely to evolve in response to changing internal and external conditions. For example, many engineering societies used to include prohibitions against competitive bidding within their codes of ethics. The NSPE code contained a provision, prior to 1979, that said the engineer "shall not solicit or submit engineering proposals on the basis of competitive bidding." The provision was intended by NSPE to protect the public by discouraging selection of firms of questionable qualifications whose work could put the public at risk. This became a contested legal issue and was subsequently declared illegal by the U.S. Supreme Court which ruled that the ethical provision could serve the self-interest of engineering firms and, therefore, harm the public (Martin and Schinzinger 2005). This is an example of an ethical provision becoming an illegal act and being removed from a code of ethics.

Incidentally, in a related and somewhat compensatory manner, the Brooks Act of 1972 prohibits federal agencies from using competitive bidding to select engineering and other similar firms for professional services. Some states have mini-Brooks Acts. Therefore, the intent of the former anti-bidding provision that was struck from the codes of ethics of engineering organizations was, in effect, offset in special circumstances by federal and state legislation (Clough et al 2005).

As another example of evolving codes of ethics, consider the former provision in the code of the National Society of Professional Engineers (NSPE) that prohibited self-laudatory advertising. This provision was intended to protect the dignity of the engineering profession. The provision has been removed from the code as result of an agreement between NSPE and the Federal Trade Commission (FTC). The effect of this change is that NSPE can no longer try to govern the nature of its members' advertising, as long as it is truthful and non-deceptive. Self-laudatory advertising, which was until recently considered unethical but legal is now, in effect, both ethical and legal (*Indiana Professional Engineer* 1993), although some consider it unethical.

Ethics Codes for Other Professions

Engineers and other technical professionals are likely to interact with members of other professions who, in turn, belong to professional societies such as the American Bar Association (ABA), the American Institute of Architects (AIA), the American Institute of Certified Planners (AICP) of the American Planning Association (APA), the American Institute of Professional Geologists (AIPG), and the Association for Computing Machinery (ACM). These groups have codes of ethics tailored to their functions and responsibilities.

Codes of non-engineering groups are likely to share some common elements with the engineering codes. Consider the AICP Code of Ethics and Professional Conduct (APA 2011). The section titled "Principles to Which We Aspire" states: "Our primary obligation is to serve the public interest and we, therefore, owe our allegiance to a conscientiously attained concept of the public interest that is formulated through continuous and open debate." This ethics provision resonates with the safety-health-welfare provision in engineering society codes. Another ethical expectation shared across various professions, such as engineering, medicine, and law, is maintaining client confidentiality (Whitbeck 1998).

A comparison of codes associated with various professions also reveals some sharp differences. "Unlike engineering codes," according to Whitbeck (1998), "the codes for law and medicine have no rule against paying or accepting bribes." Of course, codes aside, accepting bribes in generally considered unethical, code or no code. "As another example, physicians are forbidden to terminate their relationship with a patient under their care without first referring that patient to another provider," according to Whitbeck, while engineering codes do not include that requirement.

As you work with members of other professions, be sensitive to possible ethics differences. That awareness will help you empathize with and work more effectively with team members, clients, owners, customers, and stakeholders while at the same time holding to your ethical principles.

Business Codes of Ethics

Businesses often adopt codes of ethics to meet their specific needs and to complement the codes of ethics of various professional societies in which their businesses, personnel, clients, owners, and customers are members. More employers in the business sector appear to be moving toward some form of ethics code. Listed below are a few, but highly-varied, engineering-oriented businesses having ethics codes and the website addresses for access to the codes. The primary purpose of the following list, as with the earlier list of engineering society codes, is to encourage you to read a code, or a few codes, for an engineering business or businesses that interests you. In this way, you will broaden your understanding of ethics codes. From a practical perspective, you will be even better prepared, when you are seeking employment, to ask about codes and, as an employee, to use them. The codes and their websites follow:

- Bechtel, the engineering, construction, management, and development company that provides services around the globe, uses a Code of Conduct presented in six languages. This code presents the company's vision, values, and covenants which are followed by many detailed provisions explained, in part, with a question and answer format. www.bechtel.com/assets/files/PDF/CodeofConduct/Bechtel_Code of Conduct_%20Web.pdf
- Boeing, the multinational aerospace, airliner, and defense corporation, has established Ethical Business Conduct Guidelines which include the company's Code of Conduct and guidance for marketing, conflict of interest, relationships with suppliers, use of company resources, and recruiting former government employees. www.boeing.com/companyoffices/aboutus/ethics/ethics_booklet.pdf
- Skanska, a multinational construction company, created a Code of Conduct organized according to these categories: general principles, employee relations, behavior in the marketplace, and the environment. www.skanska.com/Global/Careers/SKANSKA-orig-code-of-conduct-belly-band-EU.pdf
- Wright Water Engineers, a small consulting engineering firm, developed a concise Business Ethics Policy modeled after the principles and canons of the ASCE Code of Ethics. www.wrightwater.com/documents/EthicsPolicy.pdf

A word of caution is in order when discussing formal codes of ethics within organizations that employ engineers and other primarily technical personnel. The apparent trend toward written codes notwithstanding, the absence of a written code, especially in a small organization, does not necessarily mean the absence of high ethical expectations. Exemplary action by people in leadership positions is very important.

Such action may be all that is needed, especially in a highly-communicative, small organization, to engender ethical behavior.

Government Codes of Ethics

Many federal, state, provincial, and local government agencies and entities have adopted ethics codes. As with the codes used by businesses, these codes can complement the ethics codes of relevant professional societies. A few varied government organization ethics codes are listed here along with the website addresses for access to the codes. Consider reading at least one to further enhance your understanding of and appreciation for the value of codes. The codes and their websites are:

- Standards of Ethical Conduct for Employees of the Executive Branch [of the U.S. government] www.usoge.gov/laws_regs/regulations/5cfr2635.aspx
- State of Florida Code of Ethics for Public Officers and Employees www.ethics.state.fl.us/ethics/Chapter_112.html
- City of Valparaiso, Indiana Ethics Ordinance (www.ci.valparaiso.in.us/DocumentView.aspx?DID=227)

Forewarned is Forearmed

Government ethics codes typically include some highly-restrictive requirements that could be problematic for unaware engineers and other technical professionals. The Valparaiso, IN Ethics Ordinance provides examples of some potentially problematic ethics provisions. Of particular interest are:

- "Public official" is defined as "any elected official or department head." One significant aspect of this kind of provision: As you advance in your career and, if you serve government entities, you will encounter many public officials. Therefore, be aware of ethical provisions that apply to you as a service provider and to them as representatives of the government entity.
- The conflict of interest section of the code states that "it shall be a conflict of interest . . ." for a public official "to participate in any vote or participate in any discussion in his or her public capacity on any matter if the matter has an economic benefit to the public official, his or her family member, or anything in which he or she has a financial interest." One implication of this type of provision: Be sure that no public official participating in procurement of your engineering firm's services is connected, by family or financial interest, to your firm. If that connection exists, the code may offer a remedy for the public official such as abstaining on voting on such matters.
- The Valparaiso code's conflict of interest section also indicates that a public official should not ". . . solicit or receive a gift or loan when it has been or would reasonably appear to have been solicited, received, or given with the intent to give or obtain special consideration or influence as to any action by such public official in his or her official capacity." Be cautious with gifts you may be inclined to give, no matter how little the value, as well as perceptions of gift giving.

- The public disclosure section stipulates that "...all public officials shall be required to file an annual statement disclosing the name of any outside business or occupation outside his or her duties with the city..." Significance of this kind of provision: If you, as a citizen and as an engineer employed in the business, government, academic or other sector, decide to pursue an elective office in your community, which is highly encouraged, be sure to meet disclosure requirements.

Accordingly, when beginning a relationship with a potential or new public client or owner or when considering public service, inquire about the existence of public entity's ethics codes. This advice also applies to your relationship with potential or new private sector clients because, as noted earlier in this chapter, many of them also have adopted ethics codes.

Real Complicated Real Quick

While their intent is usually worthy, ethics codes can be very complicated in practice. To illustrate this, consider the above-noted Standards of Ethical Conduct for Employees of the Executive Branch of the U.S. government. The code is based on the premise that public service is a public trust. Fourteen general principles apply to every employee and the first principle, for example, is: "Public service is a public trust, requiring employees to place loyalty to the Constitution, the laws and ethical principles above private gain." The code elaborates on the principles by setting forth many standards and then offering hypothetical examples to explain the applications of the principles and standards. For example, relative to the "private gain" issue raised in the first principle, the code states:

An employee may accept unsolicited gifts having an aggregate market value of $20 or less per occasion, provided the aggregate market value of individual gifts received from any one person under the authority of this paragraph shall not exceed $50 in a calendar year.

The federal code then gives this example to illustrate application of the preceding gift provision:

An employee of the Defense Mapping Agency has been invited by an association of cartographers to speak about his agency's role in the evolution of missile technology. At the conclusion of his speech, the association presents the employee with a framed map with a market value of $18 and a book about the history of cartography with a market value of $15. The employee may accept the map or the book, but not both, since the aggregate value of these two tangible items exceeds $20.

This example drawn from the U.S. federal ethics code only begins to suggest the amount of individual and organizational effort that may be required to conscientiously function in accordance with very detailed and complex codes of ethics that seem to assume most personnel require "watching." An attractive alternative, and one that may only be feasible in a "small" organization, is a short and strong statement of principles coupled with a very careful personnel recruitment and retention program.

University Codes of Ethics

Colleges and universities sometimes adopt ethics codes. Some are narrowly focused on academic matters, such as some honor codes, while others address a wide range of issues similar to the codes of professional societies, businesses, and government entities. Both types can serve useful functions such as helping to govern day-to-day behavior and enhancing education by sensitizing students to the need for and use of codes in various sectors of society. If you are a student in an institution with some form of ethics code, consider taking a fresh look at it in light of the ideas and information presented in this chapter.

Assuming you do not have that opportunity, listed here are a few, but varied, ethics codes from academic institutions for your perusal:

- Lawrence Technological University Academic Honor Code www.ltu.edu/currentstudents/honor_code.asp
- Purdue University Student Conduct Code www.purdue.edu/odos/osrr/studentconductcode.php
- Smith College Student Conduct and Social Responsibility section of the Student Handbook www.smith.edu/sao/handbook/socialconduct.php
- United States Military Cadet Honor Code www.usma.edu/committees/honor/Info/main.htm

The Lawrence Technological University entry is an example of an academic honor code whereas the Smith College and Purdue University documents have a broader scope. Again, I do not mean to imply that either is generally preferred; presumably both meet the needs of their respective institutions.

Academic honor codes may be narrowly focused on students. However, that is not the case with the Lawrence Technological University academic code. The code emphasizes community-wide academic responsibility by stating: "It shall be the responsibility of every faculty member, student, administrator, and staff member of the University community to uphold and maintain the academic standards and integrity of Lawrence Technological University. Any member of the University community who has reasonable grounds to believe that an infraction of the Academic Honor Code has occurred has an obligation to report the alleged violation."

University codes may be very succinct. For example, the U.S. Military Academy's Cadet Honor Code is simply stated as "A cadet will not lie, cheat, steal, or tolerate those who do." According to the USMA, "On a behavioral level, the Code represents a simple standard for all cadets. On a developmental plane, West Point expects that all cadets will strive to live far above the minimum standard of behavior and develop a commitment to ethical principles guiding moral actions." Besides it succinctness, the USMA's honor code is closely linked to the academy's core mission which is "to develop leaders of character for our Army . . . A leader of character will apply the Spirit of the Code when making decisions involving ethical dilemmas."

Codes Cannot Anticipate All Circumstances

As already mentioned, an ethics code cannot anticipate all of the ethical decisions and dilemmas a student or practitioner may encounter. For example, the City of Valparaiso code states "This chapter provides guidance for ethical conduct, but it is not intended to set forth all ethical or unethical behavior or actions." Accordingly, codes typically make reference to the principles on which they are based. For example, Standards of Ethical Conduct for Employees of the Executive Branch of the U.S. government offers this advice: "Where a situation is not covered by the standards set forth in this part, employees shall apply the principles set forth in this section in determining whether their conduct is proper." Reference to foundation principles may help you make an ethical decision or resolve an ethical dilemma.

Views of Others

Encouraging the long view, clergyman H. W. Beecher said "Expedients are for the hour; principles for the ages." Courage is often required in ethical matters, especially when dealing with friends, as suggested by author J. K. Rowling: "It takes a great deal of bravery to stand up to our enemies, but just as much to stand up to our friends." And finally be aware of those who wear their "ethics" on their sleeves. Heed the advice of minister, lecturer, and writer Ralph Waldo Emerson who said "The louder he talked about his honor, the faster we counted our spoons."

DEALING WITH ETHICAL DILEMMAS: USING CODES AND OTHER RESOURCES

Assume you, as an individual or as a group in your organization, are trying to choose among various courses of action in a challenging ethical situation. Fortunately, you have a wide variety of resources available to help you do the right thing. Consider five types of resources on which you can draw and mix and match as needed.

Ethics Codes

Reference to one or more codes of ethics may be all the guidance you need. Consider your employer's code and the code or codes of your professional societies. If you are unable to provide a specific reference to your particular situation then, as noted earlier, refer to the foundation principles on which the code or codes were built. Ethics codes are omnipresent in that they are very likely to arise as individuals and groups apply the following four additional resources.

Advice of Experienced Personnel

Included within the staff of most organizations are seasoned professionals representing various areas of technical and other experiences. They are a gold mine of wisdom in that they have faced many ethical dilemmas and made many decisions, some good and some not so good, and they learned in the process. They represent a wealth of wisdom that you, either acting alone or as a group, can draw on for guidance. Individuals who are senior in terms of breadth and wealth of their professional experience may have already encountered the very ethical dilemma you or your group face. Even if your particular situation is new to an experienced professional, he or she is still likely to be able to offer valuable guidance.

You can also find wisdom relative to your work outside your organization. Potential advisors include professional colleagues, parents, religious leaders, and former professors and other teachers. However, in contemplating seeking external advice on internal matters, be careful to not violate ethical provisions or confidentiality requirements such as revealing confidential information about a client to a third party in such a way as to violate client confidentiality requirements. For example, the NSPE Code of Ethics in Section II-1-c states that "Engineers shall not reveal facts, data, or information without the prior consent of the client or employer except as authorized or required by law or this Code."

A Nine-Step Individual or Group Process

The National Institute for Engineering Ethics at Texas Tech University (2011) provides useful guidelines. The nine-step process, more formally called, Guidelines for Facilitating Solutions to Ethical Dilemmas in Professional Practice, is quoted here, except for parenthetic comments:

1. Determine the facts in the situation—obtain all of the unbiased facts possible
2. Define the stakeholders—those with a vested interest in the outcome
3. Assess the motivations of the stakeholders—using effective communication techniques and personality assessment
4. Formulate alternative solutions—based on most complete information available, using basic ethical core values as guides (To generate ideas, consider using some of the creativity and innovation thinking techniques described in Chapter 7 such as brainstorming and mind mapping)

5. Evaluate proposed alternatives—short-list ethical solutions only; may be a potential choice between/among two or more totally ethical solutions (In evaluating alternatives, possibly apply some of the evaluation tools discussed in Chapter 7 such as multivoting and SWOT)
6. Seek additional assistance, as appropriate—engineering codes of ethics, previous cases, peers, reliance on personal experience, and prayer
7. Select the best course of action—that which satisfies the highest core ethical values
8. Implement the selected solution—take action as warranted
9. Monitor and assess the outcome—note how to improve the next time

Note reference to "codes" and "peers" in Item 6 which reinforces the two previously suggested resources, that is, referencing ethics codes and obtaining advice of experience personnel.

A Systematic Group Process

Frederick (1997) suggests a systematic group process, as illustrated in Figure 12.4. The group first creates an exhaustive list of options, again perhaps using some of whole brain tools and techniques offered in Chapter 7. Then the group seeks consensus on the options that should not be pursued. Possible solutions that are rejected by everyone in the group are considered unacceptable solutions. Advantages of this approach, at least to this point, are that it encourages identification of a wide range of potential solutions and it values consensus. All remaining possible solutions are considered "provisional solutions" or "ethically responsible" solutions because they apparently have no "glaring ethical defects."

The next step, assuming there are a significant number of options still available, is to drop any options that one or more individuals are opposed to and all remaining members are ambivalent about. In other words, one individual can "black ball" a solution. At this point, all remaining possible solutions are those that members of the group are either positive or ambivalent about. These are referred to as viable options in Figure 12.4.

Now a decision has to be made, and the process could become contentious. However, the preceding thorough and amiable process may be a solid foundation for constructive final decision making. Enable the final decision by applying some of the Chapter 7 evaluative methods such as multivoting or SWOT, and then implement it. In summary, the suggested approach is to:

- Prepare an exhaustive list of options
- Eliminate all options for which there is consensus agreement that they ought not to be done
- Screen the remaining provisional options and drop any that one or more individuals are opposed to and all others are ambivalent about
- Decide among the remaining options

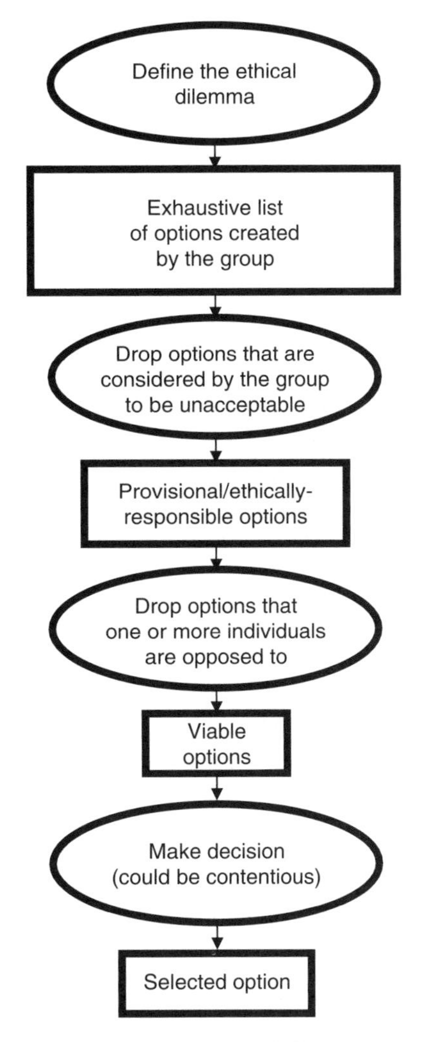

Figure 12.4 This systematic group process may help a group resolve an ethical dilemma.

Application of Moral Imagination

Telushkin (2000) suggests that one way to guide individual and corporate lives on a day-to-day basis is to develop "moral imagination." According to Telushkin, moral imagination is "...the ability to think through the implications of our actions, particularly as to how they will affect others." He goes on to observe that during the past century, our society "...has made extraordinary technological advances because of the active imaginations of our scientists and researchers." However, he concludes, "...we have been slower to advance morally because of a general unwillingness to practice imagination in the moral sphere."

The preceding suggests that another way to resolve an ethical dilemma is to apply moral imagination. The power of this approach is that it encourages individuals and groups to take the long view in pondering ethical questions. Given the demands and pressures of everyday living, including the crucial bottom line, the short view tends to be paramount. But individuals and groups have to live with the long-term consequences of decisions. Moral imagination can help make decisions that will pass the test of time.

Historic Note

Telushkin (2000) cites the example of Lee Atwater, the "highly aggressive" manager of George H. W. Bush's successful 1988 U.S. presidential campaign. Atwater learned that an opponent had suffered from depression and received electric shock treatments. Atwater released this information and, in response to objections from the opposing candidate, said that he wouldn't reply to a person "hooked up to a jumper cable." Ten years later, Atwater was dying of an inoperable brain tumor. Apparently realizing the evil of what he had done to the referenced political opponent, he wrote to him, begging forgiveness. The example illustrates the harm that can occur in the absence of moral imagination.

CASE STUDY: DISCOVERING A MAJOR DESIGN ERROR AFTER CONSTRUCTION IS COMPLETE

Design and Construction

Consider an instructive story that begins with an unusual design (Fleddermann 1999, Goldstein and Rubin 1996, Morgenstern 1996). In the 1970s, structural engineer William J. LeMessurier, and architect Hugh Stubbins, Jr., designed the 59 story, 910 foot Citicorp Center, which covers an entire city block in New York City's Manhattan. LeMessurier, a preeminent structural engineer, was responsible for the structural design. Because of the need to reserve space beneath one corner of the building for construction of a freestanding church (to replace an earlier church), the four supporting columns were placed in the middle of each façade or side of the building, rather than at the corners, as shown in Figure 12.5.

Four columns or legs supporting the Citicorp Center's frame are positioned at the middle of each side of the building. LeMessurier designed a unique system of V-shaped wind braces on each facade, arranged in a chevron pattern, to transfer loads to the column. Construction of the Citicorp Center was completed in 1977.

Post-Construction Discovery

As a result of a June 1978 question from an engineering student, who was writing a paper on the Citicorp Center, LeMessurier reviewed his original design. He discovered:

- Quartering winds, in contrast with winds perpendicular to the building facades, increased strains on some of the chevrons by 40 percent over the design calculations.

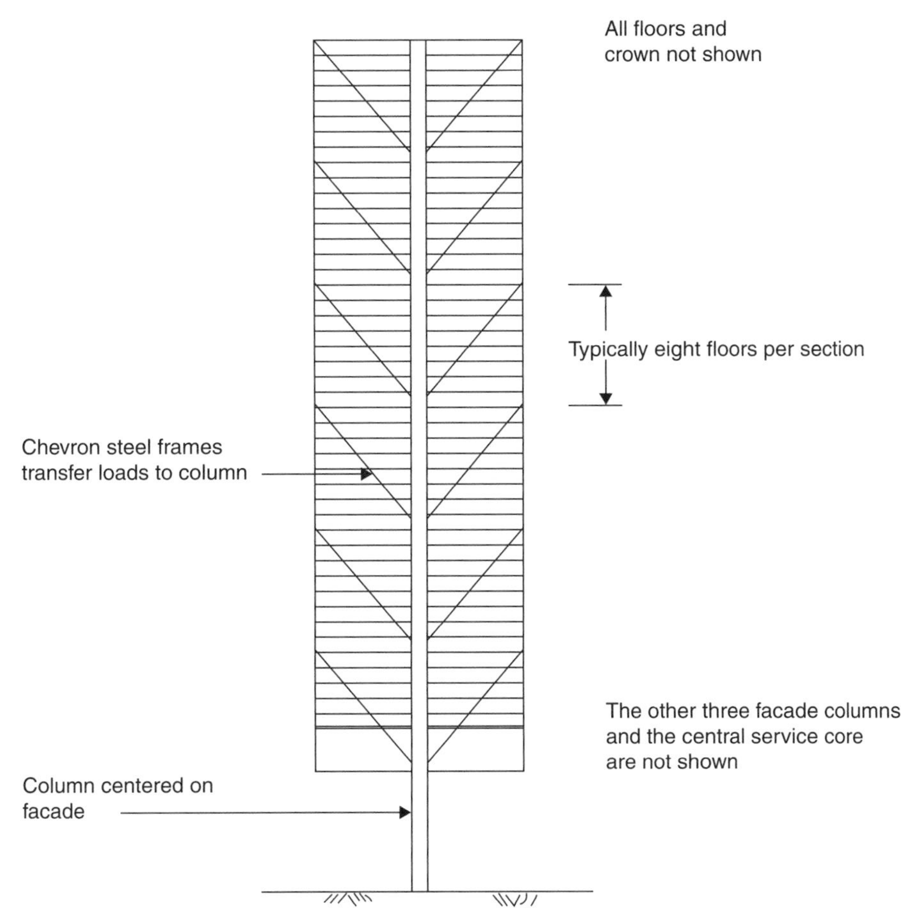

Figure 12.5 The Citicorp Center's four columns were positioned in the middle of each of the building's four sides. (Source: Adapted from Goldstein and Rubin 1996 and Whitbeck 1998.)

- The original design called for welded connections within the chevron wind bracing system. However, high-strength bolts were used during construction to avoid the cost of welding. LeMessurier had learned about this change in May 1978 and now realized its significance: the full potential strength of the connections would not be achieved.
- Although the building had presumably been designed to withstand winds corresponding to roughly a 50-year recurrence interval storm, the combination of the design error and the construction change reduced the protection to a 16-year storm.
- The hurricane season was rapidly approaching.

What did LeMessurier do? What would you do?

The Engineer's Actions

This portion of the case study is a synopsis of a detailed account provided by Whitbeck (1998) and the quotes are taken from that source. Apparently realizing his ethical responsibilities, including protection of public safety, health, and welfare, LeMessurier sought advice from Alan Davenport, a Canadian consultant who had performed the original wind-tunnel testing and now carried out additional tests. He generally confirmed LeMessurier's disturbing, post-design analysis. LeMessurier performed a preliminary design of welded "Band-Aids" over the bolted joints. The engineering solution was relatively simple.

He considered reactions of Citicorp, City officials, the profession, the press, and the general public. LeMessurier also consulted with an attorney for the lead firm that designed Citicorp building and attorneys for Citicorp's insurance companies. He then brought in Les Robertson, another structural consultant, who generally confirmed LeMessurier's analysis.

LeMessurier began to "go public" by meeting with Walter Wriston, Citicorp's Chairman who agreed to the proposed repair. LeMessurier also met with construction company representatives who concurred with the "welded Band-Aids" idea and retained meteorological experts to provide warnings of hurricanes or major storms. He led the drafting of an emergency evacuation plan for the Citicorp building and the ten-block diameter surrounding it.

He explained the situation to city officials who responded "with approval and encouragement." A press release was issued "indicating that the building was being refitted to withstand slightly higher winds." This was an understatement. It "was true to some extent, for the meteorological data suggested that the winds for that year were going to be somewhat higher than normal." A citywide press strike occurred shortly after the press release was issued which reduced press attention.

The contractor built a "plywood house" around each repair site to minimize attention to the work and performed welding at night. The first hurricane moved toward New York on September 1, 1978 while repair work was underway. However, the hurricane turned and the repair work was complete in October of 1978. Note how quickly LeMessurier moved in that the problem was discovered in June 1978 and solved in about four months.

What Happened to LeMessurier?

LeMessurier's liability insurance company paid two million dollars to Citicorp and Citicorp agreed "... to find no fault with LeMessurier's firm and to close the matter." Finally, LeMessurier's liability insurance company lowered his premiums partly because he had prevented one of the worst insurance disasters. In summary, "Far from behaving in an incompetent or devious manner LeMessurier had acted in a commendable way: He had discovered an unforeseen problem, acted immediately, appropriately, and efficiently to solve it, and solved it. LeMessurier's handling of the Citicorp situation increased his reputation as an exceptionally competent, forthright structural engineer."

CONCLUDING THOUGHTS: SEEING SERMONS

The danger of talking about ethics is that it ends there—with talk. Someone said, "I would rather see a sermon than hear one." Most of us have experienced the "do as

I say not as I do" hypocrisy. Berglas (1997) describes the importance of example—of action over talk in ethical matters. He says "...just look at how the company leader behaves and you will know with 100 percent certainty how the employees will act and feel..."

Few would argue that the "company leader," whether in business, government, academia, volunteer, or other organization needs to walk the talk. But the responsibility to set an ethical example goes well beyond the chief executive officer or other upper-level positions. In fact, everyone in an organization can lead in that individual actions – more specifically, your ethical actions – influence others. Be especially sensitive to the ethical messages your actions send within your organization. In addition, be careful with the ethical signals that you transmit to clients, owners, customers, government officials, business partners, vendors, and competitors. In many instances, you "are" your organization.

Good character is more to be praised than outstanding talent.
Most talents are, to some extent, a gift.
Good character, by contrast, is not given to us.
We have to build it piece by piece –
by thought, choice, courage, and determination.

(*John Luther, lawyer and writer*)

CITED SOURCES

American Planning Association. 2011. "American Institute of Certified Planners Code of Ethics and Professional Conduct." (www.planning.org/ethics/ethicscode.htm) May 13.

Berglas, S. 1997. "Liar, Liar, Pants on Fire." *Inc.*, August, p. 33.

Clough, R. H., G. H. Sears, and S. K. Sears. 2005. *Construction Contracting - Seventh Edition*. John Wiley and Sons: New York, NY.

Covey, S. R. 1990. *The 7 Habits of Highly-Effective People*. Simon & Schuster: New York, NY.

Delatte Jr., N. J. 2009. *Beyond Failure: Forensic Case Studies for Civil Engineers*. ASCE Press: Reston, VA.

Fleddermann, C. B. 1999. *Engineering Ethics*. Prentice Hall: Upper Saddle River, NJ.

Frederick, R. E. 1997. "How to Teach Difficult Ethical Problems." *CBE News*, Volume 6, Number 1, pp. 3–5.

Goldstein, S. H. and R. A. Rubin. 1996. "Engineering Ethics." *Civil Engineering*, American Society of Civil Engineers, October, pp. 41–44.

Indiana Professional Engineer. 1993. "NSPE Agrees to Revise Code of Ethics." *Indiana Professional Engineer*, September-October.

Martin, M. W. and R. Schinzinger. 2005. *Ethics in Engineering - Fourth Edition*. McGraw Hill Higher Education: New York, NY.

McCuen, R. H. and J. M. Wallace. (Eds.) 1987. *Social Responsibilities in Engineering and Science: A Guide for Selecting General Education Courses*. Prentice-Hall: Englewood Cliffs, NJ.

McCuen, R. H. and K. L. Gilroy. 2010. *Ethics and Professionalism in Engineering*. Broadview Press: Peterborough, ON, Canada.

Merriam-Webster. 2011. *Dictionary*. (www.merriam-webster.com/dictionary) May 13.

Morgenstern, J. 1995. "The Fifty-Nine-Story Crisis." *The New Yorker*, May 29, pp. 45–53.

Onsrud, H. J. 1987. "Approaches in Teaching Engineering Ethics." *Civil Engineering Education—ASCE Civil Engineering Division*, Fall, pp. 11–27.

Russell, J. 2004. "Choosing to Cheat." *Leadership and Management in Engineering – ASCE*, January, pp. 19–22.

Telushkin, J., 2000. "Ethics, One Day at a Time." *Imprimis*, March, pp. 1–7.

Texas Tech University. 2011. "Guidelines for Facilitating Solutions to Ethical Dilemmas in Professional Practice." Murdough Center for Engineering Professionalism, College of Engineering. (www.niee.org/case_of_the_month/ethics5.htm). May 13.

Valparaiso, City of, Indiana. 2011. "Ethics Ordinance." (www.ci.valparaiso.in.us/Document View.aspx?DID=227. May 13.

Vesilind, P. A. 1988. "Rules, Ethics and Morals in Engineering Education." *Engineering Education*, February, pp. 289–293.

Whitbeck, C. 1998. *Ethics in Engineering Practice and Research*. Cambridge University Press: Cambridge, UK.

ANNOTATED BIBLIOGRAPHY

American Society of Civil Engineers. 2008. "*Ethics Guidelines for Professional Conduct*." ASCE: Reston, VA. (This booklet was "developed for use in the day-to-day conduct of engineers' professional and business-related affairs.")

American Society of Civil Engineers. 2011. "Six Year Archive of Professional Ethics Questions Now on Line." May 13. (www.asce.org/Leadership-and-Management/Ethics/A-Question-of-Ethics). (ASCE states: "The entire archive of answers to questions submitted each month to *ASCE News'* column "A Question of Ethics" is now available online. The answers are based on recent ethical cases considered by ASCE's Committee on Professional Conduct, or ethical issues brought to the attention of ASCE's counsel.")

Hoke, T. 2011. "Defining Ethical Behavior." A Question of Ethics column, *Civil Engineering*, American Society of Civil Engineers, May, p. 48. (This column addresses the question "does the Society's Code of Ethics govern a member's private activities as well as his or her professional activities?").

Online Ethics Center for Engineering and Research. 2011. "Codes of Ethics in English." National Academy of Engineering, May 13. (www.onlineethics.org/Resources/ethcodes/EnglishCodes.aspx) (Provides quick access to many and varied ethics codes drawn from a wide array of engineering and engineering-related disciplines.)

Schwartz, A. E. 2010. "Is It Unethical to be Unaware of Current Design Standards?" Ethics column, *Structural Engineering and Design*, June, p. 20. (The question posed by this column's title is not definitively answered. The author partially answers the question with this question: "But how far is it necessary to go in order to remain current?" The column also suggests that an engineer could be found legally negligent, but not determined to be unethical, which may be at odds with Quadrant 3 in Figure 12.3. Some interesting points worthy of further discussion.)

Veach, C. M. 2006. 'There's No Such Thing as Engineering Ethics." *Leadership and Management in Engineering - ASCE*, July, pp. 97–101. (Argues that ethics is ethics. "If we engineers want to lead ethical lives, we must practice ethical behavior based on the Golden Rule or one of similar guiding principles. However . . . we must also study and learn about how to apply these ethical principles in the practice of engineering.")

Vesilind, P. A. (Editor). 2005. *Peace Engineering: When Personal Values and Engineering Careers Converge*. Lakeshore Press: Woodsville, NH. (Peace engineering is defined as "the pro-active use of engineering skills to promote a peaceful and just existence for all people." Peace engineering is relevant to this ethics chapter in that the book's editor says "Another way to promoting peace in engineering is to teach ethical skills alongside technical skills in our universities.")

Walesh, S. G. 2006. "The Inverse Talk-Action Law." *Indiana Professional Engineer*, January/February, pp. 3–5. (Addresses the sometimes unfortunate disconnect between what we say and what we do.)

EXERCISES

12.1 ETHICAL CHOICES AND WHERE DO WE STOP? This exercise (adapted from Onsrud 1987) suggests that ethical issues arise frequently and apparently harmlessly in what appear to be ethical situations and can lead to unethical and harmful situations. Furthermore, the exercise reminds us that personal value systems vary and each of us will ultimately be judged by what we did, not by what others may have suggested was acceptable. Suggested tasks are:

A. Assume you are a young engineer employed by a city. You have been placed in charge of inspecting a sewer project that is being built for the city by a private contractor. Or, assume that you are a young engineer employed by an electric utility. You have been assigned to inspect a power line project being constructed by a contractor. In either case, your responsibility is, as explained in Chapter 9, not to direct the construction but watch for departures from plans and specifications and, if observed, advise the contractor, owner, and designer and facilitate a resolution. Because of your education and field engineering experience, you could suggest techniques and procedures that save the contractor both time and money, but that is not your role.

B. Assume further that this is quitting time on a hot summer Friday afternoon. The president of the construction firm comes to the site, after quitting time, and offers a soft drink to each of his employees. The president offers you a soft drink. May you accept it? Why or why not?

C. What if the president hands each worker a pen with his company's name on it? Would you accept? Why or why not?

D. The president hands every worker, including you, a six-pack of soft drinks. Is it ethical for you to accept it? Why or why not?

E. What if the president hands every worker a can of beer and offers one to you? Can you ethically accept? Why or why not?

F. What if the president offers every worker a bottle of scotch? Would you accept? Why or why not?

G. How about a case of scotch? Would you accept? Why or why not?

H. What if the president offers a hat and jacket with the company name and logo on it? Would you accept? Why or why not?

12.2 A LEGAL AND UNETHICAL ACT: The purpose of this exercise is to encourage you to further explore the four quadrants in the legal-ethical domain illustrated in Figures 12.1, 12.2, and 12.3. Suggested tasks are:

A. Using the code of ethics of "your" engineering or other professional society and your understanding of the legal aspects of engineering and related disciplines, identify an actual, from your experience or knowledge, engineering or business situation that was/is both legal and unethical. Although not as valuable, if an actual situation is not feasible, develop a hypothetical one. If you use an actual situation, please change the names of people, places, and organizations if appropriate.

B. Write a description of the situation and explain how it is both legal and unethical.

12.3 COMPANY CODE OF ETHICS: As a result of participating in this team exercise, you will further appreciate the value of an organizational ethics code while understanding the difficulty of developing it. Suggested tasks are:

A. Recall the ethical dilemmas posed by the kinds of scenarios presented in Exercise 12.1.

B. Assume you and others are members of an ad hoc committee formed by the president of your consulting firm and are charged with formulating your company's policy on employees accepting gifts or receiving other considerations from existing or potential clients, owners, or customers; contractors; or other individuals or organizations.

C. Write the policy.

12.4 APPLICABILITY OF ETHICS CODES TO PRIVATE ACTIVITIES: This exercise's purpose is to cause you, or you and team members, to think about the possible application of professional society codes of ethics to private life. Suggested tasks are:

A. Study the article by Tara Hoke (2011) and then formulate a consensus or majority view.

B. Document your thought process and conclusions. If there is a minority view, present it.

12.5 ETHICS AND COMPETENCY: The intent of this exercise is to challenge you, or you and your team, to connect ethics and maintaining professional competence. Suggested tasks are:

A. Review the article by Arthur E. Schwartz (2010) and analyze the issue.

B. Document your thought process and conclusions. If there is a minority view, present it.

CHAPTER 13

ROLE AND SELECTION OF CONSULTANTS

The light that a man receiveth by counsel from
another is drier and purer than that which cometh
from his own understanding and judgment,
which is ever infused and drenched
in his affections and customs.

(*Francis Bacon, English philosopher and statesman*)

After defining consultant from an organizational and individual perspective and exploring reasons why consultants are retained, this chapter discusses the characteristics of successful consultants. The consultant selection process, which varies in complexity from extremely elaborate and costly to simple sole source selection, is described. It includes discussion of price-based selection (PBS) versus qualifications-based selection (QBS) of professional service firms and their impacts on clients, owners, customers, and stakeholders. A discussion of the negative impacts of PBS selection on consulting firms concludes the chapter.

CONSULTANT DEFINED AND WHY YOU SHOULD CARE

Recall the Chapter 1 discussion of the triangular client/owner/customer – engineer/other technical professional – constructor/manufacturer/implementer interaction which is illustrated in Figure 1.1. As the discussion suggested, individuals and organizations requiring planning, design, management, operations, and other engineering-related services often retain consultants to provide those services rather than having the work done "in-house."

The Meanings of Consultant

What does "consultant" mean? From a contractual, formal perspective, consultant usually refers to a consulting firm that enters into legal agreements with clients,

Table 13.1 This profile of over 5000 U.S. engineering firms that are members of the American Council of Engineering Businesses (ACEC) indicates that almost three-fourths are very small.

Size category as defined by number of employees	Percent of firms in the category
Small (1-30)	71
Medium (31-75)	16
Medium large (76-150)	6
Large (151-499)	5
Extra large (500-999)	1
Extremely large (1000+)	1

(Source: ACEC 2011)

owners, and customers for the provision of services. Data presented in Table 13.1, which is provided by the American Council of Engineering Companies (ACEC), an organization of more than 5000 engineering firms, provides a profile of the U.S. consulting industry. Consulting firms range in size from essentially one person businesses to firms with thousands of personnel. Interestingly, almost three-fourths of the firms are small, that is, have 30 or less employees.

On a personal basis, individuals within many client, owner, and customer organizations view "consultant" as a particular person, or perhaps a small group of professionals, on the staff of a consulting firm who have demonstrated competence and with whom they have established a mutually-beneficial relationship. One indication of this interpretation of "consultant" is the strong allegiance to individual professionals that clients, owners, and customers show when a particular engineer or other technical professional moves from employment in one consulting firm to employment in another firm or establishes his or her own consulting practice. Another indication of this personal view of "consultant" is when a person who works for an organization that uses consulting firms moves to another organization that uses consulting firms "takes his or her consultant along."

This second, more personal interpretation of "consultant" suggests that a trustful relationship is critical in carrying out the consulting function—expertise is necessary, but clearly not sufficient. One of the highest compliments that an individual consultant can receive for him or her, or for his or her firm, is to be retained on a sole-source basis by a prospect, that is a potential client, owner, or customer. The necessary trust is earned, in part, through consistent, long-term, no-matter-what-happens ethical behavior in keeping with the content of Chapter 12 and is often initiated by the trust-based marketing model, or similar approach, presented in Chapter 14.

Why You Should Care

As an engineering student or entry-level engineer or other technical professional on the staff of a government entity, a manufacturing organization, a contractor, or other organization that may retain consultants, you should be familiar with the role and selection of consultants. Similarly, if you are on the staff of a consulting organization, essentially all of the projects you work on will be the result of the selection process described in this chapter.

Although as an entry-level professional you will not play a major, formal role in the consultant selection process, you will have opportunities to participate in it. If, for example, you are on the professional staff of an organization that utilizes consulting firms, you should observe the manner in which various firms present themselves and, then if selected, provide services and, as opportunities arise, share your preliminary conclusions with colleagues and superiors within your organization. If you are a member of a consulting firm, you can note the variations in the expectations of your clients, owners, and customers and share that information with others. Understanding the role of consultants, being familiar with the selection process, and acting on that knowledge will enable you to be a more productive young professional regardless of where you are in the previously-mentioned triangle of interaction.

Looking beyond your early career, you may already see the possibility and desirability of starting your own business. It might be a consulting firm that provides services or a business that uses such services. Whether you are a student or young practitioner, you may be asking yourself questions like the following within a decade of completing your formal education (Walesh 2000a, 2000b, 2001):

- Is my job security fading?
- Would I prefer career security, that is, always being employed doing what I love and being fairly compensated for it?
- Is the corporate, government, academic, or other bureaucracy getting me down?
- Am I stagnating intellectually and/or emotionally?
- Am I increasingly concerned about how much I work relative to how little I earn?
- Am I tired of pay-for-performance talk?
- Do I want more autonomy?
- Could I "do better" in achieving success and significance?

Depending on your particular set of questions at that future time, you may decide that now is the time to "fly solo," to go out on your own, to start your own business. With that possibility in mind, you should be even more mindful the role and selection of consultants as described in this chapter.

WHY RETAIN A CONSULTANT? LET'S DO IT OURSELVES!

Consultants, in the form of consulting firms or individual professionals, are typically retained for one or more of the following five reasons:

1.Temporarily Acquire Necessary Expertise: In this increasingly technological world, many business, government, academic, volunteer, and other organizations, even those with engineers and other technical professionals on staff, do not have certain types of expertise. While they could develop such expertise, they are often reluctant to do so, to incur the necessary cost, unless they see a continuous need for the expertise. Accordingly, they seek a consultant who has that expertise. Executives of organizations planning to develop in-house expertise should consider contracting with one or more consultants to provide education and training in the desired area of expertise, in addition to completing the project or projects at hand.

2. Supplement In-house Personnel: Regardless of whether or not an organization has the necessary in-house expertise to carry out a project or accomplish a task, they may not have a sufficient number of staff members available at a given time to complete the project or task on schedule. Accordingly, they solve their "people power" shortage through the temporary use of consultants.

3. Provide Absolute Objectivity: A business, government, academic, volunteer, or other organization, even one with wide expertise and sufficient staff, may find itself embroiled in controversy, the resolution of which requires a high degree of objectivity which can be provided by a carefully-selected consultant.

Personal

A municipality experienced a heavy rainfall with resulting widespread flooding. This catastrophe brought into question the adequacy of storm water facilities designed by one of the city's engineering consultants and then recently constructed. The in-house professional staff, or the consultant that did the original design, could certainly have been asked to review the design. However, both were involved, to varying degrees, in the original design. Accordingly, the municipality decided that the most credible approach was to retain an outside consultant to encourage objectivity. In this case, the preferred outside consultant was the engineering firm I worked for because we had the necessary technical and other expertise, had never provided services to the municipality, and our offices were outside the community and its state. I managed that design review project. Our conclusion: The original design was consistent with the criteria selected by the city and the storm that occurred exceeded those criteria.

4. Perform Unpleasant Tasks: Carrying out unpleasant tasks or doing the "dirty work" is rarely the sole or principal purpose a consultant is retained, although it may be the principal focus of a management consultant. However, engineering and other technical projects, particularly those in the public sector, often involve unpleasant and stressful tasks. For example, the long-term and frequent failure of crucial municipal facilities and services such as water supply, wastewater, transportation, and flood control can lead to deep-seated and widespread frustration among citizens of a community. Consultants are often retained to find a planning and engineering solution to such problems. Regardless of the other reasons why the consultants might be retained—such as to provide expertise, needed staff, and objectivity—the consultants are often expected to release, deal directly with, and re-channel the pent-up frustration within the community. Consultants may also be asked to facilitate cooperation among conflicting private and public entities.

One effective mechanism for responding to stakeholder concerns is the conduct of community-wide or neighborhood information meetings shortly after the consultant is retained at which one or more representatives of the consulting organization are introduced and asked to provide an outlet for citizen frustration and to describe the intended next steps. The hope is that the meeting will culminate in citizen confidence that the problems can and will be solved and that citizens share information and their ideas. One result could be a productive, on-going program of interaction between the professionals and the public (Walesh 1993, 1999).

5. Reduce liability: This, the fifth and last reason for retaining a consultant should "raise a red flag." You and your firm may be retained partly or primarily to absorb risk that will be shifted from someone else or some other organization to your firm. Your client, owner, or customer may be a consulting firm, a contractor, or a government entity. Technical specialties within engineering that are generally considered to be more risky include geotechnical, structures, and hazardous waste. Consulting firms offering these services should be especially cautious with contract and agreement language, exercise great care in performing tasks and writing reports, and secure errors and omissions insurance. Chapter 11 provides ideas on how to minimize liability while providing services.

In summary, consulting firms or individual consultants are typically retained for one or more of the preceding five reasons. Both parties, that is, the consultant and the client, owner, or customer, should strive to identify and prioritize the reason or reasons for needing the consultant prior to entering into an agreement. Mutually-satisfying outcomes depend on a clear understanding of needs and expectations. Hopefully, as a result of the ideas and information presented here, and throughout this book, you as the aspiring or entry-level engineer will view your role and responsibility as being more than simply doing technical tasks, as important as that function is. At minimum, as you begin work on a new project, find out why the organization you are working for retained your firm to provide services. If your organization retained a consultant, determine why you are not doing the work yourselves, that is, in-house.

CHARACTERISTICS OF SUCCESSFUL CONSULTANTS

Of the various employment opportunities for engineers and other technical professionals, consulting is one of the most demanding and the most satisfying. The consulting world is typically dynamic—new problems to solve; new technologies to learn; new clients, owners, and customers to serve; and new geographic areas to work in. These characteristics, which are likely to be viewed as positives by many young professionals, must be weighed against potential negatives such as long hours, erratic schedules, extensive travel, and high levels of stress. In my view, success in consulting requires the following six characteristics and traits (Walesh 2000a):

1. Inquisitiveness and Currency of Knowledge: Recall that the consultant is often retained to provide expertise the client, owner, or customer does not possess. Consultants, as individuals or as organizations, should define their areas of expertise and remain current in them. On the surface, one might think that consultants are successful primarily because of the answers they provide based on their knowledge and skill.

However, the questions they ask those they serve and themselves, based on their knowledge, skill, and experience, are more important than the answers they give. Once key questions are asked, the consultant knows how to find the answers. The successful consultant is a perpetual student, or to use Chapter 1 terminology, always a student. The consultant is strongly oriented toward trying to find out more about his or her area of expertise, the current task or project, and the organization he or she is serving. Clearly, the consultant's inquisitiveness spans technical and non-technical subjects and topics.

2. Responsiveness to Schedules and Other Needs: Recall that the consultant may be retained because the client, owner, or customer does not have the personnel to complete a task or do a project. If the effort is late because of the consultant, the principal reason for retaining the consultant is negated. Responsiveness to client, owner, or customer needs and schedules requires that the consultant have a strong service orientation. Too many engineers think that consultants are in a technical business that just happens to provide service. A more productive perspective is to view their effort as being in a service business that happens to focus on technology. And frankly, a student could go through an entire formal engineering education and never hear the word service. You just did "hear" the word service and it is important.

3. Strong People Orientation: Although technical professionals plan, design, construct, manufacture, and care for "things," they do this for the benefit of people. The consultant is the very important part of the interface between the wants and needs of people and the possibilities of meeting those expectations with the applications of science and technology. Wants and needs, and the distinction between them, are discussed in the next chapter. Because responding to wants and needs is critical, effective consultants strive to develop excellent communication skills, as discussed in Chapter 3, with emphasis on listening, writing, and speaking. Successful consultants enjoy interacting with people—even under unpleasant circumstances. The people challenges of consulting are further complicated by frequent changes in current and potential clients, owners, or customers and their liaison persons.

4. Self-motivation: Even though an individual consultant is "working for" a client, owner, or customer, that entity's representative often does not know how to direct the consultant or have the time or the inclination to do so. Accordingly, most of what consultants do for those they serve is at the consultant's initiative within the overall framework established by the agreement between the two. The organizations being served typically assume that if they are not hearing anything from "their consultant," the consultant is proceeding with the project in a timely fashion. Moreover, the consultant will be available, on a very short notice, to answer a question, give advice, or provide a status report or other accounting of the efforts to date. Consultants must have the self-discipline to be proactive to the point of being intrusive in their relationships with those they serve.

5. Creativity and Innovation: Consultants must have the ability to be creative and innovative, to synthesize, and to see previously unforeseen patterns and possibilities. The typical technical project involves technical, regulatory, financial, economic, personnel, and other facets, all of which can be easily assembled in a variety of ways, most of which are suboptimal. A consultant's combination of knowledge, skill, varied experiences, and objectivity should enable him or her to suggest approaches and solutions not apparent to others. Recall the Chapter 8 discussion of the origin of the word "engineer," which concludes that the word has deep roots in creativity. Also reflect on the tools and techniques, presented in Chapter 7, for stimulating creativity and innovation.

6. Physical and Emotional Toughness: The successful consultant needs physical and emotional strength to withstand pressure, long hours, and travel. Some of the consultant's meetings and presentations are difficult because they occur in situations highly charged by personality conflicts, political pressure, financial concerns, and liability issues. In addition, consultants are often not selected for projects even though they believe they were the most qualified or had the best proposal. Frequent rejection can take its toll on conscientious and competent individuals, but is one of the realities of the consulting field.

CONSULTANT SELECTION PROCESS

The process by which a client, owner, or customer selects a consultant to provide planning, design, construction, manufacturing, operations, facilitation, education-training, or other services is, at the detailed level, unique to each situation. However, you can gain useful insight into consultant selection by considering the overall approach or model presented in this section.

Cost Versus Quality

As illustrated by Figure 13.1, almost always the consultant selection process is driven implicitly or explicitly by the natural tension that exists between the quality of service (conformance to all requirements as defined in Chapter 7) and the total cost of that service and what is produces or yields. You routinely encounter the same tension in your personal life when you shop for clothing, search for a home, or select a restaurant for a special occasion. In a rough way, as the proposed cost of well-defined consulting services for a given project diminish, the quality of the resulting plan, design, or other product is likely to diminish and the capital and operation and maintenance costs associated with the product are likely to increase. Whether functioning as an individual, in our personal lives, or as a clients, owners, or customers in professional life, we get what we pay for.

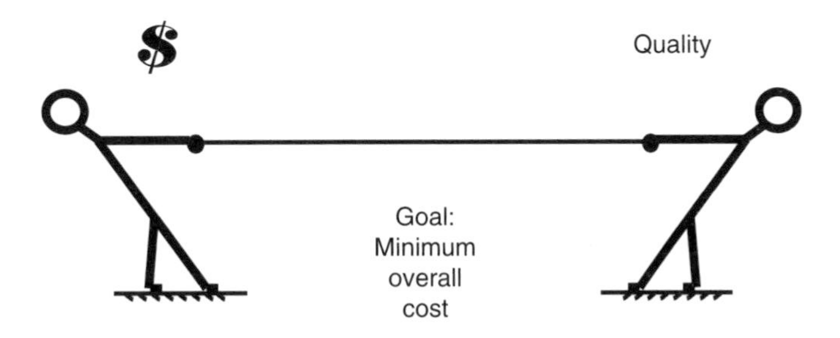

Figure 13.1 The cost-quality tension is usually present in the consultant selection process.

Price-Based Selection

The price of consulting services is certainly important and this can lead to price-based selection (PBS), that is, selecting a firm solely or mostly on the basis of price (fees plus expenses). Because price can be readily quantified relative to other selection factors such as experience, creativity and innovation, responsiveness, communication effectiveness, and productivity, it tends to assume excessive influence.

Unusually low fees proposed by consultants sometimes reflect a lack of experience and, therefore, unawareness of all necessary aspects of a project. At other times, low fees might reflect an individual consultant or a consulting firm's desire to obtain a contract for a new type of project on which they can gain valuable experience. They are, in effect, willing to "buy" (lose money or make little or no profit on) the assignment in exchange for the knowledge they will gain. This objective might be achieved at significant additional cost to the client, owner, or customer.

Most people and organizations retaining consultants know that they should avoid being penny-wise and pound-foolish. However, a completely rational approach is not always possible because of insufficient information. Consultants must put themselves "in the shoes" of the prospect's decision-makers, particularly those who are public officials subject to public scrutiny. The basis for that scrutiny could be an article in the next day's local newspaper or a feature on a news program that compares proposed fees plus expenses for various consulting firms. Officials may be hard pressed to justify a large proposed price over a small proposed price when, at least at the surface, the resulting proposed deliverables appear identical. Remember, at the time of consultant selection, the total cost of the project (consultant fee plus expenses, construction or other implementation, and long-term operation or use) is usually not known or is not perceived as an important factor.

Refer to Table 13.2, which is an accounting of all costs for a hypothetical project. Note that the consulting cost ultimately paid by the owner is a small part (2.5 percent) of the total present worth cost that will be incurred by the owner in obtaining and using the structure, facility, system, product, or process. Full appreciation of Table 13.2 requires understanding present worth and now to estimate it. Perhaps you have already learned that in an engineering economics or decision economics course, or

Table 13.2 Consulting fees are typically a very small fraction of the total life-cycle cost for an engineered project as suggested by this hypothetical accounting of costs.

Design fee proposed by consultant	$100,000
Construction/manufacturing cost to client, owner, or customer (usually not known when consultant is being selected)	2,000,000
Total initial capital cost to client, owner, or customer	2,100,000
Operation and maintenance cost (present worth) incurred by client, owner, or customer over the economic life of the structure, facility, system, product, or process (usually not known when consultant is being selected)	1,900,000
Total (present-worth) cost to client, owner, or customer (usually not known when consultant is being selected	$4,000,000

portion of a course. If not, refer to a textbook such as that written by Blank and Tarquin (2005) or Grigg (2010).

While professional service costs proposed by various consulting firms are often known before the potential project gets underway, the remainder of the total project cost, and by far the largest part of the total project cost, is largely unknown. Therefore, the decision makers within the prospect organization must make a decision without complete fiscal information. This is one reason some users of professional services slide into PBS, that is, they place too much emphasis on relative magnitudes of the proposed consulting costs and not enough emphasis on the likely total cost that they will incur as a result of the consultant they select. The wide range in consulting fees proposed on ostensibly the same project further complicates the matter.

Note in Table 13.2 that a $10,000 or a 10 percent savings in the consulting fee reduces the total cost to the prospect by only 0.25 percent for the hypothetical example. Of course, this assumes that the consulting firm billing $90,000 can produce the same quality product as the consulting firm billing $100,000. If they can't, because of insufficient knowledge, experience, or attention, the resulting increases in construction, manufacturing, operation, and maintenance costs could easily more than offset their lower up-front fees.

Even much larger savings in up-front consulting fees will tend to result in only small savings in total project costs. For example, if one consulting firm proposes to do the hypothetical project for $50,000, a 50 percent savings relative to the consulting firm proposing to do the project for $100,000, the reduction in total project cost is only 1.25 percent. This unrealistically assumes that the "low cost" firm can produce a product of similar quality to the "high cost" firm.

The preceding analysis is, of course, based on the hypothetical accounting of costs presented in Table 13.2. Nevertheless, based on my experience, the relative costs in the table are realistic and the analysis based on the table is sound.

The Ideal Selection Process

Ideally, clients, owners, and customers should select consultants based on the goal of minimizing their total costs. This ideal selection concept is illustrated in Figure 13.2. The costs or prices proposed by potential consultants A, B, and C vary widely with the largest cost being approximately twice the smallest cost. Similarly, there are significant variations, although not as dramatic in a relative sense, in the present worth of the construction, manufacturing, operation, and maintenance costs that the prospect would incur over the life of the structure, facility, system, product, or process.

Of course, as already noted, the organization seeking professional services is not likely to know these total costs, or even relative values of these costs, at the consultant selection stage. If they were known or could be known, the prospect would obviously determine the total cost associated with each of the three potential consultants and select the consultant that would offer the lowest total cost. This would be consultant B in Figure 13.2. Recall the discussion in Chapter 8 on the disproportionate impact of the design function. That is, while design accounts for a small fraction of the total project cost it is the primary determinant of that total cost. Retaining competent design

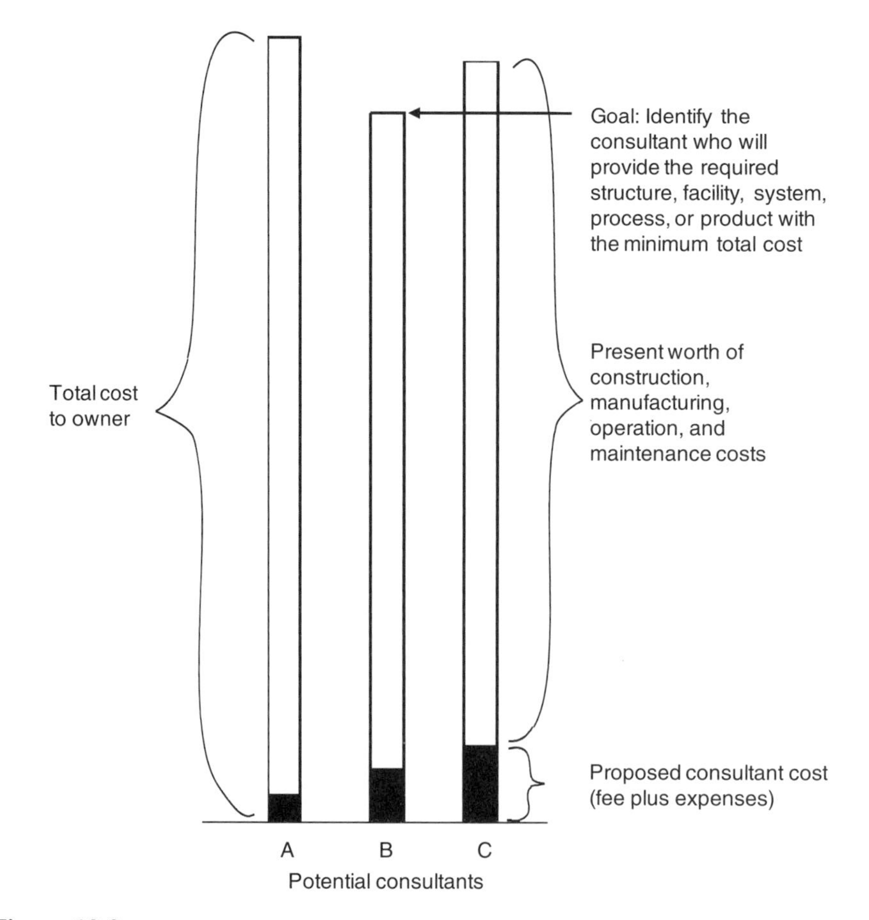

Figure 13.2 Ideally, the process used to select consultants should consider total costs.

professionals increases the probability of achieving the lowest total cost consistent with quality, that is, meeting all project requirements.

Qualifications-Based Selection

As an alternative to PBS, the consulting industry and others advocate qualifications-based selection (QBS). This consultant selection process may be described as follows (Chinowsky and Kingsley 2009): "After firms are evaluated and short-listed based on their qualifications, the top-ranked firm is selected for price negotiations, and a fair and reasonable price is reached based on a detailed scope of the project. If agreement on price cannot be reached with the most qualified firm, negotiations commence with the second most qualified firm. In the vast majority of cases the top ranked firm is selected at a price that fits the client's budget." Adding credibility to QBS is its endorsement by professional organizations that are not within the consulting industry (ACEC 2011) such as the American Bar Association (ABA) and the American Public Works Association (APWA)

Researchers recently "conducted an extensive survey of projects and analyzed the impact of QBS on project outcomes" (Chinowsky and Kingsley 2009). Key findings of the study, quoted from it, are that QBS:

- **Ensures cost-effectiveness:** Hiring the most qualified professional design services provider at a reasonable price is the best way of ensuring that the final constructed project is completed on time and on budget.
- **Lowers risk for complex projects:** Owners expressed special interest in using QBS on projects with higher risk factors and/or higher design complexity.
- **Results in better projects and highly-satisfied owners:** 93 percent of owners surveyed on QBS projects in the study rated the success of their final project as high or very high.
- **Takes account of emerging societal issues:** The team found that QBS procurements were more likely to address emerging social needs, such as sustainability, than cost-based procurements.
- **Encourages innovation, protects intellectual property:** The study confirms widely-held views that QBS promotes a higher level of innovation. In addition, there was a high degree of satisfaction on the part of design firms that the intellectual property included in the innovation was properly protected.
- **Supports owner capacity building:** QBS allowed owner organizations to gain specialized quality services from design firms as an extension of staff.

For clarity and perspective, note that the cited U.S. study was conducted at the request of the ACEC in cooperation with the APWA. Participation by ACEC does not necessarily question the study's objectivity but study users should be aware of the engineering consulting industry's involvement.

To conclude this discussion of QBS, recall the previous chapter's mention of the U. S. Brooks Act, and versions of it in some states, which prohibit competitive bidding in the selection of engineering and similar professional service firms. These laws support QBS. However, many other local governmental units in the U.S. are able to use PBS as are private sector entities. Recognizing this, engineering and similar professional service firms explicitly or implicitly decide whether or not they will participate in QBS, PBS, or both. For firms that include PBS in their business strategy, the last major section of this chapter offers some thoughts about the costs, monetary and otherwise, that they are likely to incur.

Steps in the Selection Process

As noted at the beginning of this discussion of consultant selection, the detailed process a particular client, owner or customer uses to select a consultant to provide other services, whether under PBS or QBS, is unique to each situation. However, many steps common to the consultant selection process can be identified and linked together as shown in Figure 13.3, for discussion purposes. Beginning with "Start," the most formal and involved selection process is the series of Steps 1 through 12 proceeding clockwise around the figure. Various optional shortcuts are possible and are frequently used by users of professional services, especially those in the private sector. The full and formal

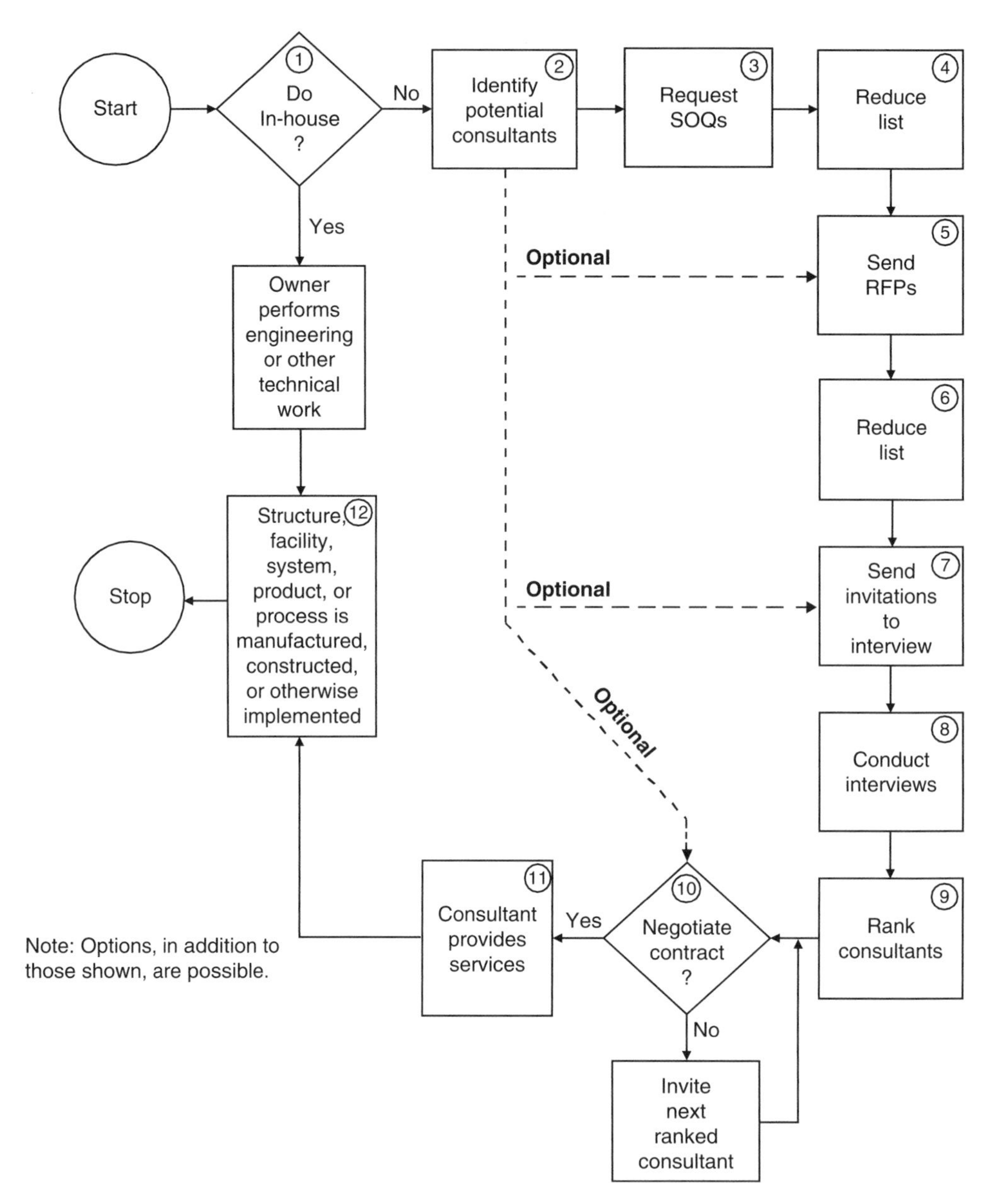

Figure 13.3 While the process used by clients-owners-customers to select consultants is unique to each situation, most selection processes are combinations of some of the steps shown here.

12-step process is described below. Note how PBS and QBS can enter into the process and influence its outcome.

Step 1-Do in-house?: The client, owner, or customer determines whether or not a consultant will be retained for a task or project. Recall the five basic reasons to retain a consultant as discussed earlier in this chapter. Assume a consultant is to be utilized.

Step 2-Indentify potential consultants: The entity needing a consultant, possibly with the assistance of a selection committee, identifies potential consultants. A list of potential consulting firms might be assembled using personal and other first-hand knowledge, referrals from colleagues at other organizations, formal listings such as those appearing in directories of engineering organizations, business cards appearing in professional publications, internet searches, and even the yellow pages in telephone books. Some firms may have been prequalified for certain service categories.

Step 3-Request SOQs: After screening the list, the client, owner, or customer requests statements of qualifications (SOQs) from consulting firms that presumably have the ability to provide the necessary services. In some cases, and they are usually government organizations, a request for SOQs is published, as in a newspaper or on website, and any firm may respond. SOQs are usually standard, "off-the-shelf" items or documents readily assembled from standard text and graphics maintained on computer systems. SOQs usually include basic information about the consulting firm such as its size; office location or locations; services offered; clients, owners, and customers served; references; experience with emphasis on projects similar to that about to be undertaken by organization receiving the SOQ; and resumes of selected professional staff on relevant projects. An SOQ typically does not address the manner in which the consultant, if selected, would approach the specific project. However, respondents may decide to include some project-specific ideas and information.

The checking of a firm's references, that is, a representative list of current and past organizations they have served, would seem to be a very effective way to screen consultants. After all, who is in a better position to comment on a firm's services than those who have received those services? Of course, and as noted, the references must be truly representative. One way to assure this is for the client, owner, or customer to ask each candidate consultant to provide the names of all organizations receiving certain services (e.g., manufacturing process, highway planning, management assistance) over the past few years, along with permission to contact any or all organizations on this list and inquire about any aspect of the services received.

Step 4-Reduce List: One or more of the prospect's professionals review the SOQs and match the perceived needs of their project with their interpretation of the experience and ability of each consulting firm. Firms judged to not have adequate qualifications are eliminated from further consideration. Other factors are likely to influence this step such as trustful or mistrustful relationships between individuals employed by the client, owner, or customer and those employed by one or more consulting firms. The next chapter stresses the importance of trust in marketing consulting services.

Step 5-Send RFPs: The organization seeking a consulting firm now invites firms remaining on the eligibility list to describe how they would complete the specific project and often asks interested firms to include an estimate of the cost of their services. RFPs typically include items such as a letter of explanation and invitation; a description of the project; an explanation of the required scope services (e.g., feasibility study, preliminary engineering, preparation of plans and specifications, construction management, start up, education and training); a project schedule; and the due date for the proposal. A list of available related reports, studies, and investigations; a description of available data and information from or known by the prospect; the name of a contact person; an indication of whether or not the proposers should provide an estimate of the cost of services; and/or a description of Minority Business Enterprise (MBE), Women's Business Enterprise (WBE), and Disadvantaged Business Enterprise (DBE) requirements may also be included. In some cases, the owner is prohibited from requesting an estimate of consultant prices. For example, recall the discussion of the Brooks Act and similar state provisions in Chapter 12. The client-owner-customer may conduct, as part of the RFP process, an information meeting for all consultants intending to submit a proposal.

Each consultant receiving the RFP typically re-visits its initial decision to pursue the project. In the interim, a firm may have learned more about the project and, based on that new knowledge, may decide to not pursue the project further. The effort required to prepare a proposal in response to the RFP is typically an order of magnitude greater than the effort required to assemble and submit the SOQ because, as noted, the SOQ is assembled from pre-prepared materials. In contrast, a proposal that is submitted in response to a RFP is a largely original document requiring considerable time and effort, including that of high level and, therefore, costly professional personnel. In fact, the likelihood of a consultant successfully and profitably completing a project, assuming it is ultimately selected, depends on the care used to prepare the proposal. The typical proposal must be prepared with a clear understanding of the project requirements and, in response, what the consultant will do, how the consultant will do it, how long it will take, and what it will cost. In a sense, the project is worked out "on paper" as part of the proposal preparation process. The proposal may be considered a very preliminary version of the project plan described in Chapter 5.

Step 6-Reduce List: Using the project-focused information provided by consultants receiving the RFPs, the prospect's personnel eliminate some consultants from further consideration. Factors may include one or more of the following: poor responsiveness to the RFP; indications of creativity and innovation, that is, including too much or too little; specific personnel to be assigned to the project; experience or lack thereof on similar projects; results of reference checks; list of deliverables; and, of course, the proposed price if it was requested under a PBS approach.

Step 7-Send invitations to interview: The selecting organizations invite the remaining consultants to interview for the project. Each firm receiving an invitation to interview is likely to accept. However, as was the case when a firm was invited to prepare a proposal in response to a RFP, a consulting firm may re-visit its initial decision to pursue the project. In the time that has passed since submitting the proposal, additional information may have been obtained about the potential project or the prospect might cause a reversal of the original decision. This re-visiting of earlier decisions to pursue the project is prudent because of the additional time that will now be required to prepare for the interview. Although, in some cases an interview is a relatively informal affair requiring minimal preparation, in other situations an interview is a formal event requiring a major investment by the consulting firm. The labor and expenses invested in assembling SOQs; preparing proposals in response to RFPs; and getting ready for, participating in, and following up on interviews all add to the consulting firm's overhead as discussed in Chapter 10. These costs are passed back to clients, owners, and customers in that they are reflected in the multiplier.

Step 8-Conduct interviews: Each interview is typically conducted in private. The consulting firm's team usually consists of a principal of the firm, the person who would manage the project, and one or more members of the designated project team possibly including a specialist with expertise specifically related to the project. Incidentally, consultants sometimes send a team to the interview that is not representative of the team, particularly the project manager and key members of the project team, that would actually work on the project. This "bait and switch" tactic, while it might enhance the interview, is a poor business practice and some would argue is unethical. The best approach to follow is "what you see is what you get." The consulting team may provide additional text, tables, figures, and other printed material prior to or at the interview. In addition, the consulting team may use audio-visual materials such as posters, computer presentations, slides, transparencies, videotapes, and equipment demonstrations. After some sort of formal presentation by the consulting team, the prospect's team typically asks questions and a general discussion ensues.

The consulting team attempts to develop rapport with the prospect's selection team and the selection team tries to determine if a good working relationship could be established with the consulting team. Although difficult to measure and sometimes denied, interpersonal "chemistry" probably becomes a significant factor at this point in the overall selection process because the selecting organizations and each potential consultant are now interacting with each other in a manner that roughly approximates the working relationship that would exist on the project. Interviews often conclude with a closing statement by someone on the consulting team.

Step 9-Rank consultants: Based largely on the interview, but perhaps on additional consideration of the proposal received prior to the interview, the client, owner, or customer representatives typically rank the competing consultants. This is a difficult task because of the voluminous amount of quantitative and qualitative information, including the personalities of participants, that is now available.

Step 10-Negotiate contract: The prospect and the first-choice consultant try to negotiate a contract. The first-ranked consultant is invited to prepare an estimate of the cost of services, or maybe a complete contract for professional services, and present it in draft form as the basis for negotiation. Typically the consultant will re-visit the proposal submitted earlier, convert it to contract language, make modifications based on ideas and information obtained during and subsequent to the interview, and submit the new document to the prospect. Somewhat self-laudatory language and other terminology that sometimes appear in proposals should not appear in a draft contract or agreement. The "Liability: Minimizing It" section of the Chapter 11 offers advice about careful use of language in order to communicate effectively.

After a draft agreement has been sent to the client, owner, or customer, typically one or more of their representatives meet with one or more representatives of the consulting firm to review the document in detail and arrive at a mutually-agreeable contract. Occasionally the two parties are not able to arrive at a mutually-acceptable agreement, in which case the organization seeking a consultant is most likely to enter into negotiation with the second-ranked consulting firm. As noted earlier, a carefully prepared proposal in response to an RFP is crucial to the successful and profitable completion of a project assuming the consultant is selected. The value of a carefully prepared proposal usually becomes evident during the contract negotiation phase because the consulting firm is in a position to convince the prospect's representatives that their needs and other project requirements will be met and that the cost of services is reasonable.

Step 11-Consultant provides services: The consulting firm now draws on its technical expertise, coupled with much of the professional knowledge, skills, and attitudes discussed in this book, to provide the agreed-upon services. This is "where the rubber hits the road" and the likelihood of future engagements, positive references, and career-long relationships is determined.

Step 12-Structure, facility, system, product, or process is constructed, manufactured, or otherwise implemented: Sometimes the consulting organization is involved in this step. For example, a mechanical engineering consulting firm that designed a new manufacturing process might be retained to supervise the installation and start-up of the process. A civil engineering consulting firm that designed a high-rise structure may be retained to monitor, but not supervise, its construction so that the structure's owner knows the degree to which the construction is conforming to the plans and specifications. Some consulting firms provide even broader services with respect to engineered structures, facilities, and systems. For example, design-build firms do both design and construction; other firms offer operation and maintenance services; and a few organizations offer all or most of the preceding plus finance services

Welcome Exceptions

As indicated earlier and illustrated with the dashed line examples in Figure 13.3, several optional, shorter, and simpler consultant selection processes are possible and often most welcome because they reduce costs for all concerned and reflect the positive effect of trustful interpersonal relationships. For example, the client, owner, or customer generally familiar with the consulting community might move directly from Step 2, identify potential consultants, to Step 5, send RFPs, thus eliminating requesting and reviewing SOQs. An even shorter version of the overall process is to move directly from Step 2, identify potential consultants, to Step 7, extend invitations to interview. This shortcut might apply in situations where a client, owner, or customer is very familiar with the qualifications of a set of consulting firms and wants to focus immediately on how any one of those firms would go about doing a particular project.

Sometimes most of the process is omitted, and this is most likely to happen in the private sector, when the experienced user of professional services goes from Step 2, identify potential consultants, to Step 10, negotiate a contract. The organization desiring professional service predetermines which consulting firm is most likely to provide the desired services at an acceptable cost and invites that firm to learn about the project and negotiate a contract. As noted at the beginning of this chapter, one of the highest compliments that an individual consultant or consulting firm can receive is to be retained on a sole-source basis. Other options, besides those shown in Figure 13.3, are possible.

Summing Up the Consultant Selection Process

Figure 13.3 might also be viewed as an intense inter-organizational and interpersonal communication process. The five components of communication (listening, speaking, writing, use of graphics, and use of mathematics) discussed in Chapter 3 are typically used throughout the consultant selection process. The most qualified firm may not be selected and the client, owner, customer and its stakeholders may be denied the most appropriate services because of inadequate communication skills on behalf of either the consulting firm or the selecting entity.

The consultant selection process is more fully understood in the context of marketing professional services. As long and complex as the selection process may be, it is only one component—and a small one at that—of marketing. Chapter 14 is a comprehensive treatment of marketing and provides ideas and tools that can enable you, as a member of an organization that uses professional services or provides them, to participate even more effectively in the consultant selection process.

Views of Others

Before giving advice as a consultant, contemplate the thought of author Harper Lee, "You never really understand a person until you consider things from his point of view—until you climb into his skin and walk around in it." Once we are ready to offer our counsel, consider this thought from English poet and philosopher, Samuel Taylor Coleridge about how to gently do so: "Advice is like snow; the softer it falls, the longer it dwells upon, and deeper it sinks into, the mind." Author Hannah Whitall Smith provides this advice that is applicable if we conclude that the receiver is not receptive to our well-intentioned consulting advice: "The true secret of giving advice is, after you have honestly given it, to be perfectly indifferent whether it is taken or not and never persist in trying to set people right." Finally, this thought from the Irish playwright, novelist, and poet Oscar Wilde, which recognizes the reluctance of some to accept counsel from consultants no matter how carefully crafted and conscientiously communicated, "The only thing to do with good advice is pass it on."

PRICE-BASED SELECTION: THREE COSTS TO THE CONSULTANT

To this point, the discussion of PBS versus QBS has focused on the users of professional services strongly suggesting that their interests are served best by QBS. Let's "look at the other side of the coin" by considering the best interests of the provider of professional services. Participating in PBS, instead of QBS, can result in up to three "costs" to the consultant and often all three. Those three, perhaps not so obvious possible costs, are explained here (Walesh 2007).

Offering Less Than We Could

The first cost occurs during the proposal and contracting phase. Because price will drive the selection, the professional must carefully and narrowly respond to the prospect's RFP or other description or knowledge of wants and needs. The proposal preparer must "sharpen his or her pencil." He or she may be tempted, as one consultant told me, to create the image of offering a "big box" while being sure there is as little as possible inside.

Some of the relevant and valuable knowledge gained through education and experience may not be used – the budget won't allow it. For example, the consultant preparing the proposal may believe that a stakeholder involvement program is needed if the designed facility is to be ultimately supported by the public and be constructed. However, stakeholder involvement is not being requested and would, therefore, add upfront costs that are likely to jeopardize selection.

Or the type of solution that the prospect wants may not, in the judgment of the professional, be the most effective in the long run. If a range of options were to be explored, which would add design costs, a solution with a lower life-cycle cost would be likely. But, the extra professional service costs would be counterproductive in the PBS environment.

You might ask what is this first cost to the consultant? It is the frustration caused by not being able to fully share the benefit of one's education and experience with the prospect. Even the ethically-disadvantaged service provider would seem to be affected, that is, to experience this frustration. The foundation of our profession is its growing body of knowledge that we strive to learn, add to, and use to serve others. Under utilizing that body of knowledge in order to win in a PBS process seems just plain wrong. Would we want our medical doctor, financial advisor, attorney, or child's teacher to draw on only part of what they know about meeting needs in exchange for reduced compensation?

Further Reduction in Profit

The second cost to the consultant, which is related to the first cost, occurs during the project. Assuming that we have been fortunate (or unfortunate?) to be selected based on our low "bid," which probably included a less-than-normal profit, we now begin to provide the promised scope and deliverables. Our intent is to do so in as "bare bones" a manner as possible.

As we proceed, however, those important project elements we omitted to yield the lowest bid, simply do not go away. They are now planted in our subconscious mind and keep creeping into our conscious mind. We try to clip them but many of us, remembering what we left out when preparing the price-base proposal and possibly driven by ethical concerns, succumb to the temptation to put some of those valuable elements back into the project.

Maybe we are compensated for some of this consultant-driven scope creep because we are able to convince the client, owner, or customer that the additions are warranted. In many cases, we unilaterally add services for which we are not compensated and lose monetarily. Our firm, in effect, further reduces its profit margin or effective multiplier on the project or our effective hourly rate declines even more. This is a monetary cost attributed to our choosing to participate in PBS; it comes out of our pockets.

Damaged Reputation

The third consultant cost arises when our portion of the project, be it a study, a plan, a design, construction, manufactured product, a process, or workshop is completed.

Maybe it begins even before our work is done. This cost emanates from the client, owner, or customer increasingly realizing that the project is deficient. Perhaps citizens and their elected representatives are expressing opposition to the project. In retrospect, stakeholder participation should have been an element of the project. Or maybe insightful decision makers are asking questions about other options considered and the real or life cycle-cost of the option that was selected from the outset. More analyses should have been conducted. Or in-house workshop participants complain that the outside facilitator was not aware of important organizational issues and, therefore, the workshop fell far short of expectations.

An ethical client, owner, or customer will accept some of the blame for what are increasingly seen, as a project progresses, to be project deficiencies. Others won't. Regardless, our professional services firm is likely to be criticized for not learning and addressing all the issues and for not appropriately drawing on our and our professions' body of knowledge. Increasingly, the client, owner, or customer and perhaps stakeholders will ask "what were we, the consultants, thinking?" Thus we arrive at the third consultant cost of PBS: Frequent and widespread allegations of deficient services. And we consultants rise or fall on our reputation.

Personal

Consulting has taken me, with some detours along the way, across the spectrum from an employee of a medium-sized consulting engineering firm to a sole proprietorship. At both ends of the spectrum, my principal functions have included marketing and project management, or what some call, the "seller-doer" role. Many lessons have been learned in this "seller-doer" role, some gently and some at high cost. One lesson is not to pursue what you view as a professional assignment if price is to be the primary selection factor. Or stated differently, if you decide to "bid," be upfront with yourself and those who will be affected, usually adversely, about why you are doing so. I realize that, as a consultant, arguing for QBS and against PBS can be interpreted as self-serving. Perhaps, however experience, within and outside of the engineering profession, suggests that there is wisdom in expressions such as "there is no such thing as a free lunch" and "you cannot get something for nothing." As someone anonymously said, "some know the price of everything and the value of nothing."

Closing Thoughts

Perhaps you agree with, or are you at least considering the merit of, the preceding three consultant costs of participating in PBS. If so, a logical follow-up question is: why do consulting firms do it, that is, why enable ourselves and our firms to incur these costs?

One reason we do this is to survive. In the short run, little or no profit seems better than none at all. Maybe we reason, as a friend and consulting firm executive said to me with tongue-in-cheek, "we will lose a little money on each project but make it up in volume." Another reason to engage in PBS is to "win" a project for the knowledge and/or contacts it may provide. I have done this, "lost my shirt" in the process as planned, and may do it again. However, these as focused and few and far between.

Maybe we participate in PBS because we are in the commodity business. We reason that many organizations do what we do, that is, the services we provide. We all do it essentially the same way. Therefore, price is the differentiator. Perhaps your firm is doing well financially and psychologically with this commodization business model. If so, ignore the preceding three-cost argument.

For the rest of us, let's include in our Go/No Go process careful probing of the probable bases for consultant selection. Study the RFP; question the prospect; review previous experience with the prospect; and ask colleagues about their experiences with the potential client, owner, or customer. If "price" is going to be a major selection factor, think of the three costs we may incur if we are "fortunate" enough to be selected.

Think also of the ethical implications of participating in a process that, from the outset, is not designed to serve the long-term best interest of those we serve and their stakeholders. Besides the three costs we are likely to incur, consider the costs that will probably be incurred by those we serve, as explained by the English philosopher John Ruskin:

It is unwise to pay too much, but it is worse to pay too little.
When you pay too much, you lose a little money, that is all.
When you pay too little, you sometimes lose everything
because the thing you bought is not capable of doing the thing it was bought to do.
The common law of business balance prohibits
paying a little and getting a lot—
it can't be done.
If you deal with the lowest bidder,
you might as well add something for the risk you will run
and if you do that,
you will have enough to pay for something better.

CONCLUSIONS ABOUT THE ROLE AND SELECTION OF CONSULTANTS

Already as a student, you should understand the role and selection of consultants because you may want to be one during some of your career and/or your organization, or even you, may use their services. Consultants are essential within the engineering profession because, depending on the situation, they provide necessary expertise, supplement in-house personnel, offer objectivity, perform unpleasant tasks, and reduce liability. Being a consultant is both demanding and satisfying. The consultant selection process, whether viewed as a member of a consulting firm or a member of an organization that uses consulting services, is complex and costly. The use of PBS or QBS inevitably influences the outcome. Use of QBS is more likely to lead to a win-win result, that is, protect the best interests of clients, owners, customers, and stakeholders and the consulting firms that serve them.

The bitterness of poor quality
remains long after the sweetness of low price is forgotten.

(*Anonymous*)

CITED SOURCES

ACEC. 2011. Personal communication with Nina S. Goldman, Director, Sales & Member Organizational Services, May 5.

ACEC. 2011. "Qualifications-Based Selection." (www.acec.org/advocacy/committees/pdf/qbsccd04.pdf) May 13.

Blank, L. and A. Tarquin. 2005. *Engineering Economy – Sixth Edition*. McGraw Hill Higher Education: New York, NY.

Chinowsky, P. S. and G. A. Kingsley. 2009. *An Analysis of Issues Pertaining to Qualifications-Based Selection*. American Council of Engineering Companies and American Public Works Association, Washington, DC. and Kansas City, MO.

Grigg, N. S. 2010. *Economics and Finance for Engineers and Planners*. ASCE Press: Reston, VA.

Walesh, S. G. 1993." Interaction with the Public and Government Officials in Urban Water Planning." Hydropolis—The Role of Water in Urban Planning. Proceedings of the International UNESCO-IHP Workshop, Wageningen, The Netherlands and Emscher Region, Germany, March-April, pp. 159–176.

Walesh, S. G. 1999. "DAD is Out, POP is In." *Journal of the American Water Resources Association*, June, pp. 535–544.

Walesh, S. G. 2000a. *Flying Solo: How to Start an Individual Practitioner Consulting Business*. Hannah Publishing: Valparaiso, IN.

Walesh, S. G. 2000b. "Seven Reasons to Consider Going Out On Your Own as a Consultant." *Journal of Management in Engineering – ASCE*, September/October, pp. 18–20.

Walesh, S. G. 2001. "Time to Fly Solo?" *The Bent of Tau Beta Pi*, Summer, pp. 21–24.

Walesh, S. G. 2007. "Price-Based Selection: Three Costs to the Consultant." *Leadership and Management in Engineering – ASCE*, July, pp. 104–105.

ANNOTATED BIBLIOGRAPHY

NSPE. 2011. "In New York, State Saves Money by Using Consulting Engineers, Study Says." *PE: The Magazine for Professional Engineers*, April, p. 5. ("The research . . . analyzed and compared the New York State Department of Transportation's employee costs to the costs for an engineer in private practice firm. The conclusion: Outsourced designs cost 15 percent less.")

Walesh, S. G. 2004. *Managing and Leading: 52 Lessons Learned for Engineers*. Lesson 12: "Are You Unemployable?", ASCE Press: Reston, VA. (This lesson recognizes that some of us reach a career point where we realize we are not meant to be an employee and must start our own business.)

Weiss, A. 2003. *Great Consulting Challenges and How to Surmount Them: Powerful Techniques for the Successful Practitioner*. Jossey-Bass/Pfeiffer: San Francisco, CA. (Discusses a wide range of individual practitioner topics including the consultant selection process as seen from the consultant's perspective.)

Zupek, R. 2010. "So You Want to Be a Consultant?" *PM Network*, December, pp. 54–59. (Offers this observation: "You will now be your own bookkeeper, marketing department, office supplier, and scheduler . . . All that work that got done by others in your previous employer will now be done by you.")

EXERCISES

13.1 CONSULTANT SELECTION PROCESS: The purpose of this exercise is to enable you and your team to further explore the consultant selection process by viewing it from the perspective of a consulting firm seeking an assignment. Suggested tasks are:

A. Assume you and your team are members of a consulting engineering firm tracking a potential project to be conducted, with heavy consultant assistance, by a client, owner, or customer, that is, a prospect. Your team just received the RFP from the prospect as a result of Step 5 in Figure 13.3. (Note: To enhance the realism of this exercise, obtain an actual RFP. One way to do this is to contact a local government entity and ask them to share an RFP with you. An already-issued RFP is a public document and, therefore, you should be able to access it. RFPs may be posted on the government entity's website. You may also be able to obtain an RFP from a private entity such as a manufacturer or land developer.)

B. Study the RFP and then write a list of questions that you would like to ask a representative of the prospect so you can prepare a responsive and helpful proposal. In developing the questions, recall the question-asking advice offered in Chapter 7 section titled "Strive to Understand Client, Owner, and Customer Wants and Needs" and look ahead at the Chapter 14 section "Ask-Ask-Ask: The Power of Questions."

C. Assume you met with a representative of the prospect, your questions were answered, and you prepared and submitted your proposal. Because of the proposal's responsiveness and thoroughness, you are invited by the prospect to an interview (Step 8 in Figure 13.3).

D. Outline the content of the presentation you and selected members of your team, and others in your firm, such as an office manager or other executive, will make at the interview. Within the outline, explain how your firm would execute the project, who will be speaking about what, supporting visual aids and/or handouts to be used, and the questions likely to be asked by representatives of the prospect during or after your presentation.

CHAPTER 14

MARKETING: A MUTUALLY-BENEFICIAL PROCESS

I don't care how much you know
until I know how much you care.

(*Anonymous*)

This chapter begins with an admonition to consider one's view of marketing and sales which is followed by review of the financial motivation for marketing. Marketing and selling are contrasted and then a three-step, win-win marketing model is presented which is applicable to the business, government, academic, and volunteer sectors. A dozen marketing tools and techniques, all of which can be used by the entry-level practitioner, are described. The chapter concludes with a summary of what works and what doesn't work in marketing.

CONSIDER YOUR VIEW OF MARKETING: ARE YOU CARRYING SOME BAGGAGE?

Perhaps the word "marketing" and the related word, "sales," engenders negative reactions or connotations. Images of brash, high-pressure car salespeople or annoying telemarketers may come to mind. Brash and high-pressure are not part of this chapter's marketing theme. High pressure does not work when marketing; persistence does.

Worse yet, marketing may suggest unscrupulous individuals. You may have a uninformed aversion to marketing based on general ethical grounds. Marketing may seem unsavory based on experiences outside your technical field because you perceive it requires misrepresentation, if not outright dishonesty. Dishonesty does not work when marketing; honesty does.

On contemplating the business, government, academic, or volunteer sectors, you, as an engineering student or young practitioner, may be repulsed by the thought of "wasting" your professional education by doing "sales" work or even being on the receiving end of any aspect of marketing. If so, try to be receptive to this chapter's positive theme and the particular marketing model that is presented. To the extent you learn to view marketing as earning trust and meeting client, owner, and customer needs, which is the essence of this chapter's approach, you may conclude that not only is marketing an ethical process, but also a satisfying and mutually-beneficial one. The same hope applies to you, as an experienced practitioner, who has been heavily involved in technical work and now, by choice or assignment, you find yourself involved in your organization's marketing efforts.

Incidentally, ethics codes, many of which are presented in Chapter 12, contain many provisions directly applicable to marketing consistent with the way marketing is presented in this chapter. For example, Fundamental Canon 2 of the ASCE Code of Ethics states that "Engineers shall perform services only in areas of their competence." For additional discussion of the crucial role of ethics in marketing of engineering services, refer to Heightchew (1999) and Snyder (1993).

The positive perspective on marketing offered here is supported by research, by the admirable successes of some organizations, and by my experiences. That win-win marketing approach is presented in this chapter with the hope that it will be of interest and value to you, the engineering student contemplating your career and the engineering graduate in the early phase of your career, regardless of the sector in which you are employed. During your career, you will have the opportunity to participate in marketing. The question is: Will that participation be positive and productive, for you and your employer, or negative and destructive?

CHAPTER'S SCOPE

This chapter focuses on the marketing of engineering and other technical services to public and private clients, owners, and customers by consulting firms. Such firms, particularly the larger ones, typically employ civil, electrical, mechanical, and other engineers as well as other technical and non-technical personnel. The range of disciplines on the consulting firm team is another reason this chapter should be of interest to all technical professionals, regardless of their area of specialization. To the extent that the government, academic, and volunteer sectors see marketing as part of their organizational responsibilities and, in my view, they should, most of the principles, the marketing model, and many of the tools and techniques set forth in this chapter are applicable to them. Marketing engineered products is not explicitly treated in this chapter. However, the underlying principles, the marketing model, and many of the tools and techniques presented here for marketing professional services are directly applicable to marketing of engineered products.

Personal

In my mid-thirties, I moved from government employment to an engineering firm. My assignment: Lead the development of a new water resources engineering service line which, among other things, required developing staff with

> the necessary expertise and obtaining new clients for the firm. I was excited and, in retrospect, I was naïve and so was my employer. I had no marketing education and training and no marketing experience. I have learned much about marketing since then while working for that consulting firm, a university, and functioning as an independent consultant. Much of what I have learned from study and experience is presented in this chapter.

THE ECONOMIC MOTIVATION FOR MARKETING PROFESSIONAL SERVICES

Income for professional services firms emanates from essentially one source: clients, owners, and customers who need and are willing to pay for services. These businesses cannot levy tax or unilaterally charge a fee. At what rate does a consulting firm's marketing program have to generate sales? The answer, of course, depends on many factors, but is heavily influenced by the size of the consulting firm in terms of total number of professional and other employees. How much should a firm invest in marketing? That, too, is influenced partly by the size of the firm. I am intentionally referring to marketing expenditures as investments, not costs, to emphasize the need to carefully plan, use, and monitor these expenditures so that the return on the investment (ROI) can be determined.

Consider the case of a small consulting firm with 10 full-time employees, most of whom are technical professionals. Assume that the overall time utilization rate (U) is 0.7, the firm's multiplier (M) is 3.0, and the overall raw labor rate is \$25 per hour. Let P_w be the total raw payroll cost per week. Then $P_w = (40 \text{ hours per week}) \times (10 \text{ employees}) \times (\$25 \text{ per hour per employee}) = \$10{,}000$. Given this weekly payroll cost, the necessary weekly net revenue is $P_w \times U \times M = (\$10{,}000) \times (0.7) \times (3.0) = \$21{,}000$. In other words, this small, hypothetical consulting firm needs to successfully market an average of \$21,000 of new contract work per week or \$4,200 per day based on a five-day work week.

Assuming seven percent of net revenue is invested in all aspects of marketing, which is a representative consulting industry value, the average daily marketing investment is \$294. The required average daily sales and the daily marketing investments for the small firm and to two larger firms are presented in Table 14.1.

Assume, for the ten person firm, that the total value of work under contract that has yet to be done is \$60,000. Given that the necessary weekly net revenue is \$21,000, the

Table 14.1 Required average daily sales and the daily marketing investments for an engineering professional services firm are large and determined partly by the size of the firm.

Size of Firm (persons)	Required Daily Sales (\$)	Daily Marketing Investment (\$)
10	4,200	294
200	84,000	5880
4000	1,680,000	117,600

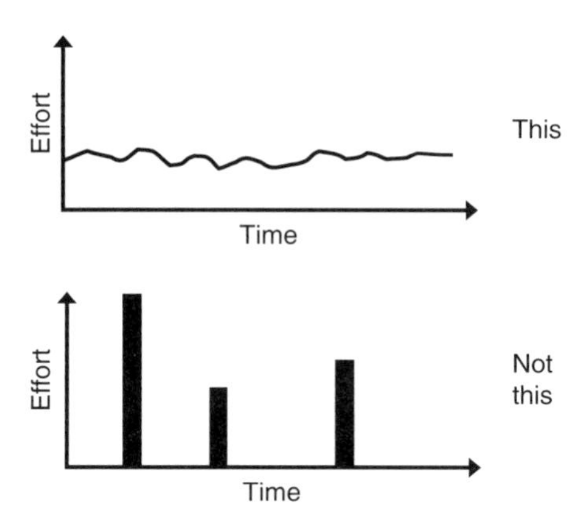

Figure 14.1 Market in a continuous, proactive manner not sporadically when work is needed.

quotient of the value of contracted work to be done and the necessary weekly net revenue is referred to as the backlog and in this case is 2.9 weeks. That is, if there are no additional contracts as a result of the firm's marketing efforts, there will be insufficient revenue to pay salaries and cover expenses three weeks from now.

Now consider the case of a large 200 person engineering firm and an extremely large 4000 person firm. Using the same values for U, M, and raw labor applied in the preceding small firm example, P_w, the required average daily sales and average daily marketing investments for the medium and large firms are shown in Table 14.1.

The preceding hypothetical, but realistic, examples illustrate the idea that marketing, to the extent that it leads to contracts for consulting services, is essential for the financial health of a professional services business. And marketing is a major annual investment on the order of a professional service firm's pre-tax profit. Therefore, the marketing effort must be carefully managed, as must other aspects of the consulting business. As illustrated in Figure 14.1, marketing should be a continuous, proactive process not a series of sporadic reactions when the firm "needs work."

While marketing is needed within all employment sectors, it is essential in the business sector. It provides the economic life blood. However, you may still be inclined, given your young age, to ask "So what has that got to do with me?" If you think marketing is that group of creative writers and graphics personnel and those individuals with out-going personalities over there on the other side of the building, than perhaps you just don't get it. As marketing consultant Henry Beckwith bluntly put it, "Marketing is not a department, it is your business."

MARKETING AND SELLING: DIFFERENT BUT RELATED

If you view, or want to view, marketing as part of your future or present "business," then let's drill down and learn more about it. Let's get more precise. What is "marketing?" Is "marketing" different than "sales?" Let's go to some credible sources for definitions of marketing. Kolter and Fox (1985) state that "Marketing is

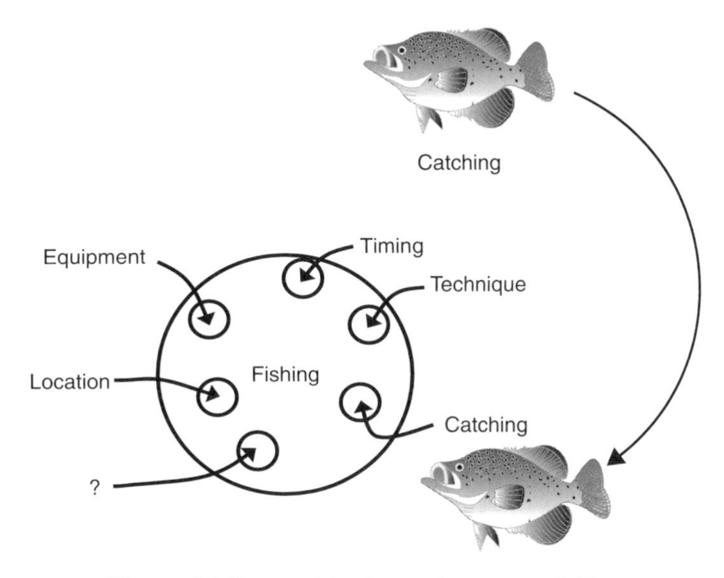

Figure 14.2 Catching is not the same as fishing.

the effective management by an institution of its exchange relations with its various markets and publics."According to Cronk, in Smallowitz and Molyneux (1987), "Marketing is creating the climate that will bring in future business." W. Coxe says "Marketing is to selling as fishing is to catching" (Smallowitz and Molyneux 1987).

Use of the word "institution" in the first definition suggests breadth of organizations that should have a marketing program which, as indicated earlier, includes businesses, government entities, educational institutions, and volunteer organizations. The second definition suggests a major, long-term effort, as in, we cannot create a climate overnight. The third definition begins to draw a sharp distinction between "marketing" and "selling." It suggests that we may be able to diminish selling; at least put it on the "back burner."

Prompted by the last definition, "marketing is to selling as fishing is to catching," which suggests a marked difference between marketing and selling, let's look further at that difference. As illustrated in Figure 14.2, catching is not the same as fishing. It is one small part of fishing.

Personal

My wife and I once lived across the street from a freshwater lake. About once a year, on a whim, one of us would say "let's catch some fish." Our fishing rods, with rusted hooks, were in the garage where we left them last year. We lacked bait, so we opened a can of kernel corn and headed out in our boat. Over the course of several years, we consistently failed to catch anything of significance.

However, others in our development were consistently successful. Why? They were "fisher-persons," they were into fishing. They had the right location, equipment, timing, and technique. We were into quick "catching," which was not successful, and they had invested in "fishing," which was successful.

So what do catching and fishing have to do with selling and marketing? The connection is illustrated in Figure 14.3. If we go out to "sell," we are not likely to be successful. We have not created context. Selling must be part of, and the result of, something much bigger and that something is marketing. Marketing is all those things we do to set the stage for selling just as fishing is all those things we do to set the stage – set the hook – for catching. An individual or organization must undertake a range of activities under the general umbrella of marketing to achieve sales, just as one must carry out many activities under the general umbrella of fishing in order to actually catch fish.

In other words, if an individual or organization attempts to simply sell without seeing selling in the context of other activities, the individual or organization is not likely to be successful. Management expert Drucker said, "The aim of marketing is to make selling superfluous" (Kolter and Fox 1985). This definition, like the one about marketing and fishing, reinforces the idea that selling is only one part of marketing and suggests that if marketing is done well, sales will almost occur naturally. Incidentally, some professionals use the term business development instead of marketing. That's fine, but you should see business development, like marketing, as being much bigger and more comprehensive than sales.

An effective marketing program is like a tasty cake because its ingredients are carefully selected and combined. And while sales are the frosting on the cake, they are supported by the cake, the cake's ingredients, and how they were combined (Galler, 2010). Get the cake's recipe right and then everything comes out right—including the frosting.

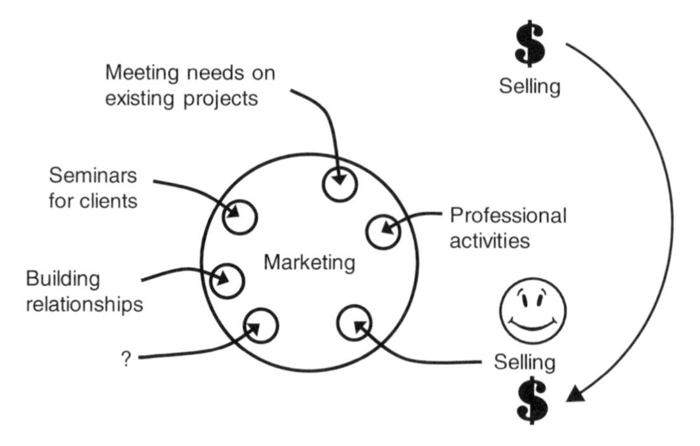

Figure 14.3 Selling is not the same as marketing.

Views of Others

Jay Conrad Levinson, marketing author and speaker, offers this broad and insightful definition of marketing: "Any contact your [organization] has with anyone who isn't part of your [organization]." Regarding earning trust and then possibly progressing to friendship, sports and celebrity marketer Mark McCormack says "All things being equal, people will do business with a friend. All things being unequal, people will still do business with a friend." Or, as stated by engineer, writer, and entrepreneur Richard Weingardt, "People do business with people they know and respect" and "competence gets firms into the game that relationships win." If you are student aspiring to achieving success and significance in engineering, you are starting to build those relationships now. Some of today's classmates could very well be your clients, owners, and customers tomorrow.

A SIMPLE, POWERFUL MARKETING MODEL

The Model

Marketing—because it is people-intensive—is complex. Individuals and organizations need a model to guide their approach. Marketing often leads to major setbacks. Individuals and organizations need a model to keep everything in perspective. In developing and using the marketing model presented here, I started with Stephen Covey's book *The 7 Habits of Highly Effective People* (1990), which I read in the early 90's soon after it was published. Covey discussed win-win interpersonal and inter-organizational relationships. I adapted his suggestions to marketing. Covey notes that the Greek philosophy, for what might now be called win/win interpersonal and inter-organizational relations, was based on ethos, pathos, and logos which are explained as follows:

- **Ethos:** Covey says "Ethos is your personal credibility, the faith people have in your integrity and competency. It's the trust that you inspire." Ethos is your ability to earn trust. "Contrary to what most people believe, trust is not some soft, illusive quality that you either have or don't have," according to motivational author Mac Anderson (2007), "rather trust is a pragmatic, tangible, actionable asset that you can create." In the ethos step, we are typically presented with a clean slate, a trust-neutral situation, that is, neither mistrust or trust. And we take it from there based on our words and actions.
- **Pathos:** Covey states that "Pathos is the empathic side—it's the feeling. It means you are in alignment with the emotional thrust of another's communication." Pathos is your connection with others to the point where they will share their wants and needs and you will understand them.
- **Logos:** "Logos is the logic, the reasoning part of the presentation," according to Covey. Logos means that you, as the offerer of professional services, and the client, owner, or customer in need of professional services, that is, the prospect, simply and openly take the logical action.

Covey emphasizes that these three elements of win/win interpersonal and inter-organizational relations must occur in the indicated order. That is, trust must be established first, then needs will be expressed and understood, and finally a logical follow-up occurs.

Applying the Model

The ethos-pathos-logos approach was originally presented almost 2000 years ago by Aristotle (Adler 1983). Nothing really new here—except the application to marketing. So, how does this become a marketing model? To begin to answer this question, refer to the sequence in the upper half of Figure 14.4. View ethos, pathos, and logos from your perspective as the offerer of professional services.

Think about what you are trying to do as you meet and interact with a potential client, owner, or customer.

- Under ethos, you want to inspire or earn trust. So you practice principled behavior.
- For pathos, you seek alignment with the other person's real wants and needs. So you ask questions and listen carefully and empathetically.
- Under logos, if there is a match between what he or she wants and/or needs and what you offer, you "sell."

Now, while looking at the lower half of Figure 14.4, view ethos, pathos, and logos from the perspective of the prospect. What is he or she looking for? What does he or she want?

- Under ethos, he or she is watching you. Can you be trusted?
- For pathos, if you can be trusted, he or she wants to explain wants and need — wants you to understand those wants and needs.
- Under logos, if there is a match between what he or she wants and what you have to offer, the potential client, owner, or customer wants to "buy."

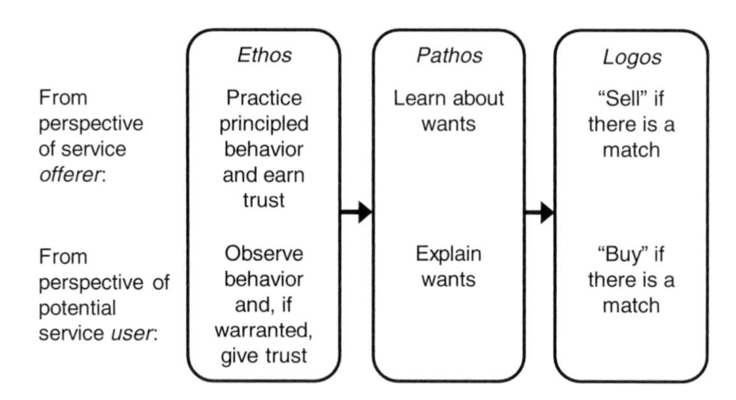

Figure 14.4 The ethos-pathos-logos steps in the marketing model may be viewed from the perspective of the service offerer and from the perspective of the potential service user.

Please appreciate that most prospects want, some desperately, to find service providers they can trust. They are increasingly faced with digging through voluminous data and information while seeking knowledge; weighing competing infrastructure needs; dealing with many vocal and varied stakeholders; satisfying demanding local, state, and federal regulations; and having too little money. Most clients, owners, and customers want a trusted advisor. Who do you want to deal with when involved in a major purchase like a condominium, a car, or a stock? Someone you can trust!

You might be thinking, this model may have potential. But, I am not comfortable with those 2000 year-old Greek words. Then describe and think about the model using six common words as follows:

- Ethos – Earn trust
- Pathos – Learn needs
- Logos – Close deal

Note that the earn trust-learn needs-close deal marketing model is a win-win process. Both parties freely participate, results are mutually-beneficial, and there is no pressure, no manipulation, no deceit, and no win-lose.

Return briefly to a topic noted in the scope section of this chapter. As stated there, most of the principles, the marketing model, and many of the tools and techniques set forth in this chapter are applicable to marketing efforts in the private, public, academic, and volunteer sectors. I reiterate that observation here to suggest that, while the preceding explanation of the marketing model assumes a developing relationship between a professional service provider and a prospect, the essence of the model could also be used by a wide range of individuals who see themselves as providing service whether they are in the private, public, academic, or volunteer sectors.

Caution: Respect the Order and Invest Time Wisely

Engineers and other technical professionals tend to be bright and analytic. We often "see" the "solution" to a problem immediately, or at least think we do. We are apt to quickly say something like "what you need is . . . " Resist that—earn trust first. Then learn wants and needs, at least as perceived by those you want to serve and want to be served. Then, and only then, offer your view as to what should be done.

Not only must trust be earned first, but it also takes, by far, the most time of the three steps as illustrated in Figure 14.5. This is the principal reason to take a rifle approach when selecting potential clients, owners, and customers—not a shot gun

Figure 14.5 Of the three steps in the marketing model, earning trust takes by far the most absolute and elapsed time.

approach. More often than not, you are going to have to work for months or years to earn trust. And investing that kind of absolute and elapsed time is worth the effort, because if we are trusted, our influence is enormous. In contrast, if we are not trusted, our influence is little or zero, regardless of our education, experience, position, or title (Tice 2010). The trust we earn is like a crystal vase. Takes a long time to craft, is highly valued, but, if shattered, impossible to restore. Of course, when using the suggested marketing model, time must be invested wisely. The next major section of this chapter describes marketing tools and techniques that can help you make the most effective and efficient use of your time.

Personal

In one extreme case, I invested eight years of low-level efforts to earn the trust of an engineering firm executive and, therefore, obtain what proved to be a large consulting assignment with a new client. In a few other situations, at the other end of the absolute and elapsed time spectrum, I was retained after two-hour meeting with strangers. Both of these situations are extreme and are intended to emphasize that significant absolute and elapsed time are normally required to earn trust.

MARKETING TECHNIQUES AND TOOLS

Many practical techniques and tools are available for implementing an individual or organizational marketing program. A dozen are described here. Working with some of these methods could enable you to contribute more to your current or future employer's marketing effort. Please look at this major section of the chapter as a smorgasbord of approaches you might peruse and selectively try in keeping with your circumstances.

Create a Personal Marketing Plan

Be proactive. Ask what you can do to enhance your organization's marketing effort? Ask to see your organization's marketing plan. How could you complement it? Think of ways you could contribute—ways you could use your education, experience, contacts, knowledge, skills, and positive attitude. For example, your organization's marketing plan calls for active membership in local or regional professional and business groups such as the American Society of Mechanical Engineers (ASME), the American Public Works Association (APWA), the Chamber of Commerce (C of C), and Rotary Club. Plan to be an active member of one appropriate organization.

Do this, not just because it is part of your organization's marketing plan, but because you want to be a contributing member of your community. While contributing to local professional and business organizations, you will have an excellent vantage point to see and hear about professional service opportunities for your organization. You, as a young or not so young, even novice marketer should get actively involved in one professional or one business organization.

Historic Note

John Dillinger, was a notorious American bank robber. As the story goes, they asked Dillinger why he robbed banks. His response: Because that's where the money is! Similarly, local professional and business organizations are where the opportunities are for your business, government, academic, or volunteer organization. Dillinger went where the money was. You need to go where the service leads are.

As part of your personal marketing plan, determine how you can learn even more about your organization with emphasis on services offered. Consider climbing out of your silo, looking outside of your box, and adopting a broader view of your job. Maybe you could walk over to the other side of your organization's offices and find out what they are doing.

You could offer to help with one or more elements of the marketing process. Maybe you could offer plan to write a portion of a SOQ or a proposal. Think about volunteering to accompany senior members of your organization who are attending an evening meeting at which they will present a proposal for professional services to a prospect. Perhaps you could assist at that meeting and learn more about your organization and marketing on the way to, at, and on the way back from the meeting. You will gain valuable insight into a process that some view as a mystery.

As part of your personal marketing plan that is aligned with your employer's marketing plan, begin to develop an "elevator speech," a "one liner," or a "tagline." So what is an "elevator speech?" It briefly, accurately, and convincingly describes your services with emphasis on the benefits to those you serve. Why "elevator?" Because we sometimes meet someone on, or as we get into an elevator, and they may ask "so what do you do?" Plan to be ready with a meaningful answer. An elevator speech is an important part of your personal marketing plan. Your elevator speech, and hopefully the elevator speeches of others in your organization, should stress the value or benefit you all provide. That is, focus less on what you "do" and more on how what you "do" might be valued by or benefit others.

Personal

While traveling, I met June Allen, who owned and operated a hair salon on Long Island, NY. As we discussed business approaches, I asked some questions and learned that she cares about her young women customers and the decisions they make. Her elevator speech: "I work on hair and change heads." That's an excellent elevator speech. What's yours?

Keep your elevator speech short and simple. Does the following text, written by scientists, sound like some of us? "The alkaline elements and vegetable fats in this product are blended in such a way as to secure the highest quality of saponification alone, with a specific gravity that keeps it on top of the water, relieving the bather of the trouble and annoyance of fishing around for it in the bottom during his ablutions." What is the message? The company's marketing people deciphered it to mean "It floats" (Brown 1999). Yes, the scientists were trying to describe an essential benefit of Ivory soap. Create an elevator speech about your organization that communicates one or more essential benefits.

Learn the Marketing Language

Your technical specialty has its terminology, so does marketing. Some terms frequently used in the consulting engineering business, and adaptable beyond, are defined below where they are listed in the approximate order they would arise or occur in the marketing process. Some of the terms are also defined in the preceding chapter and are repeated here for your convenience.

- **Prospect:** A qualified potential client, owner, or customer. That is, an organization or an individual in that organization that a professional services firm would like to serve based on the organization's or the individual's favorable profile as defined by characteristics such as the desire for assistance, reputation, stability, financial capacity, and willingness to communicate.
- **Lead:** A prospect's want or need that a professional services firm may be able to fill.
- **Request for Qualifications (RFQ):** The formal request from a prospect to a professional services firm inviting the latter to provide its Statement of Qualifications.
- **Statement of Qualifications (SOQ):** A document that presents the qualifications of a professional services firm emphasizing its experience on projects similar to that being considered by a client, owner, or customer. An SOQ typically does not address the manner in which the consultant, if selected, would approach the specific project. SOQs usually include basic information about the consulting firm such as its size; office location or locations; services offered; clients, owners, and customers served; references; experience with emphasis on projects similar to that about to be undertaken by organization receiving the SOQ; and resumes of selected professional staff which include descriptions of participation in relevant projects.
- **Request for Proposal (RFP):** The formal request from a prospect to a professional services firm inviting the latter to submit a proposal to provide services. The RFP typically includes items such as a letter of explanation and invitation; a description of the project; an explanation of the required scope of services (e.g., feasibility study, preliminary engineering, preparation of plans and specifications, construction management, start up, education and training); a project schedule; and the due date for the proposal. The RFP may also include items such as a list of available related reports, studies, and investigations; a description of data and information available from or known by the prospect; the name of a contact person; an indication of whether or not the proposers should provide an estimate of the cost of services; and a description of Minority

Business Enterprise (MBE), Women's Business Enterprise (WBE), and Disadvantaged Business Enterprise (DBE) requirements.

- **Proposal:** A document prepared by a professional services firm, in response to an RFP, and hopefully with a clear understanding of the project requirements. It describes what the firm will do, how they will do it, how long it will take, and, possibly, what it will cost. In a sense, the project is worked out "on paper" as part of the proposal preparation process.
- **Organization chart or "Org" chart:** A diagram, often included in a consultant's proposal. It typically shows key members of the prospect's organization; key members of the professional services team, including any subconsultants; the roles of all individuals; and the manner in which they will interact.
- **Price-Based Selection (PBS):** A client, owner, or customer selects a professional service firm solely or mostly on the basis of price (fees plus expenses).
- **Qualifications-based selection (QBS):** A prospect, after evaluating and short-listing firms based on their qualifications, selects the top-ranked firm for price negotiation based on a detailed project scope. If an agreement cannot be reached, which is uncommon, negotiations begin with the second most qualified firm and so on (Chinowsky and Kingsley 2009).
- **Two-envelope method:** The process by which an organization requiring consulting services invites two or more professional service firms to submit proposals. Each firm is asked to provide their proposal in two envelopes. One envelope is to contain their proposal, without their cost. The other envelope should include the cost of the service, that is, estimated fee plus expenses. On receiving the envelope pairs from each interested consulting firm, the organization seeking a consultant ostensibly opens just the proposal envelopes, evaluates the firms based on the proposals, and prioritizes the firms. Then the second envelope for the top-ranked firm is opened and, if the cost is reasonable, the prospect and the firm negotiate. If they are not successful, then the client, owner, or customer opens the second envelope for the second ranked firm and so on.
- **Go/No Go:** One of more decisions, made by a professional services firm, during the consultant selection process. At critical milestones, the firm weighs the pros and cons of proceeding further and makes a decision. Typical Go/No Go milestones, in the context of the consultant selection process discussed in the previous chapter and illustrated in Figure 13.3, are receiving an RFQ, RFP, or invitation to interview and during contract negotiations. The professional services firm may also have a Go/No Go discussion upon learning that what they thought was to be QBS will instead be PBS.
- **Platinum Rule:** This "rule" says do unto others as they would have done unto them. It contrasts with the traditional Golden Rule which is do unto others as we would have done unto us. The value of the Platinum Rule in marketing and in managing projects is that is focuses the service provider on those being served.

Schedule Marketing Tasks

So you have a personal marketing plan. Now you need to make it happen; your plan must be implemented. Notice that creating a marketing plan, the first of the dozen suggestions, is separate from this suggestion, acting on it. Your well-intentioned marketing efforts will compete with billable project work and/or other job responsibilities. Therefore, you need the self-discipline to schedule marketing tasks. Select your favorites such as calling a colleague, having lunch with a contact, attending a business or professional society meeting, writing an article, or volunteering to assist with a proposal. Schedule them and do them! How about, committing to doing one marketing task per day or two per week?

Recall the message of Figure 14.1, which is to market in a continuous, proactive manner and not sporadically when work is needed. Hopefully your organization, whether public, private, academic, or volunteer, thinks this way. Even if they don't, you can! Instead of slipping into a reactive mode of seeking sales when you or your organization needs work, try to take at least one specific marketing action every day. To aid you in being proactive, consider the following questions you might ask yourself or others to keep you tuned into marketing—to enable you to do some marketing every day:

- What is the next potential project with this client, owner, or customer? Be sure to look and listen both within and outside of your specialty area. You don't have to be an expert in a technical area to recognize a want or need and, therefore, a potential project.
- What other existing or potential clients, owners, or customers might be interested in the service we are providing to a particular client, owner, or customer? Who else might value this service?
- How might your organization leverage, re-package, or combine existing services to add value and meet evolving or new wants and needs of those you serve or want to serve?
- How could your organization "move up the food chain," that is, become more of a strategic advisor? The goal is to increasingly help those you serve decide what to do next and, of course, continue to help them do those things that you do well.
- What current wants and needs will decline and what new wants and needs will replace them? Answering this question requires a "crystal ball"—it's a tough one. Are you or your organization on top of it? You better be—change is inevitable. Accordingly, in addition to finding out what those you serve want now or need now, you should ask them what they think they will want or need five years from now.

For additional advice about questions, see the later section of this marketing techniques and tools section titled, "Ask-Ask-Ask: The Power of Questions."

Find Common Ground

Starting a relationship with a stranger—a potential client, owner, or customer—a potential friend—requires finding some common ground. You need to "connect," to develop rapport. This is not hard to do because each of us is connected, in some way, to any of us. You do not have to be an extrovert to "connect."

Common, possible points of connection include education, family, sports, and travel. There are many more such as art, aspirations, the economy, boating, investing, music, philosophy, real estate, politics, and religion. Yes, I know, that some say we are "not supposed to discuss politics and religion." However, certain people enjoy intelligent discussion of these topics. If you do, be open to finding others that share your preferences. Typically, these connections have nothing to do with engineering or whatever you do professionally. However, a little connection—a small piece of common ground—could become a part of the foundation of a relationship. Contemplate the following relevant advice, which is adapted from a statement by Howard Schultz the Starbucks founder: "We aren't in the engineering business serving people. We are in the people business serving engineering."

Personal

Consider the importance of the education connection. I recall being one member of a team making a presentation to a prospect's selection committee. The setting was informal—we sat around a table in the potential client's conference room. During a break, I pointed to the large proposal document we had provided and asked the client representative next to me what aspect of it was most important. He said something like "I go to the resumes of the proposed project team members to see where they went to college." Logical? Maybe not. Human? Yes. Where one went to school and what one studied is a common connection. If you are visiting with someone in their office, look for diplomas, certificates, mascots, and pennants.

You may think you need to do something "big" to connect, to find or create common ground. Perhaps arrange 18 holes of golf or an evening with spouses or significant others at the theater. Instead, consider little connecting acts that are directly tailored to the specific preferences of the other person. Little things mean a lot when the little things indicate that you have listened to what the other person said, that is, you're aware of some of their likes and dislikes and interests. Two actual examples:

- A consultant learns that a prospect enjoys early morning breakfast discussions. So does the consultant. They have been doing this for years and "take turns" paying. Note that both enjoy the early morning discussions—nothing superficial here.
- A consultant, who spends time in Florida, learns that a client, owner, or customer has an interest in Florida real estate. The consultant, while in Florida, visits realtors, the Chamber of Commerce, and other sources to gather information about the area, in general, and real estate, in particular. The consultant "snail mails" a large packet of material to the individual.

Neither of the above actions are "big ticket" items. Neither requires any compromise or sacrifice. The described actions are simply "little things" that "mean a lot." Such little actions indicate that you listen, care, act, and connect and you value the relationship on a person-to-person basis and not just "as business."

Earn Trust

No matter what else you do, or do not do, in the marketing arena, always place a premium on earning trust. Recall that "earn trust" is the first step in the three-step earn trust, learn needs, and close deal marketing model described earlier in this chapter. Earning trust is the foundation of the suggested model. The trust discussion associated with the marketing model stressed three points. First, the marketing process begins with earning trust. Second, trust is, to quote Mac Anderson again, "a pragmatic, tangible, actionable asset." And, finally, earning trust takes by far the most time as illustrated in the previously-introduced Figure 14.5.

So how do we "earn trust?" We have many choices and some are simple and require little effort. In fact, big trust is earned mostly by lots of small acts. And you, the novice or new comer to marketing, can unilaterally do some of them. Consider these "earn trust" examples that you can practice:

- Determine the preferred communication mode of your current or potential clients, owners, or customers. Don't assume your preferred mode is theirs.
- Share articles, emails, websites, books, and other materials with your contacts based on what you have learned about their interests.
- Compliment them on their achievements.
- Ask for their assistance because it reflects your knowledge of their competences and your confidence in them.
- Co-author and/or co-present articles and papers with existing and/or potential clients, owners, or customers.
- Admit mistakes, without excuses, and suggest remedial actions. "Admit your errors before someone else exaggerates them," is the advice of physician Andrew Mason.
- Pass tests. You will hit bumps and sometimes have accidents on the road to earning trust—view these as opportunities to demonstrate your honesty and integrity. Consider this scenario: You are earning the trust of a prospect. Finally, they have a project and ask if your organization can do it. You could cobble something together and probably complete the project. However, it is not one of your organization's strong suits. What do you do? Take on the project? Or refer them to someone else because that would be in their best interest? What's more important to you and your organization, a project or an evolving connection, a short-term financial gain or a long-term, mutually-trustful relationship?

To summarize, the first step in the suggested marketing model, earning trust, is the most challenging and, as just suggested, there are many ways to earn that trust.

Ask-Ask-Ask: The Power of Questions

The most effective "marketers" habitually ask lots of questions. These successful professionals would tend to agree with management guru Peter Drucker who said: "My greatest strength as a consultant is to be ignorant and ask a few questions." Ask questions to learn about a potential project. Also ask questions to learn more about the organization's representative. It is one way you will find common ground.

Each of your organization's existing or potential clients, owners, and customers views you and your services in a unique manner. As explained by author and management consultant Tom Peters: "[Clients] perceive service in their own unique, idiosyncratic, emotional, irrational, end-of-the-day, and totally human terms." And how can you determine any given client's, owner's, or customer's "unique, idiosyncratic, emotional, irrational, and end-of-the-day" view? Ask, listen, and ask some more!

Front End the Asking

Front end your question asking, that is, concentrate your questioning in the early part of a potential relationship or project as illustrated in Figure 14.6. The horizontal axis depicts marketing steps from prospect and lead identification through the contracting, that is, when a contract project begins. Again, the emphasis is on asking and listening, relative to telling, especially early on.

If you like to "tell," wait until later. But, you may be thinking, I like to share my ideas. I want to show the prospect how much I know and how much our organization can do. Control yourself. You can do your telling later. Heed the advice of commentator Andy Rooney who said: "In a conversation, keep in mind that you're more interested in what you have to say than anyone else."

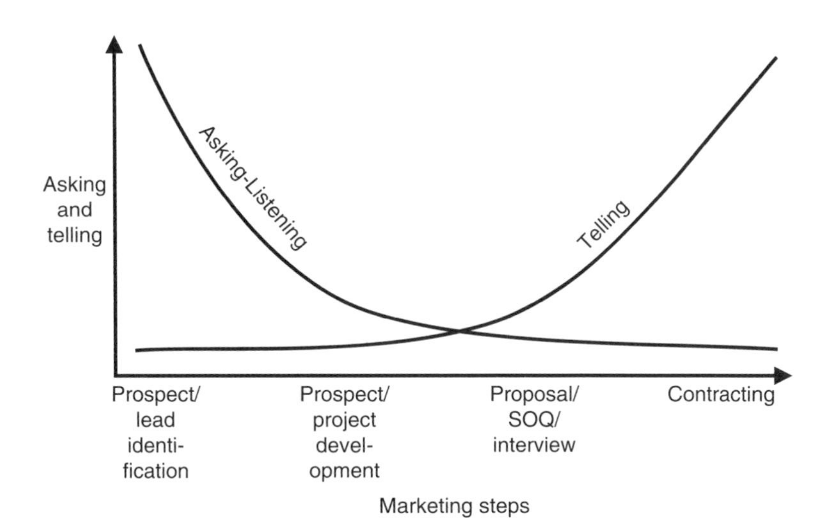

Figure 14.6 Front end the asking and listening.

Listening is hard, at least for some of us. Steven Berglas (2009), in an article titled "Learning the Art of Listening," claims that entrepreneurs, in particular, have difficulty really listening. Reasons include:

- I've "been there, done that" and know what you want or need.
- I'm the smartest person in the room. Enough of your ideas on the problem and its solution.
- I'm a doer, not a thinker. Let's get on with it.

If that describes you, slow down—slow way down!

The Power of Questions

Question asking is a powerful habit because it almost always provides the asker, and often the answerer, with many benefits. Having referred to "power," please recognize that I am not advocating using questions to misrepresent, manipulate, or exert pressure. Instead, I believe that questions enable you to clearly define wants and needs so that you can ultimately fulfill them. With that in mind, consider five "powers" of questions (adapted from Leeds 2000):

- **Questions create an obligation to respond:** Remember, in class, when a teacher or professor asked you a question and then was silent? Most people's natural inclination is to fill that silence with something and, therefore, we do our best to answer the question. Silence is uncomfortable when a question is asked. This is what I mean by saying that asking a question creates an obligation to respond. I am not suggesting that we try to cause discomfort but am simply noting a natural tendency.
- **Questions stimulate the thinking of both the asker and answerer:** Preparing and asking questions causes you to think more deeply and broadly about the organization that may use your services. Similarly, their thinking is enhanced as they respond to your questions.
- **Questions provide valuable data, information, and knowledge:** Recognize that while your questions reveal your concern for the other party and reflect your expertise, those questions also provide you with the data, information, and knowledge needed to serve that prospect, if and when you get the opportunity.
- **Questions put the asker "in the driver's seat:"** You can use questions to direct a conversation in a direction that could be potentially useful to you and the client, owner, or customer you hope to serve.
- **Questions enable people to persuade themselves:** Thoughtful questions and the thinking they stimulate tend to define issues and move all parties toward resolution. As they say, "a problem well defined is half solved." More specifically, your questions can help you and the prospect ascertain values, discover and/or elaborate on issues, reveal wants, identify needs, define milestones, formulate conceptual options, and select a course of action.

Confronting Barriers to Asking Questions

Having considered the preceding case for asking lots of questions, you may be reluctant to do so. Your reluctance may be caused by three barriers. The first is a reluctance to question authority, that is, authority figures. Recognize that most of us are authorities, just on different things. The person you interact with, who represents a potential client, owner, or customer, is an authority on something or some things. Similarly, you as a professional, are an authority. By asking questions, you are not questioning the authority of the other person. Instead, you are reflecting yours.

The second barrier is fear of appearing uninformed or poorly prepared. Consider this scenario: Tomorrow morning you awake with a pain in your chest and are rushed to the emergency room at the local hospital. The emergency room doctor asks: "What's wrong?" You answer, "chest pain," and the doctor says, "we are immediately performing triple by-pass heart surgery." The pain aside, how would you feel? Might you want the doctor to ask more questions as part of a careful diagnosis of your problem before deciding how to solve the problem? My point: Asking questions does not indicate you are uninformed or poorly prepared. It should mean just the opposite, that is, because if you are well informed, you know what to ask. The type and number of questions you ask reveal your expertise.

Personal

In preparation for an interview regarding a potential education and training project, I prepared a list of 24 questions for the prospects with the hope that I would be able to ask some of them. On arriving at their office and meeting with them, I discovered that they were not well prepared in that their leader said something like "So what would you like to know?" I asked, and they answered, essentially all of the 24 questions and they immediately retained me for their project. I suspect that my questions suggested that I was informed and prepared.

The third barrier is concern with appearing rude. Someone said "I don't care how much you know, until I know how much you care." You demonstrate civility, good manners, and care by preparing and asking thoughtful, probing questions. Of course, the questions are asked in a polite, sensitive manner.

Assume question asking, as part of your marketing contribution, makes you uncomfortable, for one or more of the reasons described here or for other reasons. Then ask the other person or persons for permission. For example, you might say "may I ask you some questions about this topic?" I have yet to have anyone say no.

Mix Question Types and Drill Down

Mixing questions and drilling down during question asking are discussed in detail in Chapter 7 in the section titled "Strive to Understand Client, Owner, and Customer Wants and Need." You are urged to review, or read for the first time, the rationale presented there for mixing closed and open-ended questions. Also note the questioning tactics that are described, namely "5 Whys" and Kipling's six honest servers, and also appreciate the need to distinguish between wants and needs.

Talk to Strangers

When we were children, our parents probably advised us not to talk to strangers. However, we are no longer children, presumably can take care of ourselves, and, therefore, may be open to this marketing advice: Talk to strangers. In her book, *How to Work a Room*, consultant Susan Roane (1988) challenges readers to "work the world." She urges us to adopt the philosophy that we are surrounded by opportunities to make contacts, ask questions, and learn. But we often have to take the initiative, whether we are at our place of work, doing personal errands in our community, sitting in an airport between flights, or attending a conference.

Will talking to strangers always provide useful information or a new contact? Roane says: "That's not the point. The point is to extend yourself to people, be open to whatever comes your way, and have a good time in the process. One never knows... The rewards go to the risk-takers, those who are willing to put their egos on the line and reach out—to other people and to a richer and fuller life for themselves." As hockey great Wayne Gretsky said, "You miss 100 percent of the shots you never take."

Personal

I obtained a rewarding, multi-year consulting arrangement as a result of approaching a speaker after his presentation at a conference. I complimented him on his presentation, mentioned that we seemed to share technical interests, and, as they say, "one thing led to another." Some of my other positive results of talking to strangers are learning about a useful personal time management tool and being appointed to a national committee. I tend toward introversion, as do the majority of engineers. Nevertheless, I've made the effort to approach strangers in a variety of professional and other settings and, as a result, have enjoyed many positive results. If so inclined, you can do the same

Stress Benefits, Not Features

Recall the earlier discussion of elevator speeches where the emphasis was on the benefits of services. That idea extends well beyond elevator speeches. We engineers and other technical professionals tend to get wrapped up in the processes and tools we use. As important as processes and tools are, think of the other party, the prospect. Stress benefits, not features, outputs, not inputs, unless the person or persons you are communicating with explicitly expresses interest in features. Anticipate the "so what?" test. Consider two examples:

- **Feature:** Our groundwater model includes both confined and unconfined aquifers. **So what?** So we can cost-effectively determine the impact of new wells on exiting wells in highly-complex situations.

- **Feature:** We have 26 offices in five countries and we offer 11 specialty areas. **So what?** So we can deliver services locally while drawing expertise globally when needed.

> **Personal**
>
> Early in my career, one of my responsibilities was marketing water resources engineering services. Our firm made heavy and effective use of digital computer hydrologic-hydraulic models. Whenever I had the opportunity, I would talk to potential clients about how our computer models worked. This was not effective, as indicated, in part, by the way their eyes glazed over. I gradually learned, based on experience and advice from a marketing professional, that instead of stressing how the models worked, I should stress the benefits the models provided. I focused on benefits and did it often enough that describing benefits became habitual and my marketing effectiveness improved. While I no longer market modeling services, I habitually focus on benefits, not features, whenever I have the opportunity to discuss my services.

Focus on Existing Clients, Owners, and Customers

While we need to continuously seek new organizations to serve, our emphasis should be on serving existing clients, owners, and customers. As a young professional, you are, as illustrated in Figure 14.7, which is based on an idea from Wahby (1993), often there—under that client tree—usually not alone, but nevertheless there.

Figure 14.7 If a project is an apple, a client is an apple tree.

The apple is your project, the project you are working on. Take care of it. Every now and then look up. If you help your project team do quality work, that is, meet project requirements—deliverables, schedule, budget, and other expectations—you will earn the right to pick additional apples. There are other orchards and many other trees, as suggested by the grove on the left side of Figure 14.7. However, first take care of the tree right above your head.

Why do I suggest concentrating on existing clients, owners, and customers? Because the cost of earning and securing a new project with an existing client, owner, or customer is probably one-tenth the cost of obtaining the first project with a new client, owner, or customer. All the more reason to carefully care for those individual and organizations you are now serving. The relatively low cost of that care goes to the bottom line.

Another way of explaining the suggestion to focus on existing clients, owners, and customers is to emphasize what might be called "farmer" marketing—taking care of those you are already serving. That is, cultivating relationships, exploring new ways to nourish them, harvesting additional projects, and reaping referrals from current clients, owners, and customers to potential new ones. Clearly, we also need to invest some resources in developing new prospects. This is what might be called "hunter" marketing and it requires loving the thrill of the hunt (Hewlett-Packard 1995). Each organization should search for the optimum mix of farmer and hunter marketing with major emphasis on the former.

Help to Establish Multiple-Level Links

The reason for creating multiple links between your organization and the organizations you serve, as illustrated in Figure 14.8, is that you want to be well connected to them. The relationship between your and their organization should not be dependent on only one link. That is too fragile. Contact personnel in your or the client, owner, or

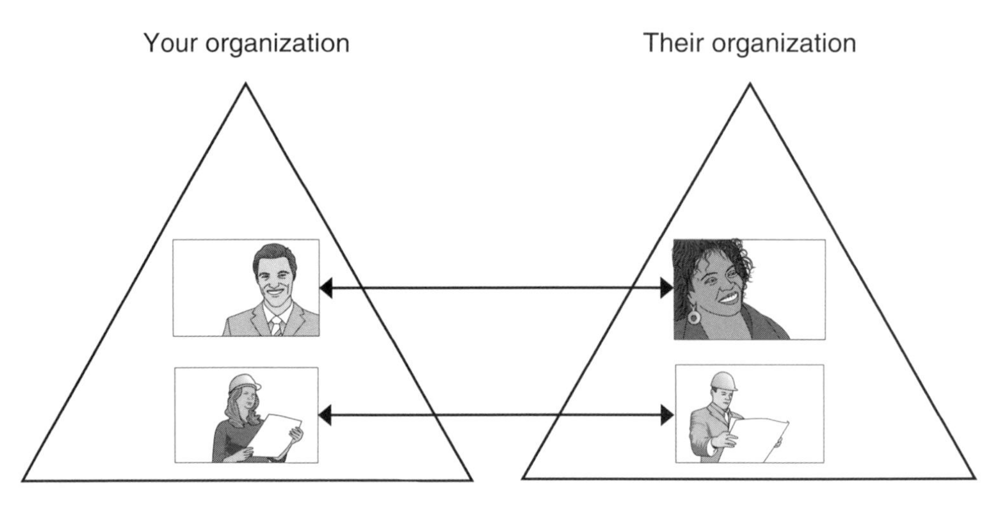

Figure 14.8 Help to create multiple links between your organization and those you serve or want to serve.

customer organization may retire, take on different functions, or move on to new employers. Seek multiple links at two or more levels.

You, as a young practitioner, can help with this link-building process. For example, the top link in Figure 14.8 may be composed of your "boss," on the left, and his/her counterpart, on the right. That could be you, on the bottom left of Figure 14.8, and your counterpart, on the lower right. With time, the two of you will move up together into decision-making positions. One or both of you are likely to "move up" in current and/or other organizations. Multiple-level links help you and your organization become even more aware of the wants and needs of those you serve and of prospects and, as a result, enable your organization to secure new projects. Links also assist in executing projects.

Personal

Consider an example of moving up together in careers. One person has used my services for over 25 years while he has been employed in three different organizations and I have been employed in two organizations and then after I established my own independent consulting business. He and I have grown professionally and moved up together.

Proactively Establish the Next Step

Assume that you, in your marketing role, have applied some of the tools and techniques discussed here. Furthermore, you've "connected" with your counterpart at the prospect organization. He or she has told you about a want or need for a future project or you suspect there is a need that your prospect isn't yet aware of. Now what? Stay in the driver's seat—nail the follow-up! Don't leave him or her in control of future communication. Instead, be assertive and indicate what you are prepared to do next to be helpful. In other words, avoid the types of neutral or reactive comments shown in the left column of Table 14.2. Instead, proactively arrange follow-ups using statements like those in the right column. Another way of saying this is to recognize that polite persistence prevails (Galler 2011) and speak accordingly. Of course, if you sense or hear push back, then adjust your approach accordingly.

Table 14.2 Proactively establish the follow-up.

No	Yes
"Please call if the material interests you"	"I will call next week to get your response"
"Let me know if you want more information"	"Now that I know more about your needs, I will send you . . . "
"Hopefully we can get together again"	"Could we meet next Wednesday when I will be back in town?"

(Source: Adapted from Weiss 2003)

Selectively Share Data, Information, and Knowledge

Most of the preceding marketing tools and techniques, any or all of which you can use, result in acquiring data, information, and knowledge about existing and potential clients, owners, and customers and their current and future wants and needs. This is good! Now what? Habitually document what you learn as you learn it. That is, unless you have an extremely good memory, always write it down. Even if you do have an unusually-effective memory, the written word is easier to share. When meeting with someone, take skeleton notes. Ask permission, if that makes you feel more comfortable. I almost always do so, that is, habitually ask permission and take rough notes.

Immediately after that discussion, elaborate on—fill in—your skeleton notes. You might do this in your car in the parking lot, at a coffee shop on your way back to the office, or on the plane. Do it before you return to the office while the discussion is fresh in your mind. Furthermore, by the time you get back to the office, other activities and tasks will demand your attention.

Then leverage what you have learned and documented by selectively sharing it. Become known as a person who, as a matter of habit, carefully shares data, information, and knowledge about prospects and possible projects. Others will reciprocate. This "document and share" habit applies across the board. It should be practiced by seller-doers, client champions, and members of management. Share laterally, down, and up. When in doubt, share—subject, of course, to not sharing what was given in confidence. Why do I stress documenting and sharing? Because it works, even though it is often not done. For example, in 1998, O'Dell and Grayson published the book *If Only We Knew What We Know*. Their message: Most organizations, to their detriment, do a poor job of sharing data, information, and knowledge.

WHAT WORKS AND WHAT DOESN'T WORK

You will observe individuals in the business, government, academic, and volunteer sectors, who are effective in the marketing arena—and some who aren't. Learn from all of them. Reflect on what you see and hear. Table 14.3 summarizes my "what works and what doesn't work" lessons learned.

Personal

My office in an engineering firm was next to that of a superb, full-time marketer—and that was an education. For example, he introduced me to the tactic of stressing benefits, not features as discussed in this chapter. Another competent marketing professional helped me appreciate the value of asking lots of questions and carefully listening to the answers, including doing so empathetically.

Table 14.3 Experience reveals "what works" and "what doesn't work" in marketing.

What Works	What Doesn't Work
Listening—to earn trust and learn needs	Talking—about what we do
Building relationships	Pursuing projects
Asking questions	Pontificating
Researching, qualifying, and ranking prospects	Viewing all prospects as being the "same"
Active involvement in targeted professional/business organizations	Passive membership in randomly-selected professional/business organizations
Keeping current—technically and otherwise	Maintaining status quo
Getting leads and following up	Getting leads and expecting others to follow-up
"Face time"	Mass mailings
What you see is what you get	Bait and switch
Illustrating benefits	Pushing features
Multiple-level contacts with organizations that being served	Single-level contact
Written materials featuring white space, graphics, photographs, color, variety	Lots of words
Suggesting program and project approaches	Reacting to RFPs
Client, owner, and customer-oriented project descriptions, SOQs, proposals, and interviews	Consultant-oriented project descriptions, SOQs, proposals, and interviews
Tailoring to clients, owners, and customers	Boilerplating from files
Defining and meeting requirements	Talking "quality" and spewing slogans
Preparing project plans and sharing with those being served	"Winging it"
Delivering locally while drawing globally	Attempting to do it all locally
Admitting errors and fixing them	Blaming others
Asking clients, owners, and customers how they want to communicate	Using firm's preferred mode of communication
Delivering draft deliverables to clients, owners, and customers throughout the project	Dumping deliverables on clients, owners, and customers at the end of the project
Caring for existing clients, owners, and customers by performing on their projects	Neglecting existing clients, owners, and customers while "chasing" new ones
"Rifle"	"Shotgun"
Making promises and delivering	Breaking promises and offering excuses
Persistence	Instant success
Saying "thank you"	—

MARKETING CONCLUDING COMMENTS

Be open to the positive perspective on marketing presented in this chapter and, if you are able to do that, then proactively participate for your and your organization's benefit. Simply stated, marketing means mutually-beneficial exchanges of wants/needs and services or products.

Recognize that marketing is much more than selling and is costly and, therefore, should be a carefully managed process that involves analysis, planning, implementation, and control; it is not a collection of random actions started when "we need the work." Adopt a marketing model that works, perhaps the one described in this chapter. As part of your marketing efforts, select from and apply marketing tools and techniques like those described here.

The game of business is very much like the game of tennis.
Those who fail to master the basics of serving well, usually lose.

(*Anonymous*)

CITED SOURCES

Adler, M. 1983. *How to Speak, How to Listen*. MacMillan Publishing Company: New York, NY.

Anderson, M. 2007. *You Can't Send a Duck To Eagle School: And Other Simple Truths of Leadership*. Simple Truths: Naperville, IL.

Berglas, S. 2009. "Learning the Art of Listening." Head Coach column, Forbes.com, July 14.

Brown, S. 1999. *How To Talk So People Will Listen*. Baker Book House: Grand Rapids, MI.

Chinowsky, P. S. and G. A. Kingsley. 2009. *An Analysis of Issues Pertaining to Qualifications-Based Selection*, American Council of Engineering Companies (ACEC) and American Public Works Association (APWA), Washington, DC. and Kansas City, MO.

Covey, S. R. 1990. *The 7 Habits of Highly-Effective People*. Simon & Schuster: New York, NY.

Galler, L. 2010. "Marketing & Baking—The Same But Different." *The Times*, Northwest Indiana, November 14.

Galler, L. 2011. "Polite Persistence Prevails for Salesman." *The Times*, Northwest Indiana, May 1.

Heightchew, Jr., R. E. 1999. "Selling With Integrity to Develop Your Engineering Practice." *Journal of Management in Engineering—ASCE*, May/June, pp. 47–51.

Hewlett-Packard. 1995. "The Baron Group: Telling Selling Secrets." *Hewlett-Packard Resolution*, Fall, pp. 6–7.

Kolter, P. and K. F.A. Fox. 1985. *Strategic Marketing for Educational Institutions*. Prentice-Hall, Inc.: Englewood Cliffs, NJ.

Leeds, D. 2000. *The 7 Powers of Questions: Secrets to Successful Communication in Life and Work*. Berkley Publishing: New York, NY.

O'Dell, C. and C. J. Grayson, Jr. 1998. *If Only We Knew What We Know*. The Free Press: New York, NY.

Roane, S. 1988. *How to Work a Room: A Guide to Successfully Managing the Mingling*. Shapolsky Publishers: New York, NY.

Smallowitz, H. and D. Molyneux. 1987. "Engineering a Marketing Plan." *Civil Engineering—ASCE*, August, pp. 70–72.

Snyder, J. 1993. *Marketing Strategies for Engineers*. American Society of Civil Engineers: New York, NY.

Tice, L. 2010. "Are You Trustworthy?," *Winner's Circle Network*, The Pacific Institute (www.thepacificinstitute.com), December 6.

Wahby, D. 1993. "Managing the A/E Firm in Turbulent Times." *Journal of Management in Engineering—ASCE*, April, pp. 122–124.

Walesh, S. G. 2000. *Engineering Your Future: The Non-Technical Side of Professional Practice in Engineering and Other Technical Fields-Second Edition*. ASCE Press: Reston, VA.

Weiss, A. 2003. *Great Consulting Challenges and How to Surmount Them*. Jossey-Bass Pfeiffer: San Francisco, CA.

ANNOTATED BIBLIOGRAPHY

Frederickson, D. 2011. "Corporate Branding From the Inside Out." *CE NEWS*, April, p. 24. (Offers an informative introduction to branding stressing "the importance of starting at the top and working it into the culture until it becomes part of the everyday business.")

Mandino, O. 1968. *The Greatest Salesman in the World*. Bantam Books: New York, NY. (This story describes the reading and use of ten leather scrolls each of which contains "a principle, a law, or a fundamental truth written in a unique style to help the reader understand its meaning. To become a master in the art of sales one must learn and practice the secret of each scroll." Mandino's book uses the word sales in the way this book uses the word marketing.)

Martin, N. 2008. *Habit: The 95% of Behavior Marketers Ignore*. Pearson Education: Upper Saddle River, NJ. (Argues that habit, that is, our unconscious minds, control up to 95 percent of our behavior. If accepted, this observation has implications for everything we do, or more specifically, want to do better, including marketing.)

Jennings, O. R. 2008. "Building 3D Client Relationships: Becoming Indispensable to Your Clients." *PE*, June, p. 16. (Urges service providers to move past project-level relationships to proactive, personal relationships. Stated differently, the author says that technical professionals should strive to be client-relationship managers.)

Walesh, S. G. 2004. "The Chimney Sweep and the Sewer Cleaner: The Importance of Style." Lesson 44 in *Managing and Leading: 52 Lessons Learned for Engineers*. ASCE Press: Reston, VA. (Engineers are advised to "Exude enthusiasm, be polite, listen carefully, speak clearly, explain thoughtfully, assist positively, dress appropriately, walk tall, and smile.")

Weingardt, R. 1998. *Forks in the Road: Impacting the World Around Us*. Palamar Publishing: Denver, CO. (In Chapter 10, "10 Commandments of Marketing," the successful engineer-author offers useful marketing guidelines for engineers.)

EXERCISES

14.1 MARKETING ELEMENT OF THE BUSINESS PLAN FOR A NEW CONSULTING BUSINESS: This exercise will help you, or you and others if a team exercise, apply some of the marketing fundamentals described in this chapter. Suggested tasks are:

A. Review the scenario described in Chapter 10, Exercise 10.2, Task A. Exercise 14.1 builds on that income statement exercise, the Chapter 10 finance exercise (Exercise 10.3), and the Chapter 11 legal form of business ownership exercise (Exercise 11.6) as part of the overall effort to develop a business plan for your new consulting business. The business plan will now be completed by preparing the marketing element.

B. Develop your marketing strategy and tactics for the period beginning now, assuming it is about October 1, and extending through the first calendar year of your new consulting business. Clearly, you are faced with many unknowns and parts of your marketing plan are bound to require revision. However, even if you markedly revise your plan later, the planning process will cause you and the other two principals, and others you may involve, to synergistically and collaboratively think of many and varied marketing facets the most important of which can be addressed proactively now, rather than reactively later.

Referring again to strategy and tactics, the former means overarching principles such as selecting your area or areas of specialization; identifying the kinds of clients, owners, and customers you want to serve such as only public, private, or both; defining quality; deciding if you will participate in PBS; and determining your initial geographic service area. Tactics, in contrast, encompass specific actions and processes examples of which are developing a web page, contacting specific individuals as references or potential clients, joining and actively participating in certain professional and business organizations, and attending selected conferences. Strategies tend to be firm over extended periods, such as a year or more, contrasted with tactics which are more ephemeral in response to changing conditions.

Be sure to address marketing tactics that you can and cannot ethically undertake during the time one or more of the three principals are still with their current employer. Also, determine if any of you are legally bound by employment contracts that may limit your participation in the new business.

CHAPTER 15

THE FUTURE AND YOU

Chance favors only the prepared mind.

(*Louis Pasteur, French chemist and microbiologist*)

This forward-looking chapter begins by noting that there are only two futures for individuals and organizations. The one the individuals and organizations create for themselves or, in the vacuum of inaction, weak action, or reaction, the one others create for them. Then the chapter offers various views of the 21st Century world in which engineers and other technical professionals will work – the stage on which they will play their roles. The traditional people-serving role of engineers and other technical professionals will not change. However, success on that world stage will require, in addition to traditional qualities, personal and group qualities such as adaptability, collaboration, creativity, empathy, entrepreneurship, innovation, synthesis, and visualization. The chapter concludes with a major section that suggests principles and offers advice to enable you and others to lead the changes needed to achieve individual and organizational vitality in the 21st Century.

WHAT DOES THE FUTURE HOLD?

What does the future hold for students of engineering and other technical professions and for young practitioners? More specifically, what does the future hold for you? To what extent can you, and other aspiring professionals, create your future, both personally and professionally? My answer: You can determine your future to the extent that it does not have to be something that happens to you and you can do so, in part, by drawing on the professional knowledge, skills, and attitudes (KSAs) discussed in the preceding chapters of this book and supplemented in this chapter.

There are only two futures for you, the one you proactively create for yourself or, in the vacuum of your inaction, weak action, or reaction, the one other's create for you.

Perhaps that is a bit strong because your future will be influenced by external conditions you cannot control. But you can adopt strategies and implement tactics that will largely determine your professional and personal future. You can engineer your future.

The same "only two futures" idea applies to your organization, whether it be the university or college at which you are student, your employer or professional society, or, some day, your own business. Of course, creating the future of an organization is much more challenging than creating your future.

In keeping with this book's managing and leading theme, this last chapter will further help you engineer your future, and that of organizations of which you are a member, and do it in the following two steps:

- First, the chapter shares views of changes likely to occur around the globe which will influence the work of engineers and other technical professionals. You should understand the changing global stage on which you will play out your career and the audience you will encounter. As noted in Chapter 1, change will increasingly be the only certainty.
- Second, building on the seven qualities of effective leaders, as also discussed in Chapter 1, this chapter offers advice about how you can be a change agent for constructive change and thus benefit your employer, your professional society, whatever communities you belong to, those you serve, and yourself. Given your education and your intelligence, you can take on the important work of effecting change. As noted by Theodore Roosevelt, 26th U.S. President, "Far and away the best prize that life offers is the chance to work hard at work worth doing." And being a change agent is "work worth doing."

THE WORLD YOU WILL WORK IN: SAME ROLE BUT NEW STAGE

From the beginning of recorded history and all over the earth, individuals we would now label engineers have met the basic needs of communal society (e.g., see Fredrich 1989, Walesh 1990, Weingardt 2005, White 1984). These needs include providing shelter, water supply, wastewater disposal, irrigation, transportation, communication, and environmental protection. Frankly, the fundamental work of these professionals hasn't changed. As discussed in Chapter 1, in the section "The Engineer as Builder," and in the Chapter 8 section "Design as a Personally-Satisfying and People-Serving Process," the common bond among engineers is designing and building for the benefit of society. Looking into the future, we can expect marked changes in tools and techniques, however, engineers and some other technical professionals will continue to be the builders; they will continue to fulfill the crucial role of meeting society's basic needs.

While the role will remain essentially the same, the stage on which that role is played will change dramatically. The following sections explore that new stage with the hope that you will leverage your education and early career experience so that you can fulfill your role on that stage to the best of your ability. If you don't play your role effectively on the 21st Century stage, the curtain may come down on your career.

After the Knowledge Age, the Conceptual Age?

Advanced societies have progressed through the agricultural and industrial ages and into the knowledge age. Daniel H. Pink (2005, 2007) argues that the present knowledge worker age, which followed the agricultural and industrial ages, is gradually being superceded, as illustrated in Figure 15.1, in the U.S. and other advanced countries by what he calls the conceptual age.

What does Pink mean by the conceptual age? Recognize that the root word is conception which suggests a new, beginning, or original idea or concept. Pink says the conceptual age is "an era in which mastery of abilities that we've overlooked and undervalued" will be required (Pink 2007). These increasingly-valued abilities emanate from the right brain and include visualization, innovation, creativity, synthesis, empathy, and helping people find meaning. (Recall the left and right-brain discussion in the "Mind Mapping" section of Chapter 7).

Functioning effectively in the knowledge age requires primarily left hemisphere or left brain abilities. Engineers and other technical professionals are prime examples of knowledge workers. They logically and sequentially collect and analyze data, calculate, and design to meet requirements. Also relying heavily on their left hemispheres, accountants prepare tax returns, lawyers research lawsuits, radiologists read diagnostic data, software experts write code, and stockbrokers execute transactions.

According to Pink, left brain abilities will be necessary, but not sufficient, in the conceptual age. A half a brain will be necessary, but not sufficient. A whole brain will be needed if one is to succeed, especially in the U.S. and other advanced countries.

Why? Because work that "can be reduced to a set of rules, routines, and instructions," the functions of the left brain, is "migrating across the oceans... Now that foreigners can do left-brain work cheaper, we in the U.S. must do right-brain work better" (Pink 2007).

Fiber-optic cables and a growing number of ambitious, smart, and English-speaking workers in India, China, the Philippines, Singapore, and other countries facilitate this outsourcing process. Pink notes that as of 2010, India is the country with the most English speakers. Tomaz Arcieszewski (2009) asserts that outsourcing occurs

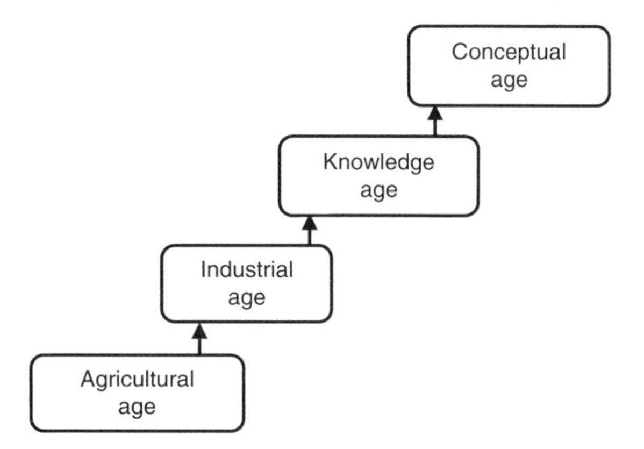

Figure 15.1 Perhaps the knowledge age is being replaced by the conceptual age.

also because engineers in some countries receive better engineering education in some areas, such as structural engineering, than in the U.S. He also notes that engineers in other countries use the same computer programs for routine design as are used in the U.S.

Accordingly, if Pink is correct, in the conceptual age leading-edge engineers will focus less on solving problems and more on finding and developing opportunities. In similar fashion, accountants will serve more as financial advisors, lawyers will concentrate more on convincing juries and mastering the nuances of negotiation, and stockbrokers will become financial advisors to help people realize their dreams.

After The Knowledge Age, the Opportunity Age?

Futurist John Naisbitt (2006) offers a related view of the future. He states: "When you're looking for the shape of the future, look for and bet on the exploiters of opportunities, not the problem solvers." He goes on to claim that individuals tend to embrace one of two poles, stasis or dynamism, stability versus evolution, predictability or surprise. His contrasts are aligned with the paradigm paralysis versus paradigm pliancy discussion later in this chapter.

Naisbitt claim suggests, as illustrated in Figure 15.2, that problem solvers tend to have one foot in the past; it's the origin of the problems they solve. In contrast, opportunity exploiters, as also illustrated in Figure 15.2, while living in the present, have one foot in the future; it's the place of promise. To restate, in a different way, a point made in the preceding discussion of Pink's ideas, most professions focus on solving problems and do a superb job. Examples are engineering, law, and medicine. Most engineering curricula emphasize problem solving and, to a lesser extent, problem prevention. Rarely, especially at the undergraduate level, would a student be explicitly exposed to finding and pursuing opportunities, as espoused by Naisbitt. How do you, as a student, or how do you, as a college graduate, fare on the problem-solving and opportunity-pursuing scale?

Figure 15.2 Problem solvers focus on the past and the present while opportunity pursuers concentrate on the present and the future.

Engineers solve well-defined problems and do it very well. This admirable ability is learned, in large part, during engineering education. This teaching-learning method is also very left-brained. For example, it is linear as in present theory, discuss theory, assign problems the solution of which requires theory, provide students with everything needed to understand the problems, use theory to solve problems, get "the answer," and discuss how the successful students got the answer.

I strongly support the preceding method of learning theories and applying them to solve problems. After all, understanding theories is essential to engineering practice and problem solving is an important aspect of engineering. However, engineers can also perform other functions, besides problem solving, that incidentally, also use theories.

Personal

I vividly recall hearing professors say, near the end of their lecture, something like, "for the next class, solve Problems 1, 3, and 9 at the end of the chapter." Later, as a professor, I said the same thing! This approach is pedagogically sound in that the principles or theories just introduced, described, and discussed in lecture are immediately applied and further understood and appreciated by doing the homework. Furthermore, a brief discussion of the homework at the beginning of the next class period could further elucidate the topic. However, we must and can do more for the benefit of students, those they serve, and their country.

After The Knowledge Age, the Solving Wicked Problems Age?

John Kao (2007), teacher, consultant, and innovation expert, is concerned that the U.S. may feel smug about its pre-eminence thinking. That is, he questions the idea that other countries will continue "to settle for being followers, mere customers, or imitators of our fabulous creations" and he asserts that "innovation has become the new currency of global competition as one country after another races toward a new high ground where the capacity of innovation is viewed as a hallmark of national success." He goes on to say that "what's at stake is nothing less than the security of our [U.S.] nation."

Kao's book (2007), *Innovation Nation: How America Is Losing Its Innovation Edge, Why It Matters, and What We Can Do To Get It Back*, diagnoses the U.S. situation, describes innovation best practices from around the globe, explains how innovation works at the national level, and proposes a U.S. strategy. That strategy is to become what he calls an Innovation Nation, that is, "a country with a widely-shared, well-understood objective of continuously improving our innovation capabilities in order to achieve world-changing goals." Clearly, Innovation Nation would, as a matter of policy, begin to teach creativity and innovation to its children and young people or, to use Kao's words, "fix the U.S. education system."

Kao envisions “a concentrated application of our vast resources to innovate on a huge scale for human benefits.” Kao wants America “to be in the wicked problems business.” By this he means taking on global issues such as “climate change, environmental degradation, communicable diseases, education, water quality, poverty, population migration, and energy sufficiency.” Creative and innovative solutions to the wicked problems are the key to making the most consequential breakthroughs of the 21st century; these solutions will generate “an enormous amount of social and economic value” and enable Innovation Nation, that is, the U.S. to do good and do well.

Additional Views of the World Stage

To supplement the views of Pink, Naisbitt, and Kao and to more fully describe the world stage on which you are likely to play, consider the thoughts of others—an engineer and a geographer. And while contemplating their thoughts, think of the implications for your role, your organization, and the engineering profession.

Ralph Peterson-Engineer

According to Ralph Peterson, former Chairman and Chief Executive Officer of the engineering firm CH2M-Hill (ASCE 2007), the next 20 years will see “a smaller portion of engineers originating from the U.S. and other industrialized countries, urbanized development and infrastructure growth concentrated in the developing world, [and] larger more youthful work forces residing in Asia, Africa, and Latin America.” He concludes by saying this about civil engineering: “One can see that the United States/developed country civil engineer in 2025 will be part of a truly global and multi-cultural profession in which there is probably no “dominant” culture of “national standards” influence. Everybody will need to learn from one another’s experience because new and exciting projects will be going on everywhere. Adaptability to and respect for different cultures will be valued in a world where global sourcing and multi-disciplinary teams are the norm.”

Assume Peterson is correct. Engineers who function on the global stage he describes will need many qualities not the least of which is a creative and innovative approach to both technical and non-technical problem solving, problem prevention, and opportunity pursuing.

Harm de Blij-Geographer

Geography professor Harm de Blij (2009), in his book *The Power of Place*, looks around at the global landscape and ahead at the troubling implications of what he sees. He says: “The Earth, physically as well as culturally, still is very rough terrain, and in crucial ways its regional compartments continue to trap billions in circumstances that spell disadvantage... The power of place still holds the majority of us in its thrall.”

He partitions the earth’s geography into the global core and the periphery. The global core is the urbanized and wealthy portion of the globe comprised of Europe,

North America, extreme East Asia, and Australia-New Zealand. Everything else is periphery, that is, roughly, Asia, Africa, and Central and South America.

Approximately 15 percent of the earth's population resides in the global core and they earn nearly 75 percent of the world's annual income. This means that the other 85 percent of the population lives in the periphery of the earth's geography and earn only 25 percent of the world's annual income. Using individual income as the indicator, the "haves" having an annual per capita income almost twenty times that of the "have nots" suggests a very "un-flat" world in contrast with that described by Friedman (2005).

Juxtaposition of the global core and the periphery creates a dynamic in that the core attracts millions of legal immigrants and asylum seekers and also illegal workers and revolutionaries. The core simultaneously spawns anger and hope. An example of hope: "The remittances sent home by just one successful mobal [a risk-taker migrant] can sustain an entire extended family in Mexico, India, China, the Philippines, or a host of other countries."

The author places Christianity and Islam, the two dominant global religions, in the context of his core-periphery view of the globe. Christianity has about 1.6 billion adherents and is diminishing while Islam has 1.3 billion followers and is growing rapidly. Furthermore, while Christianity is scattered over the globe in the global core and the periphery, Islam is in the periphery, contiguous, and concentrated in northern Africa, the Middle East, and southwest Asia. De Blij sees the potential for religious differences to drive conflict between the "disadvantaged" in the periphery and the "advantaged" in the global core. His view is consistent with recent world events and worthy of consideration.

Harm de Blij convincingly argues that the world is rough terrain for the majority of the earth's residents. Might he be describing the global stage on which you and your colleagues will work? If so, you and they will need to fulfill the role of creatively and innovatively contributing to the prevention of or solution to social, economic, environmental, and infrastructure problems.

Implications for You

Assume, for purposes of discussion, that Pink, Naisbitt, Kao, Peterson, and Blij are collectively correct in stressing that maintaining U.S. global leadership, enhancing national security, enjoying organizational vitality, and achieving personal professional success and significance will increasingly require personal and group qualities such as adaptability, collaboration, creativity, empathy, entrepreneurship, innovation, synthesis, and visualization. If these above assumptions are correct, they have serious implications for you and, by extension, your university, your country, your profession, your employer, and those you serve.

Where are you, as a student or young practitioner, in the context of your role on the world stage, on the vitality scale shown in Figure 15.3? Are you thriving, just surviving, or dying? And how about your academic, business, government, volunteer, or other organization? Hopefully, you as a student or young practitioner are working to enhance your technical knowledge, skills, and attitudes in your courses or on-the-job.

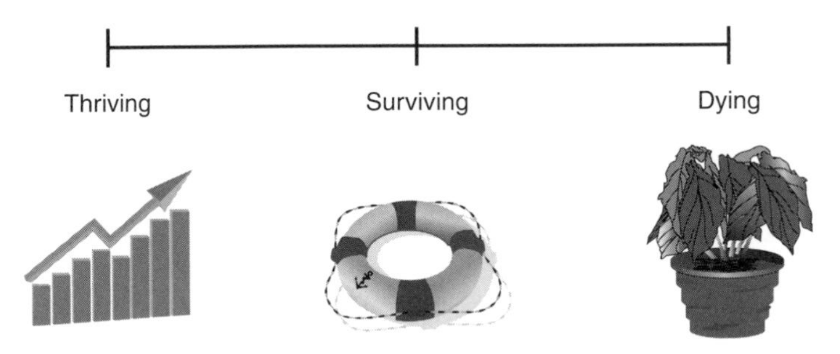

Figure 15.3 Where are you and/or your organization on the vitality scale in light of the global challenges faced by the post-knowledge age?

Furthermore, for the sake of balance, you are also developing your professional or non-technical knowledge, skills, and attitudes role, presumably using this book as one resource. I trust that you are preparing to act, or are already acting, on the global stage as described above and helping your organization to do likewise. You and your organization should strive to stay on the left side of the vitality scale.

Perhaps you are already benefitting from the movement underway, in some U.S. based engineering societies, to reform the education and prelicensure experience of engineers as discussed in this book's Preface section titled "This Book and the Body of Knowledge Movement." Stated differently, hopefully the BOK you are studying, or have studied, will prepare you for the way engineering will be practiced in the 21st Century, not the way is was practiced.

HOW TO LEAD CHANGE

In keeping with the subtitle of this book's Chapter 2, assume that you are getting and keeping "your personal house in order." Furthermore, you recognize that, while the engineer's fundamental role will not change, the stage on which that role is played will change and you will need a broader and deeper BOK to successfully play your part.

Let's revisit the topic of leading which was originally discussed in the Chapter 1 section titled "The Seven Qualities of Effective Leaders" and then view a special kind of leading, that is, leading change. Whether or not you choose to act on them, you and your colleagues will have many opportunities to lead some of the changes that will be needed for the 21st Century. My final assumption: You are receptive to leading change. How do you get started? How do you follow through? Many ideas follow with the suggestion that they be considered in approximate order in which they are presented.

Encounter a Leadership Gap

You have encountered a likely need for change and it is seems to be exasperated by a leadership gap. While you, and perhaps few others, sense a need for change you also are beginning to realize that no one is willing and/or able to lead. Perhaps you, a university student, are a member of a struggling student club, student chapter of a professional society, or a Greek organization. As a cooperative education student, intern, or summer employee you might see some technical processes that you think could be vastly improved.

Maybe you, as a young employee in the public sector, see a failing process. Or, as a member of an engineering or other professional services firm, you are increasingly aware of a dying service line. You may be in academia and be concerned about a stagnant curriculum. The change need that concerns you, and perhaps a few others, may be within your professional society, your community, or in some other setting. You, and maybe several concerned friends or colleagues, are contemplating taking the lead in closing the leadership gap and working for change.

Move Beyond Being the Thermometer: Also be the Thermostat

Recognizing the need for change, in at least a preliminary manner, and possibly seeing the related leadership gap is good; it is the important first step in effecting change. However, it is like being a thermometer, that is, measuring or defining the problem or situation. Leading change, which means doing something about the need for change and the leadership gap, is like being the thermostat. Like a thermostat, the leader in you sets, that is, defines and then moves toward the desired or envisioned condition. This thermometer-thermostat metaphor is attributed to Martin Luther King who said that, when faced with a problem, we have two options. "We can act like a thermometer and merely make a record. Or we can act like a thermostat and correct what is wrong" (Flaherty 2007). The essence of leading change is to "act like a thermostat."

As you and your change team work to fulfill both the thermometer and thermostat functions, selectively use the tools and techniques for stimulating creative and innovative thinking that are described in Chapter 7. These methods will stimulate your group to synergistically and collaboratively think more deeply and widely about where you are and where you want to go.

Define the Situation: What, Why, Who, How, and When?

Some of this book's chapters, including Chapters 3, 4, 5, 7, and 14, urge you and others to ask many and varied questions in particular situations and then listen carefully and empathetically to the responses. Do exactly that at this early stage in the contemplated change project. As a guide, refer to the twenty questions listed below which is based, in part, on Maxwell (1993) and Russell (2006). The list is provided to stimulate questions; it is not intended to be all inclusive nor is it intended to be a minimum set of questions to be asked. Based, in part, on the list, quiz yourself and many others and thus begin to address what should be changed, why it should be changed, who would be or thinks they would be affected, how might the change occur, and when.

1. Are you doing this primarily benefit the organization and those it serves or are you doing this primarily to elevate/bring attention to you?

2. What is the fundamental problem/opportunity/issue and how will you communicate it so others understand?
3. Is your commitment sufficient to deal with likely prolonged opposition and/or apathy?
4. Is the change compatible with the organization's mission and vision, or do you propose to change the organization's mission and vision?
5. Who will be positively affected by the change and what are the "benefits" to them?
6. Who will be negatively affected by the change and what are the "costs" to them?
7. What are the long-term implications for the organization of not changing, of proceeding in the current mode?
8. Who will not be impacted, positively or negatively, by the contemplated change but is likely to initially think they are a stakeholder?
9. What unexpected changes could occur as a result of the contemplated change?
10. Is the contemplated change visionary enough to excite and engage other leaders or are you aiming too low?
11. Can you confidently identify likely co-leaders and the reasons they will be supportive?
12. How will the core team learn more about the change process and how will the group be expanded?
13. Who will be the principal opposition, at least initially, and why?
14. What individuals and/or organizations outside of your organization might assist?
15. Can you point to similar or related changes made elsewhere to use as examples and/or learning experiences?
16. What messages and media will comprise your contemplated communication program?
17. What are some of the major milestones and metrics needed to achieve the change?
18. What are some small successes that will demonstrate commitment and progress?
19. How will you fund, finance, and/or obtain resources for the change effort?
20. Could the contemplated change be applied on a trial or pilot basis or, once the change begins, is it irreversible?

Recognize Widespread Resistance to Change

Your approach to effecting change must anticipate resistance to change, especially if you are suggesting a major change. The Italian politician and writer Nicolo Machiavelli (Machiavelli 1980) offered this sage advice during the Renaissance:

> There is nothing more difficult to plan,
> more doubtful of success, nor more dangerous to manage
> than the creation of a new system.
> For the initiators have the enmity of all who would profit
> by the preservation of the old institutions and
> merely lukewarm defenders in those who would gain
> by the new one.

Note, in particular, his mention of the initial "enmity" of many who oppose change contrasted with the only "lukewarm defenders" of change. Effecting change is difficult. Nevertheless, the leader in us wants change—we are dissatisfied with the present situation and can see a better one.

Why do many of use resist change? The possibility of change causes each of us to compare the way things are to the way things could be. We contrast the familiar and comfortable with the unfamiliar and uncomfortable. I believe that most of us, at the cognitive level, can see and weigh the "pros" and "cons" of a proposed change, especially if thoughtfully presented. However, at the emotional level, we fear how we are going to get from here to there. The unknown trip is scary and conditions at the destination are uncertain! In summary, when faced with change, we tend to revert to fear and other emotions, not reason.

Statements such as the following, which are adapted from Barker (1992) and Carroll (2004) who discuss change, reflect the tendency to react emotionally, in knee-jerk fashion, to a proposed change or even the suggestion to consider change. These are ways to douse change talk with water rather than fuel it with gasoline:

- We've always done it this way
- It won't work
- That's the dumbest thing I ever heard
- We can't do that
- After you are here for awhile, you will see why that cannot be done
- We tried something similar and it did not work
- Get real
- We don't know how
- The boss won't go for it
- Impossible!
- None of our competitors do it
- Let's wait until next year or next semester or next…
- Oh that it were that easy
- Let's wait for others to do it
- If that were needed, someone would already have done it
- That idea is outside of your area of responsibility

In addition to natural individual resistance to change, some organizational cultures resist specific kinds of change or, even more broadly, change in general. By culture, I mean the way things really work around here. As noted by Stephen E. Armstrong (2005), "culture wields great power over what people consider permissible and appropriate." He goes on to say "The embedded beliefs, values, and behavior patterns carry tremendous weight. The culture sends its energy into every corner of the organization, influencing virtually everything." Effecting change in organizational cultures is extremely difficult, if not impossible. The previously-mentioned 20 questions can help you and your team begin to define the culture within which you want to accomplish change.

Practice Paradigm Pliancy: Prevent Paradigm Paralysis

Definition of Paradigm

Leading change requires, using common expressions, "taking off the blinders," "getting outside of your box," and "looking over the edge of your silo" which naturally leads to the topic of paradigms. Covey (1990) defines a paradigm as "The way we 'see' the world—not in terms of our visual sense of sight, but in terms of perceiving, understanding, interpreting." According to Barker (1992), who must be credited with showing the relevance of paradigms in all aspect of our lives, "a paradigm is a set of rules and regulations (written or unwritten) that does two things: 1) it establishes or defines boundaries; and 2) it tells you how to behave inside the boundaries in order to be successful." Kriegel and Brandt (1996) suggest that a paradigm is like the sand box you played in as a child—it was your world.

Examples of Paradigms

Paradigms are all around us. Five examples within the U.S. and many other countries – and, yes, paradigms are nationalistic and cultural–include:

- Men and women participate in inter-school high school athletics
- Every engineering student and practitioner owns or has easy access to a digital computer
- Japanese products are of high quality at a competitive price, the implication being that manufacturing organizations hoping to compete globally must match or exceed the quality of Japanese products
- Packages are delivered overnight
- Most watches use the quartz crystal

So why are these today's paradigms? Haven't things always been like this? Let us defer answering that question.

Some Characteristics of Paradigms: Paralysis and Pliancy

Barker (1989) reiterates the idea that one of the characteristics of paradigms is that they are so common that they are invoked or used implicitly with little thought. Paradigms are useful, given the complexity of society. They almost always allow for more than one "right" answer. For example, there are many ways to interpret U.S. paradigms such as the nine-month school year and the 40-hour work week.

On the negative side, according to Barker (1989), paradigms tend to reverse the often quoted and relied on "seeing and believing" process. Intelligent, thoughtful people like to think that they are rational, that they "believe because they see." However, because of paradigms, people often "see because they believe." Consider, for example, some beliefs you hold and have held for a long time and highly value. Are you not likely to find many examples of situations that tend to support your belief and might you not actually be looking for them? And might you be "seeing because you believe" rather than "believing because you see?" "We do not see things as they are," according to business executive H. Jackson Brown (1988), "We see things as we are."

Personal

I studied engineering in a baccalaureate environment that linked lectures to laboratory sessions. For example, on Tuesday, in the lecture portion of a fluid mechanics course, we may have learned about the conservation of mass and conservation of mechanical energy principles. We applied those principles in class and with homework to predict what would happen to the pressure in a liquid as it flowed into, through, and out of a horizontal venturi. I still recall thinking, when I was a student, that predicting pressure would be lowest in the middle of the venturi was counter-intuitive. Then on Wednesday, in the laboratory portion of the course, we set up a similar situation and, by using manometers, we saw what happened – and it was as predicted. Over and over again, I was exposed to this pedagogical approach and I learned from it. When I subsequently taught engineering courses, including fluid mechanics for several years, I used exactly the same approach. And, to justify it, I always saw evidence supporting the wisdom of having my theoretical lecture followed by a physical laboratory experience. What if there was a more effective way to link theory and reality. Was I open to it? Was I paralyzed by my paradigm?

Unfortunately, if paradigms are held too strongly, and they often are, the holders risk paradigm paralysis. Selective paradigm pliancy is a much better strategy, according to Barker (1989), especially in turbulent times. Pliancy is a quality or state of yielding or changing. Fortunately, at least a handful of people in most organizations can change their paradigms. Even for them, paradigm pliancy is at best difficult.

The creators or advocates of new paradigms, and that may be you and a core group of change agents, are often unwelcome. Leadership writer Warren G. Bennis (1989) asserts that "Most organizations would rather risk obsolescence than make room for the nonconformists in their midst." Discussions, debates, and disagreements over old versus new paradigms do not typically pit the incompetent against the competent, the weak against the strong, and the uncaring against the caring. Paradigm paralysis adherents differ from paradigm pliancy adherents mainly in their regard, or lack thereof, for the status quo. As Heilmeier explained (1992), "History seems to indicate

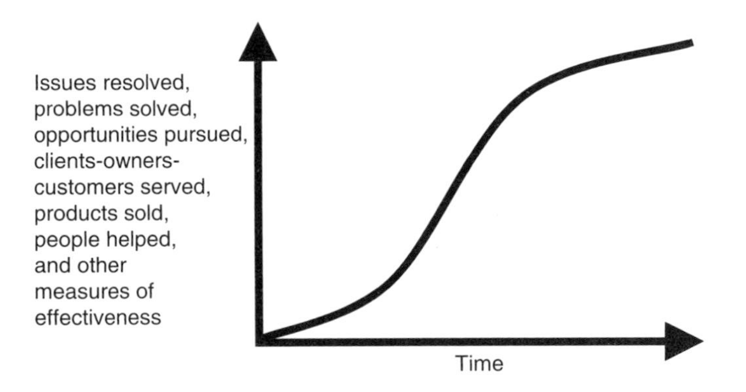

Figure 15.4 Paradigms tend to evolve following the familiar S-shaped curve. (Source: Adapted from Barker 1992)

that breakthroughs are usually the result of a small group of capable people fending off a larger group of equally capable people with a stake in the status quo."

Another characteristic of paradigms is that they evolve as suggested by the familiar "S-shaped" curve in Figure 15.4 (Barker 1992). The vertical scale represents measures of paradigm effectiveness such as issues resolved, problems solved, opportunities pursued, organizations served, products sold, people helped, and other things accomplished. Figure 15.4 illustrates the concept that the effectiveness of a paradigm is minimal early in its life and then accelerates significantly. Typically, with time, the effectiveness of a particular paradigm begins to decline. In other words, paradigms have a finite, but unknown life.

Bridges form from "old" to "new" paradigms, offering the possibility that a more effective paradigm can replace an "aging" paradigm. The series of paradigms might, for example, represent various personal computers, including the abacus, slide rule, electronic calculator, and digital computer.

The most successful individuals and organizations are those that are among the first to recognize and cross the bridges to newly developing paradigms or, better yet, create and develop new paradigms and the bridges that other individuals and organizations will eventually adopt and use.

Paradigms Shift

As noted, we shift or bridge from paradigm to paradigm. One may find many examples within engineering and other technical professions as well as throughout society. Listed here are five situations that existed several decades ago in the U.S. and many other nations. While thinking about them, compare each to its counterpart in the earlier list of today's paradigms:

- Inter-high school and college sports were essentially exclusively for males (Barker 1989).
- The slide rule was the personal computer—a digital computer with its peripherals typically filled an entire room. Ken Olsen, the then-president of Digital

Equipment Corporation, a computer manufacturer, said this in 1979: "There is no reason for any individual to have a computer in their home." Three decades earlier, IBM Chairman Thomas J. Watson said "I think there is a world market for about five computers" (Barker 1992).

- Products manufactured in Japan were inferior to U.S. products.
- Mail routinely took days to be delivered.
- Watches were mechanical, including complex gear and spring mechanisms. In the late 1960s, the Swiss mechanical watches enjoyed 60 percent of the global watch market (Barker 1989).

The preceding five paradigms were replaced by new paradigms. Consider the following five paradigm shifts, or at least major changes, that occurred mostly within the past century, with emphasis on the "shifters," "outsiders," and "odd balls," responsible for them:

- **Golden Gate Bridge:** Engineer Joseph Strauss dreamed of bridging San Francisco's Golden Gate even though he faced widespread skepticism partly because of site challenges. For two decades, he led the planning, design, and construction of the now famous bridge. The intensity of Strauss' faith in his vision is suggested by these lines from one of his poems: "Launched midst a thousand hopes and fears, damned by a thousand hostile sneers. Yet ne'er its course was stayed. But ask of those who met the foe, who stood alone when faith was low; ask them the price they paid." Strauss saw the 1937 opening of the bridge and then died approximately one year later. Strauss' persistence is recognized with a statue at the south end of the bridge dedicated to The Man Who Built the Bridge (Fredrich 1989, McGloin 2011). Today, bridges are constructed virtually anywhere.
- **Electrostatic Photography:** Chester Carlson developed the process of electrostatic photography, offered it to 43 companies in the late 1940s, and finally found one organization with foresight. The result was the development of what is now called xerography. At the time the photography paradigm consisted of film, developer, and a darkroom—there was no other way to do it (Barker 1989). At least 43 companies could not envision the now omnipresent copy machines.
- **Quality Movement:** Most American businesses ignored the late W. Edwards Deming, whose advice on how to achieve quality is presented in Chapter 7. He assisted the Japanese after World War II and, in a few decades, they set the world standard for manufactured products (Barker 1989).
- **High Jump Technique:** Dick Fosbury, a high jumper, was ridiculed for leading with his head at a time when all others jumped feet first. However, the ridiculed "Fosbury Flop" enabled Fosbury to win the high-jump gold medal at the 1968 Mexico City Olympics. Leading with the head is now the standard for world-class high jumpers (Kriegel and Brandt 1996).
- **Overnight Mail:** Fred Smith founded Federal Express in 1971 so mail, in the U.S. at least, could be routinely delivered overnight. As a Yale University

student, Smith wrote a paper proposing overnight mail delivery in the U.S. using trucks and airplanes operating within a hub and spoke system. Some say the professor gave Smith a "C" on the paper and said the idea was interesting but would never work (Barker 1989, Kriegel and Brandt 1996).

Individual and Organizational Implications of Paradigms

The most successful individuals and organizations will not permit across-the-board paradigm paralysis—they will selectively practice paradigm pliancy. Forward-looking businesses, government entities, universities, and volunteer organizations will create new paradigms and build bridges to them. Or, they will at least recognize new paradigms when they are coming down the pike and see the opportunities within them. Recall the thriving-surviving-dying spectrum depicted on Figure 15.3. Thriving individuals and organizations practice paradigm pliancy. Paradigm paralysis characterizes individuals and organizations satisfied with surviving or in the process of dying. What paradigms will you contribute to your engineering or other technical field or, what paradigms created by others will you enthusiastically embrace and advance?

Appreciate the Movers–Movables-Immovables Structure

As you further contemplate leading a change, consider putting members of the organization or group that would be affected into one of the three categories illustrated in Figure 15.5: The movers, the movables, and the immovables (Annunzio and Liesse 2001).

There is no quantifiable support for this model, but it is suggested by Annunzio and Liesse (2001). Furthermore, the categories and their relative sizes resonate with my experience and observations. Consider each of the three categories:

- The movers, comprising roughly ten percent, are predisposed to change and to leading change. Look for them. Recall Question 11 in the 20 Questions: "Can you confidently identify likely co-leaders and the reasons they will be supportive?"

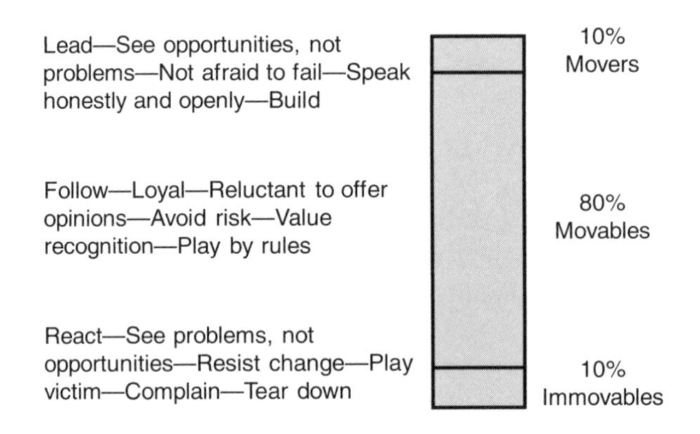

Figure 15.5 When contemplating leading change, the 10–80–10 structure is an effective way to view the mover, movable, and immovable members of the organization that would be affected.

- The movables, the large approximately eighty percent component, are predisposed to follow. They can be convinced of the need for change.
- The immovable, the remaining ten percent, tend to react and do so negatively. They are not likely to seriously consider any arguments for change. Question 13, "Who will be the principal opposition, at least initially, and why?" will facilitate identifying the immovables.

An alternative set of three terms, having essentially the same meaning, is accepters, undecided, and rejecters (Brenner 2009). The creator of this terminology notes, as I have, that the number of undecided (the above movables) is much larger, at the outset, than the acceptors (movers) and rejectors (immovables). He also states that, at the outset, the rejectors (immovables) tend to be more vociferous than the acceptors (movers).

Work Effectively With the Movers, Movables, and Immovables

Having established, at least for discussion purposes, a way to view an organization in which you and others want to lead change, move forward as shown in Figure 15.6, by devoting most of your efforts, say eighty percent, to communicating your vision and or goal and initial strategy and tactics ideas to the movers. Ask them to thoughtfully consider your ideas, refine them, and in hopefully, in principle, accept them. This could require a major effort by you and them and considerable elapsed time. However, in engaging the movers you are working with forward-looking individuals like you.

Ask the now hopefully committed movers to, in turn, communicate with and engage the movables. Neither you nor the movers should invest too much time and energy on the immovables. While you should respectfully inform them of the proposed change and invite their input, recognize their mind set. As someone anonymously noted: "Some minds are like concrete, thoroughly mixed up and permanently set."

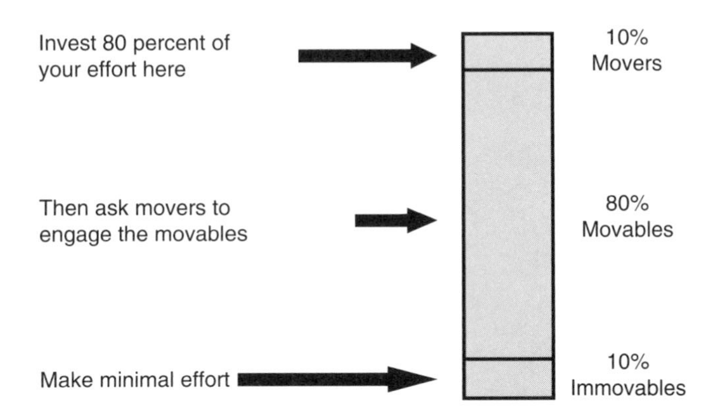

Figure 15.6 Invest most of your effort on the movers and, if they generally support the change, urge them to invest most of their effort on the movables.

Consider another way of explaining the process by which you, the advocate of change, focus first on movers and then, if successful, on the movables. Communicate the issue, the problem, or the opportunity and explain how the change you are advocating will resolve the issue, solve the problem, or seize the opportunity. We are tempted in situations like this, because of efficiency considerations, to go exclusively or mostly with mass communication such as e-mails, newsletters, memoranda, posters, banners, coffee mugs, key chains, and wallet-size cards. Don't go with mass communication—other than as a supplemental measure. Mass communication is not an efficient use of your time and energy. Instead, go one-on-one and one-on-small groups and focus on that ten percent–the movers (Smart 2007).

Personal

A participant in one of my leading webinars said: "Note that the immovables could also be called the removables – consistent resistance to change invites people to consider removing them from the organization." Given that organizations can be classified as thriving, surviving, and dying, the comment about the immovables reminds us of the need carefully recruit and retain mostly movers and movables if we want to be thriving organization.

Expect the Awareness–Understanding–Commitment–Action Cascade

Now consider the cascading awareness–understanding-commitment-action process shown in Figure 15.7, and suggestions for how to use it. It cascades in that it flows from the top to the bottom while the number of participants becomes smaller as the process proceeds. However, even so, the number of individuals remaining at the last or action level is often adequate to effect change.

As noted earlier and illustrated with typical statements, on becoming aware of a possible change, many of us react in a mostly emotional knee-jerk fashion. You, as the change leader, should anticipate and gracefully tolerate knee-jerk reactions. Simply

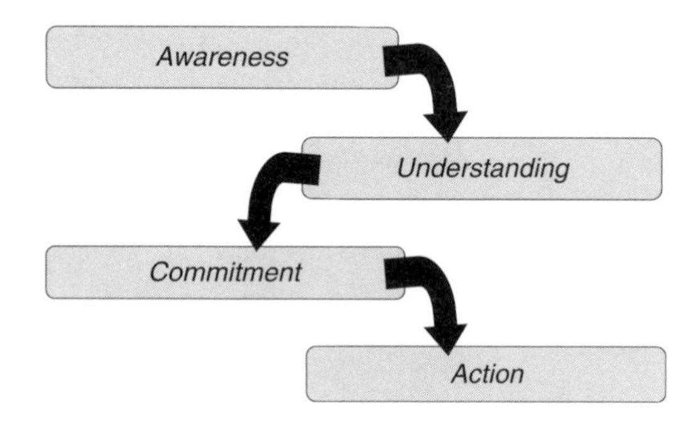

Figure 15.7 You, as the change leader, should anticipate and patiently work through the cascading awareness-understanding-commitment-action process.

ask for understanding of what is being proposed and the reasons for it. Some of the knee-jerkers will show you that courtesy. And, on understanding, a portion of them will commit. Finally, for some, that commitment will lead to action needed to advance the change effort.

AH HA! is another way of presenting the cascade process. The first "A" represents awareness, that is, we learn of a proposed change. The first "H" represents head, that is, understanding the proposed change and its features. The second "H" is for heart in that some of those who understand will commit to supporting the change. The second "A" represents action meaning that some of the committed will act to effect the proposed change.

Personal

Consider a specific example of the awareness-understanding-commitment-action process. Awhile back, my wife said something like "let's redecorate our house." My knee-jerk reaction was no!—its fine for now—we just moved in—the living room furniture is not that old. I noted the three ominous D's that were reasons for not making the suggested change. That is, we would have to deal with decisions, disruptions, and dollars. To encourage the second step, understanding, my wife reminded me that we had been in our house eight years and hadn't done a thing. Furthermore, we brought the living room furniture with us and it was eight years old then. I understood! My wife was right and, therefore, I committed to the project. This lead to action; we painted, re-carpeted, and bought some furniture. As a result of the redecorating project, our home looks and lives great! In retrospect, I'm sure glad I proposed the change.

Test Drive Terminology

The strategy and tactics employed to achieve a goal or vision should include sensitivity to how the various stakeholders might respond to the language used to describe the change. Words that seem appropriate to you and others leading the effort may be misunderstood or even viewed negatively by others. Reflect on Mark Twain's thought, "The difference between the right word and the almost right word is the difference between lightning and the lightning bug."

Recall, from the Preface, the discussion of the ASCE–led effort to reform the education and pre-licensure experience of U.S. civil engineers. This is a major change program. Let's use one aspect of that reform initiative as an example of what can happen if words are not carefully chosen.

In October 1998 the ASCE Board of Direction adopted Policy Statement 465, which began as follows: "The ASCE supports the concept of the master's degree as the First Professional Degree (FPD) for the practice of civil engineering at the professional level" (ASCE 2004). The intent was to gradually move toward a vision of more formal education for tomorrow's U.S. civil engineers. Unfortunately, the wording was interpreted by some practicing U.S. civil engineers to mean that their bachelor's degree was not a professional degree.

Partly because of that negative interpretation, the policy was reworded in 2001 to read: "The American Society of Civil Engineers (ASCE) supports the concept of the master's degree or equivalent (MOE) as a prerequisite for licensure and the practice of civil engineering at the professional level" (ASCE 2004). This version seemed to diminish some of the initial negative reaction while continuing to support the vision of more formal education for U.S. civil engineers.

In 2004, the policy was refined to begin as follows: "The ASCE supports the attainment of a body of knowledge for entry into the practice of civil engineering at the professional level" (ASCE 2008). Now the specification of a degree was replaced with body of knowledge (BOK). Finally, the acceptable terminology was achieved. Accordingly, in spite of the preceding missteps, the ASCE-led effort to reform the education and pre-licensure experience of U.S. civil engineers is moving forward. The BOK concept has proved to be an interest shared by both academics and practitioners

The lesson: After drafting a goal or vision and beginning to work on the implementation strategy and tactics, "test drive" the language and terminology before moving into wide public exposure. For example, circulate draft text, make trial presentations, and/or use focus groups.

Learn Why Change Efforts Fail

You, as a change agent or potential one, will enjoy successes and make mistakes and you will observe or study change initiatives, some of which have succeeded and some of which failed. View all of these, including the failed efforts, as sources of lessons learned. Perhaps the unsuccessful attempts were fundamentally flawed because they did not incorporate the kinds of principles and approaches discussed above. Or maybe a few key details were missed. Whatever the causes of failures, knowledge of them can help you, and your change team, be successful. "Failure is success if we learn from it," according to publisher Malcom Forbes. For example, consider the following possible reasons for failed change, adapted from leadership expert John P. Kotter (2007) and based on my observations:

- Unimaginative vision or mediocre goals
- Poor communication of vision or goals
- Shallow definition of what, why, who, how, and why
- Unimaginative identification of long term consequences with and without the proposed change
- Weak core group, that is, absence of necessary diversity of experience, knowledge, and skills
- Mistrust and/or miscommunication within the core group
- Carelessness in identifying likely stakeholders, especially proponents and opponents
- Lack of credibility by stakeholders of those leading the change

- No sense of urgency
- Inadequate financial and other resources
- Allowing obstacles to remain
- Insufficient persistence
- Failure to recognize short-term wins
- Declaring victory too soon
- Not instituting policies and procedures to make changes "stick"

Adopt Change Principles and a Change Process

Try the change principles and processes presented here, find others that suit you, or develop your own. But do have guiding principles and a supporting process to sustain the leader in you and your colleagues. Change can be tough on the leader in us. We may get "walked on." However, even that could be a good sign. As educator Roy West said: "If you're going to be a bridge, you got to be prepared to be walked on." Persist and look for signs that you are succeeding such as when a former vocal opponent agrees to engage in conversation, when a previous fence sitter asks what he or she can do to help, or you begin to hear comments like "I'm sure glad we thought of that."

Views of Others

Recognizing the often initial knee-jerk resistance to proposed change, economist John Kenneth Galbraith said "Faced with the choice between changing one's mind and proving that there is no need to do so, almost everybody gets busy on the proof." And, in a more positive vein, "All truth goes through three stages," according to German philosopher Arthur Schopenhauer, "first it is ridiculed, then it is violently opposed, finally it is accepted as self-evident." Finally, the foundational principle of change, according to anthropologist Margaret Mead: "Never doubt that a small group of committed people can change the world. It is the only thing that ever has."

CONCLUDING THOUGHTS ABOUT YOU AND THE FUTURE

You can engineer your future in that it does not have to be something that happens to you. Similarly, you and kindred spirits can create the future of your business, government, academic, volunteer, community, or other organization. You and colleagues can be the change agents needed for individuals and organizations to achieve success and significance and to thrive, and not die or merely survive, in our rapidly-changing world. How can you and others engineer the future? Partly by drawing on the professional knowledge, skills, and attitudes (KSAs) discussed in this book supplemented with the change principles and process presented in this chapter.

The great use of life is to spend it
for something that will outlast it.

(*William James, psychologist and philosopher*)

CITED SOURCES

American Society of Civil Engineers. 2004. *Civil Engineering Body of Knowledge for the 21st Century: Preparing the Civil Engineer for the Future*. ASCE Press: Reston, VA.

Armstrong, S. C. 2005. *Engineering and Product Development Management: A Holistic Approach*. Cambridge University Press: Cambridge, UK.

American Society of Civil Engineers. 2007. *The Vision for Civil Engineering in 2025*. ASCE: Reston, VA.

American Society of Civil Engineers. 2008. *Civil Engineering Body of Knowledge for the 21st Century: Preparing the Civil Engineer for the Future - Second Edition*. ASCE Press: Reston, VA.

Annunzio, S. with J. Liesse. 2001 *eLeadership: Proven Techniques for Creating an Environment of Speed and Flexibility in the Digital Economy*. The Free Press: New York, NY.

Arciszewski, T. 2009. *Successful Education: How to Educate Creative Engineers*. Successful Education LLC: Fairfax, VA.

Barker, J. A. 1989. *Discovering the Future: The Business of Paradigms*. ILI Press: St. Paul, MN.

Barker, J. A. 1992. *Paradigms: The Business of Discovering the Future*. Harper Business: New York, NY.

Bennis, W. G. 1989. *Why Leaders Can't Lead—The Unconscious Conspiracy Continues*. Jossey-Bass Publishers: San Francisco, CA.

Brenner, R. 2009. "Letting Go of the Status Quo: The Debate." *Point Lookout*, e-newsletter from Chaco Canyon Consulting, December 30.

Brown, H. J. 1988. *A Father's Book of Wisdom*. Rutledge Hill Press: Nashville, TN.

Carroll, J. 2004. "In An innovation Rut? Here Are Tip-Offs." *The Globe and Mail*, November 12.

Covey, S. R. 1990. *The 7 Habits of Highly Effective People*. Simon & Schuster: New York, NY.

de Blij, H. 2009. *The Power of Place: Geography, Destiny, and Globalization's Rough Landscape*. Oxford University Press: New York, NY.

Flaherty, M. 2007. "Let Them At Least Have Heard of Brave Knights and Heroic Courage." *Imprimis*, Vol. 36, No. 2, February, Hillsdale College, Hillsdale, MI, pp. 1–5.

Fredrich, A. J. (ed.). 1989. *Sons of Martha: Civil Engineering Readings in Modern Literature*. "Strauss Gave Me Some Pencils," ASCE: Reston, VA.

Friedman, T. L. 2005. *The World is Flat: A Brief History of the Twentieth Century*. Farrar, Straus, & Giroux: New York, NY.

Heilmeier, G. H. 1992. "Some Reflections on Innovation and Invention." *The Bridge*, National Academy of Engineering: Washington, DC. Winter.

Kao, J. 2007. *Innovation Nation: How America Is Losing Its Innovation Edge: Why It Matters, and What We Can Do To Get It back*. The Free Press: New York, NY.

Kotter, J. P. 2007. "Leading Change: Why Transformation Efforts Fail." *Harvard Business Review*, January.

Kriegel, R. and D. Brandt. 1996. *Sacred Cows Make the Best Burgers: Paradigm-Busting Strategies for Developing Change-Ready People and Organizations*. Warner Books: New York, NY.

Machiavelli, N. 1980. *The Prince*. Translated by E. R.P. Vincent. New American Library: New York, NY. (Originally published in 1537.)

Maxwell, J. C. 1993. *Developing the Leader Within You*. Nelson Business: Nashville, TN.

McGloin, J. B. 2011. "Symphonies in Steel: Bay Bridge and Golden Gate Bridge." Museum of San Francisco website. (www.sfmuseum.org/hist9/mcgloin.html). May 24.

Naisbitt, J. 2006. *Mind Set! Reset Your Thinking and See the Future*. HarperCollins: New York, NY.

Pink, D. H. 2005. *A Whole New Mind: Moving From the Knowledge Age to the Conceptual Age*. Riverhead Books: New York, NY.

Pink, D. H. 2007. "Revenge of the Right Brain." *Public Management*, July, pp. 10–13.

Russell, J. 2006. Personal communication, Professor and Chair, Civil and Environmental Engineering Department, University of Wisconsin – Madison, March 13.

Smart, G. 2007. "Strategic Planning in 5 Relatively Easy Steps." Education Session, NSPE 2007 Annual Conference, Denver CO, July.

Walesh, S. G. 1990. "Water Resources Science and Technology: Global Origins." *Urban Stormwater Quality Enhancement*. Proceedings of an Engineering Foundation Conference, Davos, Switzerland, ASCE, pp. 1–27.

Weingardt, R. G. 2005. *Engineering Legends: Great American Civil Engineers*. ASCE Press: Reston, VA.

White, K. D. 1984. *Greek and Roman Technology*. Thames and Hudson: London, UK.

ANNOTATED BIBLIOGRAPHY

Collins, J. 2001. *Good to Great*. Harper-Collins: New York, NY. (Argues that having the right people "on the bus," engaged in frank, open-minded, out-of-the-box thinking is a key to developing ideas and strategies for a successful future.)

Deutschman, A. 2005. "Change: Why Is It So Darn Hard to Change Our Way?" *FAST COMPANY*, May, pp. 53–62. (Argues two points worth considering. The first is that radical, sweeping changes may be easier to attempt than gradual changes. Reason: Big changes quickly yield benefits. The second point is that affecting change requires facts plus emotion.)

Gerber, R. 2002. *Leadership the Eleanor Roosevelt Way: Timeless Strategies from the First Lady of Courage*. Prentice-Hall Press: Upper Saddle River, NJ. (Offers leading lessons drawn from Eleanor Roosevelt's life including developing empathy, finding a mentor, taking action in crises, finding one's passion, embracing risk, and becoming an effective speaker.)

Graham, L. R. 1993. *The Ghost of the Executed Engineer: Technology and the Fall of the Soviet Union*. Harvard University Press: Cambridge, MA. (Asserts that the Soviet Union, even though it had ample natural resources and many engineers, failed to become a modern industrialized nation because of "misuse of technology and squandering of human energy," including its engineering talent. Is the U.S. practicing sound stewardship with its engineering talent?).

Levitt, S. D. and S. J. Dubner. 2005. *Freakonomics*. William Morrow: New York, NY. (The leader in us can benefit from an overriding theme in this book: Don't easily accept conventional wisdom, even when "supported" by "numbers.")

Taleb, N. N. 2007. *The Black Swan: The Impact of the Highly Improbable* Random House: New York, NY. (One message of the book is that "...we fool ourselves into thinking we know more than we actually do. We restrict our thinking to the irrelevant and inconsequential, while large events continue to surprise us and shape our world.")

Walesh, S. G. 2008. "Vision: Pie-in-the-Sky or Organizational Priority?" Engineering Your Future column, *Leadership and Management in Engineering-ASCE*, January, pp. 45–46. (Argues that visioning is a credible and valuable process subject to some qualifications).

EXERCISES

15.1 ANALYZE A CHANGE EFFORT: This exercise provides you with an opportunity to recall and reflect on a change effort, in which you were directly or indirectly involved, illuminated by the ideas and principles presented in this chapter. Suggested tasks are:

A. Recall a change effort that you participated in or were indirectly involved to the extent that you knew about it. The proposed change must have gone beyond just personal change, that is, it was intended to effect many people. The effort may have been successful or a failure and the object of the change could be in almost any aspect of society. The point is that change was advocated and you were there.

B. Determine why the change effort succeeded or failed considering the change ideas and principles discussed in this chapter. If it was successful, what were the principal success factors and if it failed, what were the main reasons?

C. Write a memorandum that describes the change effort and discusses the prime reasons for its success or failure.

15.2 LEAD A CHANGE EFFORT: The purpose of the exercise is to encourage you, and eventually a core group, to effect change. Suggested tasks are:

A. Identify a leadership gap, some cause you are passionate about.

B. Guided by the "How to Lead Change" section of this chapter, lead the change effort.

APPENDIX A

ENGINEERING YOUR FUTURE SUPPORTS ABET BASIC LEVEL CRITERION 3

When used as a textbook or reference book in an engineering program, *Engineering Your Future* provides content that supports all of ABET's Basic Level Criterion 3 non-technical or partly non-technical program outcomes in effect as of the 2011 writing of this book. Refer to ABET (www.abet.org) for the current Basic Level Criterion 3. More specifically, the following matrix connects those eight outcomes to book chapters:

Chapter title	Letter and short name of ABET Basic Level Criterion 3 non-technical or partly non-technical outcome							
	Design c	Team-work d	Problem solving e	Professional and ethical responsibility f	Comunication g	Context h	Life-long learning i	Contemporary issues j
1. Introduction: Engineering and the Engineer	X	X	X	X	X	X	X	
2. Leading and Managing: Getting Your Personal House in Order		X		X	X		X	
3. Communicating to Make Things Happen					X			
4. Developing Relationships		X			X			
5. Project Management: Planning, Executing, and Closing		X	X		X	X		
6. Project Management: Critical Path Method and Scope Creep		X	X	X	X	X		
7. Quality: What is it and How Do We Achieve It?	X	X	X		X	X		
8. Design: To Engineer is to Create	X	X	X	X	X	X		
9. Building: Constructing and Manufacturing	X	X	X		X	X		
10. Business Accounting: Tracking the Past and Planning the Future		X	X		X			
11. Legal Framework	X			X	X	X		X
12. Ethics: Dealing With Dilemmas	X	X	X	X	X	X	X	X
13. Role and Selection of Consultants		X	X	X	X	X		
14. Marketing: A Mutually-Beneficial Process		X	X	X	X			
15. The Future and You		X	X	X	X	X		X

APPENDIX B

ENGINEERING YOUR FUTURE SUPPORTS ABET PROGRAM CRITERIA FOR CIVIL AND SIMILARLY-NAMED ENGINEERING PROGRAMS

When used as a textbook or reference book in an engineering program, *Engineering Your Future* provides content that supports all of ABET's Program Criteria for Civil and Similarly-Named Engineering Programs which were effective as of the 2011 writing of this book. Refer to ABET (www.abet.org) for current program criteria. More specifically, the following matrix connects the five topics in the program criteria to the book's chapters:

Chapter title	Management	Business	Public policy	Leadership	Professional licensure
1. Introduction: Engineering and the Engineer	X	X		X	
2. Leading and Managing: Getting Your Personal House in Order	X				X
3. Communicating to Make Things Happen				X	
4. Developing Relationships	X	X		X	
5. Project Management: Planning, Executing, and Closing	X	X	X	X	
6. Project Management: Critical Path Method and Scope Creep	X	X		X	
7. Quality: What is it and How Do We Achieve It?	X	X	X	X	
8. Design: To Engineer is to Create	X	X			
9. Building: Constructing and Manufacturing	X	X			
10. Business Accounting: Tracking the Past and Planning the Future	X	X			
11. Legal Framework	X	X		X	
12. Ethics: Dealing With Dilemmas	X	X	X	X	
13. Role and Selection of Consultants	X	X	X	X	
14. Marketing: A Mutually-Beneficial Process	X	X			
15. The Future and You	X	X	X	X	

APPENDIX C

ENGINEERING YOUR FUTURE SUPPORTS THE CIVIL ENGINEERING BODY OF KNOWLEDGE

When used as a textbook or reference book in an engineering program, *Engineering Your Future* provides content that supports all nine of the Professional Outcomes in the Civil Engineering Body of Knowledge that was in effect as of the 2011 writing of this book. That BOK is documented in the 2008 ASCE report *Civil Engineering Body of Knowledge for the 21st Century-Second Edition* available at www.asce.org. Refer to that website for possible supplements or a possible third edition. The following matrix connects the nine Professional Outcomes to book chapters. The outcomes are numbered and named in the table. Refer to the BOK report for descriptions of each outcome and the prescribed level of cognitive achievement to be fulfilled through the bachelor's degree, the master's degree, and prelicensure experience.

Chapter title	Number and name of Professional Practice Outcome								
	Communication 16	Public policy 17	Business and public administration 18	Globalization 19	Leadership 20	Teamwork 21	Attitudes 22	Life-long learning 23	Professional and ethical responsibility 24
1. Introduction: Engineering and the Engineer	X		X		X	X	X	X	X
2. Leading and Managing: Getting Your Personal House in Order	X					X	X	X	X
3. Communicating to Make Things Happen	X				X		X		
4. Developing Relationships	X		X		X	X	X		
5. Project Management: Planning, Executing, and Closing	X	X	X		X	X			
6. Project Management: Critical Path Method and Scope Creep	X		X		X	X			X
7. Quality: What is it and How Do We Achieve It?	X	X	X		X	X	X		
8. Design: To Engineer is to Create	X		X			X	X		X
9. Building: Constructing and Manufacturing	X		X			X	X		
10. Business Accounting: Tracking the Past and Planning the Future	X		X			X			
11. Legal Framework	X		X		X		X		X
12. Ethics: Dealing With Dilemmas	X	X	X		X	X	X	X	X
13. Role and Selection of Consultants	X	X	X		X	X	X		X
14. Marketing: A Mutually-Beneficial Process	X		X			X	X		X
15. The Future and You	X	X	X	X	X	X	X		X

Index

About the Author

Dr. Stuart G. Walesh, PE provides management, engineering, education/training, and marketing services. He draws on more than 40 years of engineering and management experience in the government, business, and academic sectors to help individuals and organizations engineer their futures. Walesh has functioned as a project engineer, project manager, department head, discipline manager, marketer, professor, and dean of an engineering college.

Representative clients include ASCE; Boston Society of Civil Engineers; BSA Life Structures; Castilla LaMancha University; CDM; Clark Dietz; Daimler Chrysler; DLZ; Earth Tech; Harris County (TX) Flood Control District; Hinshaw & Culbertson; Indiana Department of Natural Resources; Indiana Department of Transportation/Purdue University; J. F. New; Leggette, Brashears & Graham; Midwest Geosciences Group; MSA Professional Services; PBS&J; Town of Pendleton, IN; Pennoni Associates; Taylor Associates; City of Valparaiso, IN; University of Wisconsin Engineering Professional Development; and Wright Water Engineers.

Walesh authored *Urban Surface Water Management* (Wiley 1989), *Engineering Your Future: Launching a Successful Entry-Level Technical Career in Today's Business Environment* (Prentice Hall 1995), *Flying Solo: How to Start an Individual Practitioner Consulting Business* (Hannah Publishing, 2000), *Engineering Your Future: The Non-Technical Side of Professional Practice in Engineering and Other Technical Fields-Second Edition* (ASCE Press 2000), *Managing and Leading: 52 Lessons Learned for Engineers* (ASCE Press 2004), *Managing and Leading: 44 Lessons Learned for Pharmacists* (ASHP 2008, co-authored with Paul Bush, Pharm.D), and *Engineering Your Future: The Professional Practice of Engineering-Third Edition* (Wiley and ASCE Press, 2012). Walesh is author or co-author of many articles, papers, and other publications and has facilitated or presented several hundred workshops, seminars, webinars, and meetings throughout the U.S and internationally.

Walesh is a member of ASCE's Committee on Academic Prerequisites for Professional Practice, was Special Issues Editor for ASCE's Committee on Publications, and chaired several national committees. He has been recognized for his professional contributions as follows: Public Service Award from the Consulting Engineers of Indiana (1995), Distinguished Service Citation from the College of Engineering at the University of Wisconsin (1998), Excellence in Civil Engineering Education Leadership Award presented by ASCE (2003), Distinguished Member of ASCE (2004), Diplomate of the American Academy of Water Resource Engineers (2005), Engineer of the Year by the Indiana Society of Professional Engineers (2007), Distinguished Service Award from the National Society of Professional Engineers (2007), William H. Wisely American Civil Engineer Award from ASCE for leadership in promoting engineering as a profession (2008), George K. Wadlin Distinguished Service Award from the Civil Engineering Division of the American Society for Engineering Education (2009), and Fellow Member of the National Society of Professional Engineers (2010).

Walesh received his BSCE degree from Valparaiso University, his MSE from The Johns Hopkins University, and his PhD from the University of Wisconsin-Madison.